NATO ASI Series

Advanced Science Institutes Series

A series presenting the results of activities sponsored by the NATO Science Committee, which aims at the dissemination of advanced scientific and technological knowledge, with a view to strengthening links between scientific communities.

The Series is published by an international board of publishers in conjunction with the NATO Scientific Affairs Division

A Life Sciences **B Physics**	Plenum Publishing Corporation London and New York
C Mathematical and Physical Sciences **D Behavioural and Social Sciences** **E Applied Sciences**	Kluwer Academic Publishers Dordrecht, Boston and London
F Computer and Systems Sciences **G Ecological Sciences** **H Cell Biology** **I Global Environmental Change**	Springer-Verlag Berlin Heidelberg New York London Paris Tokyo Hong Kong Barcelona Budapest

NATO-PCO DATABASE

The electronic index to the NATO ASI Series provides full bibliographical references (with keywords and/or abstracts) to more than 30000 contributions from international scientists published in all sections of the NATO ASI Series. Access to the NATO-PCO DATABASE compiled by the NATO Publication Coordination Office is possible in two ways:

- via online FILE 128 (NATO-PCO DATABASE) hosted by ESRIN, Via Galileo Galilei, I-00044 Frascati, Italy.
- via CD-ROM „NATO Science & Technology Disk" with user-friendly retrieval software in English, French and German (© WTV GmbH and DATAWARE Technologies Inc. 1992).

The CD-ROM can be ordered through any member of the Board of Publishers or through NATO-PCO, Overijse, Belgium.

Series I: Global Environmental Change, Vol. 11

The ASI Series Books Published as a Result of Activities of the Special Programme on Global Environmental Change

This book contains the proceedings of a NATO Advanced Research Workshop held within the activities of the NATO Special Programme on Global Environmental Change, which started in 1991 under the auspices of the NATO Science Committee.

The volumes published as a result of the activities of the Special Programme are:

Vol. 1: **Global Environmental Change.** Edited by R. W. Corell and P. A. Anderson. 1991.
Vol. 2: **The Last Deglaciation: Absolute and Radiocarbon Chronologies.** Edited by E. Bard and W. S. Broecker. 1992.
Vol. 3: **Start of a Glacial.** Edited by G. J. Kukla and E. Went. 1992.
Vol. 4: **Interactions of C, N, P and S Biogeochemical Cycles and Global Change.** Edited by R. Wollast, F. T. Mackenzie and L. Chou. 1993.
Vol. 5: **Energy and Water Cycles in the Climate System.** Edited by E. Raschke and D. Jacob. 1993.
Vol. 6: **Prediction of Interannual Climate Variations.** Edited by J. Shukla. 1993.
Vol. 7: **The Tropospheric Chemistry of Ozone in the Polar Regions.** Edited by H. Niki and K. H. Becker. 1993.
Vol. 8: **The Role of the Stratosphere in Global Change.** Edited by M.-L. Chanin. 1993.
Vol. 9: **High Spectral Resolution Infrared Remote Sensing for Earth's Weather and Climate Studies.** Edited by A. Chedin, M.T. Chahine and N.A. Scott. 1993.
Vol. 10: **Towards a Model of Ocean Biogeochemical Processes.** Edited by G. T. Evans and M. J. R. Fasham. 1993.
Vol. 11: **Modelling Oceanic Climate Interactions.** Edited by J. Willebrand and D. L. T. Anderson. 1993.
Vol. 12: **Ice in the Climate System.** Edited by W. Richard Peltier. 1993.
Vol. 13: **Atmospheric Methane: Sources, Sinks, and Role in Global Change.** Edited by M. A. K. Khalil. 1993.
Vol. 14: **The Role of Regional Organizations in the Context of Climate Change.** Edited by M. H. Glantz. 1993.

Modelling Oceanic Climate Interactions

Edited by

Jürgen Willebrand

Institut für Meereskunde
Universität Kiel
Düsternbrooker Weg 20
D-24015 Kiel, Germany

David L.T. Anderson

Department of Atmospheric Physics
Clarendon Laboratory
University of Oxford
Oxford OX1 3PU, U.K.

Springer-Verlag
Berlin Heidelberg New York London Paris Tokyo
Hong Kong Barcelona Budapest
Published in cooperation with NATO Scientific Affairs Division

Proceedings of the NATO Advanced Study Institute on Modelling of Oceanic Climate Interactions, held at Les Houches, France, February 17–28, 1992

Additional material to this book can be downloaded from http://extra.springer.com.

ISBN-13:978-3-642-84977-0 e-ISBN-13:978-3-642-84975-6
DOI: 10.1007/ 978-3-642-84975-6

Library of Congress Cataloging-in-Publication Data
Modelling oceanic climate interactions / edited by David L. T. Anderson, Jürgen Willebrand. p. cm. (NATO ASI series. Series I, Global environmental change; vol. 11)
"Proceedings of the NATO Advanced Study Institute on Modelling of Oceanic Climate Interactions, held at Les Houches, France, February 17-28, 1992" – T.p. verso. Includes index.
ISBN-13:978-3-642-84977-0

1. Ocean-atmosphere interaction–Mathematical model–Congresses. 2. Climatology–Mathematical models–Congresses. I. Anderson, D. L. T. (David L. T.) II. Willebrand, J. (Jürgen), 1941- . III. NATO Advanced Study Institute on Modelling of Oceanic Climate Interactions (1992: Les Houches, Haute Savoie, France) IV. Series. GC190.2.M63 1993 551.5'246–dc20 93-26021

Softcover reprint of the hardcover 1st edition 1993

Typesetting: Camera ready by authors
31/3145 - 5 4 3 2 1 0 - Printed on acid-free paper

Preface

The ocean plays a central role in determining the climate of the earth. The oceanic circulation largely controls the temporal evolution of climate changes resulting from human activities such as the increase of greenhouse gases in the atmosphere, and also affects the magnitude and regional distribution of those changes. On interannual and longer time scales the ocean is, through its interaction with the atmosphere, a source of important natural climate variations which we are only now beginning to recognise but whose cause has yet to be properly determined. Chemical and biological processes in the ocean are linked to climate change, particularly through interaction with the global carbon cycle.

A quantitative understanding of the oceanic role in the climate system requires models which include many complex processes and interactions, and which are systematically verified with observations. This is the objective of global research programs such as TOGA, WOCE, and JGOFS. Coupled numerical models of the oceanic and atmospheric circulation constitute the basis of every climate simulation. Increasingly it is recognized that in addition a biological/chemical component is necessary to capture the pathways of carbon and other trace gases. The development of such coupled models is a challenging task which needs scientists who must be cognizant of several other disciplines beyond their own specialty.

The purpose of the NATO Advanced Study Institute which was held from February 17-28, 1992 in Les Houches, France, was to introduce young scientists (53 student participants from 18 countries) into the current state of modelling oceanic climate interactions. The lectures covered a range of topics from the fields of physical, chemical and biological oceanography as well as meteorology, with the principal focus being on ocean circulation dynamics, hydrological processes in the atmosphere, tropical ocean-atmosphere interactions, and the oceanic carbon cycle. Lecturers not only shared their expert knowledge with the participants, but highlighted the uncertainties in present models.

Primary support for the Study Institute was provided by NATO under the auspices of the Scientific and Environmental Affairs Division. Additional support was provided by the Les Houches Centre de Physique Winter Programme, and by the National Science Foundation. It is a great pleasure to thank these organisations and in particular Dr. L.V. da Cunha and Professor Nino Boccara who helped to make everything run smoothly.

We would like to thank Shona Anderson who coped with the endless technical problems in producing this book in Latex; and Andreas Oschlies, Stefan Rahmstorf and a number of reviewers for helping to improve the quality of the final manuscripts. Finally, we would like to thank all contributors who have given a stimulating series of lectures which is recorded in this volume.

David Anderson

Jurgen Willebrand

April 1993

CONTENTS

Participants

Myles R. Allen, Department of Physics, Atmospheric, Oceanic and Planetary Physics, Clarendon Laboratory, Parks Road, Oxford, OX1 3PU, U.K.

Ms. M. Alonso Balmaseda, Dpto. de Fisica, Universidad de Alcalà de Henares, 28880 Madrid, Spain

Keith Alverson, EAPS Department, M.I.T. 54-1419, Cambridge, MA 02139, USA

Dr. David L.T. Anderson, University of Oxford, Dept. of Atmospheric Physics, Clarendon Laboratory, Parks Road, Oxford OX1 3PU, U.K.

Norman H. Barth, Climate Modeling & Global, Change Team - CERFACS - 42, Ave Gustave Coriolis, 31057 Toulouse Cedex, France

Jérôme Benveniste, Massachusetts Institute, of Technol. Earth, Atmosph. and Planetary Sciences, Center for Meteorology, and Oceanography, Room 54-1514, Cambridge, MA 02139, USA

Eric Blayo, Institut de Mécanique de Grenoble, Domaine Universitaire, B.P. 53X, 38041 Grenoble Cedex, France

Pascale Braconnot, Laboratoire d'Océanographie, Dynamique et de Climatologie, Université P.& M. Curie, Tour 14 - 2e étage, 4 Place Jussieu, Paris Cedex 05, France

R. Budich, Max-Planck-Institut für, Meteorologie, Bundesstr. 55, D-2000 Hamburg 13, Germany

Gerrit Burgers, Ministry of Transport and, Works, Royal Netherlands Meteorological Institute, Dept. of Oceanogr. Research, P.O.Box 201, 3730 AE de bilt, The Netherlands

J.-M. Campin, Institut d'Astronomie et de, Géophysique G. Lemaître, Université Catholique de Louvain, chemin du cyclotron, 2, 1348 Louvain La Neuve, Belgique

Dr. A. Colin de Verdière, IFREMER, Departement Etudes Oceaniques, BP 70, 29263 Plouzane, France

Dr. J.C. de Munck, The Netherlands Institute, for Sea Research, P.O.Box 59, 1790 AB Den Burg, The Netherlands

Deng-Hua Wu, Atmospheric, Oceanic and Planetary Physics, Dept. of Physics, University of Oxford, Clarendon Laboratory, Parks Road, Oxford, OX1 3PU, UK

Prof. Dr. H.F. Durukanoglu, Karadeniz Technical University, Surmene Higher School of Marine Sciences and Technology, 61080 Trabzon, Turkey

Matthew England, Physical Oceanography, F05, The University of Sydney, Sydney NSW 2006, Australia

Augustus F. Fanning, 379 Hamilton Avenue, St. John's NF, Canada, A1E-1K4

Christoph Frei, Atmospheric Physics ETH, Hönggerberg HPP, CH-8093 Zürich, Switzerland

Dr. A. Gargett, Institute of Ocean Sciences, 9860 W. Saanich Road, P.O. Box 6000, Sidney, B.C. V8L 4B2, Canada

Peter J. Gleckler, Program for Climate Model, Diagnosis and Intercomparison, Lawrence Livermore National Laboratory, P.O. Box 808, L-264, Livermore, CA 94550, USA

Dr. Catherine Goyet, Woods Hole, Oceanographic Institution, Woods Hole, MA 02543, U.S.A.

Dr. R.J. Haarsma, KNMI, P.O.Box 201, 3730 AE de Bilt, The Netherlands

Susan L. Hautala, School of Oceanography, WB-10, University of Wahington, Seattle, Washington 98195, USA

Dr. M. Heimann, Max-Planck-Institut für Meteorologie, Bundesstr. 55, 2000 Hamburg 13, Germany

Dr. A. Hollingsworth, European Centre for Medium Range Weather Forecasts (ECMWF), Shinfield Park, Reading, Berkshire RG2 9AX, UK

Tertia Hughes, Department of Meteorology, McGill University, 805 Sherbrooke Street West, Montreal, Quebec, Canada H3A 2K6

Dr. V.M. Kattsov, A.I. Voeikov Main Geophysical Observatory, Divis. of Dynamical Meteorology, 7, Karbyshev Str., 194018 Leningrad, USSR

Dr. Jens Krüger, Alfred-Wegener-Institut für Polar- und Meeresforschung, Postfach 12 01 61, Columbusstraße, D-2850 Bremerhaven, Germany

Paul Kushner, Department of Physics, University of Toronto, Toronto, Ontario, Canada, M5S 1A7

Laura L. Landrum, School of Oceanography, University of Washington, WB-10, Seattle, WA 98195, USA

Pascal Le Grand, MIT, room 54-1523, Cambridge, MA 02139, USA

Piero Lionello, c/o Prof. Hasselmann, Max-Planck-Institut für Meteorologie, Bundesstr. 55, D-2000 Hamburg 13, Germany

Kjetil Lygre, Nansen Environmental and Remote Sensing Center, Edvard Griegsvei 3a, N-5037 Solheimsvik, Norway

Dr. Jochem Marotzke, Dept. of Earth & Planetary Sciences, M.I.T., Room 54-1528, Cambridge, MA 02139, U.S.A.

Olivier Marti, Laboratoire d'Océanographie Dynamique et de Climatologie, Univ. Pierre et Marie Curie, Tour 26-0, 4 place Jussieu, 75252 Paris Cedex 05, France

Dr. R.P. Matano, College of Oceanography, Oregon State University, Corvallis, OR 97331-5503, USA

Mike McCann, Department of Oceanography, Naval Postgraduate School, Monterey, CA 93943, USA

Ian McConnell, Department of Applied and, Computational Mathematics, P.O.Box 597, Sheffield S10 2UN, UK

Prof. Dr. P. Lemke, Alfred-Wegener-Institut für Polar- und Meeresforschung, Columbusstraße, Postfach 12 01 61, 2850 Bremerhaven, Germany

Dr. Ch. LeProvost, Institut de Mécanique de Grenoble, Domaine Universitaire, B.P. n 68, 38402 Saint-Martin d'Héres Cedex, France

Dr. E. Maier-Reimer, Max-Planck-Institut für Meteorologie, Bundesstr. 55, 2000 Hamburg 13, Germany

Christophe Menkes, ORSTOM, Centre de Noumea, Boite Postale A 5 Noumea, Nouvelle Calédonie

Dr. M. Miller, European Centre for Medium Range Weather Forecasts (ECMWF), Shinfield Park, Reading, Berkshire RG2 9AX, UK

R. Miller, Atmospheric and Planetary Sciences, NASA Goddard Institute for Space Studies, 2880 Broadway, New York, NY, 10025, USA

A. Oschlies, Institut für Meereskunde an der Universität Kiel, Düsternbrooker Weg 20, 2300 Kiel, Germany

Prof. Dr. D. Olbers, Alfred-Wegener-Institut für Polar- u. Meeresforschung, Columbuscenter, 2850 Bremerhaven, Germany

Tamás Palvölgyi, Magyar Köztársaság, Meteorologiai Szolgálata, Meteorological Service of the Republic of Hungary, P.O.Box 38, H-1525 Budapest, Hungary

Dr. G. Philander, NOAA/GFDL, Princeton University, P.O. Box 308, Princeton, NJ 08540, U.S.A.

Dr. S. Rahmstorf, Institut für Meereskunde an der Universität Kiel, Düsternbrooker Weg 20, 2300 Kiel, Germany

Jacyra Ramos Soares, Department of Oceanography, University of Southampton, Highfield, Southampton SO9 5NH, UK

Urmas Raudsepp, Institute of Ecology and Marine Research, 1 Paldiski Str., 200031 Tallinn, Estonia

Prof. Dr. P. Rhines, WV-10, Dept. of Oceanography, University of Washington, Seattle, WA 98195, U.S.A.

Paul Robbins, Woods Hole Oceanographic Institution, Dept. of Physical Oceanography, Woods Hole, MA 02543, USA

A. Russo, Instituto per le Ricerche sulla Pesca Marittima, (IRPEM) - CNR, Molo Mandracchio, 60100 Ancona, Italy

Pawel Schlichtholz, Laboratoire d'Océanographie et de Climatologie (LODYC), Tour 14 - 2e étage, 4, Place Jusieu, 75 252 Paris Cedex 05, France

Dr. D.B. Stephenson, University of Reading, Dept. of Meteorology, 2 Earley Gate, Whiteknights, Reading RG6 2AU, U.K.

David Straub, Robert Hooke Institute for Atmospheric Research, The Observatory, Clarendon Laboratory, Parks Road, Oxford OX1 3PU, U.K.

Rémi Tailleux, LMCE DSM-Batiment 522, CEN Saclay, 91191 GIF-sur-Yvette Cedex, France

Dr. A. Taylor, Plymouth Marine Laboratory, Citadelle Hill, Plymouth PL1 2PB, UK

Dr. Laurent Terray, Global Change and Modelling Team, C.E.R.F.A.C.S., 42 Avenue G. Coriolis, 31057 Toulouse Cedex, France

S.R. Thompson, IOS, Institute of Oceanographic Sciences, Brook Road Wormley, Godalming, Surrey, GU8 5UB, U.K.

Dr. LuAnne Thompson, School of Oceanography WB-10, University of Washington, Seattle, WA 98195, USA

Christoph Völker, Straßburgerstr. 35, 2800 Bremen 1, Germany

Anthony T. Weaver, Atmospheric, Oceanic and Planetary Physics, Department of Physics, Clarendon Laboratory, Parks Rd, Oxford, OX1 3PU, U.K.

Dr. M. Whitfield, Plymouth Marine Laboratory, Citadelle Hill, Plymouth PL1 2PB, UK

Prof. Dr. J. Willebrand, Institut für Meereskunde, an der Universität Kiel, Düsternbrooker Weg 20, 2300 Kiel, Germany

Goro Yamanaka, Meteorological Research Institute, 1-1, Nagamine, Tsukuba, Ibaraki, 305 Japan

Dr. J. P. v. Ypersele, Institut d'Astronomie et de Géophysique, Georges Lemaitre, Université Catholique de Louvain, Chemin du Cyclotron 2
B-1348 Louvain-La-Neuve, Belgium

Lisan Yu, Mesoscale Air-Sea Interaction Group, B-174, 020 Love Building, The Florida State University, Tallahassee, Fl 32306-3041, USA

THE MODELLING OF HYDROLOGICAL PROCESSES IN THE ATMOSPHERE

M J MILLER
European Centre for Medium-Range Weather Forecasts
Reading, Berkshire

Abstract

The global hydrological cycle and associated energy fluxes are briefly reviewed and the role of large-scale numerical models in studying these is examined. A selective discussion of individual physical processes follows together with consideration of the parametrization problems that these processes pose. The fundamental role of physical parametrization in studying climate and climate change is emphasised particularly with regard to the complex interactions between the major processes.

1 Introduction

The importance of large scale numerical models of the atmosphere for studying the earth's climate and its variability has increased dramatically during the last decade and there is a growing emphasis on climate studies with coupled General Circulation Models (GCM) of the atmosphere and oceans. A major and fundamental component of these studies is the modelling of the hydrological cycle and associated energy fluxes by suitable physical models. Since these fluxes are due to motions on scales smaller than are resolved by the numerical models, the fluxes must be represented in terms of resolved variables, i.e. parametrized. Much of the progress made in this area has been through the continuing development and refinement of NWP models which provide the weather forecast products worldwide. Increasingly the distinction between global forecast models and climate models is one of application rather than of major differences in model design and assumptions. The determination of the observed hydrological cycle and its energy fluxes depends critically on global analysis systems which can process and combine a variety of data types; this is optimally achieved through the use of data assimilation systems which combine the available data in whatever form with a short-range forecast from an operational forecast model. The forecast model, by advecting information into data sparse regions, enables a high quality global analysis to be carried out. This in turn makes determination of the global hydrological budget practical.

NATO ASI Series, Vol. I 11
Modelling Oceanic Climate Interactions
Edited by J. Willebrand and D. L. T. Anderson

Several of the GEWEX (Global Energy and Water Cycle Experiment) objectives recognise the central role of the models in *determining* the hydrological cycle, in *modelling* the hydrological cycle and *predicting* its variation on global and regional scales in response to environmental change.

Following a review of the components of the global budgets and their magnitude, the following sections consider with varying levels of detail how they are parametrized in current models. No exhaustive exposition of any is attempted but rather some emphasis on the interactions between the various components and the demonstrated sensitivity of models to particular schemes etc.

For reasons of brevity only, the representation of the land surface and its associated fluxes of heat, moisture and momentum is not discussed here. Likewise the parametrization of the planetary boundary layer (PBL) is only considered in the context of the relationship of the PBL to convective processes. Current land surface parametrizations are reviewed in Schumgge and Andre (1991) and Wood (1991), while the parametrization of the PBL is reviewed in Sommeria (1988) and Stull (1988).

Most of the model results used in this paper are taken from operational or research forecasts made with the ECMWF model at relatively high resolutions. In general it is intended that these results are illustrative of general modelling problems and sensitivities, some of which could have been generated by long climate-type simulations at much lower resolutions.

2 Global Budgets

Before considering the energy fluxes it is pertinent to stress the importance of water in the earth/atmosphere system. In vapour form it is the most variable radiatively active constituent and the principal contributor to the greenhouse effect; as cloud it controls the planetary albedo and hence the radiative energy reaching the earth's surface; as liquid or ice its availability is crucial to life of all kinds. It also modulates deep ocean circulation and the buoyancy driving of the upper ocean. Estimates of the components of the water cycle are shown in Fig. 1 (GEWEX, 1991). As well as stressing the importance of the ocean fluxes, it can be seen that the mean cycling time of water substance is relatively rapid ($\sim$ 10 days) underlining the requirement for careful modelling of moist processes.

Globally integrated climatological estimates of the principal fluxes of energy (and water) are summarized in Fig. 2. Athough the values of the radiative fluxes at the top of the atmosphere are now well known from satellite data, surface fluxes and those of water substance still have

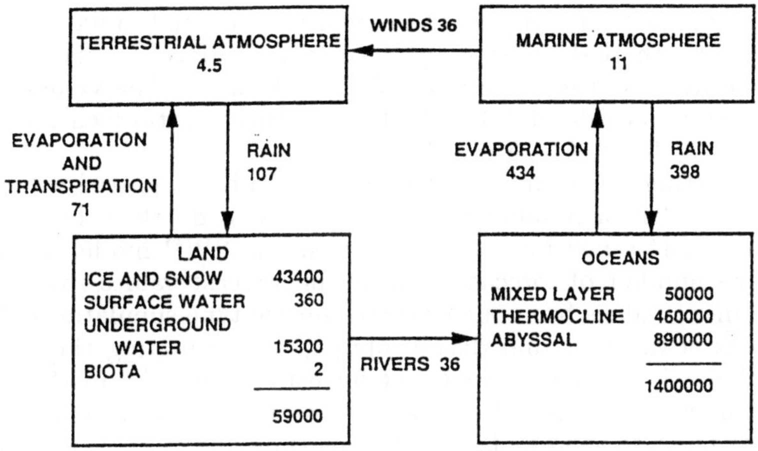

Figure 1: The global water cycle and its reservoirs. Units are 10^{15} kg. yr^{-1} for the water cycle and 10^{15} kg for the reservoirs.

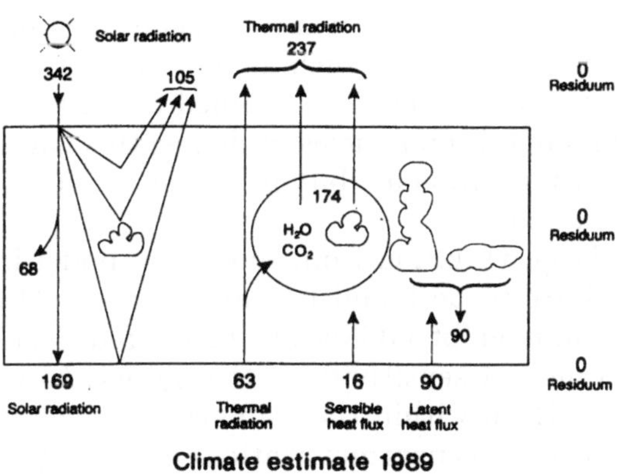

Figure 2: Climatological estimates of the global annual mean energy and water cycle. Units: W/m^2 (Ramanathan *et al.*, 1989).

uncertainties of several percent. The net surface thermal radiation is the difference between that emitted by the earth's surface ($\sim$ 390 Wm^{-2}) and that emitted by the atmosphere ($\sim$ 327 Wm^{-2}). The values given in fig.2 define an albedo of 0.30 ± 0.01 and a Bowen ratio (sensible heat flux/latent heat flux) of 0.18 ± 0.02.

It is interesting to see the corresponding globally-integrated values for a GCM. There are a number of possible ways to do this: by looking at the very short-range forecast values from an NWP model averaged over a large number of forecasts, the mean forecast values say at day 10, or the mean values from one or more climate-type simulations. Each of these has advantages and disadvantages; for estimating the 'truth', the first choice is to be preferred, but for establishing the model's own hydrological cycle the other choices are necessary. Fig. 3 shows ECMWF operational forecast values (days 9-10), for three complete years during which model changes were made.

The two main changes relevant to this figure are a major change of radiation scheme and convection scheme in May 1989 and an important change to the parametrization of ocean evaporation and convective cloud cover in summer 1990. The impact of these changes can clearly be seen in comparing Fig. 3(a) and (b), and Fig. 3(b) and (c). By comparison with Fig. 2 it is clear that Fig. 3(c) is generally closer to the climatological estimates. Since there are small residual fluxes at the boundaries the model has not achieved total thermal equilibrium by this forecast range. Results from longer climate runs typically achieve consistent but relatively wrong global distributions and magnitudes of these fluxes by internal mutual adjustments and compensations between processes under the overall conservation constraints such as those on total energy, water, angular momentum etc. In this context coupled ocean-atmosphere models tend to magnify model discrepancies in fluxes, this can be considered advantageous in exposing problems but makes such models more problematic in their application.

Examples of the large variation from model to model in basic quantities such as rainfall are to be found in Randall *et al.* (1992) in which fluxes from a large number of GCMs are compared. This study also examines the wide range of sensitivity of the models to changes in sea surface temperatures. Even within the framework of one model, significant differences can be seen as parametrizations are changed or numerical techniques refined. Fig. (4) shows such differences for recent versions of the ECMWF model run at T63 for a season. Each wave corresponds to a version of the model used opperationally or considered for operational use.

The precise details are not relevant, only that this variability is char-

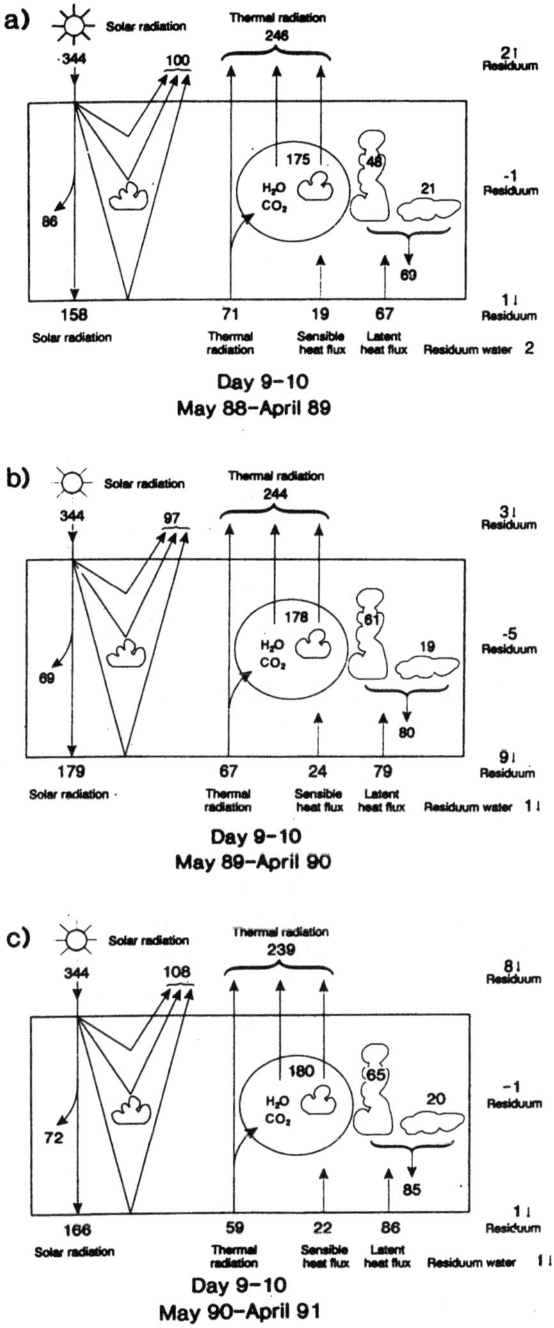

Figure 3: Global annual mean energy and water cycles derived from day 9-10 forecasts for three consecutive years. (Klinker, 1991)

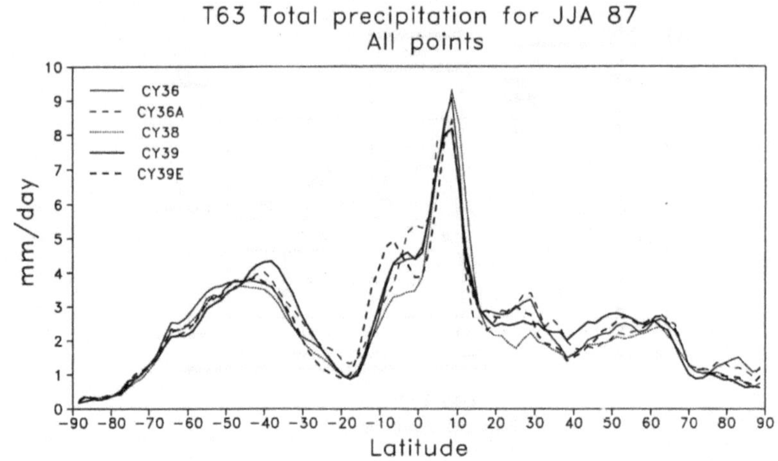

Figure 4: Zonal mean precipitation from five 120 day T63 summer integrations starting from 1/5/87 and averaged over the period June - August. Each curve identifies a different version of the model physical parametrization schemes or numerical algorithms.

acteristic of typical model changes. Further insight into the basic mechanisms of the hydrological cycle can be gained by looking at zonal mean cross-sections from model simulations. Fig. 5 shows results of three-monthly average tendencies of various physical processes calculated from a T63 winter simulation. The precise partitioning between processes is a function of the model in question and its associated physics. However the role of each process is clearly shown as is the nature of the balance between diabatic and adiabatic processes in Fig. 5a. The relative latitudinal importance of convective and stratiform (gridscale) condensation is shown in Fig. 5b and the role of boundary layer turbulent fluxes and shallow (predominantly) non- precipitating convection illustrated in Fig. 5c. In the tropical belt, the radiative cooling in the free atmosphere and the convective heating are close to balancing, as shown in Fig. 5d. This balance is basic to the observed quasi-equilibrium temperature and moisture profiles which characterise much of the tropical troposphere (see Section 5).

2.1 Basic Equations

Provided the scales of motion to be parametrized are distinctly separated from the resolved scales, typical model equations for dry static energy ($\overline{s}$)

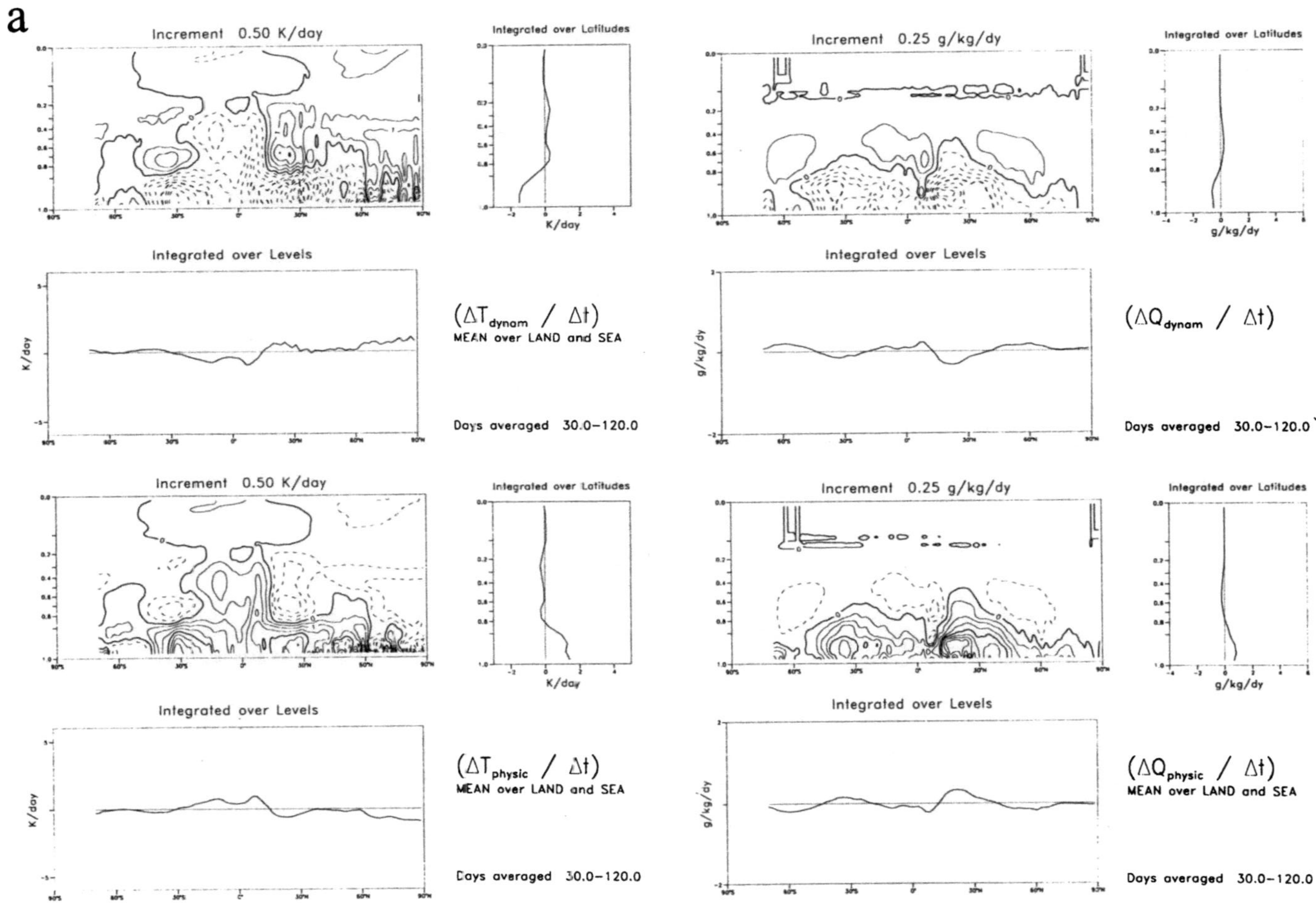
a
Increment 0.50 K/day
Integrated over Latitudes
K/day
Integrated over Levels
K/day
(ΔT_dynam / Δt)
MEAN over LAND and SEA
Days averaged 30.0–120.0
Increment 0.25 g/kg/dy
Integrated over Latitudes
g/kg/dy
Integrated over Levels
g/kg/dy
(ΔQ_dynam / Δt)
Days averaged 30.0–120.0
Increment 0.50 K/day
Integrated over Latitudes
K/day
Integrated over Levels
K/day
(ΔT_physic / Δt)
MEAN over LAND and SEA
Days averaged 30.0–120.0
Increment 0.25 g/kg/dy
Integrated over Latitudes
g/kg/dy
Integrated over Levels
g/kg/dy
(ΔQ_physic / Δt)
MEAN over LAND and SEA
Days averaged 30.0–120.0

b

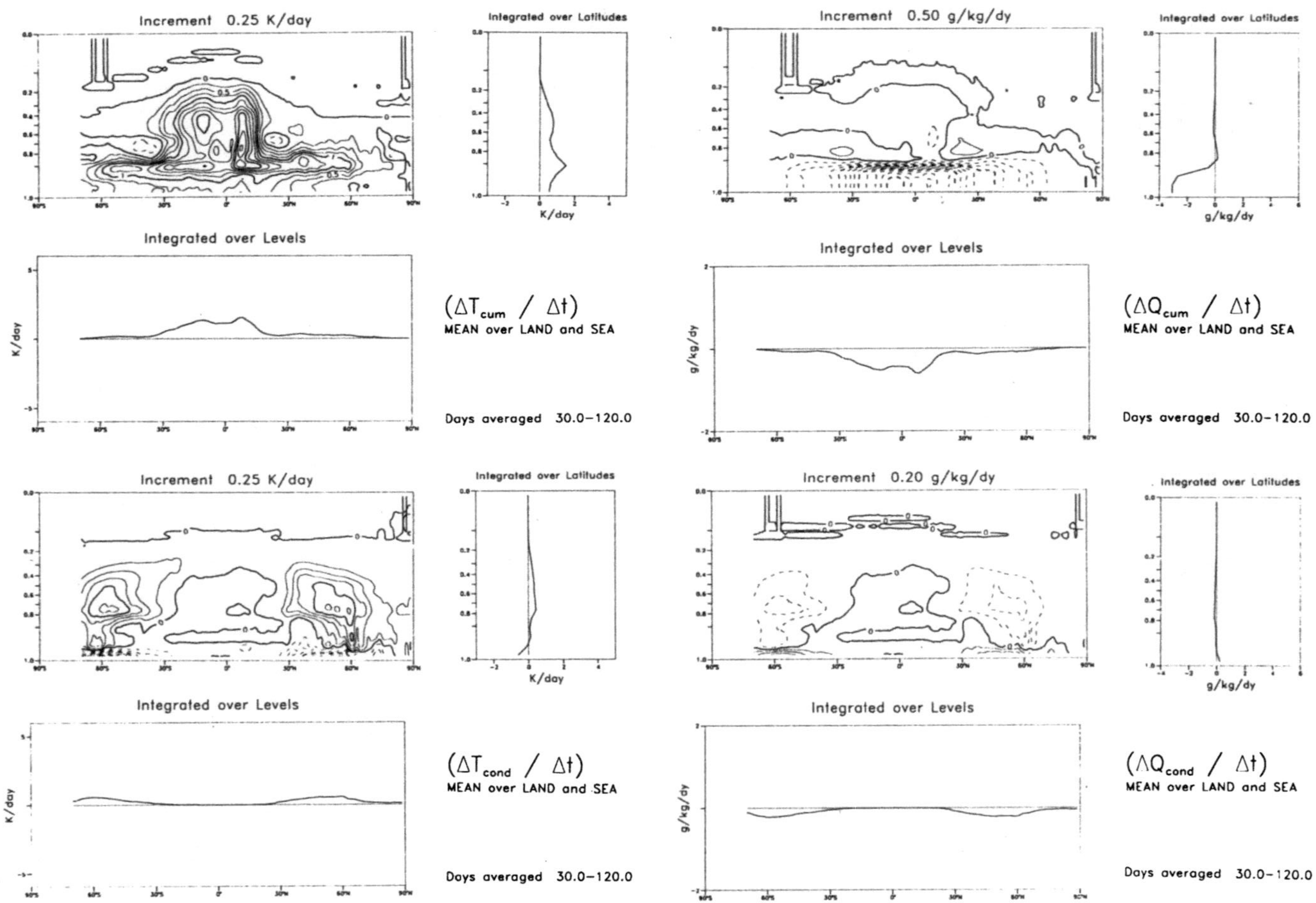
Increment 0.25 K/day
Integrated over Latitudes
K/day
Integrated over Levels
K/day
$(\Delta T_{cum} / \Delta t)$
MEAN over LAND and SEA
Days averaged 30.0–120.0
Increment 0.50 g/kg/dy
Integrated over Latitudes
g/kg/dy
Integrated over Levels
g/kg/dy
$(\Delta Q_{cum} / \Delta t)$
MEAN over LAND and SEA
Days averaged 30.0–120.0
Increment 0.25 K/day
Integrated over Latitudes
K/day
Integrated over Levels
K/day
$(\Delta T_{cond} / \Delta t)$
MEAN over LAND and SEA
Days averaged 30.0–120.0
Increment 0.20 g/kg/dy
Integrated over Latitudes
g/kg/dy
Integrated over Levels
g/kg/dy
$(\Delta Q_{cond} / \Delta t)$
MEAN over LAND and SEA
Days averaged 30.0–120.0

c

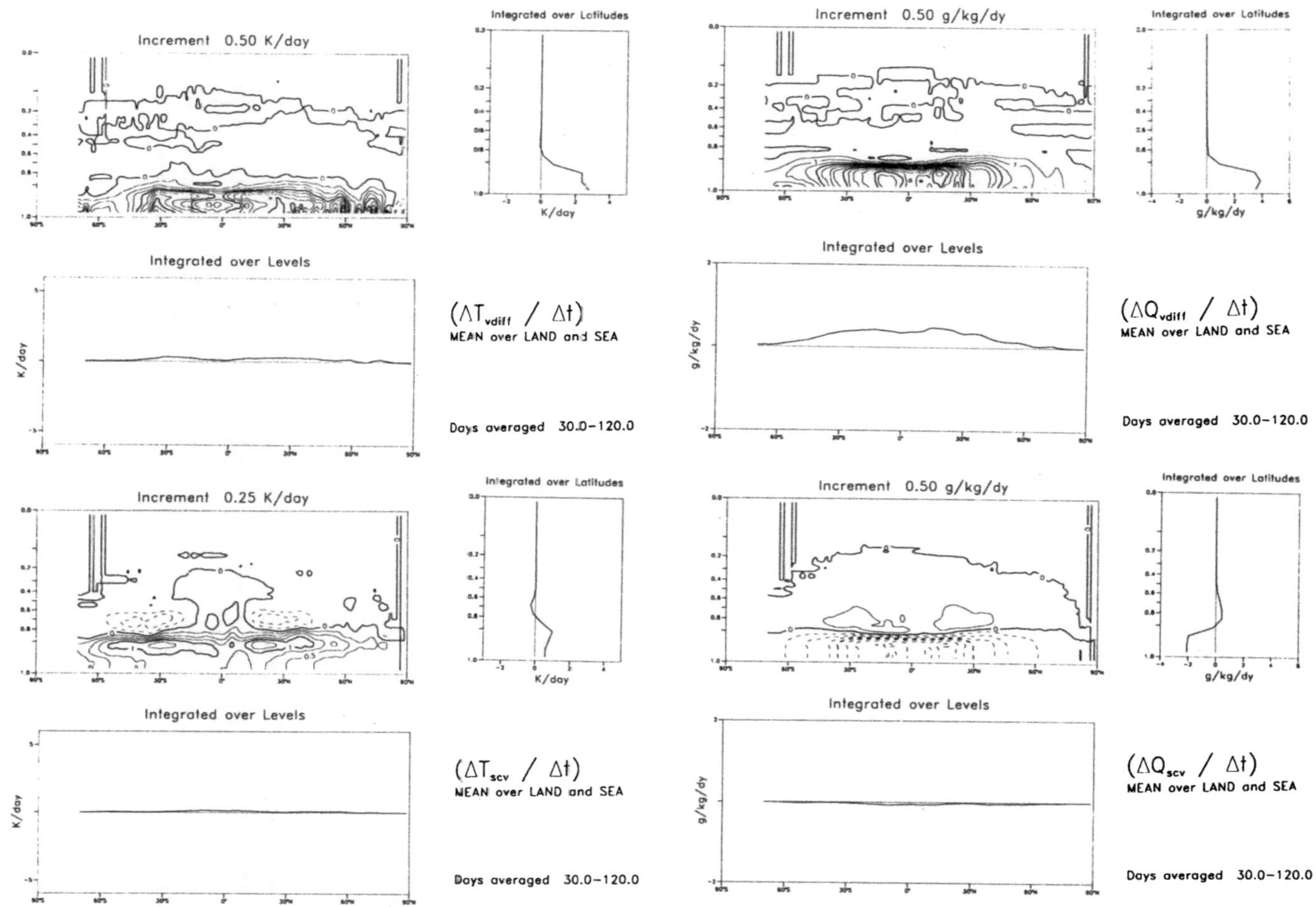
Increment 0.50 K/day
Integrated over Latitudes
K/day
Integrated over Levels
K/day
(ΔT_vdiff / Δt)
MEAN over LAND and SEA
Days averaged 30.0-120.0
Increment 0.50 g/kg/dy
Integrated over Latitudes
g/kg/dy
Integrated over Levels
g/kg/dy
(ΔQ_vdiff / Δt)
MEAN over LAND and SEA
Days averaged 30.0-120.0
Increment 0.25 K/day
Integrated over Latitudes
K/day
Integrated over Levels
K/day
(ΔT_scv / Δt)
MEAN over LAND and SEA
Days averaged 30.0-120.0
Increment 0.50 g/kg/dy
Integrated over Latitudes
g/kg/dy
Integrated over Levels
g/kg/dy
(ΔQ_scv / Δt)
MEAN over LAND and SEA
Days averaged 30.0-120.0

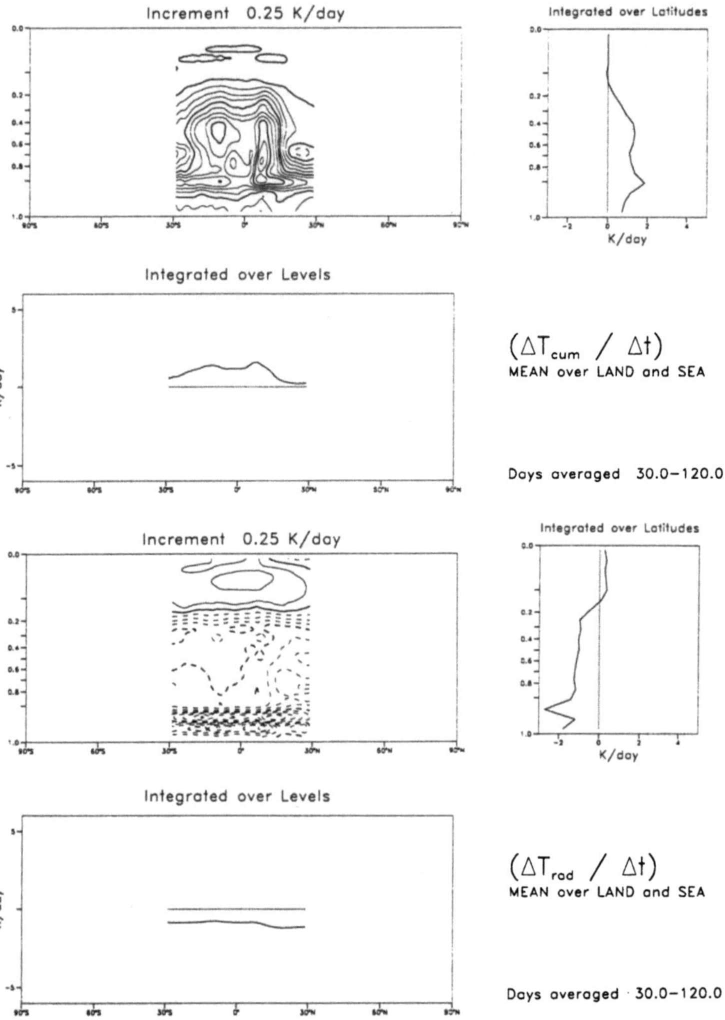

Figure 5: Zonally averaged three-month (Dec/Jan/Feb) mean values of temperature and moisture tendencies (units °K/day and g/kg/day) from a T63 simulation (initial date 1/11/90). a) Total dynamics (upper) and total physical processes (lower) b) Total convection (upper) and large-scale condensation (lower) c) Vertical diffusion (upper) and shallow convection (lower) d) Convection (upper) and Radiation (lower) (temperature tendencies for tropical latitudes only).

(or alternatives) and specific humidity ($\overline{q}$) can be written as

$$\frac{\partial \overline{s}}{\partial t} + \overline{v}.\nabla \overline{s} + \overline{w}\frac{\partial \overline{s}}{\partial z} = -\frac{1}{\overline{\rho}}\frac{\partial}{\partial z}(\overline{\rho w' s'}) + L(\overline{C} - \overline{E}) + C_P Q_{RAD} \tag{1}$$

where $s = C_p T + gz$ and

$$\frac{\partial \overline{q}}{\partial t} + \overline{v}.\nabla \overline{q} + \overline{w}\frac{\partial \overline{q}}{\partial z} = -\frac{1}{\overline{\rho}}\frac{\partial}{\partial z}(\overline{\rho w' q'}) - (\overline{C} - \overline{E}) \tag{2}$$

Primed variables are deviations from gridscale means and horizontal derivatives of prime terms are neglected. $\overline{C}, \overline{E}$ are condensation and evaporation rates and Q_{RAD} contains all radiative terms. Global integration and simple manipulations provide the formal relationships implied in the preceding budget discussion. Representation of the terms on the right hand side of equations (1) and (2) is considered in the following sections.

3 Radiation and Cloud Interactions

As the primary energy source for the atmosphere/ocean climate system, it is self-evident that as accurate a representation of radiative fluxes as possible is needed in any large-scale model. The main hypotheses for climatic variation and change pivot upon radiative sensitivities either through internal atmospheric properties or external variations in solar input. Fortunately the basic radiative transfer equation and its dependencies on atmospheric constituents is well understood, and very accurate models are possible; however the computational requirements are prohibitive in, for example, long climate simulations and considerable simplification and approximation are necessary to produce a radiation scheme for such a model. The calculation of the radiative fluxes requires detailed knowledge of the water vapour distribution, the cloud amount and its vertical distribution, and the water/ice content of cloud. The often large uncertainties in these translate into inaccuracies in radiative fluxes. A detailed discussion of radiative transfer, its role and parametrization in climate models can be found in Fouquart (1988).

The importance of cloud/radiation interactions has long been recognized and the role of cloud feedbacks in climate change scenarios much debated in the literature (e.g. Cess, 1976, Arking, 1991). Estimates of cloud/radiative feedback (Slingo, 1990) indicate that the direct greenhouse warming due to doubling the atmospheric CO2 can be compensated by i) increasing lowlevel cloud amounts from, for example, 25% to 30%; ii) reducing cloud droplet radii from, for example, 10μ m to 8μ m; iii) increasing the liquid water content of the low clouds from, for example,

50 g m^{-2} to 65 g m^{-2}; all very much within observed physical limits. More recent debate has focused on the sensitivity of the climate system through radiative processes to the water vapour distribution especially in the middle/upper tropical troposphere (Lindzen, 1990a,b; Betts, 1990). Since this water vapour distribution is not well-known in nature and still rather poorly treated in models, this will continue to be an important area of research for some time.

Calculation of radiative fluxes in a 3-D cloud field is in its infancy, and currently clouds are treated as occupying one or more model layers within a grid column, with a fractional area coverage and an associated liquid/ice water content. Since model layers may have different fractional cloud areas further assumptions must be made to overlap the layer values. This then becomes a model artefact which depends on vertical and horizontal resolution. The importance of the vertical distribution of cloud is simply illustrated in Fig. (6). In this example low cloud increases the infrared cooling whilst high cloud reduces it. The availability of radiation and cloud data from programmes such as the Earth Radiation Budget Experiment (ERBE) and International Satellite Cloud Climatology Project (ISCCP) means that verification of radiative parametrizations (and associated cloud schemes) is possible to a much greater degree than other parametrizations such as convection or surface fluxes for example.

An example of ERBE data produced by J-J Morcrette (personal communication) (Fig. 7), based on Ramanathan *et al.* (1989) shows the globally averaged long and short wave fluxes at the top of the atmosphere and the corresponding clear sky values extracted selectively from the dataset. Also shown are the net fluxes (i.e. SW minus LW). Of particular interest in the climate problem is the quantification of the global role of clouds. The figure shows the SW and LW cloud forcing (i.e. total minus clear SW fluxes and the clear minus total LW fluxes) and the net cloud forcing (i.e. SW cloud forcing plus LW cloud forcing). The net effect of cloud is thus negative, i.e. a cooling relative to the clear-sky case.

4 Clouds

The importance of clouds has already been stressed. However the lack of knowledge of cloud processes and their complex interactions with turbulence, radiation and advection has made rapid progress in cloud parametrization impossible. Climate model simulations with varying cloud representations have underlined the inadequacies in cloud parametrizations and clouds remain one of the major uncertainties in any climate sensitivity scenarios (e.g. Mitchell, 1989; Cess *et al.*, 1989).

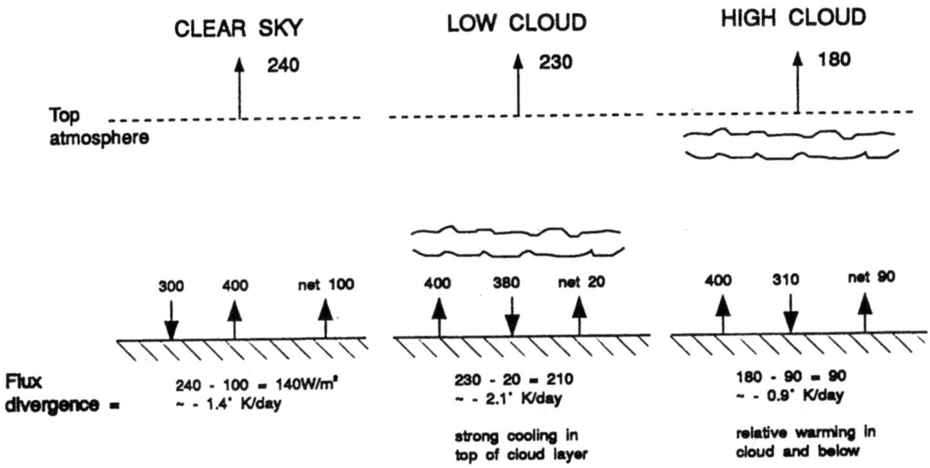

Figure 6: A simplified illustration of how a layer of low or high cloud modifies the thermal radiative flux divergence and hence the vertically integrated cooling of an atmospheric column.

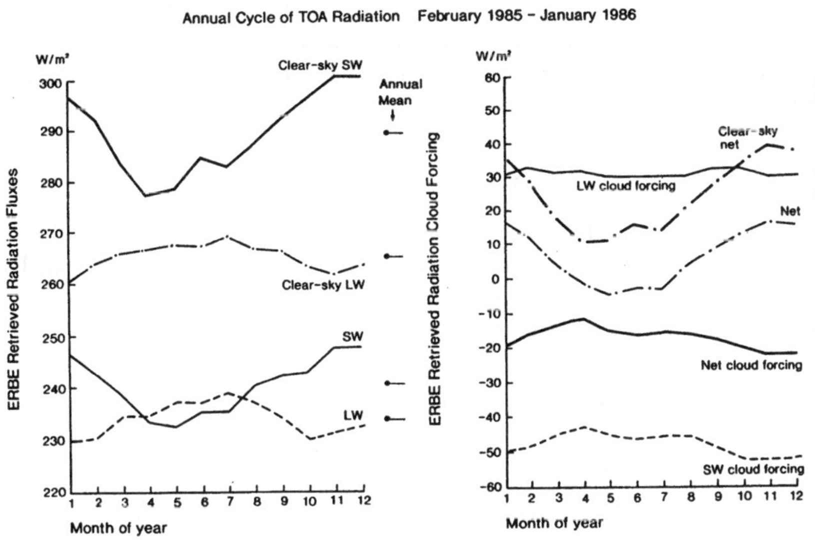

Figure 7: The annual cycle of radiative fluxes at the top of the atmosphere deduced from ERBE data (Units: W/m^2).

In the last decade parametrizations have been based predominantly on *diagnosing* cloud cover from atmospheric parameters such as relative humidity, vertical velocity and stability (e.g. Slingo, 1987). Such schemes have quite successfully produced the main cloud distribution but it is necessary and difficult to prescribe liquid water contents and associated optical characteristics. Also there are considerable problems in handling boundary layer clouds in particular.

Since such diagnostic schemes are well documented in the literature, attention is now focused on the development of *prognostic* cloud schemes which attempt to unify cloud processes within a physically consistent treatment of water and its phase changes. Such parametrizations have been developed and used in cloud resolving models for many years but adaptation for large-scale models was pioneered by Sundquist (1978). This approach requires the realistic treatment of cloud microphysics, turbulence and cloud optical properties, and many difficulties remain in these areas.

Perhaps the most difficult problem is the treatment of convective clouds and their anvils. It is recognised that large thick anvil clouds provide a significant fraction of the precipitation associated with convection particularly when organized in squall lines and clusters.

A further complication in prognostic schemes is the need for modelling the horizontal advection of cloud parameters, especially in higher resolution models and for ice clouds.

4.1 Prognostic Cloud Scheme

Tiedtke (1992) considers the basic problems of representing clouds by two prognostic variables, namely cloud water and cloud area, and their sources and sinks in the following manner.

Equations for the prognostic variables can be written as:

$$\frac{\partial l}{\partial t} = A(l) + S_{CV} + S_{BL} + C - E - G_P$$

$$\frac{\partial a}{\partial t} = A(a) + S(a)_{CV} + S(a)_{BL} + S(a)_C - D(a)$$

where $A()$ represents the advective changes, S_{CV} and S_{BL} sources from convection and boundary layer turbulence, C and E condensation/evaporation of cloud water, G_P the rate of generation of precipitation and $S(a)_c$ and D are the increase and decrease of cloud due to condensation or evaporation respectively.

Equations for specific humidity q, dry static energy $S(= C_P T + qz)$ and precipitation flux, P complete the equation set for moist processes i.e.

$$\frac{\partial q}{\partial t} = A(q) - S_{BL} - C + E + E_P \tag{3}$$

$$\frac{\partial s}{\partial t} = A(s) + L(S_{BL} + C - E - E_P) - L_F M \tag{4}$$

$$P = \int (G_P - E_P)\rho\, dz \tag{5}$$

where L and L_F are the latent heat of condensation and freezing respectively and M is the rate of snowmelt.

Some terms due to cumulus convection no longer appear explicitly in the equations for q and S in this unified approach. Discussion of the source terms $S_{CV}, S(a)_{CV}, S_{BL}$ and $S(a)_{BL}$ can be found in Tiedtke (loc sit). An important role of deep convection is described by this formulation in that the evolution of anvil clouds and their significance as sources of hydrometeors are represented and this coupling of convective and stratiform clouds is an undoubted strength of such an approach. However there are many difficulties and uncertainties in defining source terms and there is much scope for improvement and increased sophistication. It remains to be seen how much better such prognostic schemes can represent boundary layer clouds under a wide range of conditions such as marine stratocumulus, arctic stratus or fog.

5 Cumulus Convection

Cumulus convection is a major process in determining the atmospheric circulation and its thermodynamic structure. It influences the atmosphere through diabatic heating due to latent heat release, turbulent transports of heat, moisture and momentum and through the interaction of convective clouds and radiation. Depending on the time and space scales, each of these can play a dominant role. Diabatic heating is fundamental to the tropical heat balance and the large-scale tropical circulations but it is also a key component in many smaller scale systems such as hurricanes, clusters and squall lines. On the other hand the mean thermodynamic structure and characteristics of the tradewind regions are crucially dependent on the turbulent fluxes of heat and moisture from shallow mostly non-precipitating clouds. These shallow cumulus clouds, the much deeper cumulonimbus clouds with their often extensive anvils and the stratocumulus decks over large areas of ocean, together play a major role in determining the earth's radiation balance and hence its climate. Although convection is predominant in the tropics and subtropics, convection in the extratropics can be equally important, for example in modulating the diurnal cycle of cloud, precipitation and temperature over continents and in the rapid and extensive warming and moistening of cold dry continental and polar air as it flows out over the oceans in wintertime.

Representing these effects must therefore be a key requirement in any detailed atmospheric model. However since convection is almost exclusively subgridscale, the ensemble of these varied effects must be parametrized in terms of the resolved parameters describing the atmosphere in which convection occurs. The relationships between the type and magnitude of convective activity and the large-scale flow parameters, known as the 'closure' problem, is fundamental, and relating the magnitude of the convective fluxes to larger-scale parameters is central to the parametrization problem. Further, the complexity of microphysical processes, precipitation efficiency and subcloudscale mixing and evaporation ensure that no simple answer to the closure problem exists. However observational studies have confirmed that, at least in lower latitudes, basic constraints on the atmospheric thermodynamic structure exist, and enable progress to be made. Fig. (8) from Tiedtke (1990) shows how some well-known schemes relate to each other in terms of closure. The basis for computing the vertical distribution of convective fluxes varies in complexity from direct adjustment of profiles to attempting to define the spectrum of clouds within each grid box. Ideally a parametrization scheme should represent the subgridscale fluxes due to all the processes that characterise convective clouds and their organization forms. Because our understanding of convection is still rather basic and limited by theoretical and observational difficulties, current schemes are far from comprehensive.

Detailed consideration of the 'closure' problem can be found in Arakawa and Chen (1986) and also in Tiedtke (1988); this latter reference also provides an excellent and extensive discussion of the main problems of convective parametrization and the formulation of the schemes in common use in large-scale models. Although many models still use schemes such as the Kuo scheme (Kuo, 1974) or even simpler moist adiabatic adjustment (Manabe *et al*, 1965); it is widely accepted now that schemes based on a massflux approach or more realistic adjustment techniques are superior. The following briefly summarises the most salient features of two such schemes in current use in state-of-the-art GCM and regional models. One, the massflux scheme developed by Tiedtke (1989), the

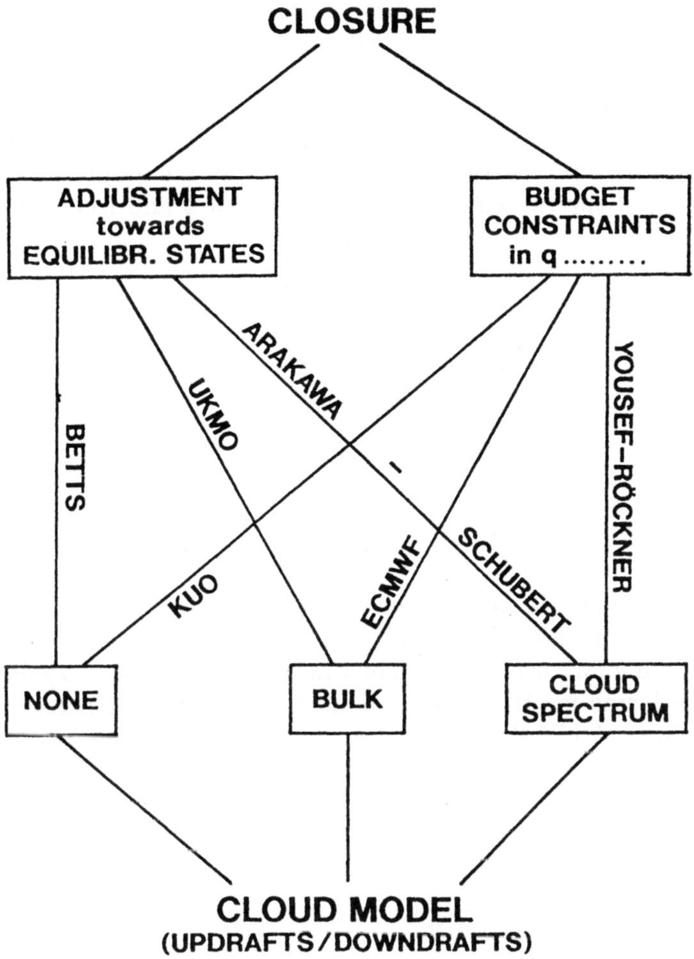

Figure 8: Classification of conventional cumulus parametrization schemes (Tiedtke, 1990).

other an adjustment scheme developed by Betts and Miller (1986, 1992). Full details can be found in these references. The massflux approach was pioneered by Ooyama (1971) and Arakawa and Schubert (1974). By assuming a 1-dimensional cloud model, and with suitable simplification and averaging of the updraught equations for mass continuity and flux forms of the equations for dry static energy (s), water vapour (q) and cloud liquid water (l); the massflux M and fluxes of s, q and l can be written

$$\frac{\partial M_c}{\partial z} = E - D \tag{6}$$

$$\frac{\partial}{\partial z}(M_c s_c) = E\overline{s} - Ds_c + L\overline{\rho}C \tag{7}$$

$$\frac{\partial}{\partial z}(M_c q_c) = E\overline{q} - Dq_c - \overline{\rho}C \tag{8}$$

$$\frac{\partial}{\partial z}(M_c l) = -Dl + \overline{\rho}C - \overline{\rho}G_p \tag{9}$$

where $\overline{()}$ are environment values and $()_c$ are cloud values.

Provided the profile of M_c can be calculated (from a closure assumption such as a moisture budget constraint on the boundary layer) and appropriate forms for the entrainment and detrainment rates E, D and the conversion of cloudwater to precipitation (G_p) are assumed, then these equations can be solved, layer by layer, for the relevant fluxes.

The eddy flux terms in equations (1) and (2) can now be calculated since

$$(\frac{\partial s}{\partial t})_{CONV} = -\frac{1}{\overline{\rho}}\frac{\partial}{\partial z}(\overline{\overline{\rho}w's'}) \simeq -\frac{1}{\overline{\rho}}\frac{\partial}{\partial z}\{M_c(s_c - \overline{s})\} \tag{10}$$

and

$$(\frac{\partial q}{\partial t})_{CONV} = -\frac{1}{\overline{\rho}}\frac{\partial}{\partial z}(\overline{\overline{\rho}w'q'}) \simeq -\frac{1}{\overline{\rho}}\frac{\partial}{\partial z}\{M_c(q_c - \overline{q})\}. \tag{11}$$

Similar computations can be performed for convective downdraughts. In principle, it is possible to represent various types of convective motion by extending equations (10,11). The effects of windshear and convective organisation are clearly problematic in this approach however.

By utilizing the observed quasi-equilibrium of temperature and moisture structures in a convective atmosphere, Betts and Miller (1986) developed a generalised adjustment scheme which relaxes the atmosphere towards profiles characterising the observed quasi-balance of convection and large-scale forcing. The analogues of equations (10) and (11) can then be written as

$$(\frac{\partial T}{\partial t})_{CONV} = (T_{ref} - \overline{T})/\tau \tag{12}$$

$$(\frac{\partial q}{\partial t})_{CONV} = (q_{ref} - \overline{q})/\tau \tag{13}$$

where the subscript '*ref*' denotes the reference profiles to be computed and τ is a relaxation timescale. Details of the construction of these reference profiles can be found in the afore-mentioned reference, crucially these profiles differ from a wet adiabat in being subsaturated and more unstable in the lower half of the troposphere and more stable above. In principle, it is possible to represent convection characteristic of different regions and of different types, by adjusting towards different reference profiles. Although much simpler than massflux schemes, additional assumptions and computation are required if byproducts such as massfluxes or cloudwater content are required.

6 Shallow Convection

The preceding sections have focused primarily on deep, precipitating convection. The importance of shallow convection has, until recently, been underestimated. Furthermore the rather shallow cloud layers are difficult to represent in low vertical resolution models. Shallow cumulus clouds occupy very large regions of the globe; over much of the tropical oceans throughout the year, substantial areas of the tropical and 'summer' land masses and large areas of the winter oceans in cold air outbreaks. Shallow convection helps to maintain the PBL structure in the Tradewind areas by providing cooling and moistening to balance the drying and warming effect of subsiding largescale motion. Without parametrizing this process the PBL becomes very shallow and moist with a very strong inversion above. Consequently, ocean evaporation is underestimated and the supply of moisture into the deep convection areas such as the ITCZ is limited. The modelled hydrological cycle is then too weak. Over land the diurnal cycle of shallow convective cloud cover modulates the available solar radiation reaching the surface and hence the surface temperatures and rate of evaporation. In the last few years large-scale models have included parametrization of the fluxes (and cloud cover) due to these shallow clouds.

The most commonly used approach is to treat the convective fluxes as an extension of the turbulence parametrization, either an eddy diffusion scheme or higher order closure scheme. Tiedtke *et al.* (1988) show the substantial impact of an enhanced eddy diffusion approach and this has been adopted in several of the main models. An alternative approach using the adjustment concept was developed by Betts and Miller (1986). This is basically as already discussed for deep convection; however, the adjustment uses the clear observational evidence that shallow convection is a moist mixing process between cloudlayer air and the drier air above. Since this mixing process is well defined by the mixing-line concepts proposed by Betts (1982), this adjustment approach requires no other observational input such as observed lapse rates and is therefore self-contained. A third type of scheme is the logical extension of the massflux approach already discussed for deep convection. To close this application of the massflux approach, the cloudbase massflux is determined primarily by the turbulent moisture flux at the surface. This closure is supported by observations, particularly for maritime convection.

The preceding discussions have not referenced the difficult problem of parametrizing convective transports of momentum or vorticity. In this context the use of 1-D cloud models and/or adjustment ideas are no longer such useful simplifications and although some attempts have been made to incorporate momentum fluxes in an analogous way to thermodynamic ones, this remains highly problematic and parametrization of such fluxes remains very much an open question.

The extent to which the full complexity of convective processes is parametrized identifies these two basic approaches to parametrization. Either one tries to represent as much of the physics of clouds as possible or one bypasses the use of cloud models and adjusts the thermodynamic profiles directly. Use of one or more cloud models is physically realistic but is limited by the simplicity and uncertainties of the prototype cloud models used. Direct adjustment neatly avoids this but obscures the relative importance of separate cloud processes. Nevertheless, experience with both massflux-type and generalised adjustment schemes suggests that both are capable of modelling the basic diabatic forcing due to tropical convection. In the extratropics, the important role of windshear and the inappropriateness of quasi-equilibrium and moisture budget closures ensure that existing convective parametrizations are rather less satisfactory and leave considerable scope for improvement.

7 Surface Fluxes

As was discussed in section 2, the available solar energy reaching the earth's surface and its partitioning into sensible and latent heat fluxes is very sensitive to model parametrizations of cloud and surface characteristics such as availability of moisture from the ocean surface.

The evaporation plays a fundamental role in the energetics and dynamics of the atmosphere. This is especially so in the Tropics where the sea surface temperature (SST) is relatively high and where the moisture from the ocean is a direct energy source for convection. Consequently variations in SST modulate convective activity, and the ability of GCMs to represent the tropical circulations and to simulate the sensitivity of the atmosphere to SST variations (such as 'El Nino' or such as might occur in a changing climate) depends critically on the parametrization of surface latent heat fluxes. Recently it was discovered that the ECMWF model was deficient in its treatment of evaporation at very low windspeeds such as are found in the tropical 'warm pool' areas. The following discussion and results summarizes parts of Miller *et al.* (1992). Although this refers to a change in surface flux parametrization, the impacts are very significant on the other components of the model's hydrological cycle and model systematic errors, hence its inclusion here.

The sensitivity to the evaporation parametrization was discovered indirectly through the results of an SST experiment in which the SST was enhanced by 1°K in the region bounded by (95E-140E; 20N-10S). The rainfall distribution of this experiment (a T42, 90 day simulation with the ECMWF model) is shown in Fig. (9). A more realistic rainfall climatology and an improved Indian monsoon circulation were generated. Improvements to zonal mean wind, temperature and moisture errors were also apparent. In general the positive impact on the mass and wind field from the SST enhancement was consistent with large scale dynamical responses to increased diabatic heating in the vicinity of the area of SST change. Fig. (10) shows large scale difference maps of this response. As this area encompassed the warm pool regions of very low windspeed, the parametrization of surface fluxes was re-examined. Generally models all use surface flux equations of the general form $F_\psi = \rho C_\psi |v| (\psi_1 - \psi_s)$ where ψ_s is the value of ψ at the surface and subscript 1 refers to the lowest model level. C_ψ is the transfer coefficient for ψ and contains dependencies of the turbulent fluxes on atmospheric stability, windshear and surface characteristics such as roughness.

In the ECMWF model C_ψ is determined using Monin-Obukhov similarity theory plus surface roughness lengths for momentum, heat and moisture. These roughness lengths are set equal to each other with that of momentum calculated according to the Charnock relation

$$z_0 = Max\{0.018\frac{u_*^2}{g}, 1.5 \times 10^{-5}\} \tag{14}$$

where u_* represents the friction velocity.

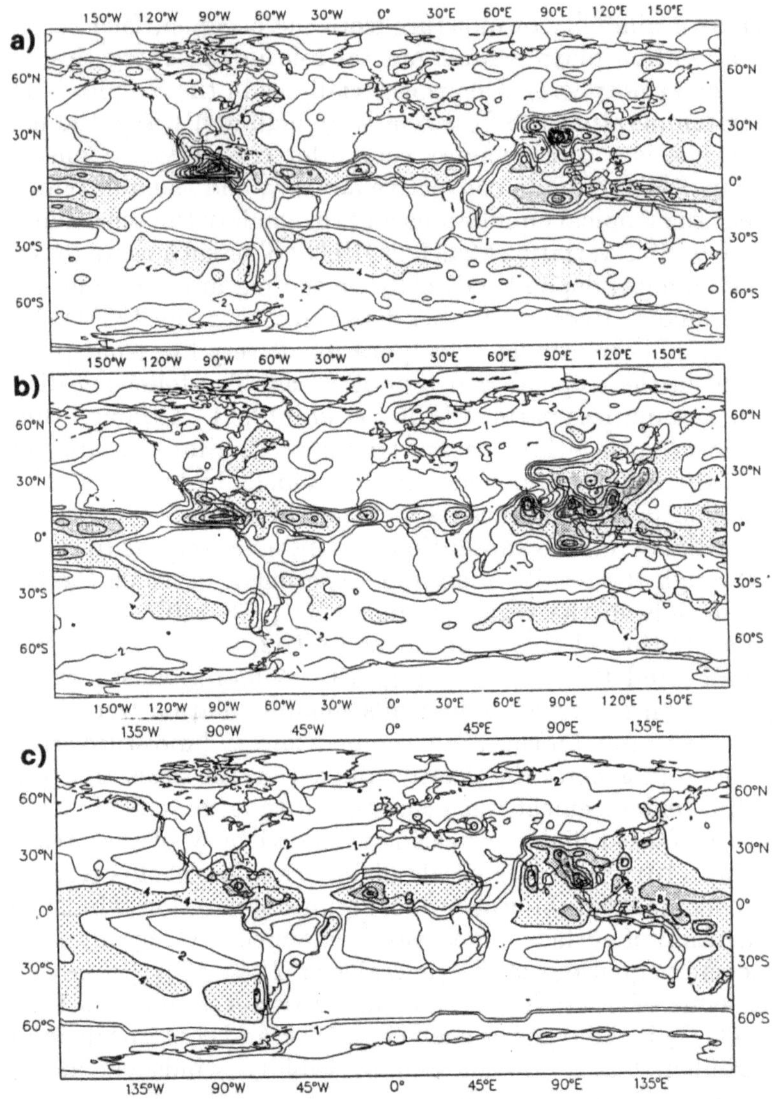

Figure 9: Rainfall averaged over 90 days from a T42 90-day experiment with observed SST (June, July and August). Rainfall for a similar run with the SST artificially increased by 1K in the area between 10°S and 20°N and between 95° and 140°E is shown in b). The climatology by Jaeger for June, July and August is given in c). Contour intervals are: 1, 2, 4, 8, 12, 16 ... millimetres per day, and areas with rainfall larger than 4 mm per day are stippled.

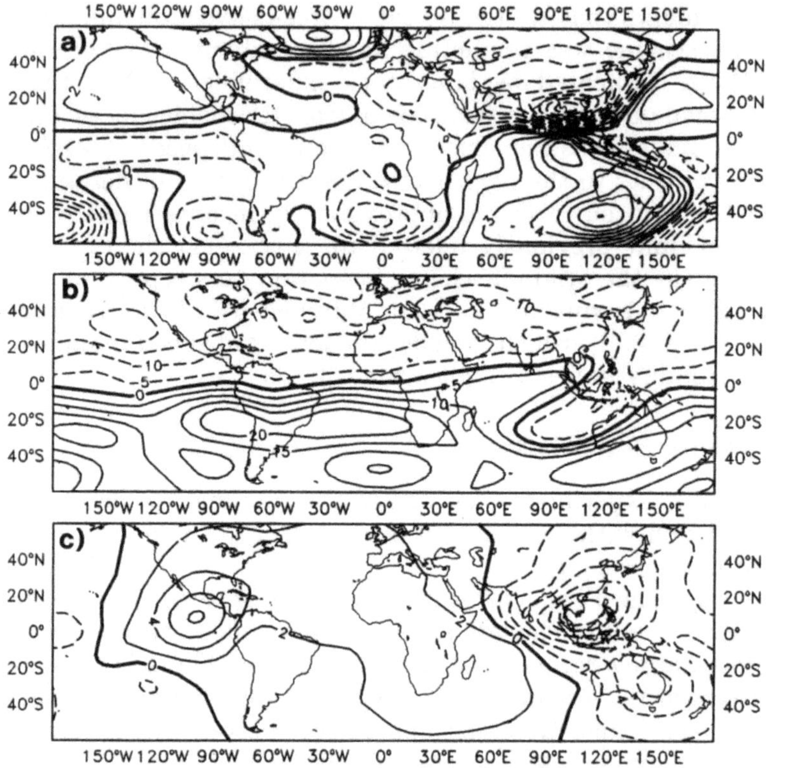

Figure 10: The 90-day mean streamfunction difference at 850 mb a) (contour interval 1 x 10^6 m^2 s^{-1}) and 200 mb b) (contour interval 5 x 106 m^2 s^{-1}) and the velocity potential difference at 200 mb c) (contour interval 2 x 10^6 m^2 s^{-1}) between the integration with 1K SST anomaly and the control integration.

It was found that at the so-called free-convection limit (windspeed $\rightarrow 0$) the scheme was predicting fluxes of about 5 W/m^2 rather than 35-40 W/m^2 indicated by laboratory experiments. Also the available observational data for the variation of transfer coefficient with windspeed did not provide measurements at very low windspeeds $< 4m/s$.

Improvements to the parametrization (discussed in detail in Miller *et al.* (1992) resulted in fluxes approximately in agreement with the free convection limit previously referred to. Fig. (11) shows the improved scheme versus the original scheme for parameters typical of the tropical ocean. The following results were obtained using the 'revised' scheme indicated in Fig. 11 which includes the low wind speed modifications but does not modify fluxes at higher windspeeds.

7.1 Model Sensitivity

Results from experiments for winter and summer 90 day periods with T42 resolution with and without the revision to the ocean surface fluxes are now discussed.

By comparing Figs. (12) and (10) it can be seen that the pattern of response to the flux change is very similar indeed to the SST anomaly experiment albeit rather smaller in magnitude. There is increased upper level divergence over the warm pool region and remote descent over the Eastern Pacific and increased upper level westerly flow, clearly a large scale remote dynamical response. There is some increase in evaporation over large areas of the tropical ocean and especially the warm pool regions (Fig. (13)). The crucial interaction between evaporation changes and the convection (whereby latent energy is converted into diabatic heating) is illustrated in Fig. 14a,b and c. Large increases in the mean boundary layer equivalent potential temperature are reflected in a corresponding warmer and moister troposphere as the mean thermodynamic structure is adjusted and maintained by the deep convection.

Consistent with the SST experiment, there are improvements to the rainfall and monsoon circulations (not shown). Perhaps the most striking effect of the evaporation change can be seen in the reduction of zonal mean wind and temperature errors for both winter and summer, Fig. (15). As referred to previously, there is a marked increase in (tropical) upper troposphere westerlies giving a major improvement in the pre-existing easterly bias.

As well as impacting on the systematic errors of the model, experiments with the observed SST anomalies from different years (Palmer *et al.*, 1992) showed that the response of the model to these interannual anomalies was strongly sensitive to this flux parametrization change; and

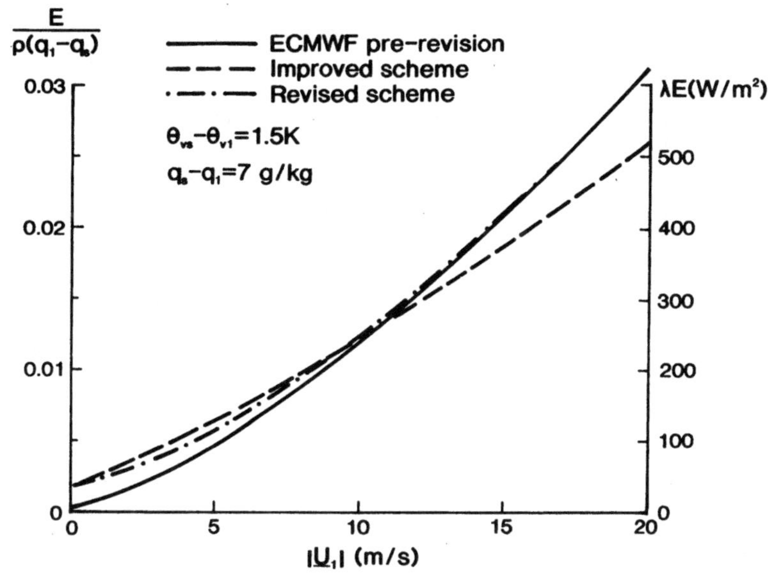

Figure 11: The "conductivity" of the lowest model layer for moisture transfer as a function of wind speed. The reference height is 32 m, which is the height of the lowest level in the ECMWF model. A typical tropical situation has been chosen with an SST of 30°C and a virtual potential temperature difference over the lowest model layer of 1.5 K. The right hand scale indicates the latent heat flux for a specific humidity difference of 7 g kg^{-1}. The revised scheme can be seen as an empirical fit of the Monin Obukhov scheme at low wind speeds.

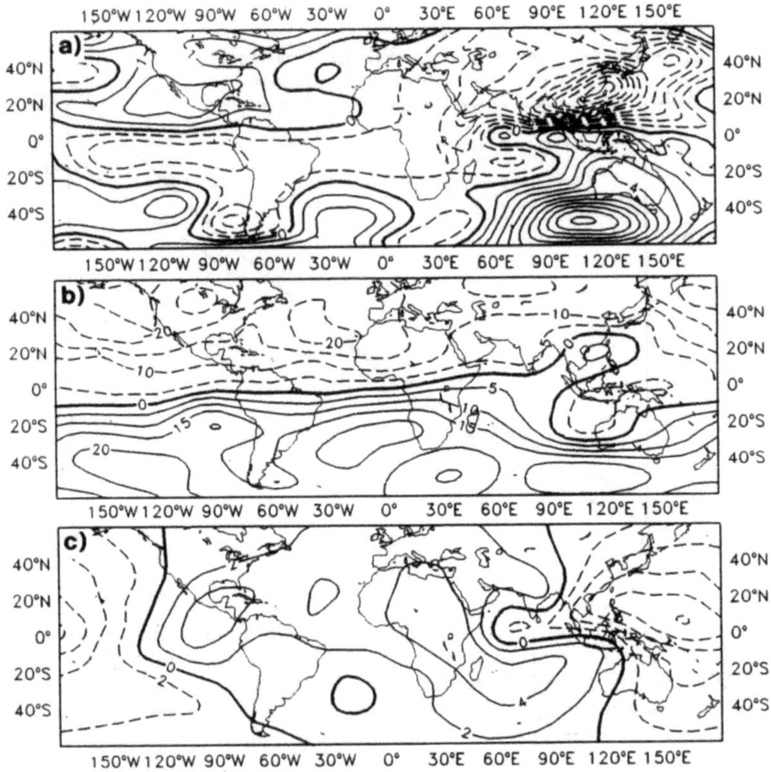

Figure 12: The 90-day mean streamfunction difference at 850 mb a) and 200 mb b) and the velocity potential difference at 200 mb c) between integrations with and without the revised flux scheme. Contour intervals as in Fig. 10.

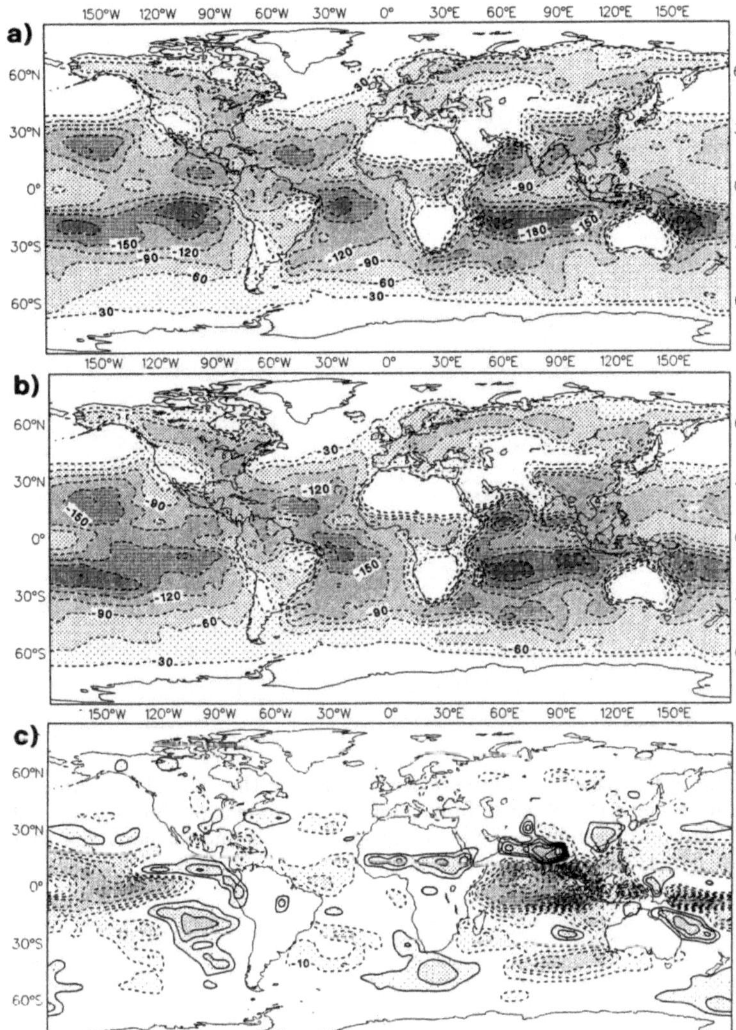

Figure 13: The 90-day average of the latent heat flux at the surface for June-July-August integrations with (a), the old and (b) the revised flux schemes. Dashed contours indicate negative values that correspond to upward fluxes (by model convention). Panel (c) gives the corresponding differences (revised minus old) with contour intervals of 10 W m^{-2}.

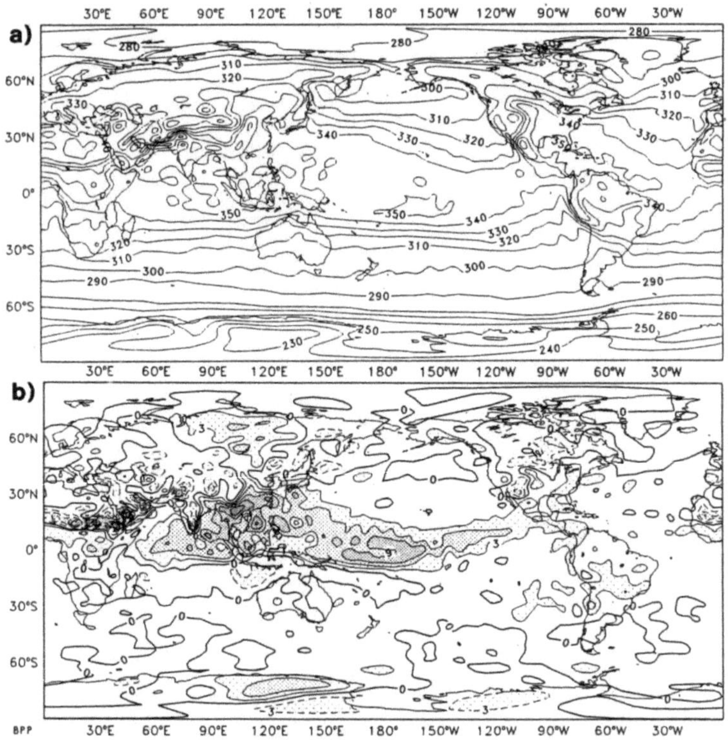

Figure 14: The upper panel (a) represents a θ_e map for the control experiment without the revised flux scheme.(b) Mean difference of the equivalent potential temperature (θ_e) at 1000 mb between June-July-August integrations with and without the revised flux scheme (lower panel) (contour interval of 3°).

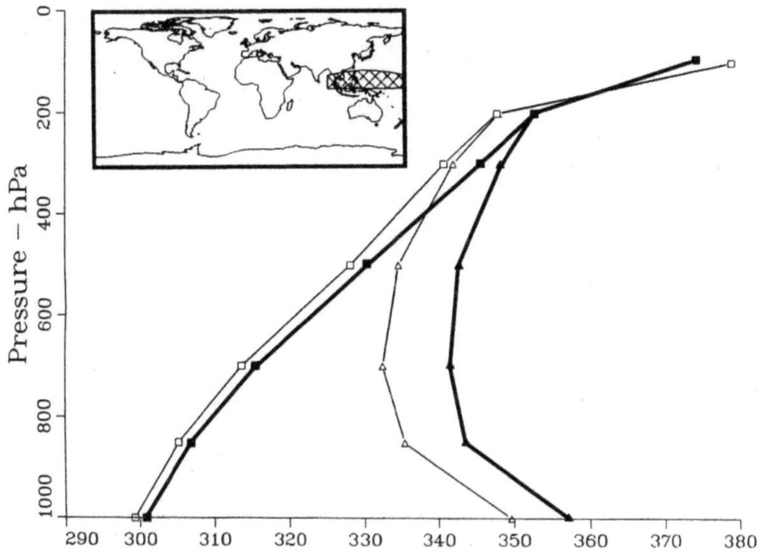

Figure 14: (c) Area-mean profiles of $\theta(\square)$ and $\theta_e(\triangle)$ and for control (thin line) and revised fluxes (thick line). Area marked by hatching in insert.

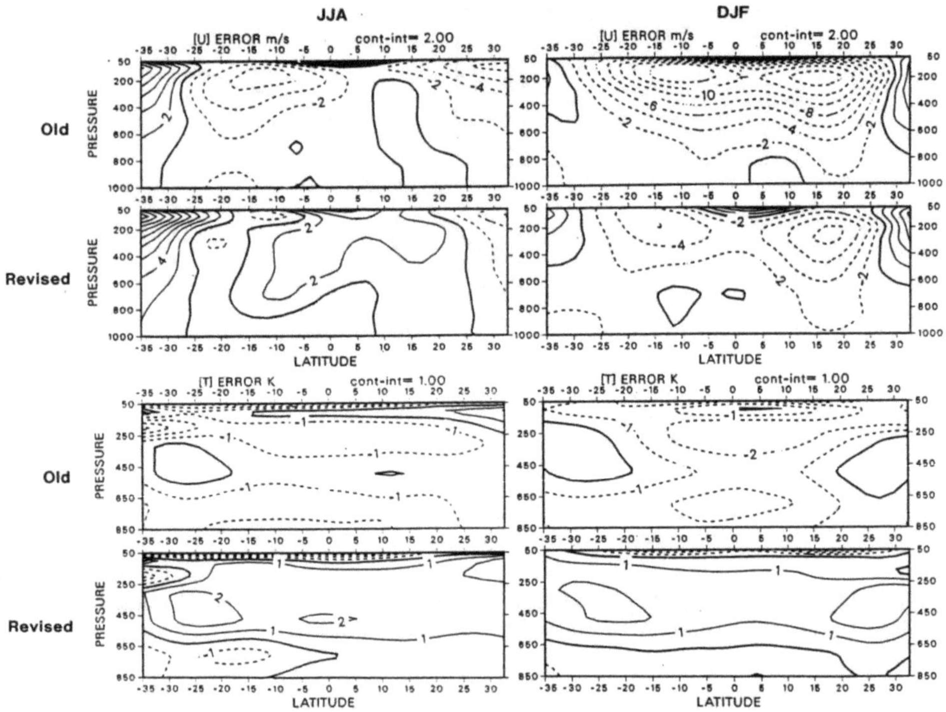

Figure 15: Zonal-mean zonal wind and temperature errors (difference between model result and analysis) averaged over the last 45 days of the 90-day integrations for June-July-August (JJA) and December-January-February (DJF), with the old and the revised scheme.

without it the model showed very little response at all. One further encouraging result in these simulations was evidence of tropical intraseasonal variability in the form of low frequency eastward propagation indicative of the Madden and Julian Oscillation (Madden and Julian, 1971) previously lacking.

The sensitivities discussed above underline the need for a physically accurate description of the boundary conditions in long simulations. In coupled experiments inaccuracies in surface fluxes manifest themselves in strong climate drifts such as in the SST of the warm pools (e.g. Gordon and Hunt, 1989).

8 Conclusions

The efficacy of GCM models and their use in studying the hydrological cycle whether in forecast or assimilation mode or in climate simulations, depends on the physical parametrizations. This review has selected some of the more crucial processes and their interactions and discussed more recent parametrization improvements and ideas relevant to climate simulation and sensitivity.

The close couplings between radiation, clouds, convection and the boundary layer fluxes are such that significant improvements in our ability to model the earth's climate system and its sensitivity will come primarily from improvements in our knowledge and understanding of moist processes, their representation by parametrization schemes and accurate modelling of their interactions.

References

ARAKAWA, A. and W.H. SCHUBERT, 1974: Interaction of a cumulus cloud ensemble with the large-scale environment: Part I. *J.Atmos.Sci.,* **31,** 674-701.

ARAKAWA, A. and J-M. CHEN, 1986: Closure assumptions in the Cumulus Parametrization Problem. Presented at the WMO/IUGG NWP symposium, Tokyo, 4-8 August 1986. *Journal of Met.Soc. of Japan,* pp 107-132.

ARKING, A., 1991: The radiative effects of clouds and their impact on climate. *Bull.Amer.Meteor.Soc.,* **72,** 795-.

BETTS, A.K., 1982: Saturation point analysis of moist convective overturning. *J.Atmos.Sci.,* **39,** 1484-1505.

BETTS, A.K. and M.J. MILLER, 1986: A new convective adjustment scheme. Part II: Single column tests using GATE wave, BOMEX, ATEX and arctic air-mass data sets. *Quart.J.R.Met.Soc.,* **112,** 693-709.

BETTS, A.K., 1990: Greenhouse warming and the tropical water budget. *Bull.Amer.Met.Soc.,* **71,** 1464-1465.

CESS, R.D., 1976: Climate change: An appraisal of atmospheric feedback mechanisms employing zonal climatology. *J.Atmos.Sci.,* **33,** 1831-1843.

CESS, R.D. and 18 AUTHORS, 1989: Intercomparison of cloud-climate feedback as produced by 14 atmospheric general circulation models. *Science,* **245,** 513-516.

FOUQUART, 1988: Physically-based modelling and simulation of climate and climatic change. Part I Edited by M.E. Schlesinger NATO ASI Series, *223-284.*

GEWEX (1991) UK National Plan for GEWEX: *The Royal Society Nov. 1991.*

GORDON, C. and D.L. HUNT, 1989: Surface evaporation during free convection. *Dyn. Clim. Tech.* Note No. 80, Met.Off. 20, Bracknell, UK.

KLINKER, E., 1993 Aspects of large-scale modelling. In: Energy and water cycles in the climate system. E. Raschke and D. Jacob (Eds.), *Springer Verlag NATO ASI Series.*

KUO, H.L., 1974: Further studies of the parametrization of the influence of cumulus convection on large-scale flow. *J. Atmos Sci,* **31,** 1232-1240.

LINDZEN, 1990a: Some coolness concerning global warming. *Bull.Amer.Met.Soc.,* **71,** 288-299.

LINDZEN, 1990b: Response to greenhouse warming and the tropical water budget *Bull.Amer.Met.Soc.,* **71,** 1465-1467.

MADDEN, R.A. and P.R. JULIAN, 1971: Detection of a 40-50 day oscillation in the zonal wind in the Tropical Pacific. *J.Atmos.Sci.,* **28,** 707-708.

MANABE, S., J. SMAGORINSKY, and R.F. STRICKLER, 1965: Simulated climatology of a general circulation model with a hydrologic cycle. *Mon.Wea.Rev.,* **93,** 769-798.

MILLER, M., A.C.M. BELJAARS and T.N. PALMER, 1992: The sensitivity of the ECMWF model to the parametrization of evaporation from the tropical oceans. *J.Climate,* **5,** pp 418-434.

MITCHELL, J., 1989: The greenhouse effect and climate change. *Rev.Geophys.,* **27,** 115-140.

OOYAMA, K., 1971: A theory of parametrization of cumulus convection. *J.Meteor.Soc. Japan,* **49,** 744-756.

PALMER, T.N., P. VITERBO and M.J. MILLER, 1992: Modelling interannual variations of summer monsoons. *J.Climate,* **5,** 399-417.

RAMANATHAN, V., R.D. CESS, E.F. HARRISON, P. MINNIS, B.R. BARKSTROM, E. AHMAD and D. HARTMANN, 1989: Cloud-radiative forcing and climate: Results from the Earth Radiation Budget Experiment. *Science,* **243,** 57-63.

RANDALL, D.A. and 30 AUTHORS, 1992: Intercomparison and interpretation of surface energy fluxes in atmospheric general circulation models. *Journal of Geophys.Res.,* **97,** 3711-3724.

SCHMUGGE, T.J. and J-C ANDRÉ (EDITORS), 1991: Land surface evaporation: measurement and parametrization. *Papers presented at a workshop held in Banyuls, France, 10-21 Oct., 1988. Springer-Verlag, pp 424.*

SLINGO, J.M., 1987: The development and verification of a cloud prediction scheme for the ECMWF model. *Quart.J.R.Met.Soc.,* **113,** 899- 927.

SLINGO, J.M., 1990: Sensitivity of the Earth's radiation budget to changes in low clouds. *Nature,* **343,** No. 6253, pp 49-51.

SOMMERIA, G., 1988: Physically-based modelling and simulation of climate and climatic change. *Part I Edited by M.E. Schlesinger NATO ASI Series, pp 331-374.*

STULL, B.B., 1988: An introduction to boundary layer meteorology. *Kluwer Academic Publishers, p 666.*

SUNDQUIST, H., 1978: A parametrization scheme for non-convective condensation including prediction of cloud water content. *Quart.J.R.Met.Soc.*, **104**, 677-690.

SUNDQUIST, H., 1988: Physically-based modelling and simulation of climate and climatic change. *Part I Edited by M.E. Schlesinger NATO ASI Series, 433-464.*

TIEDTKE, M., 1988: Physically-based modelling and simulation of climate and climatic change. *Part I Edited by M.E. Schlesinger NATO ASI Series, 375-432.*

TIEDTKE, M., W.A. HECKLEY and J. SLINGO, 1988: Tropical forecasting at ECMWF: The influence of physical parametrization on the mean structure of forecasts and analyses. *Quart.J.R.Met.Soc.*, **114**, 639-664.

TIEDTKE, M., 1989: A comprehensive mass flux scheme for cumulus parametrization in large-scale models. *Mon.Wea.Rev.*, **117**, 1779-1800.

TIEDTKE, M., 1990: Aspects of cumulus parametrization. *ECMWF Seminar on Tropical extra-tropical interactions, Reading, UK, 10-14 Sept. 1990, 441-446. ECMWF, Shinfield Park, Reading, Berkshire, UK.*

TIEDTKE, M., 1992: Representation of clouds in large-scale models (in preparation).

WOOD, E.F. (EDITOR), 1991: Land surface, atmosphere interactions for climate modelling: observations, models and analysis. *Kluwer Academic Publishers, pp 314.*

OCEAN-ATMOSPHERE INTERACTIONS IN THE TROPICS

GEORGE PHILANDER
Princeton University
Princeton, New Jersey

1 Introduction

When a group of experts met in Princeton, New Jersey in October 1982 to discuss plans for an international program to study El Niño, no one was aware that the most intense and devastating El Niño episode of the past century was developing at that time. Less than a decade later scientists were able to predict months in advance that El Niño conditions would develop towards the end of 1991. This forecast proved correct: during the early months of 1992 the eastern tropical Pacific experienced abnormally high sea surface temperatures, a cessation of equatorial and coastal upwelling, and torrential rains. Even Texas and southern California suffered devastating floods. The success of this prediction is convincing evidence of rapid progress during the 1980s in our ability to monitor and predict conditions in the tropical Pacific. This chapter briefly summarizes recent developments in the study of ocean-atmosphere interactions in the tropics, and the earlier achievements that made the rapid progress possible. For a detailed discussion the reader is referred to Philander (1990). The crucial aspect of large-scale ocean-atmosphere interactions is a circularity: sea surface temperatures strongly influence atmospheric conditions and are also determined by those atmospheric conditions, especially by the surface winds in the tropics. It is therefore necessary to start with a discussion of the atmospheric response to sea surface temperature changes (section 2) and the oceanic response to changes in the surface winds (section 3) before proceeding to interactions between the ocean and atmosphere (section 4).

2 The Atmosphere

Atmospheric variability at high frequencies, up to periods of days and weeks, is associated primarily with instabilities of the time-mean atmospheric motion. On time-scales of months and longer a different source of variability becomes of paramount importance: changes in the oceanic, land and ice surfaces below the atmosphere induce large-scale changes

NATO ASI Series, Vol. I 11
Modelling Oceanic Climate Interactions
Edited by J. Willebrand and D. L. T. Anderson

in atmospheric conditions. This is especially true of the tropics where direct thermal circulations play a central role. A well-known example of such a circulation is the monsoons over India. During summer the Indian subcontinent attains such high temperatures that the air there starts to rise, causing low level winds to converge onto that region. The moisture that these winds carry condenses where the air rises and this release of latent heat to some extent drives the winds. India during the northern summer is but one tropical region of high surface temperatures, rising, convective motion, low surface pressures, and heavy rainfall. Other convective regions in the tropics, and their seasonal north-south movements, are clearly evident in figure 1 which shows contours of outgoing long wave radiation (from the tops of tall clouds). Of special interest is the convective zone over the "maritime" continent of southeastern Asia and northern Australia. This zone migrates eastward interannually, during El Niño, because it is over an area of high sea surface temperatures and that area expands eastward during El Niño, into the central equatorial Pacific. Such a movement of the region of heavy precipitation implies a rainfall deficit over the western tropical Pacific and torrential floods over the otherwise arid and cold central and eastern tropical Pacific. On such occasions the Trade Winds near the equator collapse. When, after El Niño, the warm surface waters recede westward, the winds that converge onto the smaller area of high sea surface temperatures in the west include intense Trade Winds. Figure 2 illustrates the spatial structure of interannual sea surface temperature changes. Similar changes occur seasonally between September, (when a tongue of cold surface water is prominent along the equator) and March when that tongue is practically absent.

These arguments suggest that, for the prediction of low frequency variability in the tropics, it is necessary to develop models that calculate the atmospheric response to changes in surface boundary conditions, in sea surface temperatures for example. A hierarchy of models have been developed, to explain and simulate this response. The simplest, a shallow water model, captures the essence of the meteorological response to a heat source Q:

$$AU - fV + \phi_x = 0 \qquad (1)$$

$$AV + fU + \phi_y = 0 \qquad (2)$$

$$B\phi + c^2(U_x + V_y) = Q \qquad (3)$$

Here U and V should be regarded as the horizontal mass transport in the atmospheric boundary layer, ϕ is the geopotential, A and B are

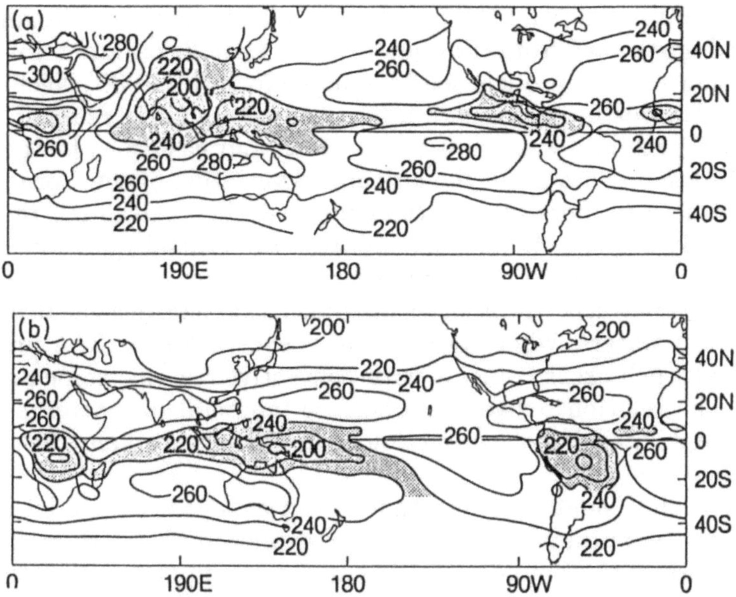

Figure 1: Mean outgoing long-wave radiation (watts per square meter) (a) for June to August and (b) for December to February as measured by polar orbiting satellites. Tropical regions with values less than 240 W m^2 (shaded) are regions of strong convection and heavy rainfall.

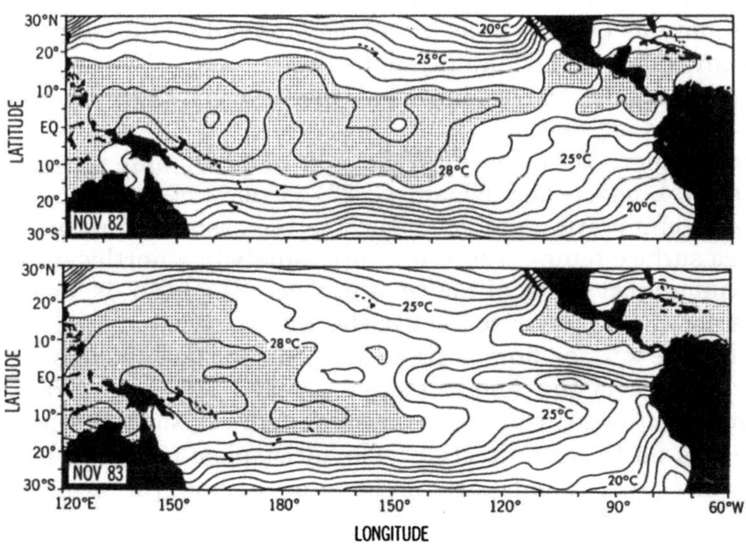

Figure 2: Sea surface temperatures in November 1982 during El Niño and one year later during La Niña.

damping parameters, and the coordinates x and y measure distance eastward, and northward from the equator. The Coriolis parameter is f. (For a discussion of such models see Matsuno 1966, Gill 1980, Neelin 1988). The model winds in response to a heat source are fairly realistic and, on the equator, include westerlies to the west, and easterlies to the east of the heat source. Because of the latitudinal variation of the Coriolis force, the westerlies are more intense than the easterlies. Although this model is reasonably realistic it does not address the crucial question concerning the relation between the atmospheric heat source and the sea surface temperature pattern. It is often assumed that convection occurs over the warmest waters, where temperatures exceed $28°C$. The problem is that not all such areas coincide with convective regions; some are free of clouds. The relation between sea surface temperature patterns and the location and intensity of convection is a difficult problem that has not been solved yet.

Lindzen and Nigam (1987) proposed an atmospheric model in which the role of sea surface temperatures is explicit. They argue that mixing processes in the boundary layer cause air temperatures in that layer to mimic those of the ocean so that pressure gradients, proportional to the sea surface temperature gradients, arise in the boundary layer and drive low level winds. These winds happen to converge onto the regions of maximum sea surface temperature but the condensation and release to latent heat over these regions are at altitudes that are too great to influence the boundary layer. Mathematically such a model differs from that of equations (1)-(3), in as far as the forcing function appears in the momentum equations (1) and (2), not in equation (3). A simple transformation converts the one set into the other (Neelin 1991) but physically the two models are distinct.

In reality there is evidence (Wallace *et al* 1989) that the mechanism proposed by Lindzen and Nigam (1987) is relevant to regions of large sea surface temperature gradients such as the eastern equatorial Pacific. There, sea surface temperatures increase rapidly in a northward direction to a maximum just north of the equator where the ITCZ is located. This creates a northward pressure force in the atmospheric boundary layer and accelerates the surface flow northwards. To the south of this equatorial zone, over much of the southeastern tropical Pacific, the Trade Winds also have a northward component even though sea surface temperatures decrease in a northward direction. The mechanism proposed by Lindzen and Nigam (1987) is but part of the story of how the atmosphere responds to sea surface temperature patterns. A factor that remains to be explored more extensively is the effect of sea surface temperatures on the vertical stability of the atmosphere. Wallace, Mitchell, and Deser

(1989) and Hayes, McPhadden and Wallace (1989) find that the surface wind speed over eastern equatorial Pacific can increase with an increase in the surface temperature. Apparently the vertical shear of the wind is considerable in that region so that enhanced vertical mixing associated with higher surface temperatures, brings momentum to the surface. The very shallow atmospheric boundary layer in this region is also associated with extensive stratus cloud decks because the descending air aloft inhibits convection. This large-scale subsiding motion can be regarded as part of the atmospheric response to the heat source associated with the deep convection over the warm western tropical Pacific Ocean. In that region of small sea surface temperature gradient the atmospheric boundary layer is very deep. In other words, it appears that the model described by equations (1)-(3) account for certain features of the atmospheric motion, those associated with deep convection, while the Lindzen and Nigam (1987) is relevant to boundary layer motion over parts of the eastern tropical Pacific. An improved simple model of the atmosphere will have to take into account changes in its vertical structure between the western and eastern tropical Pacific.

The atmospheric models described thus far isolate a few physical processes and yield insight into atmospheric dynamics but, because of numerous neglected processes, are not capable of realistic simulations of reality. To achieve that end we have to turn to General Circulation Models of the atmosphere. They usually include parameterizations of boundary layer processes and convection that permit them to calculate the atmospheric response to changing sea surface temperature patterns. One of these models has been used to reproduce atmospheric variability over several 15 year periods (Lau 1985). The first experiment, in which climatological sea surface temperatures were specified as a lower boundary condition, succeeded in reproducing the statistics of atmospheric variability in mid latitudes with reasonable accuracy but failed to produce any significant low-frequency variability in the tropics other than the seasonal cycle. The second experiment was a repetition of the first except that, in the tropical Pacific, the specified sea surface temperatures corresponded to those observed during the period 1962 to 1976. This calculation simulated the Southern Oscillation, the interannual fluctuation between El Niño and its complement La Niña reasonably well. In other words, given the sea surface temperature patterns, the model is capable of predicting the occurrence of El Niño episodes with great precision, indefinitely into the future. The catch is that future sea surface temperatures have to be known.

Although the success of atmospheric GCM's when sea surface temperatures are specified is encouraging, further improvements are vital. This quickly becomes evident when the models are put to far more stringent tests in which they are coupled to oceanic models so that interactions between the two media determine sea surface temperature. We return to this topic in section 4.

3 The Oceans

From an atmospheric point of view climate fluctuations are, in part, caused by changing sea surface temperature patterns. What causes temperatures of the surface waters to change? In the tropics, sea surface temperature patterns reflect the subsurface thermal structure. For example, the surface waters are cold in the southeastern tropical Pacific where the thermocline is shallow and are warm in the western tropical Pacific where the thermocline is deep. When temperatures in the southeastern tropical Pacific increase during El Niño, the thermocline there deepens. To explain changes in sea surface temperature we have to explain changes in the thermocline topography and hence in the oceanic currents, all in response to changes in the winds.

The tropical currents are subject to considerable variability: some currents reverse direction seasonally, the North Equatorial Counter Current in the Atlantic Ocean, for example (Garzoli and Katz, 1983) and others disappear occasionally, the Equatorial Undercurrent during the 1982 El Niño, for example (Firing *et al.*, 1983; Halpern *et al.*, 1983). This variability implies large changes in the oceanic thermal field. The associated horizontal redistribution of heat in the upper ocean can completely change sea-surface temperature patterns as during El Niño events, for example. The character of this oceanic variability depends on the time scale of the wind changes, and also on time scales intrinsic to the ocean. Of fundamental importance is the time it takes the ocean to adjust to a sudden and permanent change in wind conditions. Lighthill's (1969) seminal study of this problem was motivated by the generation of the Somali Current after the sudden onset of the southwest monsoons. He demonstrated that the adjustment time of the ocean decreases with decreasing latitude, so that the Somali Current can be generated within a matter of weeks, whereas it would take on the order of a decade to generate the Gulf Stream from a state of rest in mid-latitudes (Veronis and Stommel, 1956). It is the rapid adjustment of the tropical ocean and the speed with which heat can be redistributed horizontally in the upper ocean that makes possible phenomena such as the interannual El Niño events in the Pacific, and the considerable seasonal changes in the circulations of the tropical Atlantic and Indian Oceans. The similarities and differences between the three tropical ocean basins provide us

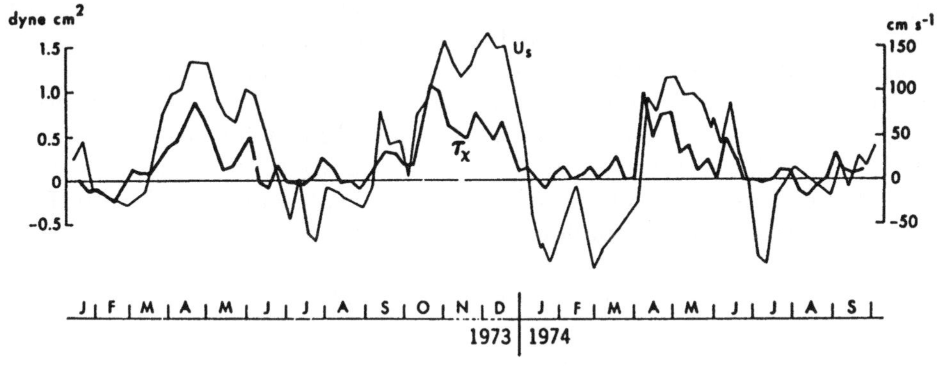

Figure 3: The zonal velocity component at the ocean surface and the zonal component of the wind stress as measured on the equator near Gan in the Indian Ocean 170°E). After Knox (1976).

with a wealth of information about the manner in which changes in the geometry of the ocean basin and changes in the forcing function affect the response. This response can be described deterministically because variability in the tropical oceans is primarily forced by the winds. Instabilities of the mean currents contribute to variability only in a narrow band of frequencies near a period of 3 weeks.

For idealized mathematical studies the response to winds that abruptly start to blow at a certain time is of special interest. The eastward winds that suddenly start to blow over the central equatorial Indian Ocean at the time of the equinoxes, behave in approximately this manner. They generate the intense oceanic jet shown in Fig. 5 (Wyrtki, 1973). This current has a number of curious features. Though there is nothing exceptional about the structure of the winds near the equator, the jet is only a few hundred kilometers wide and is centered on the equator. It accelerates at first but becomes steady after a few weeks even though the winds continue to provide eastward momentum to the ocean. Subsequently, the jet decelerates and reverses direction, though the winds never reverse direction.

To explain the behavior of the jet, an appealingly simple model of the tropical ocean will suffice. In the model, the interface between two layers of immiscible fluid, each of constant density, simulates the sharp and shallow tropical thermocline that separates the warm surface waters from the cold waters of the deep ocean. The upper layer has density ρ_1, has a mean depth H, and is bounded above by a rigid lid. The infinitely deep, motionless lower layer has density ρ_2 . Linear hydrostatic motion in the upper layer is driven by the zonal wind stress τ^x, which acts as a body force. This motion is associated with a displacement of the interface and is described by the shallow-water equations

$$u_t - fv + g'\eta_x = \tau^x/H \tag{4}$$

$$v_t + fu + g'\eta_y = 0 \tag{5}$$

$$g'\eta_t + C^2(u_x + v_y) = 0 \tag{6}$$

Here x and y measure distance in the eastward and northward (from the equator) directions; u and v denote the corresponding velocity components; t measures time and

$$f = \beta y, \quad g' = [(\rho_2 - \rho_1)/\rho_2]g, \quad C^2 = g'H$$

The latitudinal derivative of the Coriolis parameter is β, and g is the gravitational acceleration. Reasonable numerical values are

$$(\rho_2 - \rho_1)/\rho_2 = 0.002, \quad H = 100m, \quad h = 20cm, \quad C = 140cm/s^{-1} \tag{7}$$

Equations (4)-(6), can be reduced to a single equation

$$(v_{xx} + v_{yy})_t + \beta v_x - \frac{1}{C^2}v_{ttt} - \frac{f^2}{C^2}v_t = \frac{f}{C^2}(\frac{\tau^x}{H})_t - (\frac{\tau^x}{H})_{xy} \tag{8}$$

The motion induced by the sudden onset of spatially uniform zonal winds, which then remain steady, is initially independent of longitude provided that attention is confined to regions distant from the coasts. Far from the equator there is Ekman drift

$$v = \tau^x/fH \tag{9}$$

after an inertial period. If the winds are eastward then this motion at a right angle to the wind converges on the equator, implying that there is a distinctive equatorial region in which downwelling is intense. The width of this region can readily be inferred from a scale analysis of (8) and is the equatorial radius of deformation

$$\lambda = \sqrt{C/\beta} \sim 250km \tag{10}$$

Since Eq. (4) reduces to

$$u_t = \tau^x/H \tag{11}$$

at the equator-recall that zonal variations are negligible-it follows that the equatorial zone is characterized by an accelerating jet in the direction of the wind. This jet is in geostrophic balance; its acceleration matches the steadily increasing latitudinal density gradient associated with the deepening of the equatorial thermocline caused by the convergent Ekman flow (Yoshida, 1959).

The motion described thus far is independent of longitude. To satisfy the boundary conditions at the meridional coasts $x = 0$ and L it is necessary to superimpose on the jet the free modes of oscillation of the ocean. Assume that the modes are of the form

$$v = V(y)e^{i(kx-\sigma t)}$$

then

$$V_{yy} + (\beta^2/C^2)(Y^2 - y^2)V = 0 \tag{12}$$

where

$$Y^2 = [(\sigma^2/C^2) - k^2 - (\beta k/\sigma)](C^2/\beta^2)$$

Solutions to this equation are wavelike (oscillatory) in an equatorial zone of half-width Y, but are exponentially decaying poleward of latitudes $\pm Y$. In other words, wave propagation is possible only in an equatorial waveguide whose width Y has a maximum value

$$Y^2_{max} = \sigma^2/\beta^2 + C^2/4\sigma^2 \tag{13}$$

Figure 4 depicts this curve and shows that in extraequatorial latitudes inertia-gravity waves are separated from Rossby waves by a frequency band in which no free waves are possible. Near the equator there are waves at all frequencies.

An appropriate superposition of latitudinally propagating waves can form standing modes that are described by Hermite functions and whose dispersion relation is (Matsuno, 1966)

$$[(\sigma^2/C^2) - k^2 - (\beta k/\sigma)](C/\beta) = 2r + 1, \quad r = 0, 1, 2 \tag{14}$$

Particularly important in the oceanic adjustment are the low-frequency nondispersive Rossby modes with westward group velocities for which the dispersion relation (14) simplifies to

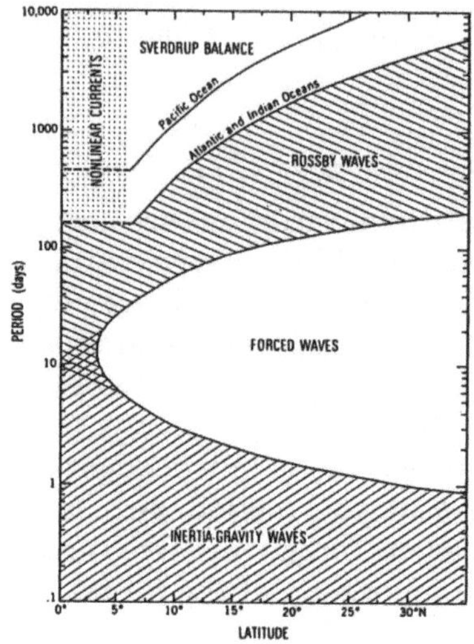

Figure 4: Periods, as a function of latitude, at which inertia-gravity, Rossby, and no free waves-only forced waves are possible. At very long periods, which depend on the zonal extent of the ocean basin, the equilibrium response to the wind is a Sverdrup balance except near the equator, where the currents are highly nonlinear.

$$\sigma = -kC/(2r+1), \qquad r = 1, 2, 3.... \tag{15}$$

The most rapid waves, associated with small values of r, are described by Hermite Functions of a low order and are strongly trapped about the equator. Those modes that extend into high latitudes are associated with large values of r, have low frequencies, and have small group velocities.

In addition to the waves described by the dispersion relation (14), there is a Kelvin wave that travels eastward nondispersively:

$$u = F(x - Ct)exp(-y^2/2\lambda^2), \tag{16}$$

where F is an arbitrary function.

Let us return to the accelerating equatorial jet. It can be expressed in terms of the complete set of Hermite Functions that describe the natural modes of the ocean. These are the modes that ensure that the zonal velocity component vanishes at the coasts $x = 0$ and L. The modes excited at the western coast ($x = 0$) must have eastward group velocities. The short dispersive Rossby waves with this property play a negligible role in the oceanic adjustment. Of paramount importance is the Kelvin wave that, at $x = 0$, must have a latitudinally integrated mass flux equal and opposite to that of the jet (Cane and Sarachik, 1976, 1977). This condition determines the function

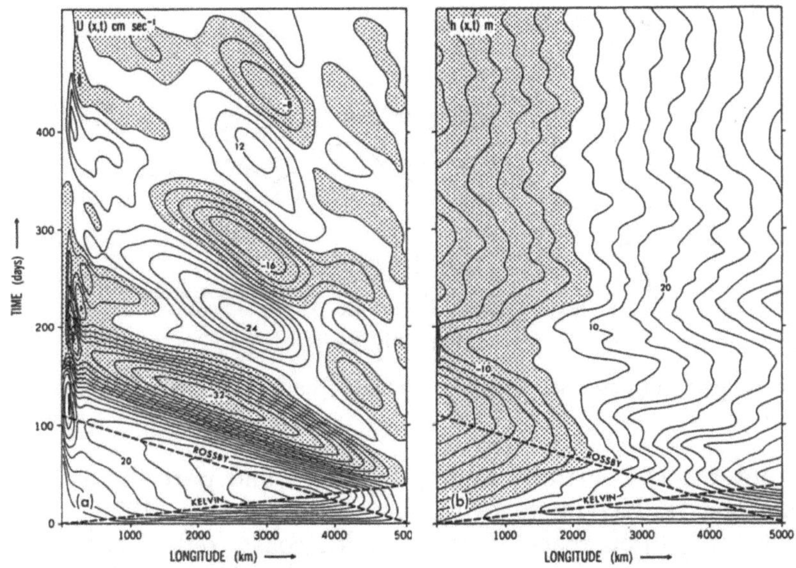

Figure 5: (a) Zonal velocity component and (b) thermocline depth variations along the equator after the sudden onset of spatially uniform eastward winds. Shaded regions indicate westward flow in (a) or a shoaling of the thermocline above it original depth of 100 m in (b).

$$u^{KELVIN} = 0, \qquad for \quad t < x/C$$

$$u^{KELVIN} = \frac{0.84\tau^x}{C_p H}(x - Ct)exp(-y^2/2\lambda^2), \qquad for \quad t > x/C$$

This equation describes a front, or bore, that is excited at the coast $x = 0$ at time $t = 0$ and that travels eastward at speed C. After it passes a point in the interior of the basin, the acceleration of the wind-driven jet, and the steady deepening of the thermocline, decreases abruptly. This happens because the Kelvin front introduces zonal gradients: in the momentum Equation (4) for the conservation of mass, divergence of the zonal current now balances the meridional convergence. Figure 5 shows how dramatically the Kelvin wave front affects the flow.

The long nondispersive Rossby waves excited initially at the eastern coast $x = L$ travel westward at speeds C/3, C/7, C/11, . . . and have an effect on the flow similar to that of the Kelvin wave. The immediate neighborhood of the equator is influenced primarily by the most rapid wave ($r = 1$), which is also the most strongly trapped about the equator. This wave meets the Kelvin wave at $t = 3L/4C$, whereafter the equatorial jet starts to decelerate. Upon reaching the eastern coast the Kelvin

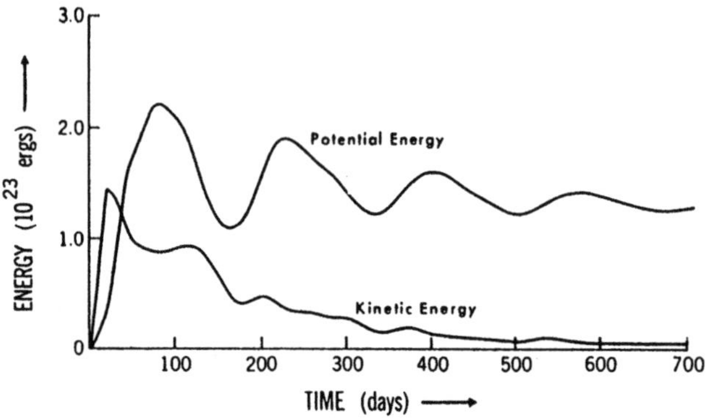

Figure 6: Kinetic and potential energy in the band from 6°N to 6°S for the motion shown in figure 5.

wave front reflects partially as Rossby waves and partially as poleward traveling coastal Kelvin waves (Moore and Philander, 1977). The long Rossby waves, upon reaching the western coast, reflect as Kelvin and short Rossby waves. All these waves, and their further reflections, contribute to the oceanic adjustment, but from Fig. 6 it is evident that the adjustment within a radius of deformation of the equator is accomplished essentially in the time it takes a Kelvin wave to propagate eastward across the basin plus the time it takes the reflected Rossby wave to cross the basin, L/C + 3L/C = 4L/C (Cane, 1979).

The Kelvin wave is confined to the narrow equatorial zone so that the oceanic adjustment outside this zone proceeds entirely from the eastern coast and is effected by Rossby waves. Far from the equator their speed is approximately, $\beta C^2/f^2$ and at a given latitude the basin is in an adjusted state after a time $Lf^2/\beta C^2$ (Anderson and Gill, 1975). This time increases with increasing latitude and with an increase in the width of the ocean basin. Its value is shown in Fig. 4.

In an inviscid ocean the equilibrium response to spatially uniform zonal winds is a state of no motion with a zonal pressure that balances the winds: $\eta_x = \tau_x/C^2$. (In the absence of friction, oscillations about this state persist.) For mean currents to be present in the equilibrium state, processes that represent the mixing of heat must be taken into account (Yamagata and Philander, 1985). It is insufficient to include only the mixing of momentum. The dissipation attenuates the waves that effect the oceanic adjustment so that mean currents are present in the equilibrium state.

The results described here suggest explanations for the measurements in Fig. 3. The initial acceleration of the jet probably stops when the passage of a Kelvin wave establishes zonal pressure gradients that balance the wind stress. The passage of a Rossby wave will cause a deceleration at a later stage. Measurements with drifting buoys in the eastern Indian Ocean do indeed seem to confirm this wave (Gonella *et al.*, 1981). In the central Indian Ocean, however, the deceleration can also be attributed to the relaxation of the eastward wind. If the winds turn off abruptly, the westward pressure force that the winds had maintained is unbalanced and accelerates the flow westward (Cane, 1980). Something similar happens in the Pacific Ocean when the westward trade winds, which normally maintain an eastward pressure force, relax. The unbalanced pressure force accelerates the equatorial currents eastward so that they transport warm water into the eastern Pacific (McCreary, 1976; Hurlburt *et al.*, 1976).

In general, wind fluctuations are not abrupt but are gradual. Suppose that the winds vary on a time scale long compared to the adjustment time of the ocean. The response will then be in equilibrium with the winds at all times. An example of such a response is the seasonal change in the zonal density gradient along the equator in the Atlantic that is practically in phase with the seasonally varying winds (Katz *et al.*, 1977). This suggests that the adjustment time for the equatorial Atlantic is less than the seasonal time scale, in agreement with the result in Fig. 4.

The shallow water model [Eqs. (4)-(6)] is a powerful tool for studying the adjustment of tropical oceans. It is even capable of simulating, with reasonable success, observed sea-level fluctuations when forced with realistic winds (Busallacchi and O'Brien, 1981). (The sea level reflects changes in the depth of the thermocline). The model has severe limitations, however, because of its simple vertical structure and gives only limited information about sea surface temperature. To learn more about that variable we have to turn to a continuously stratified model.

In a shallow water model waves excited at coasts establish the zonal density gradients. In a continuously stratified ocean these waves propagate horizontally and downward and can establish density gradients at depths greater than that of the wind-driven surface currents.

The horizontal structure of waves in a continuously stratified fluid is described by Eqs. (4)-(6) provided that the gravity wave speed C is regarded as a constant of separation that also appears in the equation for the vertical structure of the waves:

$$W_{zz} + (N^2/C^2)W = 0 \tag{17}$$

Here W is the vertical structure of the vertical velocity component, (all variables are assumed to be separable) and N is the BruntVaisala frequency of the fluid (Moore and Philander, 1977). Given the boundary conditions for the vertical velocity at the flat ocean floor and at the ocean surface, Eq. (17) yields an infinite number of eigenvalues C_n (n = 0, 1, 2, . . .) and eigenfunctions W_n known as the baroclinic plus barotropic modes. For the barotropic mode the stratification of the ocean is unimportant: the horizontal currents are independent of depth and the gravity wave speed is $C_o = (gH_o)^{1/2}$, where H_o is the total depth of the ocean. For the baroclinic modes, C_n decreases as n increases. The small values of C_n-for the first mode $C_1 \sim 2.5ms^{-1}$- reflect the very weak stratification of the ocean. The horizontal structure of each vertical mode is described by Eq. (2), where C assumes the appropriate eigenvalue. It follows that although each mode has the same waves, with the same dispersion relation, the temporal scale and the spatial scale [which is the radius of deformation of Eq. (10)] are different for each mode because they depend on the values for C_n.

The abrupt onset of zonal winds that act as a body force in a surface layer of depth H of a continuously stratified ocean drives meridional Ekman flow far from the equator and an accelerating zonal jet in an equatorial zone with $(ND/\beta)^{1/2}$. Here D is the depth of the thermocline. This motion is confined to the surface layers of the ocean and is very similar to the wind-driven jet described earlier. To satisfy boundary conditions at the coasts $x = 0$ and L, waves are excited. The amplitudes of the waves are determined by projecting the wind-driven jet onto the vertical modes of the ocean and then onto the latitudinal modes associated with each vertical mode. The most important waves for the adjustment are long nondispersive Rossby waves excited at the coast $x = L$, and Kelvin wave fronts excited at the coast $x = 0$. These waves introduce zonal density gradients in the surface layers and at greater depth. The existence of zonal pressure gradients below the surface layers permits the generation of an Equatorial Undercurrent.

The adjustment of the upper layers of a stratified ocean is essentially the same as that of a shallow-water model because the first two modes, especially the second, accomplish most of the adjustment (Philander and Pacanowski, 1980). Hence, the time it takes for basinwide density gradients in the thermocline and the Equatorial Undercurrent to be established is approximately the time it takes a second-mode baroclinic Kelvin wave to propagate across the basin. For the Atlantic Ocean this is on the order of a month. Below the equatorial thermocline, low-frequency motion corresponds to downward-propagating waves that travel at shallow angles to the horizontal and give rise to zonal jets with short vertical scales (Luyten and Swallow, 1976; McCreary, 1984).

Simplified, idealized models are excellent for exploring the physical processes that determine the oceanic response to changing winds. For realistic simulations it is necessary to turn to a model that incorporates all these processes simultaneously. The model that does that best is the General Circulation Model (GCM). It has been used extensively to reproduce the seasonal cycle of each of the three tropical oceans and to simulate the interannual fluctuations between El Niño and La Niña in the Pacific. One of the most significant results is that the low frequency variability in the tropics can be reproduced realistically, and deterministically, provided the surface winds are specified. Instabilities of the oceanic currents account for oceanic variability primarily at periods near three weeks and less. On time-scale of seasons and longer, instabilities are unimportant and oceanic variability in low latitudes is associated predominantly with variations in the forcing functions. Existing GCM's can simulate that variability realistically provided the forcing functions are known.

At present, a GCM of the tropical Pacific is being run operationally at the National Meteorological Center, Washington and each month produces detailed pictures of the surface and subsurface oceanic fields, the counterpoint of the daily weather charts (Leetmaa and Ji 1989). This model assimilates the available oceanographic measurements, the XBTs that are on the Global Telecommunication System for example, and is forced with winds, for the previous month, from the operational meteorological model. Because of this model, oceanographers are on a par with meteorologists in that, for the tropical Pacific at least, they have a network of instruments that routinely transmit data, in real-time, to a center where a General Circulation Model uses and interpolates the data to generate an accurate description of current conditions. The next step is for such a model to be coupled to an atmospheric model in order to predict future climate fluctuations.

4 Ocean-Atmosphere Interactions

The arguments advanced above to explain the response of the ocean to anomalous winds and the response of the atmosphere to changes in sea surface temperature patterns are circular and suggest that interactions between the ocean and atmosphere are unstable. Suppose, for example, that a modest disturbance causes a slight relaxation of the Trade Winds that drive the warm surface waters westward in the equatorial Pacific. The winds no longer balance the eastward oceanic pressure force associated with the slope of the thermocline so that the warm waters of the western Pacific surges eastward. This causes a change in the sea surface temperature pattern, an eastward displacement of the atmospheric convergence zone and hence a further relaxation of the Trades. In other words, the initial perturbation, a relaxation of the Trades, introduces an oceanic response that affects the atmosphere in such a way as to reinforce the weakening of the Trades. This is essentially how El Niño develops. The existence of unstable ocean atmosphere interactions raises numerous questions, including the following:

1. which factors determine the degree to which the interactions are unstable?
2. what are the dispersive properties of the spectrum of possible unstable modes?
3. does the Southern Oscillation correspond to the most unstable mode of oscillation?
4. does the seasonal variation in solar radiation force some of the unstable modes?

A hierarchy of coupled ocean atmosphere models have been developed to address these questions. The most complex ones are coupled General Circulation Models of the ocean and atmosphere. Neelin *et al* (1992) review the performance of several of these models which behave in a bewildering variety of ways: some fail to simulate any low frequency variability; some succeed in reproducing oscillations which, in certain respects resemble the Southern Oscillation but which in other respects differ from the measurements. The simulated oscillations generally differ from each other. To shed light on this matter, and to find ways to improve the coupled GCM's, it pays to turn to simpler coupled models. In the simplest of these, the atmospheric component is described by equations (1), (2) and (3) the oceanic component by equations (2). The ocean is driven by stress τ that is proportional to wind speed

$$\tau = \gamma(U, V)$$

where γ is a constant. It is far more difficult to relate sea surface temperature, the only oceanic parameter that affects the atmosphere significantly, to the oceanic currents and thermocline depth. An ad hoc approach is to write

$$T = a\eta + b\overline{T}_x \tag{18}$$

where a and b are constants, $\overline{T}_x$ is a specified zonal temperature gradient and η is thermocline depth perturbation. The case in which $b = 0$ so that sea surface temperature changes are directly proportional to changes in thermocline depth, illustrate what is special about the tropics. Because the Corolis parameter vanishes at the equator winds that converge onto a heat source over warm equatorial water drive oceanic currents that converge onto that region, thus deepening the thermocline there, increasing sea surface temperatures and intensifying the winds. This feedback can be realized when the earth is not rotating but not when the Coriolis parameter f has a constant value. In the latter case convergent winds drive divergent currents (Philander, Yamagata and Pacanowski 1984).

Hirst (1986, 1988) and Battisti and Hirst (1988) developed a simple coupled model with sea surface temperature described by equation (18), and analyzed the stability properties of various modes in detail. Modes can be stationary, or can propagate eastward or westward depending on the relative importance of advection and thermocline displacements on sea surface temperatures. The atmosphere is assumed to respond instantaneously to sea surface temperature changes so that two oceanic time-scales are of central importance: the advective time-scale, and the dynamic adjustment time which depends on the wave speed c. Neelin (1991) studied coupled modes with a time-scale very large in comparison with that on which the waves bring the ocean to a state of adjustment. (The speed c is assumed to be high). In that case the explicit time dependence disappears from the momentum equations (4) and (5) for the ocean. The atmosphere, equations 1, is assumed to have no "memory" and to adjust immediately to changes in sea surface temperature. The crucial timescale is now an advective oceanic one that controls sea surface temperatures so that equations (18) represent too simple a model. Neelin (1991) adapts the model of Zebiak and Cane (1987) who imbed, within the one-layer ocean, a surface mixed layer of fixed depth in which temperature is described by the equation

$$T_t + uT_x + vT_y + wT_z = Mixing\ processes$$

This equation can be linearized about a mean state with known currents and sea surface temperature gradients. The effects of upwelling enters only when the vertical velocity w, calculated from the divergence of the surface currents, is positive in which case the factor T_z is approximated by $(T - T_{sub})/H$ where T_{sub} depends on the thermocline depth. Neelin's linearized model can be studied analytically and yields insight into the manner in which different processes (upwelling, zonal advection, meridional advection) affect the properties of unstable ocean-atmosphere modes. Of special interest is a westward traveling mode that seems to exist in some of the coupled GCM's, specifically those of Meehl (1990) and Lau *et al* (1992). In such a mode, anomalously warm surface water in a certain region gives rise to westerly winds to the west of the anomaly. The westerlies induce convergent oceanic currents, a deepening of the thermocline and higher sea surface temperatures so that the initial warm anomaly is displaced westward. Note that this westward propagating mode is independent of the existence of coasts. The crucial phase lags that permit it to be continual is the lag, at a fixed point, between the winds and sea surface temperature. There is no explicit evidence of oceanic waves because the ocean is, dynamically, always in an adjusted state.

There exists another coupled mode in which the memory of the ocean plays an explicit role. It is referred to as the delayed oscillator mode, was first proposed in connection with El Niño by Schopf and Suarez (1988) and in its simplest form is described by the equation

$$T_t = aT + bT(t - \tau) \tag{19}$$

The first term on the right-hand-side represents the positive feedback between the ocean and atmosphere (for $a > o$) and leads to exponential growth. The second term takes into account that the ocean, at a certain time, responds not only to the winds that prevail at that time but is still in the process of adjusting to winds that prevailed at an earlier time τ. The delayed response can counter the positive feedback, and can even reverse it so that continual oscillations are possible. In other words, equation(19), for certain values of the constants a and b, can have oscillatory solutions. There is evidence that this mode is relevant to the observed Southern Oscillation which has the structure shown schematically in figure 7. During El Niño, the westerly wind anomalies are in the form of a narrow equatorial jet that is confined to the western side of the Pacific Ocean. This jet deepens the thermocline in the eastern equatorial Pacific, at the expense of the western Pacific, especially the off-equatorial regions where the shear of the wind induces divergent, oceanic Ekman drift that elevates the thermocline. The deep thermocline in the east is associated with high sea surface temperatures which in turn are responsible for the atmospheric jet. This is not an equilibrium state because the off-equatorial thermocline displacements in the west disperse into oceanic waves which ultimately travel eastward along the equator, causing an elevation of the thermocline, a change in sea surface temperatures and a change in the winds that leads to the decline of El Niño and the appearance of La Niña. The oceanic waves on the equator are Kelvin waves that travel with speed c. However, in the western Pacific they are superimposed on currents driven by continually changing surface winds. This superposition gives rise to phase propagation at a speed far slower than c. Rather then 150 cm/sec approximately, the speed in the western equatorial Pacific is near 15 cm/sec as shown in figure 8. That figure shows low-pass filtered data from a realistic GCM forced with observed winds (Chao and Philander 1992). Equatorial waves are of central importance to this coupled ocean-atmosphere mode, but so are continually changing winds that drive fluctuating currents. The waves are not present in isolation, and the properties of isolated waves are therefore not evident. Rather, waves can only be inferred from complex phase patterns such as those in figure 8 attributable to a superposition of waves and currents. Measurements need to be analyzed appropriately to determine whether similar phase patterns occur in reality. They do appear in certain coupled GCM's, those of Kitamura *et al* (1992), Philander *et al* (1992) and Latif and Villwock (1990), and are also excited in the simpler coupled models of Cane and Zebiak (1985), Schopf and Suarez (1988) and Battisti (1988).

Stability analyses of ocean-atmosphere interactions indicate that there are at least two types of modes. In one, the memory of the ocean is of central importance and the critical time-scale is that associated with the propagation of oceanic waves. If c is the speed of long gravity waves and L is the zonal dimension of the ocean basin then L/c is a convenient time-scale. The other unstable mode is possible even when $c = \infty$ so that the ocean instantaneously adjusts to a change in the winds and has no memory. In such a situation a time-scale associated with advection (by mean oceanic currents for example) come into play. Coupled GCM's with very coarse resolution, and hence high dissipation, in the oceanic component, seriously distort and damp the waves that bring about oceanic adjustment. In such models the ocean has little memory and, if the coupled models reproduce interannual oscillations at all, the ocean-atmosphere mode corresponds to the westward propagating type that depends on advection and that Neelin (1991) discussed. This type of interannual oscillation is present in the models of Meehl (1990) and Lau *et al* (1992). The latter study is of special interest because the calculations were repeated but with an important change. The spatial resolution of the oceanic component was increased by an order of magnitude thus permitting much better resolution of oceanic waves, and much lower dissipation. With this improved oceanic component the character of the interannual fluctuations changed to that of the delayed oscillator (Philander *et al* 1992).

Some coupled GCM's have no interannual oscillations at all. To explain their behavior it is necessary to investigate the factors that control the extent to which interactions between the ocean and atmosphere are unstable. One such factor is the mean depth of the thermocline. If the thermocline is very deep then changes in its depth, in response to changes in the winds, may not have a significant effect on sea surface temperatures so that the feedback on the wind is minimal. Zebiak and Cane

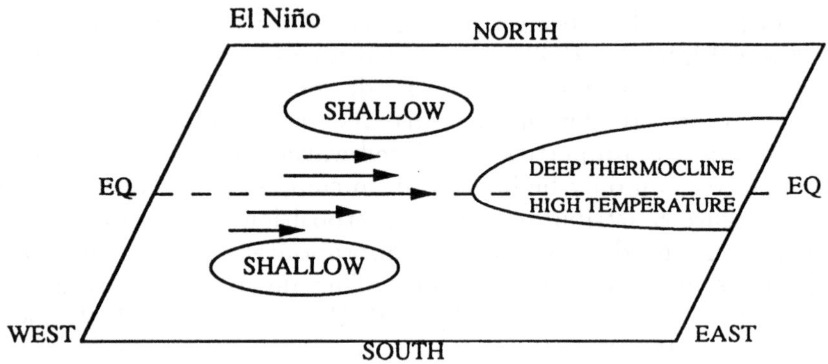

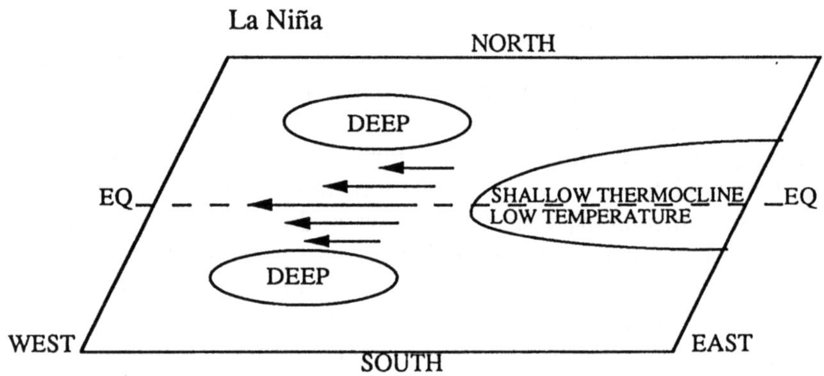

Figure 7: A schematic diagram showing spatial distributions of surface wind stress, sea surface temperature primarily off the equator, in the areas indicated by contours, because of their curl.

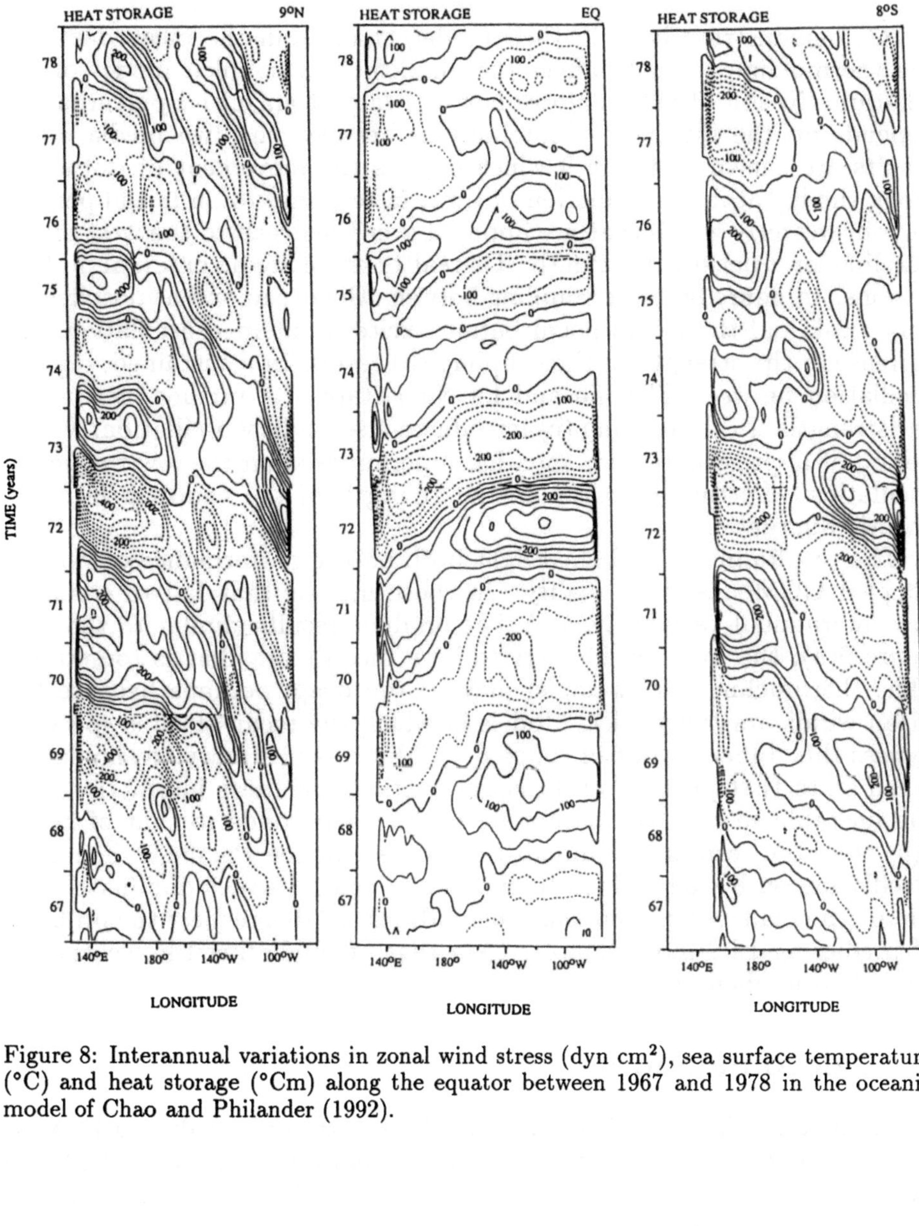

Figure 8: Interannual variations in zonal wind stress (dyn cm^2), sea surface temperature (°C) and heat storage (°Cm) along the equator between 1967 and 1978 in the oceanic model of Chao and Philander (1992).

(1987) demonstrate how, in their model, changes in the mean depth of the thermocline strongly affect the amplitude of coupled modes of oscillation. Neelin (1990) explores how the fate of an initial perturbation depends on the intensity of interactions between the two media. The results are shown in figure 9 which depicts the movement of isotherms along the equator, as a function of time, in response to an initial perturbation, westerly winds that are imposed over the ocean to the west of the dateline for a period of a month. Thereafter there are no external perturbations. In figure 9(a) the coupling between the ocean and atmosphere is very weak and the initial perturbation excites a damped oscillation. An increase in the strength of the coupling (b) results in a sustained, regular oscillation between El Niño and La Niña states. A further increase in the strength of the coupling (c) introduces a secondary oscillation, with a period of approximately six months, superimposed on the low frequency oscillation. Further increases in the strength of the coupling leads to chaotic fluctuations.

The results in figure 9 shed light both on the performance of various coupled GCM's, and on reality. In those models that fail to simulate any interannual variability, and that produce results similar to those in figure 9(a), the problem is likely to be a mean state that is too stable. (The thermocline, for example, may be too deep, perhaps because the mean winds are too weak). Models in which the mean state is specified bypass this difficulty but in coupled GCM's it is necessary to simulate a mean state that permits unstable interactions. If these interactions are too unstable so that the parameter range corresponds to that of figure 9(c) then the results are unrealistic because, in the Pacific, a Southern Oscillation with well-defined temporal and spatial scales dominates and there is no evidence of secondary instabilities. In this respect the problem of predicting El Niño is simpler than that of predicting the weather, a very chaotic phenomenon.

5 Teleconnections

The anomalous heating of the tropical Pacific atmosphere during El Niño episodes affects the global atmospheric circulation. Bjerknes (1969) was the first to examine the tropical-extratropical links in a dynamical context, although the existence of such relationships was implied by Walker's Southern Oscillation correlation patterns (Walker and Bliss,1932). Bjerknes proposed that the observed intensification of the Aleutian Low during an El Niño winter results from an enhanced regional Hadley Circulation, which increases the northward transport of angular momentum in the upper troposphere, leading to stronger westerlies and an enhanced

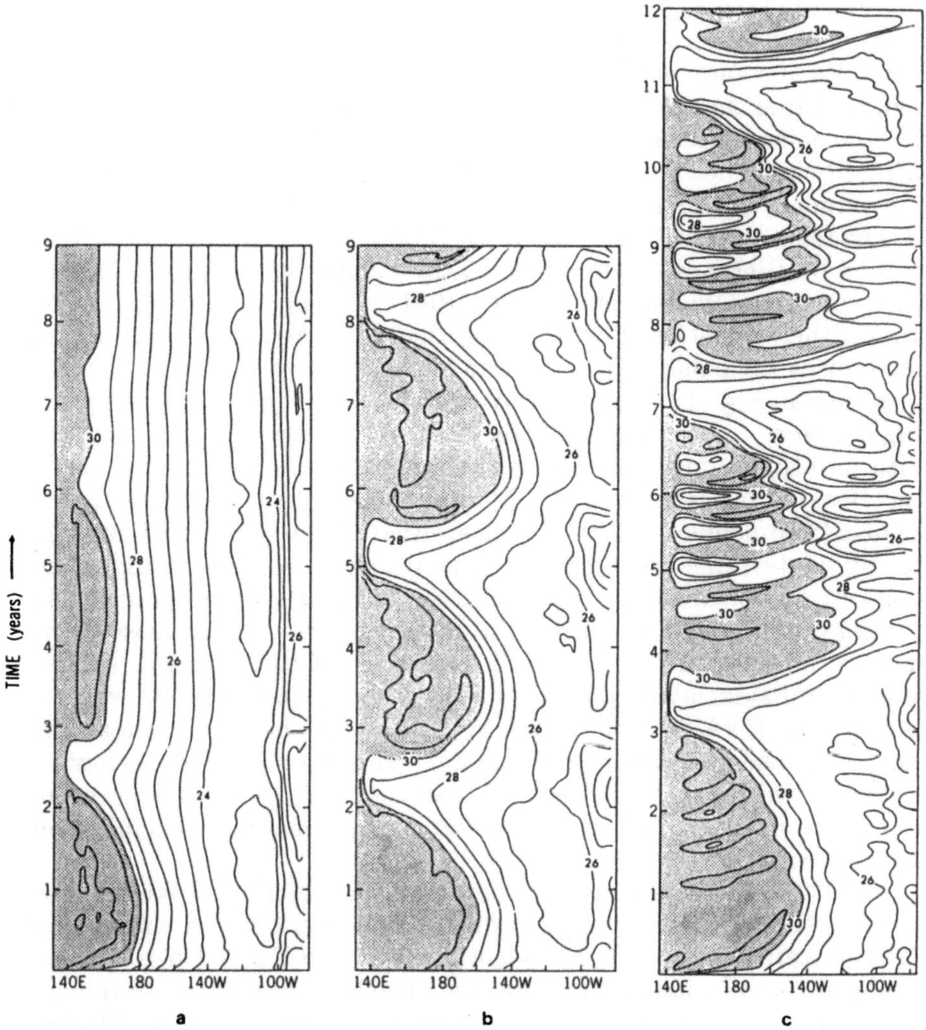

Figure 9: Isotherms (°C) along the equator over a period of years in the coupled ocean-atmosphere model of Neelin (1990). Westerly winds are imposed over the western side of the basin for a month whereafter ocean atmosphere interactions, which are weak (a), moderately strong (b), and very strong (c), control further developments. The strength of the interactions is increasing the drag coefficient that converts wind speed to stress.

meridional pressure gradient. The downstream circulation anomalies over western and eastern North America were viewed as a response to the North Pacific anomaly. Within this conceptual framework, Arkin (1983) demonstrated that the enhanced upper-tropospheric westerlies during an El Niño episode are associated with increased poleward ageostrophic flow, primarily into the winter hemisphere. This flow emanates from the region of divergence in the neighborhood of the enhanced equatorial convection.

The triplet of North Pacific-North American circulation anomalies described by Bjerknes is referred to as the Pacific-North American (PNA) teleconnection (Wallace and Gutzler,1981). During an "average" El Niño winter, these "centers of action", together with the anticyclonic anomaly center near the equator, trace out an apparent wave train with the polarity shown in Fig. 10. Horel and Wallace (1981) suggest that this pattern might reflect a Rossby-wave train that propagates along a great circle path into higher latitudes from the equatorial region of anomalous forcing (Hoskins and Karoly, 1981). More recent results from simple dynamic models (Simmons *et al.*, 1983) indicate that in the Northern Hemisphere during wintertime, when the stationary waves are very strong, the response to tropical forcing may take the form of a geographically fixed normal mode, which resembles the PNA pattern. Under these circumstances, an eastward or westward movement of the region of enhanced rainfall and associated upper-troposphere anticyclonic couplet will result in the same PNA pattern of response, although the amplitude and perhaps even the polarity might change. This normal mode behavior is associated with barotropic instability of the mean flow pattern and may be excited in many different ways; forcing by equatorial precipitation is just one way.

In 1982-1983 the anomalous convection in the tropical Pacific moved much farther east than during earlier El Niño episodes. A characteristic PNA pattern was nonetheless identifiable in the December-to February averaged 200-mb circulation anomalies. The PNA pattern showed little or no systematic eastward displacement from its normal position. Blackmon *et al.*(1983) have recently performed an experiment using the National Center for Atmospheric Research (NCAR) spectral general circulation model in which the equatorial SST anomaly was moved progressively farther east of its typical position. In the model, the PNA response remained geographically locked in position, but its amplitude decreased the farther east the SST anomaly was placed. These results seem to support the hypothesis that the PNA pattern is an unstable mode of the midlatitude troposphere (Simmons *et al.*, 1983).

Figure 10: Schematic diagram of upper-tropospheric height anomaly pattern during the mature phase (northern Hemisphere winter) of El Niño. (After Horel and Wallace, 1981).

6 Discussion

Certain phenomena associated with climate variability, the Southern Oscillation for example, correspond to modes of oscillation of the coupled ocean-atmosphere system in the same way that weather corresponds to natural modes of the atmosphere. Interactions between the tropical ocean and atmosphere permit a broad spectrum of modes of different types. The Southern Oscillation appears to correspond to the most unstable of a class of modes referred to as delayed oscillator modes in which the memory that the ocean has of earlier wind conditions is of central importance. Coupled ocean-atmosphere models, in which the mean state (and seasonal cycle) are specified so that the models deal only with departures from those states, reproduce realistic interannual fluctuations and have been used to predict El Niño successfully (Barnett *et al* 1988). Some of the more sophisticated coupled GCM's, which attempt to reproduce the mean state and interannual fluctuations, are realistic in their simulations but can not be used for predictions until they incorporate the seasonal cycle. This step is a challenge because the relation between the annual and interannual variations is poorly understood. For example, if the Southern Oscillation corresponds to one of the most unstable delayed oscillator modes then it is reasonable to assume that the annual cycle is one of the less unstable modes that is present because it is forced by the annual variations in solar radiation. However, the measurements in figure 11 indicate that the annual cycle involves, not the delayed oscillator mode, but one of the westward propagating modes, discussed by Neelin (1991), in which the memory of the ocean plays a secondary role. Exactly how this mode is forced is not entirely clear. Its structure is symmetrical about the equator so that the annual variation in radiation on the equator because of the ellipticity of the earths orbit is a possible forcing function. The amplitude of this annual forcing is much smaller than that of the semiannual forcing so that the absence of a significant semiannual signal in much of the tropics then becomes a puzzle. Another possibility involves the component of the annual forcing that is antisymmetrical about the equator. The measurements analyzed by Mitchell and Wallace (1991) indicate that ocean-atmosphere interactions determine the response to this forcing in the eastern tropical Pacific. Cross-equatorial winds that blow towards the Intertropical Convergence Zone, which is over the warmest water, keep the ITCZ in a northerly position because the winds, through upwelling, keep the waters south of the equator cold. In other words, the sea surface temperature patterns both determine and are determined by the wind. Such ocean-atmosphere interactions are yet to be explored quantitatively.

There clearly is a host of questions to be answered in connection with the seasonal cycle. These questions are of relevance to the interannual fluctuations too because the different frequencies are linked. The phase of the Southern Oscillation, for example, is strongly tied to that of the annual cycle. Off the coast of Peru El Niño tends to start, not in an arbitrary month, but in March or April, the warm phase of the annual cycle. Sometimes El Niño starts in the western rather than eastern Pacific and on such occasions it starts towards the end of the northern summer. In attempts to predict El Niño it is found that the coupled models have skill for several months if the predictions start in June or July but have far less skill if the predictions start in January or February. The reasons for these links between the annual and interannual fluctuations are, at present, unclear. What is clear is that coupled models have to take the seasonal cycle into account for them to be used as predictors. Thus far only models in which the seasonal cycle is specified are available for this purpose. The available coupled GCM's can reproduce either the delayed oscillator mode or the one that depends on advection but not both. In reality both types of mode are present, one on seasonal, the other on interannual time-scale.

El Niño is but one example of a climate fluctuation that is attributable to interactions between the ocean and atmosphere. The Atlantic experiences a similar phenomenon (during which the normally cold and arid coast of southwestern Africa experiences high sea surface temper-

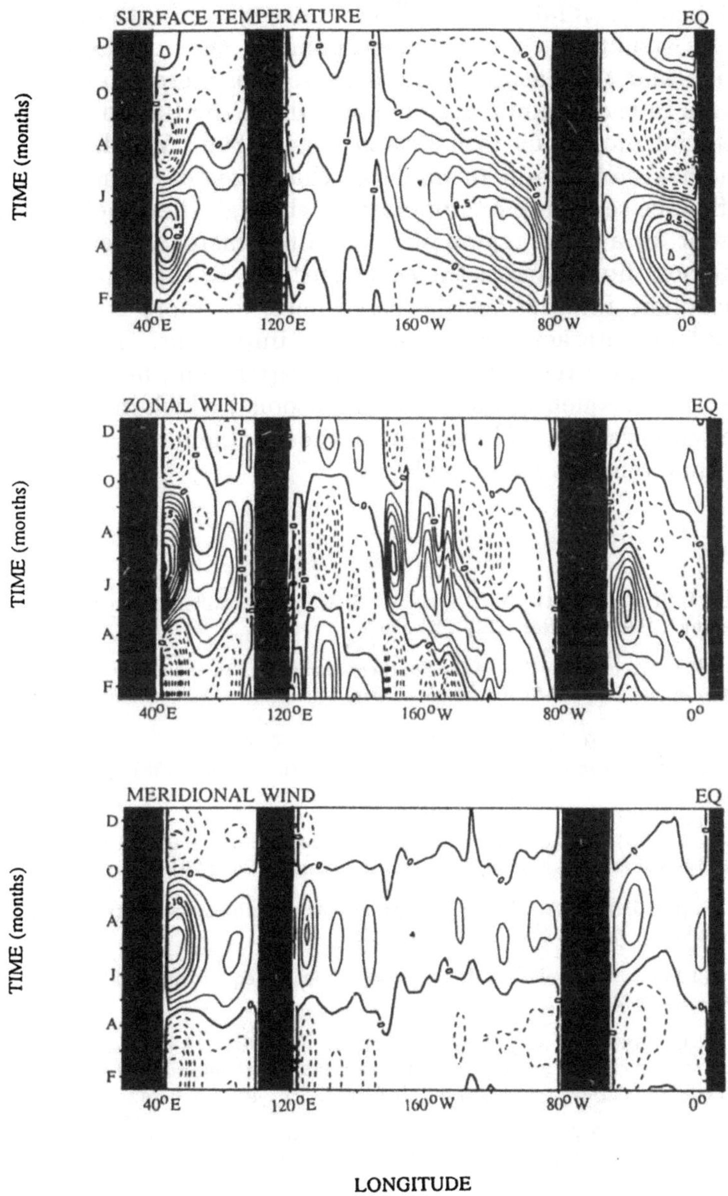

Figure 11: Climatological monthly mean anomalies, along the equator, of sea surface temperature (contour interval: 0.25°C), zonal component of the surface wind (contour interval: 0.01 dyn cm^2) and meridional component of the surface wind (coutour interval: 0.025 dyn cm^2). Dashed contours indicate negative, easterly, and northerly anomalies, respectively.

atures and heavy rainfall) but it is far more sporadic than the Pacific episodes. Its structure does not correspond to that of the delayed oscillator mode and the nature of the ocean-atmosphere interactions that cause El Niño in the Atlantic are unexplored. In the Indian Ocean where cross-equatorial monsoons rather than zonal Trade Winds prevail, the thermocline has such a large mean depth that changes in the wind do not directly influence sea surface temperatures through thermocline displacements (except for a small region off Africa) but instead affect sea surface through the heat fluxes across the ocean surface. Rather than the type of ocean-atmosphere interactions discussed here, far more complex ocean-atmosphere-land interactions appear to be important. For example, a change in sea surface temperature affects evaporation, hence the rainfall over land and consequently the winds (monsoons) which in turn influence sea surface temperatures. This is yet another problem concerning climate variability that remains to be explored.

References

ANDERSON, D. L. T., and GILL, A. E. (1975). Spin up a stratified ocean with applications to upwelling. *Deep-Sea Res.* **22**, 583-596.

ARKIN, F. (1983). An examination of the Southern Oscillation in the upper tropospheric tropical and subtropical wind-field. *Ph.D. Thesis, University of Maryland, College Park.*

BARNETT, T. P., GRAHAM, N., CANE, M. A., ZEBIAK, S. E., DOLAN, S., O'BRIEN, J., and LEGLER, D. (1988). On the prediction of El Niño of 1986-1987.; *Science* **241**, 192-196.

BATTISTI, D. S. (1988). The dynamics and thermodynamics of a warming event in a coupled tropical atmosphere-ocean model. *J. Atmos. Sci.* **45**, 2889-2919.

BATTISTI, D. S. and HIRST, A. C. (1988). Interannual variability in the tropical atmosphere ocean system *J. Atmos. Sci.* **45**, 1687-1712.

BJERKNES, J. (1969). Atmospheric teleconnections from the equatorial Pacific. *Mon. Weather Rev.* **97**, 163-172.

BLACKMON, M. L., GEISLER, J. E. and PITCHER, E. J. (1983). A general circulation model study of January climate anomaly patterns associated with interannual variations of equatorial Pacific sea surface temperature. *J. Atmos. Sci.* **40**, 1410-1425.

BUSALACCHI, A. J. and O'BRIEN, J. J. (1981). Interannual variability of the equatorial Pacific in the 1960s. *J. Geophys. Res.* **86**, 10901-10907.

CANE, M. A. (1979). The response of an equatorial ocean to simple wind stress patterns. I. Model formulation and analytic results. *J. Mar. Res. A* **37**, 233-252.

CANE, M. A. (1980). On the dynamics of equatorial currents with application to the Indian Ocean. *Deep-Sea Res.* **27**, 525-544.

CANE, M. A., and SARACHIK, E. S. (1976). Forced baroclinic ocean motions. I. *J. Mar. Res.* **34(4)**, 629-665.

CANE, M. A., and SARACHIK, E. S. (1977). Forced baroclinic ocean motions. II. *J. Mar. Res.* **35**, 395-432.

CANE, M. A. and ZEBIAK, S. E. (1985). A theory for El Niño and the Southern Oscillation. *Science* **228**, 1084-1087.

CHAO, YI and PHILANDER, S. G. H. (1992). On the structure of the Southern Oscillation. *J. of Climate,* in press.

FIRING, E., LUKAS, R., SADES, J., and WYRTKI, K. (1983). Equatorial Undercurrent disappears during 1982-1983 El Niño. *Science* **222**, 1121-1123.

GARZOLI, S. L., and KATZ, E. J. (1983). The forced annual reversal of the Atlantic North Equatorial Countercurrent. *J. Phys. Oceanogr.* **13**, 2082-2090.

GILL, A. E. (1980). Some simple solutions for heat induced tropical circulation. *Quart. J. Roy. Met. Soc* . **106**, 447-462.

GONELLA, J., FIEUX, M., and PHILANDER, G. (1981). Mise en évidence d'ondes Rossby équatoriales dans l'océan Indian an moyen de bouées derivantes. *C.R. Hebd. Seances Acad. Sci., Ser. B* **292**, 1397-1399.

HALPERN, D., HAYES, S., LEETMAA, A., HANSEN, D., and PHILANDER, G. (1983). Oceanographic observations of the 1982 warming of the tropical Pacific. *Science* **221**, 1173-1175.

HAYES, S., MCPHADDEN, M., and WALLACE, J. M. (1989). The influence of sea surface temperature upon surface wind in the eastern equatorial Pacific: weekly to monthly variability. *J. Climate* **2**, 1500-1506.

HIRST, A. C. (1986). Unstable and damped equatorial modes in simple coupled ocean- atmosphere models. *J. Atmos Sci.* **43**, 606-630.

HIRST, A. C. (1988). Slow instabilities in tropical ocean basin-global atmosphere models. *J. Atmos. Sci.* **45**, 830-852.

HOREL, J. D., and WALLACE, J. M. (1981). Planetary scale atmospheric phenomena associated with the Southern Oscillation. *Mon. Weather Rev.* **109**, 813-829.

HOSKINS, B. J., and KAROLY, D. J. (1981). The steady linear response of a spherical atmosphere in thermal and orographic forcing. *J. Atmos. Sci.* **38**, 1179-1196.

HURLBURT, H. E., KINDLE, J. C., and OBRIEN, J. J. (1976). A numerical simulation of the onset of El Niño. *J. Phys. Oceanogr.* **6**, 621-631.

KATZ, E., BELEVITSCH, R., BRUCE, J., BUBNOV, V., COCHRANE, J., DUING, W., HISARD, P., DE MESQUITA, A., MILLER, L., RYBNIKOV, A. (1977). Zonal pressure gradient along the equatorial Atlantic. *J. Mar. Res.* **35**(2), 293-307.

KITAMURA, (1990). Simulation of the annual and interannual variation of the tropical Pacific Ocean. *J. Mar. System* **1**, 169-181.

KNOX, R. A., and HALPERN, D. (1982). Long Range Kelvin wave propagation of transport variations in Pacific Ocean equatorial currents. *J. Mar. Res.* **40**, Suppl., 329- 339.

LATIF, J. and VILLWOCK, A. (1990). Interannual variability as simulated in coupled ocean-atmosphere models. *J. Marine Systems* **1**, 51-60.

LAU, N. C. (1985). Modelling the seasonal dependence of the atmospheric response to observed El Niños in 1962-76. *Mon. Wea. Rev.* **113**, 1970-1996.

LAU, N. C., PHILANDER, S. G. H. and NATH, M. J. (1992). Simulation of ENSO-like phenomena with a low-resolution coupled GCM of the global ocean and atmosphere. *J. Climate* **5**, 284-307.

LEETMAA, A. AND JI M. (1989). Operational hindcasting of the tropical Pacific *Dyn. Atmos. Oceans,* **13**, 465-490.

LIGHTHILL, M. J. (1969). Dynamic response of the Indian Ocean to the onset of the southwest monsoon. *Philos. Trans. R. Soc. London, Ser. A* **265**, 45-93.

LINDZEN, R. S. and NIGAM, S. (1987). On the role of sea surface temperature gradients in forcing low level winds and convergence in the tropics. *J. Atmos Sci* **44**, 2418-2436.

LUYTEN, J., and SWALLOW, J. (1976). Equatorial Undercurrents. *Deep-Sea Res.* **23**, 1005-1007.

MCCREARY, J. P. (1976). Eastern tropical ocean response to changing wind systems with application to El Niño. *J. Phys. Oceanogr.* **6**, 632-645.

MCCREARY, J. P. (1984). Equatorial beams. *J. Mar. Res.* **42**, 395-430.

MATSUNO, T. (1966). Quasi-geostrophic motions in equatorial areas. *J. Meteorol. Soc. Jpn.* **2**, 25-43.

MEEHL, G. A. (1990). Seasonal cycling forcing El Niño-Southern Oscillation in a global coupled ocean-atmopshere GCM. *J. Climate* **3**, 72-98.

MITCHELL, T., and WALLACE, J. (1992). On the annual cycle in equatorial convection and sea surface temperature. *J. Climate ; 5,; 1140-1156.*

MOORE, D. W., and PHILANDER, S. G. H. (1977). Modeling of the tropical oceanic circulation. *In "The Sea" (M. N. Hill, ed.), Vol. 6, pp.319-361. Wiley (Interscience), New York.*

NEELIN, J. D. (1988). A simple model for surface stress and low-level flow in the tropical atmosphere driven by prescribed heating. *Quart J. Roy. Met. Soc.* **114**, 747-770.

NEELIN, J. D. (1990). A hybrid coupled General Circulation Model for El Niño Studies. *J. Atmos. Sci.* **47**, 674-693.

NEELIN, J. D. (1991). The slow sea surface temperature mode and the fastwave limit: Analytic theory for tropical interannual oscillations and experiments in a hybrid coupled model. *J Atmos. Sci.* **48**, 584-606.

NEELIN, J. D., LATIF, M., ALLAART, M. A. F., CANE, M. A., CUBASCH, U., GATES, W. L., GENT, P. R., GHIL, M., GORDON, C., LAU, N. C., MECHOSO, C. R., MEEHL, G. A., OBERHUBER, J. M., PHILANDER, S. G. H., SCHOPF, P. S., SPERBER, K. R., STERL, A., TOKIOKA, T., TRIBBIA, J., and ZEBIAK, S. E. (1992). Tropical air-sea interactions in general circulation models, *Climate Dynamics* **7**, 73-104.

PHILANDER, S. G. H., and PACANOWSKI, R. C. (1980). The generation of equatorial currents, *J. Geophys. Res.,* **85**, 1123-1136.

PHILANDER, S. G. H., YAMAGATA J. and PACANOWSKI, R. C. (1984). Unstable air sea interactions in the tropics. *J. Atmos. Sci.* **41**, 604-613.

PHILANDER, S. G. H, (1990). El Niño, La Niña and The Southern Oscillation. *Academic Press* 289 pp.

PHILANDER, S. G. H., PACANOWSKI, R. C., LAU, N. C., and NATH, M. J. (1992). Simulation of ENSO with a global atmospheric GCM coupled to a high resolution, tropical Pacific ocean GCM. *J. Climate* **5**, 308-329.

SCHOPF, P. S. and SUAREZ, M. J. (1988). Vacillations in a coupled ocean-atmosphere model. *J. Atmos. Sci.* **45**, 549-566.

SIMMONS, A. J., WALLACE, J. M., and BRANSTATER, G. W. (1983). Barotropic wave propagation and instability, and atmospheric teleconnection patterns. *J. Atmos. Sci.* **40**, 1363-1392.

VERONIS, G., and STOMMEL, H. (1956). The action of variable wind-stress on a stratified ocean. *J. Mar. Res.* **15**, 43-69.

WALKER, G. T., and BLISS, E. W. (1932). World weather. *Y. Mem. R. Meteorol. Soc.* **4**, 53- 84.

WALLACE, J. M., GUTZLER, D. S. (1981). Teleconnections in the geopotential height field during the northern Hemisphere winter. *Mon. Weather Rev.* **109**, 784-812.

WALLACE, J. M., MITCHELL, T. P., and DESER, C. (1989). The influence of sea surface temperature on surface wind in the eastern equatorial Pacific: seasonal and interannual variability. *J. Climate* **2**, 1492-1499.

WYRTKI, K. (1973). An equatorial jet in the Indian Ocean. *Science* **181**, 262-264.

YAMAGATA, J., and PHILANDER, S. G. H. (1985). The role of damped equatorial waves in the oceanic response to winds. *J. Oceanogr. Soc. Jpn. ; 41; 345-357.*

YOSHIDA, K. (1959). A theory of the Cromwell Current and equatorial upwelling. *J. Oceanogr. Soc. Jpn.* **15**, 154-170.

ZEBIAK, S. E. and CANE, M. A. (1987). A Model El Niño-Southern Oscillation. *Mon. We. Rev.* **115**, 2262-2278.

OCEANIC GENERAL CIRCULATION: WAVE and ADVECTION DYNAMICS

PETER B. RHINES
School of Oceanography
University of Washington

Abstract

This is a discussion of the oceanic general circulation, both wind-driven and buoyancy driven. We start with basic ideas about fluids 'stiffened' by planetary rotation and Sverdrup-Rossby dynamics, describe the 'non-Doppler' effect that allows one to solve a large class of wind-driven circulations, and 'arrested wave' theory that leads to many linear models of water-mass development and circulation. This borders on 'Rossby hydraulics' in which advection and wave propagation effects are fully competitive. Modern debate over potential vorticity dynamics and 'warm' and 'cold' subduction layers into the thermocline are discussed. The production of vertical stratification that is the essence of subduction can occur from warming-induced restratification in spring or from cooling and dynamical restratification (conversion of horizontal-density gradient into vertical-density gradient). Wind-driven circulation involves active potential vorticity advection and stirring, and the ratio of advection to mesoscale-eddy diffusion in the gyres (the Peclet number) is of order 3 to 5, which is not large. In addition to the global overturning modes seen in global circulation models, the deep circulation involves smaller, faster, and more quickly responding branches of circulation which occur with topographic basins and ridges, and with fast boundary-current physics. These may be eddy forced, and shaped by topography. The capacity of basin topography to reverse the Stommel-Arons circulation (by its 'hypsometry') is described. Some of the 'missing physics' that challenges numerical ocean models is discussed, and some promotion of the value of laboratory experiments given.

1 Perspective

In preparing these lectures, two people come particularly to mind: Adrian Gill and Henry Stommel. Neither is with us any longer. Adrian was much in evidence at the predecessor to this workshop, also organized by David Anderson and Juergen Willebrand at Les Houches, in February 1985. Adrian's textbook, Atmosphere-Ocean Dynamics, demonstrates

NATO ASI Series, Vol. I 11
Modelling Oceanic Climate Interactions
Edited by J. Willebrand and D. L. T. Anderson

the power of simple models that capture the spirit of a part of our observed, complex geophysical flows. He was a quiet and unassuming person, and his contribution is ever more evident in the longevity of his insights.

Henry Stommel, who died unexpectedly a month before this meeting, had many ideas about the ocean circulation, and was also a master of 'penetrating simplification'. Twenty-four years ago, at the Woods Hole summer GFD (geophysical fluid dynamics) program of 1968, he gave a course of 5 lectures on the circulation which visited many of the problems to which he had contributed (figure 1). They are recorded in the bound lecture notes from that program. Here we shall revisit these same topics, trying to describe basic principles and some modern detail. Much modern detail was in fact provided by Stommel himself, in his remarkably productive years after returning to Woods Hole in 1978.

2 Introduction

The setting of research in climate and biogeochemical cycles is very broad and complex, and a basic tool for exploration is the general circulation model (gcm). Yet, some of the most important regions affecting biogeochemistry and climate are not yet well described by the gcm. In particular the upper mixed layer, zones of deep convection, polar fronts, the global pattern of upwelling from the abyss, shallow upwelling regions at the Equator, eastern boundaries and fronts, and high-latitude ice-laden seas all present difficulties and need concentrated study.

The basics of ocean circulation involve many issues of physical dynamics, thermodynamics, the influence of complex boundary conditions and atmospheric feedback. In 1968, at the time of Stommel's review lectures, the principal emphasis was the statistical mean circulation, upon which were superimposed harmless transients. While climatologists were keenly aware of interannual variability, dynamicists tended not to be. The process of building the mosaic of hydrographic data by gradual accumulation of cruises assumed there is a dominant underlying mean. Evidence of interannual change began to come both from el Niño and from high-latitude variability, as in the shutting down of Labrador Sea deep convection in the late 1960's.

As with C.D Keeling's Mauna Loa record of atmospheric CO_2, it is remarkable how few long time-series we have in the ocean, yet how much is recorded in what we do have. The richness of interannual change has been more and more evident in weather ship data, repeated hydrographic sections, climatological SST data, and the few purposeful time series like the Panulirus Station at Bermuda. At the same time the meteorologists have

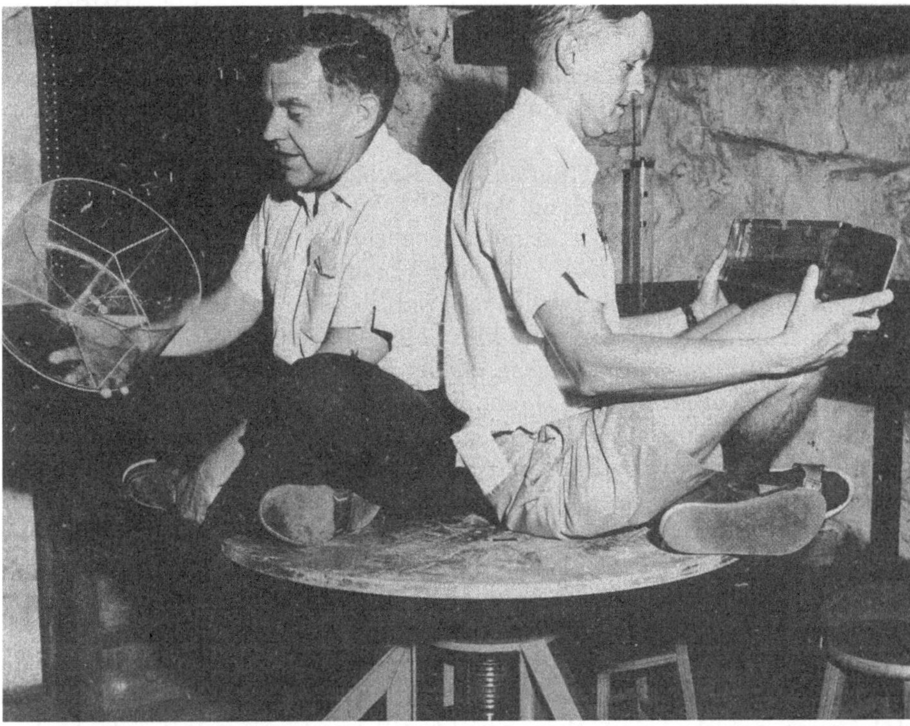

Figure 1: Henry Stommel and L.N. Howard, lecturers at the 1968 summer school of geophysical fluid dynamics in Woods Hole. Stommel is holding a 'pie' sliced plexiglas cylinder, for modelling source-driven deep circulation.

been quicker to focus on interannual change, for example in ENSO, the early work of Jerome Namias on teleconnections, and the early analysis of the 'North Atlantic Oscillation' and 'seasaw' in temperature between Greenland and Europe by Harry Van Loon, Andrew Bunker and now many others.

Sus Tabata (Tabata, Thomas and Ramsden 1986) has meticulously recorded climate change along 'line P' from the Canadian coast to station Papa (50°N, 145°W), and more recently Syd Levitus' analysis of the 15-year change in the North Atlantic (Levitus 1989) between the mid 1950's and early 1970's showed a mappable change of much of the North Atlantic, from the surface to middle depths. Dickson *et al.* (1988) trace the 'great salinity anomaly' seen earlier by transient tracer expeditions of 1980-81 (TTO group, 1983) as a freshening of the whole water column of the subpolar North Atlantic. Lazier (1992) traces the history of freshening and cooling of the Labrador Sea through the 1960's to the present, including a hiatus in deep convection in the late 1960's. The excitement now is in incorporating the full force of the atmospheric database to show how the coupled system developed during these anomalous periods, and their relationship with the development of global change over longer periods (see, e.g., Kushnir (1992), Wallace and Zhang (1992)).

This discussion emphasizes how transients are at work in the ocean, and they impel us to look at the dynamics of variability. Even if the globe were not changing, there would still be strong impetus to look at time-dependent ocean circulation, for it is a way to see cause and effect at work. Thus, the spin-up of the ocean from rest, with the onset of wind-forcing, can tell us much about the balances working in the steady state. We mean here to complement two earlier broad discussions: Rhines 1986a (for the 1985 NATO ASI very like this one) and 1986b. These centered on potential vorticity in theory and models, on wave-mean flow interaction, and on wind-driven circulation.

3 β-Plane Basics

3.1 Force Balance.

Scales of motion larger than a few km, varying with time scales greater than a day, tend to be nearly geostrophic and hydrostatic. The pressure increase with depth is nearly equal to the weight of the water overhead (neglecting vertical acceleration or Coriolis force), and the horizontal pressure variation is nearly in balance with the horizontal Coriolis force. The closeness of balance is expressed by the nondimensional Rossby number,

$$Ro \equiv U/fL$$

which is the ratio of typical nonlinear acceleration terms to Coriolis forces; U is a typical horizontal velocity corresponding to horizontal length scale L, and f is the Coriolis frequency, $2\Omega \sin\phi$, where ϕ is latitude. Its crucial variation with latitude is evident. The Ekman number,

$$E \equiv \nu/fH^2$$

measures the relative strength of viscous and Coriolis forces; ν is the effective vertical viscosity provided by turbulence in the top and bottom boundary layers, and H is the fluid depth. The nondimensional frequency

$$\omega \equiv (fT)^{-1}$$

measures temporal acceleration against Coriolis force; T is the timescale of the motion. Providing that Ro, E and ω are all much smaller than unity in the region of interest, the horizontal momentum balance is

$$\mathbf{f} \times \mathbf{u} = \nabla p/\rho; \tag{1}$$

p is pressure and ρ the density. If the aspect ratio of the motion, H/L, is small then the vertical force balance is often nearly hydrostatic,

$$p_z = -g\rho, \tag{2}$$

the 'gravity', $g \equiv |\mathbf{g}|$, depending upon the gradient of the geopotential,

$$\mathbf{g} = -\nabla\Phi$$

$$\Phi = -\frac{1}{2}(GM/r^2 + \Omega^2 r_1^2);$$

G is the gravitational constant, M the mass of the Earth, r the spherical radius, and r_1 the cylindrical radius $r\cos\phi$ (though the full integration of the gravity field is more complex than this approximation). This is not just the mean vertical force balance, but also that associated with moving fluid. The 'heaviness' of the stratification is summed up in the buoyancy frequency, N

$$N = \left[-g\left(\frac{\rho_z}{\rho} + \frac{g}{c^2}\right)\right]^{\frac{1}{2}}.$$

where the sound-speed squared, $c^2 = \partial p/\partial \rho$ at constant entropy. A related parameter that measures the relative strengths of buoyancy forces and Coriolis forces is the 'Prandtl ratio' or Burger number,

$$B \equiv NH/fL$$

It defines aspect ratio of motions that have equal Coriolis and buoyancy forces to be just $H/L \cong f/N$. If H is taken to be the full depth of the fluid then L is the Rossby radius of deformation, say $L_\rho, \equiv NH/f$. L_ρ ranges from 50 km or so at mid-latitude to less than 10 km at high latitude, and it is the scale of the most energetic motions in the ocean (other than surface and inertial waves). The corresponding scale in the Equatorial zone, where f changes rapidly, is just $(NH/\beta)^{\frac{1}{2}}$.

It has long been known (e.g., Phillips, 1964) that this state of balance allows too many possible motions, even when reasonable boundary conditions are specified. The geostrophic state is underdetermined, and therefore small acceleration or friction terms must be included to fix the circulation uniquely. There are two ways to approach this problem: the momentum balance can be expanded in powers of the Rossby number, Ekman number, and non-dimensional frequency, as a (singular) mathematical perturbation problem. Or, more directly, one can work with the curl of this balance, which is the vorticity equation. The curl of ∇p vanishes identically, so that the dominant force balance is supressed, and the remaining 'quasi' geostrophic dynamics elevated in importance. In particular, acceleration effects or viscous effects capable of determining a single, unique solution (or subset of solutions) for given boundary forcing, are now fully active in the vorticity equation. The complete vertical component of the vorticity equation is

$$\zeta_t + \mathbf{u} \cdot \nabla(\zeta + f) = (\zeta + f)w_z + F - D \tag{3}$$

where ζ is the vertical vorticity, $\nabla \times \mathbf{u}|_z$, w the vertical velocity, F is the vertical component of the curl of external forces, and D is dissipation. This equation tells us that the vorticity or 'spin plus shear' of fluid with respect to a local vertical axis can be changed by vertical stretching of the fluid (1st righthand term), external forcing F or dissipation D. In equation (3) the vorticity of the fluid is dwarfed by the vorticity of the Earth (by a factor of Ro^{-1}). It is the absolute vorticity, $\zeta + f$ that figures in the dynamics.

With spherical geometry, the radial (vertical) component of (3) exhibits the β-effect, the change in planetary vorticity seen by a fluid parcel moving north or south, as

$$\mathbf{u} \cdot \nabla f \equiv \beta v$$

where $\beta = df/dy = 2\Omega\cos\phi/a$, a being the planetary radius (which is nearly uniform).

It is particularly significant that (3) has no contribution from the density field: the twisting by buoyancy forces, $\nabla p \times \nabla\rho$, is very nearly horizontal, since ∇p and $\nabla\rho$ are themselves are nearly vertical in a hydrostatic ocean. The horizontal vorticity equation exhibits this effect, in which tipping of the planetary vortex lines is approximately cancelled by buoyancy twisting forces:

$$f\frac{\partial \mathbf{u}}{\partial z} = -(\frac{\mathbf{g}}{\rho}) \times \nabla\rho \tag{4}$$

The rate of change of horizontal vorticity is missing from (4) because it is smaller by factors Ro and ω. Very like the precession of a rapidly spinning gyroscope, the change in vorticity is dominated by the tipping of the Earth-imparted vorticity.

This standard approach to large-scale dynamics hides the irrotational and non-geostrophic motions that occur at many scales in the fluid. Their interactions with the quasigeostrophic subset occur in fronts, boundary layers and boundary waves, breaking internal waves, hydraulic flows, and generally wherever small-scale turbulence is found.

To this Newtonian dynamics one must add thermodynamics. This may be minimal, as in the case of shallow-water models of quasigeostrophic layered circulation models. Or, it may be elaborate, as in fully three-dimensional primitive-equation models. The level of thermodynamics, more than anything else, determines the degree of 'sophistication' of the model. It enters through the density field in the thermal wind equation and in the hydrostatic relation (2), in mass conservation,

$$\frac{\partial\rho}{\partial t} = -\nabla\cdot(\rho\mathbf{u})$$

and equations of conservation for temperature T, salt S and tracer species;

$$\frac{\partial(\rho C)}{\partial t} = -\nabla\cdot(\rho C\mathbf{u} - \rho\kappa_c\nabla C + F_c) \tag{5}$$

where κ_c is the molecular diffusivity of the concentration (per unit volume of mixture) of the property C and F_c the turbulent flux of C (formally redundant with $\rho C\mathbf{u}$, it appears when C and $\mathbf{u}$ are taken to be averages over time or space). Given the 10 decades of length-scale in planetary circulations, it is useful to recognize that small and intermediate scales, denoted by primes, may contribute turbulent fluxes $\langle\rho u'c'\rangle$. Finally, an equation of state relating ρ, p, S, T is incorporated.

The thermodynamic side of general circulation is poorly understood, and poorly represented in theoretical and numerical models. Quite simply, deep turbulent convection, mixing and fronts are not readily parameterized.

To get from ζ back to familiar pressure and velocity components one inverts a Poisson-type equation, for under geostrophic balance, equation (1) identifies p as a stream-function for horizontal velocity:

$$\zeta = (\rho f)^{-1}\nabla^2 p$$

to order $[Ro + \omega + L/a]$. The vertical velocity, w, is trickier, requiring the use of the density equation to 'follow isopycnal surfaces' and infer w from the mass field, very much as one does in practice with ocean data. The classical approach is through the "omega" equation (e.g., Gill (1982) p.543). It is another Poisson-type equation,

$$f^2 w_{zz} + N^2\nabla^2 w = (\frac{\beta}{f\rho})p_{xz} + 2\nabla \cdot \mathbf{Q}$$

where $\mathbf{Q} = g\hat{\mathbf{z}} \times J(\mathbf{u}_g, \rho)$, $\mathbf{u}_g$ is the geostrophic vector velocity, and $\hat{\mathbf{z}}$ a vertical unit vector. A simpler expression for w, at the expense of including time derivatives, is just the density equation with a hydrostatic expression for ρ:

$$w = \frac{1}{\rho N^2}\left[-\frac{D(p_z)}{Dt} + \mathcal{H}\right]$$

where $\mathcal{H}$ is the buoyancy flux convergence owing to heating, cooling, mixing, or salinity change and

$$\frac{D}{Dt} \equiv \frac{\partial}{\partial t} + \frac{1}{\rho f}J(p,\ ).$$

Vertical velocity is a crucial and difficult part of the dynamics: it is so small ($O((Ro+\omega)H/L)$) times the horizontal velocity) that it is difficult to calculate or measure in the ocean, yet is a crucial part of vorticity dynamics.

3.2 Rotational Stiffness.

Some feeling for the dynamics of rotating fluids can be gained from a few rules, which have intuitive basis. First consider what 'rapid' rotation does to a slowly moving fluid. Imagine holding in your hands a balloon filled with uniform-density water. If you squeeze the balloon, the water responds freely (in the 'laziest' response with the least kinetic energy that it can arrange, according to potential theory), though without noticeable compression of the fluid.

Now suppose that you and the balloon and the fluid are on a rapidly rotating turntable, with angular velocity Ω, and once again squeeze the balloon. Because the angular momentum (measured in a nonrotating frame of reference) has a large component due to the turntable rotation, squeezing the balloon inward (allowing it to expand upward at the center) will result in a great spin of the fluid, as you see it. If horizontal cross-sections of the balloon remain circular, a ring of fluid at radius r that is squeezed into a smaller ring of radius $r + \delta r$ (where δr is < 0) will develop an angular velocity ω (seen by the rotating observer)

$$-\frac{2\Omega\delta r}{r}$$

in order that total angular momementum $mr^2(\omega + \Omega)$ is conserved, where m is the mass of the ring. In the rotating frame this corresponds to the appearance of new kinetic energy (KE)

$$2m\Omega^2(\delta r)^2 \tag{6}$$

measured in the rest frame. Now the source of this energy is your squeezing work; if Ω is large you will have to squeeze very hard (with a force of order $m\Omega\delta r$) to make a given change in shape: rapid rotation *stiffens* a fluid. Because of the central role of system rotation, the stiffening occurs principally for motions that change the area of fluid elements, projected normal to the the rotation axis; that is, horizontally divergent motions. This is how the vertical stretching, $\partial w/\partial z$ on the righthand side of (3) causes vertical vorticity, ζ to change. Though a simple idea, its energetics may puzzle one: note that the change in energy depends upon reference frame. A non-rotating observer would see a larger change in KE of $-2m\Omega^2 r\delta r$ rather than (6). If the fluid is at rest in the rotating system initially, either sign of r causes a positive KE in that system, yet the non-rotating observer sees the absolute KE increase for $\delta r < 0$ and decrease for $\delta r > 0$. The puzzle is settled by realizing that the energy equation has a potential energy term

$$\rho \mathbf{u} \cdot \nabla \Phi \equiv \rho \frac{D\Phi}{Dt}$$

incorporating the work done against the centrifugal force, which of course the two observers see differently because of the Ω-dependence of Φ (more exactly, $\delta KE_{rot} = \delta KE_{rest} + m\delta\Phi + \mathbf{\Omega} \times \mathbf{r} \cdot \mathbf{F}\delta t$ where $\mathbf{F}$ is the pressure gradient, plus any external or viscous forces).

The stiffness property makes understandable the Taylor-Proudman approximation, derived by differentiating (1) in the vertical, yielding simply

$$\frac{\partial(u,v)}{\partial z} = 0 \tag{7}$$

for the horizontal velocity (u, v). The best way to appreciate and teach this fact is to put a beaker of fluid on a turntable and give it a stir, marking it with food coloring or ink. You will see the remarkable 'curtains' of dye that represent nearly two-dimensional flow: the fluid is so stiffened that vertical columns will not bend over. The two-dimensional motions do not change the projected area of fluid elements. They do not cause the 'super-spin' due to squeezing, and hence can occur freely.

Equation (7), written for uniform density fluid, is a special case of the thermal-wind equation (4) which shows that tipping over of fluid columns can be accommodated if twisting by buoyancy forces is available. (4) is the horizontal vorticity equation under geostrophic scaling. But even with this great new freedom for the horizontal circulation to be different at different depths, the vertical vorticity balance (3) still feels stiff!

3.3 Sverdrup Balance.

For slow, nearly steady motions for which Ro, ω, $E << L/a$ where a is the Earth's radius, (3) is dominated by

$$\mathbf{u} \cdot \boldsymbol{\nabla} f = f\frac{\partial w}{\partial z}$$

which is known as Sverdrup balance. Notice the interesting parameter $(Ro)a/L \equiv U/\beta L^2$ which comes up frequently as a relative measure of advection of planetary and relative vorticity, respectively. Sverdrup balance is somewhat ironic, because the full vector vorticity equation involves $\mathbf{u} \cdot \boldsymbol{\nabla}(\omega + 2\Omega)$, where Ω is the constant Earth's rotation vector. Although $\mathbf{u} \cdot \boldsymbol{\nabla}(2\Omega)$ vanishes, the vertical (radial) component of this term does not. In Earth coordinates, with v the meridional velocity, it is just

$$\beta v = f\frac{\partial w}{\partial z} \tag{8}$$

In terms of latitude ϕ, (8) becomes

$$v = \tan\phi\frac{\partial w}{\partial z} \tag{9}$$

which has a geometric feel to it. Sverdrup balance is remarkable for many reasons. It is thought to describe the dynamical balance of the

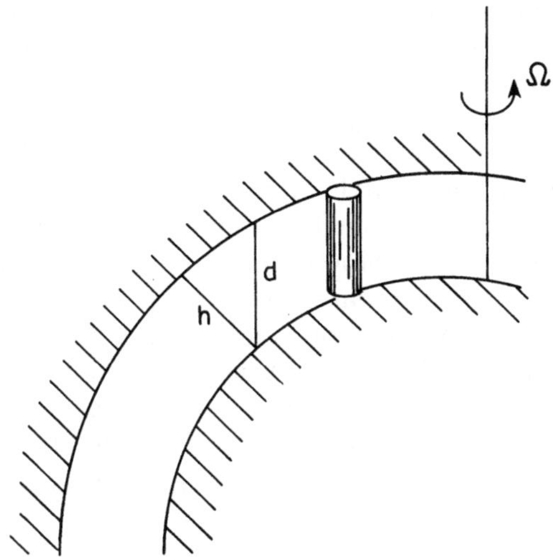

a

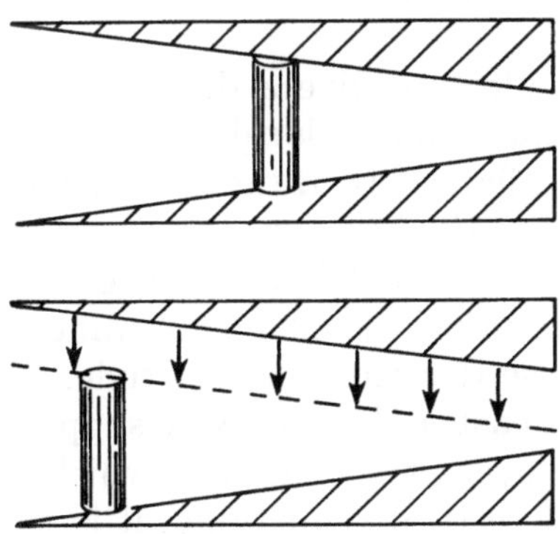

b

Figure 2: (a), Geometry of a spherical shell, in which the thickness measured along the rotation axis, d, increases Equatorward if h is constant. (b), β-plane version in which the wedge-shaped geometry also represents a north-south potential vorticity gradient. Downward vertical velocity forced at the surface drives southward flow, such as to avoid vertical stretching.

ocean interior at large scales *even for unsteady baroclinic waves.* Though it has a linear look, it can combine with a nonlinear density equation and thermodynamics to describe fully nonlinear circulation. That density equation can involve eddy contributions also, so that (8) can hold even with eddy-driven branches of deep circulation. Equation (9) has a geometric interpretation: if little metal rods aligned parallel with Ω move north or south in a spherical shell, without tipping over or changing their length, they will obey (9). Moving northward, while remaining parallel with Ω, they will occupy a greater distance in the local vertical direction (see sketch, figure 2), and thus one end or the other must move through a geopotential 'level' surface. It is this microscopic rigidity of the fluid in the Ω-direction which the Sverdrup relation conveys. Remarkably the stiffness constraint is still active when the fluid has strong density stratification, or experiences strong diabatic change in density, owing to mixing, convection, or direct heat flux.

In an earlier discussion of this effect (Rhines 1977), it was shown how both steady, wind-driven and unsteady adjustment of the circulation in baroclinic Rossby waves follow directly from this rule, (8). The other elements necessary to the model are the thermal-wind balance and the Ekman flux. Wind-stress acting on the ocean surface causes Ekman layer mass-flux whose horizontal divergence forces vertical velocity at the base of the Ekman layer.

Let us revisit this idea, to give a description of the basic responses of the large-scale ocean. Viewed locally, the spherical shell looks like a wedge-shaped domain, figure 2b. If the shell has uniform thickness h (no topographic features) its thickness measured parallel with the polar axis is $d \cong h/\cos\phi$, approximated for a thin shell. If the fluid is homogeneous, fluid columns marked parallel to Ω want to remain parallel, and not to stretch. Because d varies with latitude. the only 'free' mode of flow is zonal, directed east and west. Now suppose that the wind drives an Ekman boundary layer at the surface, which pumps fluid down into the interior with local vertical velocity w_e; it is as if the outer boundary of the shell were locally pressed downward. The fluid can respond without changing its dynamical length by moving Equatorward; the pumping velocity (projected onto Ω) just fills in the gap created by the change in d, following the fluid meridionally:

$$\begin{aligned}\frac{w_e}{\cos\phi} &= -\frac{Dd}{Dt}\\ &= -\frac{D\phi}{Dt}\cdot\frac{\partial d}{\partial\phi}\\ &= -\frac{D\phi}{Dt}\Big(-\frac{h\sin\phi}{\cos^2\phi}\Big)\\ &= \frac{v}{a}\Big(\frac{h\sin\phi}{\cos^2\phi}\Big).\end{aligned}$$

Thus

$$\frac{v}{a}\cos\phi = \frac{w_e}{h}\sin\phi$$

or

$$\beta v = f\frac{\partial w}{\partial z}$$

identically agreeing with (8),(9).

Now stable density stratification might seem to upset this picture, but it really does not. First note that the vertical vorticity equation, which yielded (9), has a very small twisting contribution from buoyancy (the slight vertical part of $g(\nabla p \times \nabla\rho)$ makes a relative contribution, compared to βv, of order $(f^2L^2/gH)(Ro)(a/L) \equiv (L/L_g)^2(U/\beta L^2)$, where $L_g = (gH)^{\frac{1}{2}}/f$ is the *external* Rossby deformation radius relating to the oceanic free surface; taking $U \cong 1$ cm sec^{-1}, $L \cong 10^3$ km $\cong \frac{1}{2}L_g$, $\beta = 2\times10^{-11}$ m^{-1}s^{-1} we have a negligible effect, of order 10^{-4}). Alternatively, if we imagine an ocean of many constant density layers, then the stiffness rule along the polar axis rule applies within each, leading again to (8).

3.4 Time-Dependence.

It is remarkable that the Sverdrup equation encompasses time-dependent motions important to the setup of baroclinic general circulation, as well as its steady state. If the wind starts blowing over a resting ocean, first of all fast barotropic Rossby waves (which are so fast that they can break down the 'stiffness' of the fluid) set up the initial, depth-independent response of the ocean. The fast waves carry information to the western boundary where a boundary current is set up to accept the Sverdrup circulation incident from the east. But, subsequently, baroclinic Rossby waves compete with the advection of the flow itself, to set the permanent depth-dependent flow, and its associated mass-field. We can see exactly these dynamics working in the analogous laboratory model (fig.2b) in which the top and bottom boundaries are tilted, and rotation is about a local vertical axis. There, if the angle of tilt is α (so that the top and bottom provide a wedge angle 2α), north-south motion is driven either by Ekman pumping at the top or by physically pressing down on the lid. The 'inclined plane' relation is simply

$$2\alpha v = w_e,$$

which is in fact the Sverdrup relation (8) because $2\alpha = \beta H/f$ (H is the total mean depth). Now, in a two-layer stratified ocean, the fluid has additional freedom. Layers 1 and 2, of equal mean depth but differing in density by the small amount $\Delta\rho$, have horizontal flow that differs by Margules relation (the discrete form of the thermal wind relation),

$$\mathbf{u}_1 - \mathbf{u}_2 = \frac{\mathbf{g}'}{f} \times \nabla\eta \tag{10}$$

where $\mathbf{g}' \equiv \mathbf{g}\Delta\rho/\rho$ is the reduced gravity vector, and η the interface displacement. But within each layer vertical 'stiffness' is felt, and so the average meridional displacement must just accommodate the vertical velocity imposed at the top. We have

$$\alpha(v_1 + v_2) = w_e \tag{11}$$

This is the 2-layer form of the classic Sverdrup relation of wind-driven circulation,

$$\begin{aligned} \beta \textstyle\int v dz &= f w_e \\ &= \nabla \times \boldsymbol{\tau}|_z \end{aligned}$$

the latter being the vertical velocity arising from the curl of the horizontal wind-stress, $\boldsymbol{\tau}$.

But, notice that with no forcing, $w_e = 0$, there is now non-trivial meridional flow allowed, provided that

$$v_1 = -v_2.$$

If such antisymmetric north-south flow occurs, it will clearly affect the 'thermocline', for as each layer moves laterally, the sloping boundary forces it to move vertically. Imagine how a trough in the interface will react to this meridional flow (figure 3); the southward upper-layer flow

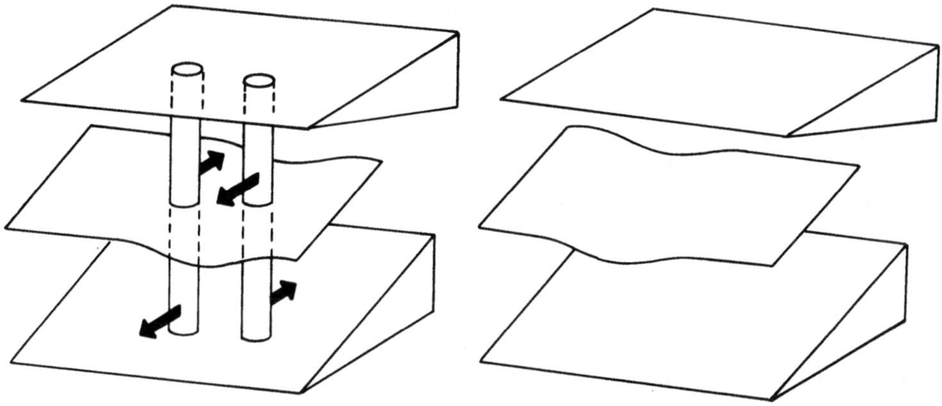

Figure 3: View from the south of a baroclinic Rossby wave; the north-south movement of 'stiff' fluid columns, allowed the thermal-wind balance drives the interface vertically, causing the pattern to propagate westward.

on the eastern side of the trough forces the interface upward, in complete agreement with the northward flow forced by the opposite-cambered wedge beneath the interface. Meanwhile the northward upper-layer flow on the western side of the trough depresses the interface. The result is propagation of the trough westward.

The details are as follows. The complete problem consists of the vorticity equation (11), the thermal wind equation (10) and the continuity equation

$$\frac{D\eta}{Dt} = \boldsymbol{\nabla} \cdot (h_1 \mathbf{u}_1) = -\boldsymbol{\nabla} \cdot (h_2 \mathbf{u}_2)$$

where $h_1 = \frac{1}{2}H - \eta$, $h_2 = \frac{1}{2}H + \eta$. A linearized form is

$$\begin{aligned} \frac{\partial \eta}{\partial t} &= -v_1 \frac{\partial h_1}{\partial y} \\ &= v_2 \frac{\partial h_2}{\partial y} \\ &\cong \alpha v_1 \quad \mathit{or\ from}\ (11), \quad -\alpha v_2 \end{aligned}$$

Combining with equation (10) we have simply

$$\frac{\partial \eta}{\partial t} + c_0 \frac{\partial \eta}{\partial x} = 0 \tag{12}$$

where $c_0 = \beta g'/f$, which is a 1st-order equation in x and t for the propagation of internal Rossby waves. Because our wedge angle 2α is a surrogate for the beta effect, the wave-speed c_0 may be written

$$c_0 = \frac{\beta g' H}{4 f^2}$$

which is just β times by the square of the Rossby radius. The general solution of (12) is simply a westward propagation without distortion, of the initial conditions, $\eta_0(x, y)$:

$$\eta = \eta_0(x + c_0 t, y).$$

The key connection here is that on the Earth and in the 'wedge' analogy the **potential vorticity** of an isopycnal fluid layer, $H(f + \zeta)/h$ has a north-south gradient: with the fluid at rest or very gently moving ζ is negligible and the gradient is β in one case and

$$(\frac{f}{h})\frac{\partial h}{\partial y} \equiv \frac{2 f \alpha}{H}$$

in the other. We might build a laboratory model in which the top and bottom boundary are rigid, yet tilted, with rotation about the local vertical (indeed, John Hart (e.g. Ohlsen and Hart, 1989) has done so, and made meticulous examinations of baroclinic waves on a polar β-plane. Scott Condie has produced a deep-ocean β-plane with topography in our laboratory (Condie and Kawase, 1992). Ross Griffiths (personal communication) has constructed a 2-layer 'sliced cylinder' laboratory β-plane which is aimed at boundary currents and wind-driven gyres. Unfortunately, it is not yet quite possible to create a fully stratified β-fluid in the laboratory, with an interior pv gradient at rest, though we are attempting now to do so, using ferromagnetic fluids.

4 Wind-Driven Circulation

In general we want to include windstress forcing, which adds a term $\frac{1}{2}w_e$ to the righthand side of (12). Using β for familiarity, write

$$\eta_t + \frac{\beta g' h}{4 f^2} \eta_x = \frac{1}{2} w_e. \tag{13}$$

The general solution is the sum of homogeneous and particular solutions of (13) which may be written

$$\eta = \frac{4f^2}{\beta g' h}\left[\int_x^{x_e} w_e(x',y) - w_e(x'-c_0 t, y)dx'\right].$$

where x_e is the eastern boundary of the fluid. It is quite nice, the way the pieces of the Sverdrup puzzle fit together: check the special cases $v_1 = v_2$ (steady, barotropic Sverdrup flow, $\partial\eta/\partial x = 0$ yet η deflecting at $\frac{1}{2}w_e$); $v_1 = -v_2$ (baroclinic Rossby wave); $v_2 = 0$ (steady baroclinic Sverdrup flow). As the two parts of the integrand 'slide apart' with time, the Sverdrup solution is left behind beneath the forcing while its image propagates westward, seeking a western boundary or eventually dissipating. If the integral $\int w_E dx$ over the whole domain of x vanishes, we have what might be called 'mid-ocean' forcing. Then, the two parts of the solution are disconnected and the steady circulation is a closed gyre in the middle of the ocean; yet, if the integrated Ekman pumping does not vanish, an east-west trough in the thermocline connects the interior (say northward-) circulation with its transient 'forerunner' (say southward flow) reaching in search of a western boundary, where a boundary current will develop.

After the wind is switched on, the interface, initially flat, supports no vertical shear. The velocity established by fast barotropic dyamics (in a few days, for a typical ocean) is independent of depth. Then, gradually, the developing thermocline tilt cancels out the circulation in the lower layer, concentrating the (constant) Sverdrup transport in the upper layer.

Many generalizations (to include Equatorial dynamics and finite amplitude, for example) have been made. But the basis of modern general circulation theory is just this systematic Rossby wave propagation, in competition with advection. 'Nature abhors an east-west tilting thermocline' one might say, for if such a tilt is to remain in place it must be held there by eastward flow, direct Ekman pumping or mixing/convection; for further exploration of the idea, see Anderson and Gill (1975), who include western and eastern boundary current evolution, and Rhines (1986b).

4.1 Advection by Baroclinic Circulation and the 'non-Doppler' Effect.

The set-up of the general circulation thus follows equation (13) in a linearized, 2-layer framework equation, together with important western boundary concentration effects (the famous $Jo((xt)^{\frac{1}{2}})$ solution of Rossby) and eastern boundary upwelling dynamics. Because of what may be called the 'Anderson and Gill catastrophe', the linear, non-diffusive problem ultimately has NO circulation beneath the upper mixed layer ($\beta v \equiv 0$)

once it is spun up. A sequence of nonlinear theoretical models had to build on this dynamical framework (Rhines and Young 1982a,b; Luyten, Pedlosky and Stommel 1983). These provide a basis for analysing the rather complicated general circulation found in numerical models and in reality. A simple property of the equations for a model with a 'few' constant-density layers has made much of this possible. This is (Rhines and Young, 1982b) that the advection of Rossby-wave 'signals' by the baroclinic mode of circulation vanishes for some choices of model geometry: the advective competition for Rossby waves is predominantly that of the (known) Sverdrup depth-averaged flow.

To see how this may work, consider the following situation: a 2-layer model with a very deep lower layer has a thermocline tilt corresponding to the shear from the upper layer to the deep layer, which is nearly at rest. Suppose that a zonal steady current exists, with speed U, from west to east. From the above arguments, it is apparent that the north-south potential vorticity gradient is increased by this flow, and hence the intrinsic westward wave-speed of the long baroclinic Rossby mode is increased. But, in equal measure, the wave is advected eastward by the mean flow. The nonlinear pv conservation equation ($pv \equiv Q = f/(H+\eta)$) for interface displacement η is *independent* of U:

$$\eta_t + \frac{\beta g'}{f^2}(H+\eta)\eta_x = 0$$

in which the UQ_x term has been just cancelled by the mean-flow induced contribution to $v\eta_y$. The westward phase/group speed is

$$\beta L_\rho^2$$

where $L_\rho^2 = g'(H+\eta)/f^2$. This interesting compensation removes any possible 'hydraulic structure' in which waves would compete with advection speeds. If there is also a barotropic component of mean east-west flow, extending through both layers, it does Doppler-shift the waves and lead to hydraulic effects. More generally, if relative vorticity is included, the QG potential vorticity in terms of the stream function ψ is

$$Q = \nabla^2\psi + L_\rho^{-2}\psi + (\beta + L_\rho^{-2}U)y$$

and linear QG waves of frequency ω, wavevector (k, ℓ) obey

$$\omega = k[U(k^2+\ell^2) - \beta]/[k^2+\ell^2+L_\rho^{-2}]$$

which shows a Doppler shift of the relative vorticity, but not the 'stiffness' part of the potential vorticity. Stationary waves, $\omega = 0$, determine a circle of possible waves $k^2+\ell^2=\beta/U$, plus upstream and downstream blocking plumes, $k=0$ (see Lighthill, 1968). Ironically, this system has Doppler shifting yet stationary waves propagate independent of stratification!

In a similar way the characteristic problem for the 2- and 3-layer QG wind-driven models of Rhines and Young (1982b) became a linear problem of integrating along characteristics determined solely by the wind field. In a 2-layer QG model, the nonlinear advection of interface height

$$\mathbf{u}_1 \cdot \nabla \eta$$

is just $J(\psi_1,\eta)=J(\psi_1,\psi_1-\psi_2)=J(\psi_1,\Psi_B+2\psi_1)=J(\psi_1,\Psi_B)$ where $\Psi_B=\psi_1+\psi_2$. Thus the advection becomes a linear advection by the barotropic Sverdrup transport, Ψ_B. The self-advection of the baroclinic mode vanishes owing to the non-Doppler effect, leaving only advection by a known field.

In a model with three equal layers, and two interfaces with equal g', the 3 Rossby-wave modes are Φ_0,Φ_1,Φ_2. The wavespeeds for the two baroclinic modes are $\beta L_\rho^2/3$ and $\beta L_\rho^2/9$. The evolution equation for the 2d baroclinic mode, Φ_2,is

$$(\frac{\partial}{\partial t}+(\beta L_\rho^2/9)\frac{\partial}{\partial x})\Phi_2+J(\Phi_2,\Phi_0)=0$$

with $L_\rho^2=g'H/f^2$ based on the total depth H. The barotropic mode is the Sverdrup transport function

$$\Phi_0=\frac{1}{3}(\psi_1+\psi_2+\psi_3)=(1/\beta H)\int \nabla\times\boldsymbol{\tau}|_z dx$$

since barotropic waves are absent (have infinite speed) in the PG model. Φ_0 is normalized so that it represents the velocity of barotropic advection. Look at the $J(\cdot)$ term: the $2d$ baroclinic mode (where the two interfaces are out of phase) is advected neither by the 1st mode (interfaces in phase) nor by itself: this is the 'non-Doppler' effect for these inter-mode interactions. Hence, again, the nonlinear advection of pv becomes the linear advection by the barotropic Sverdrup flow.

It is interesting that this result and thus the characteristics determining the whole baroclinic circulation depend upon the 2d baroclinic mode wavespeed, plus advection by the barotropic Sverdrup flow. The faster, first baroclinic mode propagating from the eastern boundary reduces the deep-ocean velocity and increases the shallow velocity, but it is the $2d$ mode which is capable of finally shutting down the circulation in both layers 2 and 3; hence it is this mode whose Doppler shift by the Sverdrup circulation is all-important.

It is remarkable that the same result occurs in the non-QG case of 2 active layers with arbitrarily large interface displacement, as followed by Rhines (1986), Luyten and Stommel (1986). With additional forcing, for example by simulated buoyancy driving, the problem becomes one of integrating along known characteristics, which are set by the wind field. Buoyancy forcing was incorporated by Rhines, Holland and Chow (1985), Luyten and Stommel (1986), and Pedlosky (1986). There are many new effects, but it remains true that the speed of waves carrying information to determine the circulation is independent of the slopes of the density interfaces, and hence the problem remains a first-order (though nonlinear) equation solved by integrating along characteristics. The barotropic, depth-averaged circulation, which is already known from the Sverdrup relation, exerts tremendous control on the problem, steering the circulation whether wind- or buoyancy driven. The 'non-Doppler' effect has arisen in other problems, see Held (1983), Chang and Philander (1989), Kessler (1990).

The time-dependent equations for interannually varying circulation retain much of this simplicity (for 2 layers, or 3 equal-depth layers in a QG model, or 2 active layers in a PG model). Dewar *et al* (1984), Rhines (1986), Dewar (1991) and Liu (1992) have explored the time-dependent adjustment of the QG and PG systems, some with nonlinear steepening and outcropping density surfaces. Finite amplitude effects which turn the above derivation into one capable of allowing outcropping and subduction effects, simply require accurate treatment of the continuity equation, yielding in the simplest case of a two-layer model

$$\frac{\partial h_1}{\partial t} + \mathbf{U}_B \cdot \nabla h_1 + C(h_1)\frac{\partial h_1}{\partial x} = -(1 - \frac{h_1}{H})w_e \tag{14}$$

where the Rossby wavespeed $C(h_1) = -\beta g' h_1(H - h_1)/f^2 H$. h_1 is the upper layer thickness ($\equiv H_1 - \eta$) and H the total depth. Thus, a pulse of elevated thermocline steepens at its eastward trailing side, and flattens on its western face, for the usual circumstance of a mean upper layer thinner than the lower. The field of characteristics describes both the steady circulation with steady forcing, and the response to interannual changes in forcing. Equation (14) exhibits outcropping under appropriate upward Ekman pumping. As above, the 'non-Doppler' effect is at work, for the advection of the baroclinic mode by itself vanishes, $(\mathbf{u}_1 - \mathbf{u}_2) \cdot \nabla h_1 = 0$. Nonlinear effects that are present involve wavespeed dependence on amplitude (i.e., coefficients depending upon h_1) rather than advection which would upset the hyperbolic nature of (14).

It is fair warning that despite the many papers written on 'non-Doppler' models, the general case (Welander's 'M-equation') does not admit this simplification. One can locally examine the general equation in terms of Rossby propagation and advection, but the sole dependence on the Sverdrup barotropic circulation is gone, and mutual advection of baroclinic modes returns. In going to a continuous density distribution we have relied upon ideas like the Prandl-Batchelor theorem for recirculating gyres, (Rhines *et al.* 1982) and assumptions about the functional dependence of potential vorticity on stream function and density in inviscid idealizations (e.g., Huang 1986, Janowitz 1986, Killworth 1987). These have discontinuous behaviour (e.g., Young and Ierly 1986) and resolution of conditions at both eastern and western boundaries is not yet completely understood.

In particular, the large 'shadow zones' that appear in the early theory are regions of stagnant ocean forced to be so by eastward radiation of zero-velocity conditions from the eastern boundary. It is indeed interesting that the oxygen minimum regions of the several tropical oceans seem to 'fit' into these shadow zones; unusually high biological productivity has been one explanation, but cfc-tracer sections (Warner 1992) show that they are 'freon holes' as well: the water is very old and poorly ventilated. However, we know also that damped Kelvin waves from the Equator can establish boundary conditions along the eastern boundary, which may be other than simply no-flow (e.g., Davey, Hsieh and Wasjowicz 1983). In some general circulation models, a specified surface meridional temperature gradient forces flow into the eastern boundary which must mix and exit at another depth (Cox 1985), radiating 'β-plumes' of shallow circulation westward. Adequate treatment of these non-geostrophic boundary signals (and their quasi-geostrophic shelf- slope- and bottom-trapped cousins over realistic topography) is essential to a complete representation of the general circulation. Buoyancy-forced circulation, in the simple models, appears to be rigidly controlled by the free-wave characteristics of the circulation driven only by wind. As above, however, models with continuous stratification or more layers will begin to lose the non-Doppler property: water-masses will advect and propagate, but they also will interact with one another. A classic case of this is readily seen in the laboratory, and in recent gcms aimed at glaciation-time-scale climate: abyssal, deep, and intermediate waters of the world ocean must compete for space. When sea-surface conditions change (for example, with sea-ice extending southward during glaciation), the attendant change in northern deep water production can shift the balance in favor of Southern Hemisphere sinking, with Antarctic Bottom Water taking over the deep ocean (e.g., Cox 1989, Imbrie *et al.* 1992).

4.2 Numerical Example.

Layered models permit one to plot fields of characteristics (showing the group velocity of the relevant Rossby wave mode, as altered by the Sverdrup flow; for a 3-equal layer QG model this is the 2*d* baroclinic mode). Some general properties are evident. The wavespeed c_0 is large in the tropics (O(10 cm sec^{-1})) yet decreases rapidly with increasing latitude (to less than 1 cm sec^{-1} at 40°N). Yet, though the surface flow of the general circulation often far exceeds 1 cm sec^{-1} it is the depth-averaged velocity that advects the density field, which will be far weaker.

A plot of these characteristics for four configurations of wind forcing is shown in figure 4, from Rhines and Schopp (1991). In this particular study we allowed for a non-symmetric pattern of westerly/easterly winds. The line of vanishing wind-curl was tilted to resemble the observed state, where the North Atlantic storm-track runs southwest-northeast from Cape Hatteras toward Ireland. These are thus the 'information' propagation lines for the middle layer of a QG three-layer model; for steady flow without buoyancy forcing or mixing they also become the streamline field, $\Psi_2(x, y)$. Notice that with the symmetric wind pattern (Ekman downpumping south of uppumping, with the zero line running east and west) the shadow zones are also symmetric, and are separated by the saddlepoint and stagnation characteristic from the recirculating gyres.

As the zero-pumping line is rotated so that it runs southwest-northeast, the subpolar gyre expands while the subtropical gyre contracts. For large 'tilt' of the wind-field, some characteristics run first westward, then snake around the subpolar gyre to high latitude. In doing so they pass beneath the boundary jet, where it flows eastward toward midocean, and can thus be energized by eddy activity there. A sort of 'Rossby propellor' occurs in which eddy flux of potential vorticity drives a large new branch of circulation. This did not happen in the symmetrically forced runs, for there the boundary current eddies and the shadow-zone characteristics are geographically separate.

The time-mean circulation from these simulations, one of which is shown in figure 5, is very close in structure to the predictions of the theory: theory, that is, modified in extreme cases to admit the added circulation due to eddy forcing near the boundary jet.

But there is much more to the ocean than PG theories would admit. The flooding of the gyre from the western boundary is influential in both spin-up and steady state flow. As the circulation develops from rest (figure 6), it agrees with the barotropic Sverdrup pattern after a few days. Then one sees both baroclinic Rossby wave modes moving westward to

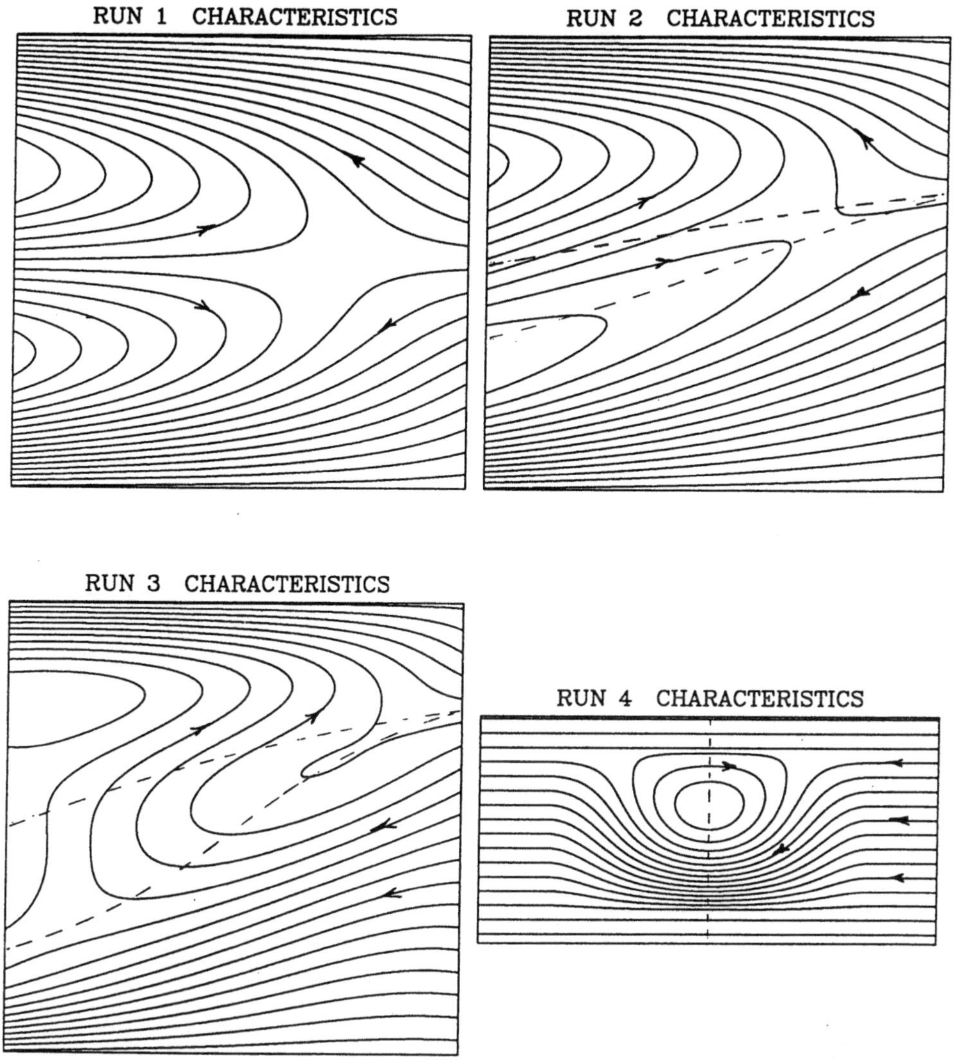

Figure 4: Characteristics for the quasi-geostrophic wind-driven circulation (Rhines and Schopp, 1991), for four patterns of wind. The first is for the familiar symmetric westerly/easterly winds; in the upper-right the zero-Ekman-pumping line (southernmost dashed line) is rotated to run SE-NW (the Sverdrup gyre boundary is the northern dashed line); in the lower left it is further rotated; in the lower right it runs north and south.

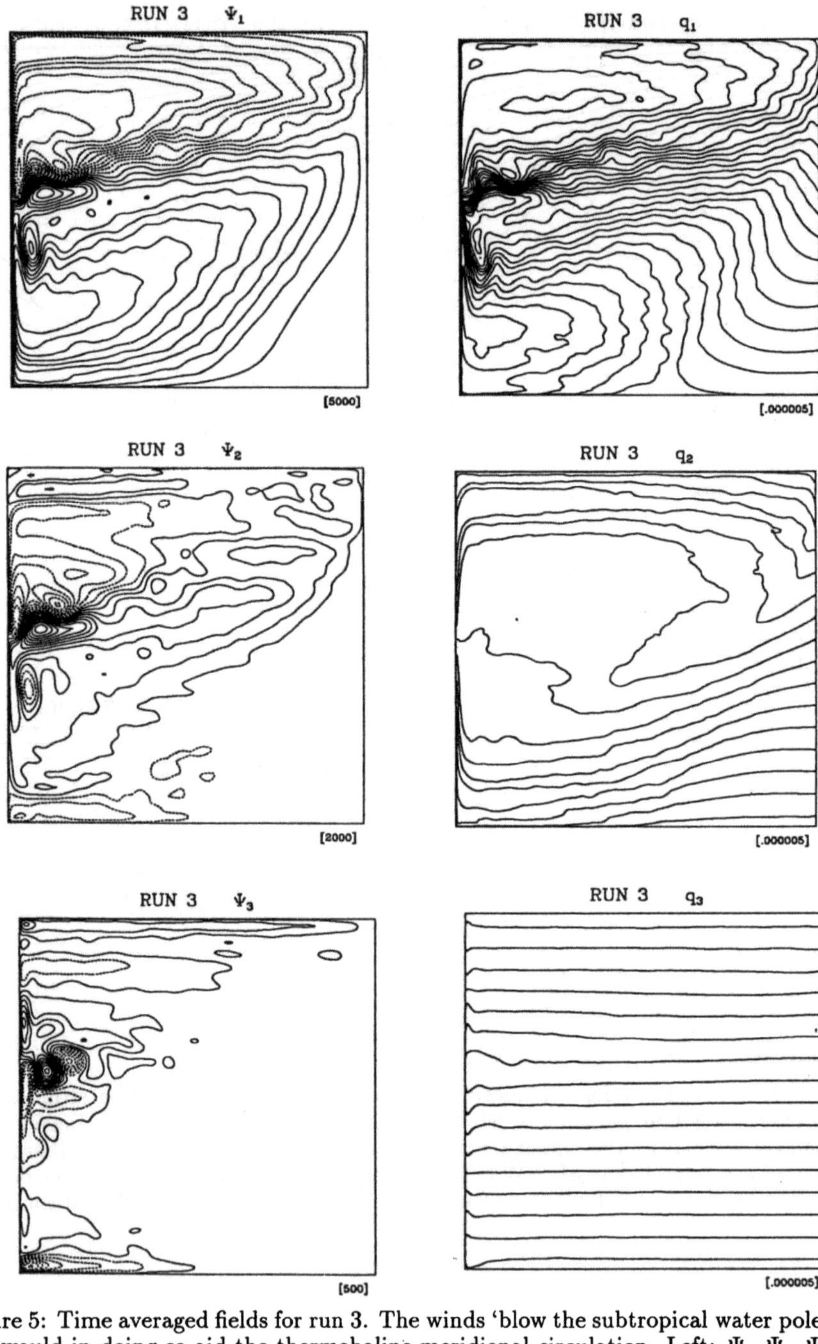

Figure 5: Time averaged fields for run 3. The winds 'blow the subtropical water poleward' and would in doing so aid the thermohaline meridional circulation. Left: Ψ_1, Ψ_2, Ψ_3 and right: Q_1, Q_2, Q_3. Eddy pv-flux propels the mid-depth circulation, as well as smaller deep gyres. Pv is well-mixed in the mid-depth gyres.

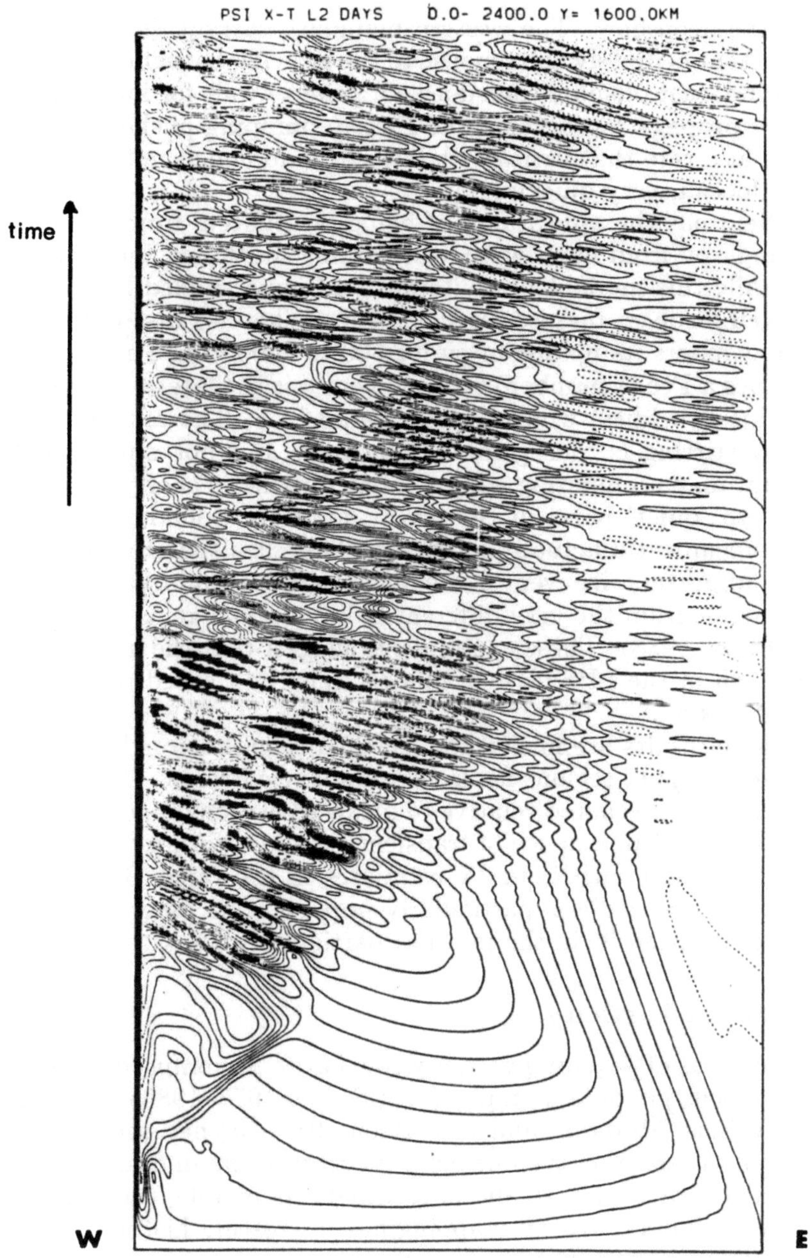

Figure 6: Time-longitude diagram for the middle-layer streamfunction, for the first 2400 days of spinup from rest, Run 2, 400km south of the mid-line of the 4000km basin. The 1st baroclinic Rossby mode reduces the amplitude and the slower 2d mode shuts off the large-scale flow in the east, while eddies and western boundary current activity invade from the west and north.

shut down the deep flow. The western boundary current develops baroclinity almost immediately, in response to the growing baroclinity directly forced by wind, just offshore (one does not have to wait for long-distance Rossby wave propagation because the Ekman pumping pushes down the thermocline at a rate that varies with latitude, and hence creates vertical shear in midocean, very quickly).

The pv field has the appearance of being 'ventilated' from the northeast, with high pv from high latitude streaming southwest into the thermocline. While this model lacks outcrops and upper mixed layer, it has instead a well-resolved mesoscale eddy field. What we discover is that the fluid remembers its upstream pv for 500 to 1000 km. The tongue of high pv defines a Peclet number (UL/κ, where κ is the lateral Lagrangian diffusivity due to eddies) which ranges from 3 to 5. The ratio of advection to eddy stirring is thus greater than unity, but not much greater. In agreement, maps comparing eddy- and mean kinetic energy show that as one goes to 'quiet' parts of the ocean, far from boundary currents, the eddy-to-mean ratio actually increases. This says that the mixing of pv remains important in even the 'eddy desert' of midocean. Similar conclusions come from the very different 3-dimensional gcms (Cox, 1985, Boning and Cox, 1988). These experiments suggest that the non-diffusive approach taken in the LPS class of models must be revised to recognize potent eddy mixing.

Unless these results are misleading us, the ocean is strongly laterally-diffusive, even in quiet regions far from boundary currents. Eddy mixing of this magnitude (measured relative to mean advection) is supported by the broad, diffusive appearance of transient tracers entering from the the sea-surface, consistent with our Pe estimates in the range 3 to 5. The vertical dependence of Pe is particularly important to ascertain, as the deep mean circulation is, in the interior, particularly weak.

4.3 Potential Vorticity.

Earlier discussions of potential vorticity (pv, not to be confused with pressure times velocity) in oceanography (e.g., Rhines 1977, 1986a,b) and meterology (Hoskins, McIntyre and Robertson 1985, Haynes and McIntyre 1987) have emphasized on the one hand its central role in dynamics, and on the other its tracer nature and surprisingly mappable, tangible presence in observational data. pv mapping has interesting history. Writing $Q(x,t)$ for potential vorticity, its conservation equation is

$$\rho\frac{DQ}{Dt} = -\nabla \cdot \mathbf{N} \tag{15}$$

where

$$\begin{aligned} Q &= \rho^{-1}(\boldsymbol{\omega} + 2\boldsymbol{\Omega}) \cdot \nabla \sigma \\ \mathbf{N} &= -\mathcal{H}(\boldsymbol{\omega} + 2\boldsymbol{\Omega}) - \mathbf{F} \times \nabla \sigma, \end{aligned}$$

$\mathcal{H}$ is buoyancy forcing rate (from heating, cooling, mixing). It is particularly important that $\mathcal{H}$ is the *convergence* of the buoyancy flux $\boldsymbol{G}$, say $\mathcal{H} \equiv -\nabla \cdot \boldsymbol{G}$ rather than the flux itself; σ is potential density, $\boldsymbol{\omega} + 2\boldsymbol{\Omega}$ the absolute vorticity vector (don't confuse vector ω with the non-dimensional frequency $(\mathrm{fT})^{-1}$ used earlier), and $\mathbf{F}$ is a viscous or external body force per unit mass (e.g., Haynes and McIntyre 1990). For the scaling appropriate to ocean circulation

$$Q \cong \rho_0^{-1}(f + \zeta)\frac{\partial \sigma}{\partial z}$$

where f and ζ are vertical components of $\boldsymbol{\omega}$ and $2\boldsymbol{\Omega}$ respectively. Q is the pv per unit mass, and ρQ the pv per unit volume. At scales greater than L_ρ the contribution of ζ, other than to the barotropic mode, is smaller than the 'stiffness' term and

$$Q \cong \rho_0^{-1} f \frac{\partial \sigma}{\partial z}$$

At this level of approximation the righthand side forcing may be written

$$\frac{f}{\rho_0}\frac{\partial \mathcal{H}}{\partial z} - (\boldsymbol{F} \times \hat{\mathbf{z}})\sigma_z.$$

Because the righthand side forcing appears as the divergence of a flux N, the integrated pv cannot be created or destroyed away from boundaries. The view of Haynes *et al* (op. cit.) is that buoyancy or mechanical forcing cause flux, concentration and dilution of pv rather than their creation or destruction. If one produces a region of low pv in the middle of a fluid by mixing its mass-density, the sharpened stratification at its edges has increased pv, in equal measure: there has been no net creation. Or if one exerts a twisting force on the fluid, regions of positive and negative vorticity are produced, which cancel in a volume integral.

It is somewhat a semantic argument however. Owing to the flux form of the righthand side of (15), the volume integral of ρQ between two isopycnal surfaces is conserved even in the presence of diabatic (heating, cooling, mixing) processes and internal or external body forces, as Rhines and Holland (1979) pointed out. Just as salt is the ionic substance whose density per unit mass is salinity, Haynes and McIntyre define 'pv substance' as the conservative property whose density is Q (per unit mass) or ρQ (per unit volume). 'pv substance' is closely related to Kelvin's (absolute) circulation

$$\int\int\int \rho Q dV = \int\int \sigma(\boldsymbol{\omega} + 2\boldsymbol{\Omega}) \cdot \hat{\mathbf{n}} dA$$

taken over the bounding surface, with outward normal $\hat{\mathbf{n}}$. If it for example is bounded by constant potential density surfaces, plus a perimeter lying parallel to $\boldsymbol{\omega} + 2\boldsymbol{\Omega}$, then the integral is just

$$[\sigma(\int \mathbf{u} \cdot d\boldsymbol{\ell} + 2\Omega A)]_{\sigma_2}^{\sigma_1}$$

(including the planetary contribution $2\Omega A$, where $\boldsymbol{\Omega} = \Omega\hat{\boldsymbol{k}}$ and A is the area projected normal to $\boldsymbol{\Omega}$; $d\boldsymbol{\ell}$ is a vector line segment defining the perimeter). The robust conservation of absolute circulation thus accounts for the conservation of integrated pv. But, we shall show below that the volume integral of ρQ is *not* at all robustly conserved when the volume is the entire body of a fluid contained in insulating, shear-stress-free boundaries.

A robust quantity may also be an uninformative one. Net pv conservation, after integrating vertically between σ-surfaces can hide important dipole-like local contributions. Knowing the Kelvin circulation about a contour of fluid particles, one may not in fact know either the Eulerian or Lagrangian mean circulation around that same contour, nor any detail of the dynamics inside the contour (Rhines and Holland 1979, p307). There are also subtle problems when stirring and mixing are involved, which has led to much debate in the atmospheric literature (e.g., Danielson, 1990, Haynes and McIntyre, 1990). Take a parcel of fluid from a heavily stratified, high pv region and suddenly insert it below in a layer that is weakly stratified, with low pv. Equation (15) requires that the pv (or we might say 'Kelvin density') move with the fluid. It doesn't violate Haynes *et al*'s idea of flux along isopycnal surfaces, for those surfaces are torn apart and later reconnected by diffusion. Quite subtly, by placing two blobs of different density next to one another one creates, at their common boundaries, sheets of pv equalling, with opposite sign, all of that present in the newly moved blob! These δ-functions of locally large pv strain the imagination, as to where they come from. Dissipation and some intesting topology are clearly involved. Regardless, the pv of the interior blob was transported vertically, and leaves its mark. But it carries around itself a compensating sheath of pv that cancels in the integrated Kelvin circulation. After further mixing, the pv is diluted and net (vertically integrated) pv within the layer, is as before.

What we are wrestling with is a tracer that carries with it strong implications to momentum (or angular momentum) and vorticity balance, and can hardly be expected to move about as simply as a cloud of dye. It remains useful to think of mixing and buoyancy sources as source-sink distributions of pv-density in physical space, recognizing that they will often occur in dipolar pairs; see my example in Haynes *et al.* (1987) for further discussion.

4.4 Global pv Budgets.

In addition, pv is a tracer whose global conservation properties may *not* be terribly useful to examine, because one can upset the entire global balance by merely changing a boundary layer here and there (as shown above). Consider an east-west section across a one-layer model ocean basin: the net, integrated pv in a western boundary current is equal to the net pv in the whole remainder of the section. We have made such calculations for a 3-layer QG general circulation model, in which the wind-input of pv in the upper layer rearranges the ambient pv of the resting ocean (the 'ramp' $Q = \beta y$), and there is significant flux through the western boundary. Lozier and Riser (1989) plot the distribution of flux from this calculation, and carry it further by looking at pv balance on individual fluid parcels moving through the circulation.

Beyond its usefulness as an ocean tracer, the dynamical role of pv relates to 'invertibility': the statement that with knowledge of the complete pv distribution one can calculate velocity and density fields. Of course, to do so one needs boundary conditions, adherence to a balance condition (e.g., geostrophic balance), and knowledge of the distribution of mass as a function of potential density. Non-geostrophic motions (some of them irrotational) like surface or internal waves, or Kelvin waves cause problems for invertibility, which are not always obvious. Indeed, the current interest in fronts and in the interaction between internal waves and vortical modes in the ocean challenges the 'completeness' of pv as a description of the motion.

An early derivation, and mapping for the atmosphere, is given by Rossby (1940). In part because of greater stratification in the atmosphere ($L_\rho \cong$ 1000's of km), the relative vorticity dominates pv over more of the atmospheric spectrum than is the case in the ocean (where $L_\rho \cong$ ten's of km). The 'stiffness' component of pv, closely approximated by $f\partial\sigma/\partial z$ where σ is a locally referenced potential density, dominates in oceanic circulations, but for Rossby-radius scale jets and mesoscale eddies.

David Behringer, working with Myrl Hendershott, made two maps of South Atlantic pv from Meteor data, as part of his Ph.D. dissertation (Behringer, 1972). The paper reporting these results was never published, apparently the referees thinking it unimportant! After Bill Young and I had written up our theory of the wind-driven circulation, which so hinged on pv dynamics, and regions of homogenization due to mesoscale eddies, Scott McDowell and I plotted pv in the North Atlantic from a nice collection of hydrographic data he was working with. The result (McDowell, Rhines and Keffer (1982), followed by Holland, Keffer and Rhines (1984), Keffer 1985, Rhines 1986a) showed the remarkable information content of this 'dynamical' tracer, and how simply it is recoverable from hydrographic data. Regions of homogenized pv appear, great tongues of pv injected at outcrops, and islands and ridges of pv created by buoyancy- and wind-forcing, and pv minima associated with 'water mass' cores all were evident. Water masses are now routinely mapped in terms of pv minima (e.g., McCartney 1982), complementary to their identification by N^2-minima or Wustian tracer extrema.

Mixing toward homogenization of pv appeared subsequently in the stratosphere as well as the ocean. McIntyre and Palmer (1983) portray stratospheric sudden warmings as enormous breaking Rossby waves. They homogenize much of the hemispheric pv in mid-latitudes, picturesquely called a 'surf-zone', while concentrating the north-south pv gradient in a narrow band surrounding the polar vortex. This concentrated pv gradient provides a stable surface that helps to maintain the isolation of the ozone hole. There was significant interaction between the ocean- and atmosphere workers at this time, each learning from the other. In particular, this was a 'paradigm shift' for the atmospheric dynamicists, who on the one hand viewed the sudden warming in zonally averaged sense, where it appears a formless mixing event, and on the other hand speculated about 'generalized Lagrangian mean' circulations in the stratosphere in which fluid particles had bounded displacement from latitude circles and pv mixing was disallowed.

The basis for homogenization is that a region of fluid without sources or sinks of pv, but with eddy energy, will stir and mix until the large-scale pv gradient is much reduced. Some sort of confinement of the mixing process, by a mid-latitude gyre or nearly closed streamline region, is necessary to prevent endless resupply of the pv-tracer from higher or lower latitudes. In the dominant oceanic circumpolar flow, the Antarctic Circumpolar Current, there is strong evidence for pv homogenization in Keffer's original maps, in the FRAM numerical model, and in that of Dirk Olbers and collaborators, reported in this volume. Although, it is also clear that as one rises along σ surfaces toward the ocean surface, there is increased stratification, and a band of high pv. Circumpolar flows are tricky however, if the region is not simply connected (Rhines and Young, 1982). The presence of an active boundary (i.e., the Antarctic coast and the outcrop lines at the sea surface) can cause a continual flux of pv northward, for example by radiating nearly-standing Rossby waves, or by turbulent-viscous eddy shedding accompanied by an enstrophy cascade (recall that boundaries, even under free-slip conditions, can be sources of pv). This may help account for the change from a homogenized band to the strong pv gradient in the outcrop regions.

4.5 Angular Momentum.

Ocean modelling needs, in addition to global budgets for heat, fresh water, mass and energy, a signed or vector momentum-like quantity to describe the general circulation, in somewhat the way that meteorologists look at the y-z structure of the x-averaged zonal momentum of the atmosphere (more exactly, the polar component of angular momentum density $\rho r x u|_{\Omega}$, where r is the spherical radius vector). Rhines and MacCready (1989), Holloway and Rhines (1991) look at the angular momentum of an idealized 'ocean gyre' in a 2-dimensional fluid on a plane, which may be written

$$\int\int \rho \mathbf{r} \times \mathbf{u}|_z dV \equiv -2 \int\int (\psi - \psi_b) dV$$

where ψ is the geostrophic streamfunction, and ψ_B is its value at the basin boundary. The idea is that the 'spin' of the gyre is a fairly robust quantity, proportional to the integrated stream function, much more robust than the integrated pv, which in the same circumstance describes the Kelvin circulation $\int \mathbf{u} \cdot d\boldsymbol{\ell}$ integrated about the rim of the basin (along arc length $d\ell$). When the ocean responds to forcing and develops a gyre of circulation, however, its center of mass moves, and this represents another very significant component of angular momentum change.

The interesting point is that with free-slip boundary conditions (the vanishing of the tangential, or long-shore component of viscous stress), there still can be a balance between wind-driven angular momentum and torque exerted by the boundary. One possible part of this is the inviscid pressure torque (a sort of lateral Rossby-wave drag), but in the simple square basin described in the papers referenced above, this vanishes. It is in fact the *normal* component of viscous stress that exerts a torque, and will spin down a gyre to rest if the forcing is removed. Normal viscous stress is somewhat unfamiliar, yet it is very tangible when one pours maple syrup from a bottle: it is the force holding the syrup column together.

Angular momentum and vorticity interact in interesting ways: they are sometimes mistaken for one another, but they often can have opposite sense. Note the mathematical identity:

$$|\mathbf{r}|^2\boldsymbol{\omega} = -2\mathbf{r}\times\mathbf{u} + \nabla\times(|\mathbf{r}|^2\mathbf{u})$$

Here ω is the vector vorticity and r a 3-dimensional position vector. If this relation is integrated over a volume of fluid one has

$$\int\int\int|\mathbf{r}|^2\boldsymbol{\omega}dV = -2\int\int\int\mathbf{r}\times\mathbf{u}dV - \int\int|r|^2\mathbf{u}\times\hat{\mathbf{n}}dS$$

where $\hat{\mathbf{n}}$ is a unit vector normal (outward) to the bounding surface S. If $\mathbf{u}$ vanishes at S (or if we integrate to infinity, where $\mathbf{u}$ vanishes), then

$$\int\int\int|\mathbf{r}|^2\boldsymbol{\omega}dV = -2\int\int\int\mathbf{r}\times\mathbf{u}dV;$$

the second spatial moment of vorticity is (minus) twice the angular momentum! Relations like this, and the analysis of 'impulse' (a generalized kind of linear momentum) found in Lighthill (1986) lead to important conservation ideas which may help in future analyses of the general circulation. For a discussion of polar angular momentum in the ACC see Straub (1992).

The use of pv ideas to describe induction of mean circulation has been described earlier (e.g., Bretherton and Haidvogel 1976, Rhines 1977, Rhines and Holland 1979, Haidvogel and Rhines 1983). Stirring and mixing by mesoscale eddies leads to locally intense gyres and branches of circulation, and also can act persistently to drive larger regions of ocean gyres toward plateaus of mixed pv, away from the north-south β gradient. What is less understood is the entry and exit of pv from boundaries: now we will discuss several aspects of this.

4.6 Subduction; pv Entry into the Ocean.

This is the process by which near-surface fluid flows from the violently stirred, convected and stressed mixed-layer down into the much quieter 'geostrophic' interior. The volume transport of fluid driven by wind in the Ekman layer is one of the most reliable predictions we have. Its divergence drives vertical motion contributing to subduction. Subduction can also occur due to horizontal velocity in presence of a sloping mixed layer base, properties which will be discussed below (e.g., Woods and Barkmann, 1986). Buoyancy forcing is constantly active at the ocean surface, and it can either help or hinder Ekman pumping. Even in total absence of Ekman pumping, a cooled fluid is capable of convectively sinking, although the shape and form of purely convective ventilation is not well understood. Quite often Ekman pumping opposes buoyancy forcing, with 'densification' (cooling dominating precipiation) at high latitude in the regions of upward Ekman pumping, yet 'lightening' of the surface water by warming dominating evaporation, in the subtropics just where Ekman pumping is downward. The general circulation must respond to this tug-of-war appropriately, and horizontal advection of density constituents in the mixed layer is a key part of the response. It will be discussed further in section 4.

We need much better knowledge of the Ekman flux, its vertical profile, and its Lagrangian trajectories. In the far North, the veered Ekman flux outward from the center of the Irminger gyre must also accommodate the pattern of density outcrops. To the south, near the Gulf Stream, the strong Equatorward Ekman flux beneath the winter storm track running northeastward from Cape Hatteras, forces heavy fluid over light, and this must be partly responsible for the deep mixed layers found between 40°N and 50°N (that is, advection of cold water southward in addition to local one-dimensional mixed-layer deepening). In the tropics Chereskin and Roemmich (1991) find a striking tendency for the strong poleward Ekman transport to extend below the base of the mixed layer (as measured by an acoustic Doppler current profiler (adcp) combined with hydrography).

Pv budgeting is interesting in this area, because one has near-zero pv in the mixed layer, very large pv at its base where the density gradient is large, and somehow the fluid escaping the mixed layer through this base takes on a climatologically smooth value of pv (e.g. Williams (1991)). It is not a smooth process, but in part occurs after the deepest convection of winter is healed over by spring warming (Stommel 1979b). Various models have been proposed, often involving a nondissipative, pv-conserving interior and a simple, schematic mixed layer in which is embedded an Ekman layer (note the contradiction with Chereskin *et al.*'s direct measurements in the tropics, in which the Ekman flux reached below the mixed layer). Nurser and Marshall (1991), Marshall and Nurser (1992) give a formula for volume flux of subduction, into the geostrophic interior (defined as the rate at which fluid moves across the mixed-layer base):

$$\begin{aligned} S &= \frac{D_b \rho_m}{Dt} / \frac{\partial \rho_0}{\partial z} \\ &= \frac{f(\mathcal{H} - \mathcal{H}_{Ek})}{g \rho Q_b} \end{aligned} \qquad (16)$$

where

$$S \equiv -w - \frac{D_b h}{Dt}$$

is the rate of vertical movement of the water relative to the (possibly sloping) mixed-layer base; $D_b\rho_m/Dt$ is the rate of change of mixed-layer density ρ_m following the flow at the mixed-layer base, $\partial\rho_0/\partial z$ is the density gradient just below the mixed layer, h is the mixed layer depth, $Q_b = -f\rho^{-1}\partial\rho_0/\partial z$ the pv just below $z = -h$, $\mathcal{H}$ is the buoyancy forcing of the mixed layer (that is, divergence of the vector buoyancy flux, assumed uniform in the layer: thus $\mathcal{H}$ = air-sea buoyancy flux/h); $\mathcal{H}_{\mathcal{E}\|}$ is the Ekman advection of buoyancy, $g(\hat{\mathbf{z}} \times \boldsymbol{\tau}) \cdot \boldsymbol{\nabla}\rho$. The flux of pv into the ocean, at its surface is given as

$$\begin{aligned} J_z &= \rho S Q \\ &= -f(\mathcal{H} - \mathcal{H}_{Ek})/g + O(Ro) \end{aligned}$$

The formulas are surprising in their apparent lack of dependence on Ekman pumping, $\boldsymbol{\nabla} \times (\boldsymbol{\tau}/\rho f)$ where $\boldsymbol{\tau}$ is wind-stress.

Heat-balance based formulas like (16) depend upon a key assumption that in the upper ocean there is *no jump* in either density or velocity at the mixed layer base, and neither mixed-layer entrainment nor diapycnal diffusion below the mixed layer contributes to the density balance. The mixed-layer density marches up and down the ambient $\rho(z)$ without any 'step' at its base (hence without turbulent buoyancy flux at its base). Accordingly, their mixed layer is in geostrophic thermal-wind balance, with a thinner Ekman layer at the surface. Subduction occurs when the sea is heated or freshened by the atmosphere, mixed-layer fluid thus becoming interior fluid as the mixed-layer shoals (in space or time, relative to the fluid, which may be moving downward and laterally). This models the restratification in springtime, and leaves a well-defined pv in the newly stratified fluid, consistent with Stommel's (1979b) 'Ekman Demon'. Conversely, passage of fluid from stratified interior to mixed layer ('entrainment', $S < 0$) is associated with cooling/evaporation flux at the sea surface. The contribution to $\mathcal{H}_{net}$ from the lateral buoyancy flux by Ekman transport introduces a wind effect on subduction, and could allow cooling of the ocean by the atmosphere to promote subduction, $S > 0$; yet it does not explicitly account for the major effect of windstress curl and Ekman pumping.

Marshall and Nurser's model is an important step in looking at the buoyancy balance of the upper ocean; a complementary discussion by Speer and Tziperman is described below. Yet, problems abound. $\partial\rho/\partial z$ just beneath the mixed layer, which figures prominently in (16), is a very uncertain quantity, with strong seasonality; velocity does vary rapidly with depth near the mixed-layer base; subduction induced by springtime warming occurs in regions (e.g., eastern subtropical North Atlantic) where the annual mean buoyancy gain by the sea is negative (cooling wins); some important modes of ocean circulation seem to be missed by the model. One example of such modes is the classic 1-dimensional advective-diffusive model of the thermocline. In regions of oceanic heat gain continuing upwelling ($w > 0, S < 0$) can be balanced by downward turbulent heat diffusion. Yet in such a region the formula predicts subduction, $S > 0$ (downward w, say, if h is steady in time). Diffusive pv forcing beneath the upper mixed layer is evident (see, e.g., the strengthening subsurface tongue of pv at $\sigma_\theta = 26.5$ in the tropical North Pacific, where wind-curl and precipitation are both large, (Rhines, 1986a)). Another example is deep convection itself, which associates cooling with sinking and entry into the deep geostrophic circulation.

The essence of subduction is in the production of vertical stratification. Marshall and Nurser and the 'Ekman demon' invoke springtime warming, which thus might be called 'Warm Subduction'. Yet cooling of the ocean surface may be intense and spatially variable. When it is so, strong horizontal density gradients are produced, and with them unbalanced horizontal pressure gradients. In classic Rossby adjustment, the fluid slumps under gravity and horizontal stratification becomes vertical stratification. We might call this 'Cold Subduction' for it is capable of producing geostrophic water masses beneath the prevailing seasonal thermocline. Although Rossby adjustment does not allow very great horizontal excursion of the fluid in the 0(1 day) available, further slumping by baroclinic instability will occur, and with it further production of vertical stratification.

A marvelous model of this process was posed by Henry Stommel in conversation one day at Woods Hole. Suppose the upper ocean to be mixed by wind, so as to have a uniform initial horizontal stratification, say N_h^2 (expressed as a frequency squared). In a 4-line solution we found that after adjustment the vertical stratification, expressed as a buoyancy frequency, N, is

$$2^{-\frac{1}{2}} N_h^2 / f,$$

with thermal wind balancing the tilted isopycnals, plus some reverberation in inertial oscillations. This particular day Stommel was not after 'cold subduction' so much as poleward heat flux, for that is what follows a repeated sequence of such events. While this model does not in a single event produce large subduction, its continued action together with more violent kinds of deep convection and more extense mesoscale slumping, give compelling candidates for subduction.

In the North Atlantic formula (16) predicts subduction principally in the tropics where heat gain from both atmosphere and poleward Ekman flux is greatest. Yet consider the region further north. The line of zero mean buoyancy gain runs southwest-northeast from about 30°N to 50°N (Schmitt *et al.* 1989). In this region the westerly wind storm-track drives cold water Equatorward-eastward. Thus $\mathcal{H} - \mathcal{H}_{Ek}$ is negative on both accounts, and yet downward Ekman pumping combined with the sloping mixed layer base lead to mean subduction, as the fluid turns Equatorward, encounters stronger pumping, and shallower mixed layer depth. Models similar to Marshall *et al*'s (e.g., Huang 1990, Williams 1991) tend to diagnose heat gain by the sea rather than the observed heat loss in these middle latitudes.

4.7 Pv Entry and Escape from the Ocean.

It is important to realize that pv influx (and probably subduction itself) are in general not simply coupled to transfer of buoyancy and stress at the ocean boundaries. This is clear from the defining conservation equation (15). The rate of change of ρQ integrated from top to bottom (assuming no influx through the sides) is

$$[\mathcal{H}(\zeta + f) + \mathbf{F} \times \nabla\sigma|_z - \rho Q w]_{bottom}^{top}.$$

neglecting possible slope effects at the bottom. The final term vanishes (but for a contribution from evaporation-precipitation). Ekman/mixed-layer dynamics has to be carefully considered before such a formula can be exploited, however. The second term expresses the effects of turbulent stress τ_{ij} *divergence* ($\boldsymbol{F}_i = \partial\tau_{ij}/\partial x_j$) at top and bottom, which rely on horizontal density gradients within the boundary layers. These cause outcropping such that the top or bottom stress divergence gives a twisting force at the 'rim' of an outcropped isopycnal layer, which can alter the circulation along the outcrop, hence contributing to the integrated absolute vorticity component normal to the isopycnal. A sloping bottom introduces a stronger 2d term, once again because the bottom stress *divergence* can then twist the absolute vertical vorticity, acting rather as a sidewall.

The first term, approximately $f\mathcal{H}$, involves not the air-sea buoyancy flux but again its divergence $\mathcal{H}$. Only by ignoring the flux at the mixed-layer base could Nurser and Marshall (1991) succeed in relating $\mathcal{H}$ to air-sea interaction.

To show that there are many possibilities, recall how the 'impermeability theorem' of Haynes and McIntyre shows that pv flux by advection and mixing cannot move net pv across isopycnal surfaces. They take this to imply 'robust conservation' but it means, in fact, that pv can move through the insulating boundary of a fluid with alacrity. Consider two examples:

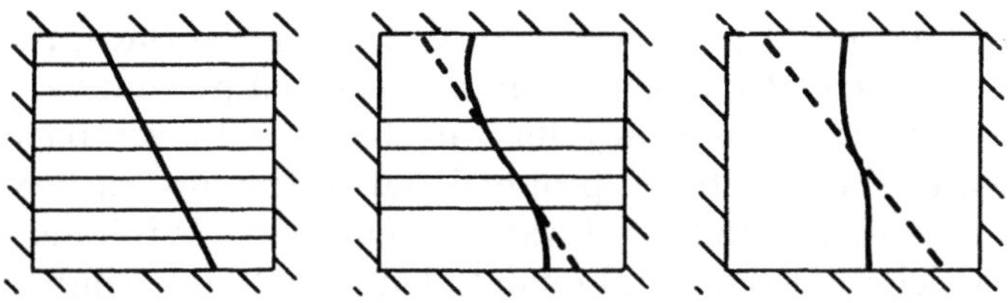

Figure 7: Potential vorticity escaping an insulated box of motionless fluid, purely by diffusion. The pv 'follows the isotherms' out of the box.

(i) A motionless box of stratified fluid. Take a thermally insulated box of temperature-stratified, motionless water on a rotating table. The initial pv is essentially $f\rho_z$. Because the boundary condition on temperature is that its normal gradient vanish on the walls, the profile is 'eaten away' by diffusion from the top and bottom (figure 7). Eventually it is isothermal and pv =0! Where did the pv go? Through the top and bottom boundaries. Using Haynes-McIntyre ideas we see that pv 'follows the isotherms' which themselves escape the box, so there is no contradiction.

(ii) A barotropic gyre with sidewalls. Take a simple gyre of circulation in a 1-layer fluid in a square box with free-slip sidewalls, discussed above in connection with angular momentum. This boundary condition means that shear-stress vanishes. For straight walls (only!: see Holloway and Rhines, 1991) this means that the vertical vorticity vanishes at the wall. Now in 2D the vorticity ζ obeys a diffusion equation,

$$\frac{D\zeta}{Dt} = \nu\nabla^2\zeta$$

with $\zeta = 0$ on the boundary. Thus by analogy with heat diffusion, ζ diffuses away, out of the box and with it the pv, energy and angular momentum of the fluid. This is surprising owing to the free-slip condition, but again it is normal viscous stress at the walls (not pressure torque!) that does it.

These two examples show that boundaries 'impermeable' to buoyancy and linear momentum are in fact 'transparent' to potential vorticity. Integrated pv (Kelvin circulation) is not very robust in contained fluids, even with insulating boundaries. *The corollary is that by specifying the boundary shear-stress and buoyancy flux, we cannot uniquely specify the pv flux into the ocean* (without solving for the entire circulation itself). It is the convergence of the buoyancy flux at the sea surface, not the buoyancy flux itself, that drives the volume integrated pv change: in Marshall *et al.* the neglect of buoyancy flux at the mixed-layer base forces the convergence of the flux to be proportional to the air-sea flux itself. For predicting Q itself in subducted waters, models like Williams' (1992) give estimates which indeed depend on detailed mixed-layer configuration. He writes an estimate for Q in terms of the difference $\Delta\rho$ of density between a recently subducted parcel and the mixed layer overhead, and its depth Δz relative to the mixed-layer base:

$$
\begin{aligned}
Q &= f\frac{(\Delta\rho/\Delta t)}{\rho_0}\left(\frac{\Delta z}{\Delta t}\right) \\
&= \frac{f}{\rho_0}\frac{D_b\rho_m}{Dt}\left(w_b + \frac{D_b h}{Dt}\right)
\end{aligned}
$$

where as above the $D_b \cdot /Dt$ operator follows the horizontal velocity at the base of the mixed layer, of depth h, and ρ and u are assumed continuous at the mixed-layer base.

Further progress with subduction needs further modelling and, desperately, better observations. The emphasis given to lateral-advective subduction (Equatorward flow into regions of shallower mixed-layers) (e.g., Woods and Barkman (1986), Huang (1990), Williams (1992)) requires better knowledge of the sheared, turbulent mixed layer and its Lagrangian pathways of circulation. Estimates from sea-surface drifters and climatological mixed-layer maps may overestimate the effect.

5 Thermohaline and Deep Circulation

How can a global numerical ocean circulation model, with typical horizontal resolution of 100 km or greater, produce vertical motion fields that in some way capture the complex, fine-scale processes of convection, sinking, entrainment, boundary layers, and interior mixing seen in the ocean? The models, particularly the Cox-Bryan primitive equation code, are widely in use in both dyanamical study, and by chemists, biologists and atmospheric scientists as part of the grander scheme of global change. Yet their internal dynamical structure is not often examined, or tested against known solutions of geophysical fluid dynamics. Only recently have the thornier issues of numerical accuracy begun to be examined with care.

The classic thermohaline circulation equations (Robinson and Stommel (1959)) (summed up in the M-equation, (Welander 1971)), gave hope of predicting the basic stratification (or 'mass-field') as well as the circulation and thermal-wind balanced density field. But they proved so difficult to solve that advances have primarily relied upon simpler models of the deep circulation which illustrate partial, yet crucial physics. The first was the series of papers beginning with Stommel and Arons (1960), which charted a global scheme of thermohaline circulation for a simplified deep ocean, driven by point sources of abyssal sinking with assumed uniform upwelling: the global convection problem was not addressed. Stommel (1962) illustrated a key property of the convection problem in which a distributed, large-scale heating-cooling pattern at the surface of a simple basin of water leads to very small regions of sinking fluid, with most of the fluid rising gently into the thermal boundary layer beneath the surface. The source-sink circulation on a spherical Earth was elegantly modelled in the laboratory by Stommel, Arons and Faller (1958) (using the pie-slice domains shown in figure 1). The Stommel-Arons theory is a classic prediction, against which observations and other models have been placed ever since. Warren's (1981) review of the deep circulation shows how many of the key deep western boundary currents have been successfully found, yet there is not a sense that the predicted poleward interior flow (say, averaged vertically beneath the thermocline) has been verified, and the topographic and baroclinic modifications of the classic picture are numerous and complex. Mesoscale eddies are so energetic that they drive deep circulation gyres of moderate size, which combine with the large-scale planetary circulation (Hogg, 1983).

The basic assumption of spatially uniform upwelling seems to be supported in simple diffusive numerical models, yet there are reasons to suspect that special sites like the Southern Ocean and the Equator contribute a great deal to the upwelling, and that it may be otherwise rather unevenly distributed in space. A parcel of water, wandering about the deep ocean looking for a way to warm and rise, might achieve this by rubbing along the base of the thermocline (enhancing vertical mixing with vertical shear converting lateral property gradients into vertical property gradients: the advective rate of change of the vertical gradient DC_z/Dt, and hence the vertical diffusive flux, is just

$$\mathbf{u}_z \cdot \nabla C,$$

using equation (5)). During this shear-dispersion process fluid would move obliquely upward, with a minimal penalty of potential energy. The (diapycnally) downward diffusive fluxes of heat and salt in the subtropics would then be balanced by vertical advection, but in the presence of much greater lateral advection. It is very likely that, in addition, T- and S transport by boundary currents into and out of the subtropical gyres is a major component of the net budgets of these regions in and above the thermocline, in competition with diapycnal mixing. Direct measurement of basin-scale upwelling should be acheivable using tracers, but so far has proved difficult. There are promising candidates like primordial He-3 which vents from the spreading center ridge crests, and has essentially zero concentration at the sea surface. The distinct patterns of silicate, nutrients and salinity, plus particularly radiocarbon, have all but demonstrated that the deep waters of the North Pacific rise only to mid-depth ($\cong$ 2500m), and thence flow southward toward further rising sites in the Southern Ocean (e.g., Fiadeiro, 1982, Toggweiler and Samuels, 1992). This incidentally resolves L.V. Worthington's (1977) classic paradox, in which he showed that upwelling to the surface of the Pacific would require such enormous biological consumption and particle rain of silica as to be unbelievable. The extensive high-latitude outcropping region in the Southern Ocean may be a dominant site for upwelling from great depth. One has strong, widespread upward Ekman pumping to assist the process, and one observes low dissolved oxygen concentration close to the surface, a sign of old circumpolar waters being uplifted.

5.1 Sinking.

The small sinking regions of classic theory are summarized by Killworth (1983). Some ingredients for successful sinking seem to recur: it usually takes place at or near a boundary, encouraged by the 'cooling chamber' effect of a continental shelf (which limits the depth of convection, enhancing the density change), it seems to involve entrapping a water mass so that densification can proceed for long enough (on the shelf, where there is topographic control, or in an inertial gyre as with the 18° water of the North Atlantic, or in a 'Taylor column' trapped by topography (the Rhone abyssal fan with the Mediterranean sinking region), or in a basin-scale gyre (the Labrador Sea). Preconditioning of the water is important, and the whole of the North Atlantic involves the creation of high salinity by evaporation. This light yet potentially heavy water moves north to the subsequent loss of the (more mobile) heat at high latitude to finally enable deep mixing. Finally, involvement of ice at high latitude is important, whether in the Ronne and Filchner Ice Shelf production region for Weddell Sea bottom water, or in offshore polynas and their maintenance of ice-free regions by convective heating from below.

Entrainment during sinking is particulary crucial to the entire circulation. It is one of the most frighteningly unpredictable processes in fluid dynamics, one that challenges both numerical and laboratory models. In our laboratory we have found that thermohaline circulations can take on a wide range of patterns of abyssal- and intermediate water formation, deep density, and deep stratification, depending simply upon the detailed turbulent entrainment rate at the sinking plume (which in turn is controlled by fine details of that region). Barringer and Price (1990) have persuasively argued that the depth of sinking is very sensitive to the density difference $\Delta\rho$ between the sinking plume and its environment. Compare the Mediterranean Water, which reaches about 1200m depth before coming to equilibrium, with the Denmark Strait Overflow Water, which sinks as a western boundary current all the way to the abyss. The Mediterranean water initially has a density that exceeds that of the Denmark Strait water but it sinks through a much lighter environment of warm thermocline water ($\sigma_\theta \cong 25 - 27$), and hence entrains violently. The water entering the Denmark Strait has density of about $\sigma_\theta = 28.5$, yet its environment has densities differing from this by less than one ppt. Barringer *et al.*(1990) are able to reproduce these differences using simple parameterization of bottom stress and entrainment rate. MacCready and Rhines (1991,1993) suggest that the bottom stress, Ekman pumping and spindown of such currents can be greatly inhibited by stable stratification and the slope ∇h of the bottom, in a manner which is very sensitive to N and ∇h (see Garrett, MacCready and Rhines, 1993). The potential energy of the sinking plume can be great enough to maintain the flow for great horizontal distances (MacCready, 1992). Fluid from the bottom boundary layer can be pumped into the fluid interior, both transforming water properities and altering the downstream momentum there. This is intimately connected to entrainment processes. A relevant study from the Faeroe-Bank Channel and its analog in the laboratory (Johnson and Olsen 1992) show, perhaps for the first time, how bottom stress and 'interfacial Ekman-layer stress' at the top of the current collaborate in the net momentum balance of a deep jet, and in the development of pockets of mixed fluid by transverse circulation. In other laboratory studies we have found enormous sensitivity of entrainment rate to the details of the sinking branch (orientation, topography, velocity and density of surrounding fluid), when carried out at scales for which diffusion was very weak.

In general circulation models, the representation of entrainment of sinking waters is *schematic* only, with a combination of the convective adjustment scheme and lateral diffusion yielding the calculated flow. They substitute a nearly continual numerical convective adjustment, over broad regions, for the actual small regions of convection and sinking. As a result, for example, the gcm's water column at high southern latitudes is mixed down to 1000m or so, whereas one sees nothing like this in observations south of the circumpolar front, wind-induced upwelling seems to draw old, low-oxygen circumpolar water up to within 150m of the surface, before it becomes ventilated by the sea surface. The many global thermohaline models must be regarded as 'thought' experiments, with the realization that greatly different circulations are possible with small changes in entrainment physics.

5.2 $\theta - S$ Models.

The distinct contributions of temperature and salinity to the fluid buoyancy, have led to a rich set of recent models, owing to the partially independent boundary conditions for heat and fresh water. Evaporation at the sea surface dominates the cooling of the fluid, and in parallel its salinity increase. Solar radiation and precipitation, on the other hand, independently affect temperature and salinity, respectively. Ice formation and melting is yet more complex: the seemingly zero-net-gain annual cycle of freezing and thawing of sea ice actually is a formidable 'brine pump', for the salt rejected on freezing is not recaptured later by the light, fresh melt water. The surface of the high-latitude ocean is awash with light melt water in summer, which gives rise to striking mirages and other optical effects [an iceberg, plus its inverted image, can take on a monstrous appearance, and was feared by early seafarers: it is sometimes called the 'Norse Merman'].

It is revealing to carry out simple laboratory experiments with heat, salt and ice. Shine an infrared lamp at the surface of a container of fluid, initially uniform in temperature and salinity (and dyed so that the heat is absorbed near the surface. The result is remarkably complex. At first, the heating lightens the surface layer, producing a stable, warm layer. This then increases the evaporation rate, which limits the warming, and increases the salinity of the surface layer. A shallow cell of circulation develops. Salt fingering takes hold, draining some of the excess salt downward and stabilizing the density profile. But eventually the evaporation may win out, salinity build up, and the fluid roll over in a massive meridional circulation. This is closely analogous to the 'massive flush' that occurs in numerical models of the onset of the thermohaline circulation from rest, for example seen by Winton and Sarachik (1992).

The classic approach would be to look at the linear stability of a fluid with mixed T and S boundary conditions. Welander (1989) shows that, even in total absence of double diffusion (i.e., if diffusivity of heat and salt are identical), mixed boundary conditions lead to strong instability near the boundary of the fluid. He focuses on an overstable instability connected with surface boundary conditions. Consider a simpler case: a fluid with a lateral boundary at which temperature is specified (say,

$$T = \alpha z \quad \text{at} \quad x = 0$$

yet salt flux vanishes:

$$\frac{\partial S}{\partial x} = 0 \quad \text{at} \quad x = 0.$$

Suppose that initial temperature and salinity gradients compensate, so that the density is uniform. If a fluid parcel near the boundary is displaced upward it gains heat from the wall, yet with no flux of salt. It is thus lightened and continues to rise. Numerous other microphysical oscillators have been discovered, to illuminate the physics of heat-salt-ice (Welander, 1986), yet much analysis and experimentation remains to be carried out.

These are microscopic versions of the heat-salt engines that are now numerous. Stommel (1961), Rooth (1982) and Bryan (1986) particularly established both the multiple steady solutions, and competition between sinking regions leading to pole-to-pole overturning circulations even when surface boundary conditions are maintained symmetric with respect to the Equator. The insulating effect of light, fresh waters at high latitude can prevent sinking, and reduce further buoyancy flux. It is speculated to have caused short-term climatic fluctuation during deglaciations, and is a prominant feature in debates about the observed rapid fluctuations in North Atlantic climate which sea-floor sediment and ice-core stratigraphy are showing with ever finer time-resolution (e.g. Keigwin, 1992). Such oscillations have appeared in coarse-resolution gcms (e.g., Weaver and Sarachik, 1991). It may also be relevant to more recent climate: the cessation of Labrador Sea convection in the late 1960's; the 'great salinity anomaly' tracked around the North Atlantic in the early 1980's (Brewer *et al.* 1983, Dickson *et al.* 1988). and other major, sudden changes in SST seen in the subarctic regions.

Modelling of interannual-to-millenial climate fluctuations in thermohaline circulation demands much of models and theory. It is good for their validity that we now have a wide spectrum of models in use, from box models, to 2D non-rotating 'slots', to coarse 3D gcms, to eddy resolving 3D models which can be run only for times of order decades. Because of the boundary layers and many other under-resolved physical effects, it is particularly important to develop 3D simulations with simplified physics, for which finite-differencing errors are provably small. Salmon (1986) has suggested the use of Rayleigh-damped momentum and density fields to this end (yet leaving in the usual vertical diffusion terms); for climate simulations with this approach see Winton *et al.* (*op.cit.*). At the longer timescales of glaciations and tectonic evolution, ocean models face still greater challenges as greater changes in the physical, biological, chemical, and geological environment come into play. Not an area for the timid, it is nonetheless 'high science' because of the remarkable observational histories being discovered in ice-cores, sediment cores, corals, and hard geological features. See Imbrie *et al.* (1992) for a proposed sequence of ocean-atmosphere events accompanying glacial cycles. We are reminded of the 1970's when many atmospheric scientists claimed that 3D gcms then represented the 'essential' atmosphere, with only details to be cleaned up. Subsequently it has been found that the viscous, low resolution nature of these models prevented them from representing many dynamical features (e.g., the quasi-biennial oscillation of the Equatorial stratosphere, ENSO, and mid-latitude response to SST anomalies). By demanding more of the models, the community has begun to ferret out their weaknesses.

5.3 Resolution Effects.

Because the resolution of 3D models is so restrictive, particularly in cases that must be run for many thousands of years, the physics is by nature very simplified. It is particularly important to identify finite-representation errors so that they may not be confused with real physics. The vertical velocity is sensitive: are the Equatorial jets and massive upwelling at the western boundary realistic, in coarse-resolution Cox-Bryan numerical models (e.g.,Cummins *et al.*, 1990, Suginohara and Aoki, 1991)? Such features can be caused, respectively, by numerical diffusion and the 'Veronis effect', (Veronis, 1975) in which huge Cartesian-horizontal diffusion in the presence of strongly tilted density surfaces yields strong diapycnal mixing, and massive vertical motion; see for example, Cummins *et al* (1986), Weaver and Sarachik (1990). In this circumstance (figure 8) the approximate density balance in the boundary current is

$$w\rho_z = \kappa_h \rho_{xx}$$

and the upwelling induced there can drive downwelling in the ocean interior. The continued presence of this kind of numerically controlled upwelling, 20 years after its diagnosis, begs for improvement. Rotating the diffusivity tensor is one approach, yet it seems intractable in an eddy-resolving model. It is an extreme model of variable diapycnal mixing, one which may have far-reaching effect on the meridional circulation.

For this reason, laboratory simulations of this process are also of value, and can demonstrate sensitivity to deep convection and mixing processes. The technique shows how crucial is the entrainment process in which the sinking branches of the circulation are diluted by, and dilute, their surroundings. As Barringer *et al.* (1990) suggest for the ocean, a sinking plume entrains at a rate dependent on the density of the surrounding fluid, as well as topography and velocity fields. Michael Cox, near the end of his life, also speculated that 'physics' problems often become worse rather than better as resolution is increased. He was increasing the vertical resolution of his 3D model to look at topographic effects on the deep circulation. His confidence in the solutions decreased, however, as the inclusion of more and more small topographic features seemed to cause eccentric and unrealistic flow patterns.

Gerdes (1991) has taken the approach of testing a gcm against known analytical solutions. He tested a 3D primitive equation model with topographic Rossby waves, of wavelenth $\cong$ 200 km. The experiment was sobering, with the 'stair-case' treatment of the lower boundary in 15- and 32-level x 20km grid models causing major errors in the solution (figure 9). Here $N^2 = 6.3x10^{-6}\text{sec}^{-2}$. The 'staircase' in the two runs has, respectively, 4 and 11 steps. While the horizontal resolution is not ultra-fine, it is better than can be expected for any global or basin-scale model, based on the 10km to 50km Rossby radius. A bottom-following coordinate scheme was developed that minimized these errors (figure 9a), yet created some difficulties in the interior, owing to diffusive errors with the non-Cartesian interior grid. Cross-sections of the perturbation density (figure 10) show that the difficulty persists even when the staircase is quite fine. Better implementation of the bottom boundary condition can likely be found, based more directly on $w = -u \cdot \nabla h$ plus an Ekman layer correction, at $z = -h$. See, e.g., Ishizaki, 1992.

5.4 Diagnosis from air-sea Flux and Surface Properties.

Given that many of the crucial elements of the circulation are very difficult to model numerically or in the laboratory, it makes sense to keep aware of a number of global properties of the circulation. Many recent works (building on early efforts of Budyko, 1963 and Bunker and Worthington, 1976) have begun to pin down the net meridional flux of heat, fresh water and mass by the general circulation.

Other than using geometric 'boxes' and control surfaces, however, one may also use potential density surfaces (referenced locally becoming 'neutral' surfaces). Their use was pioneered by Montgomery (1938) and Reid (e.g., Reid and Lynn 1971); the gentle flow of the large-scale ocean can move along these surfaces without paying a potential energy penalty. Water masses marked by tracers are seen to extend enormous distances along these neutral surfaces, and hence they have become our most valu-

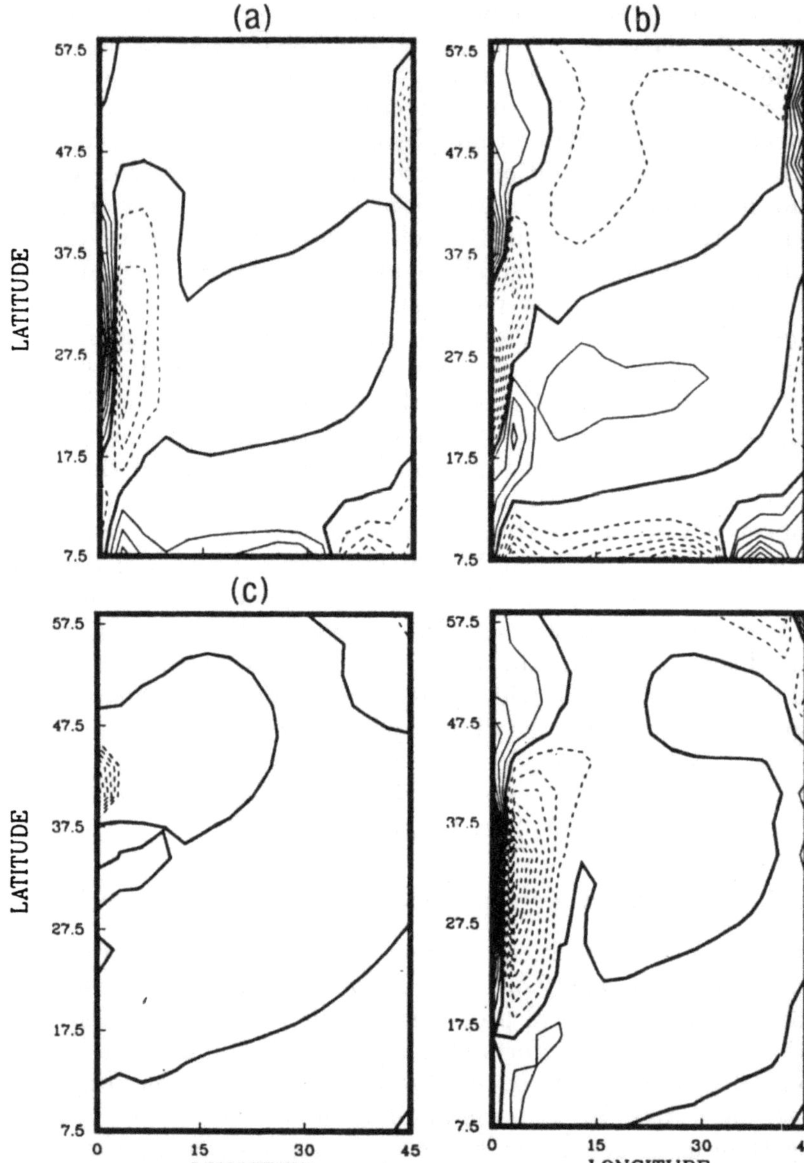

Figure 8: Balance of terms at 430m in the density equation, from Cummins *et al.*, 1990. Upper left: $w\rho_z$; upper right: $\mathbf{u}_h \cdot \boldsymbol{\nabla}\rho$; lower left: $\kappa_v \rho_{zz}$; lower right: $\kappa_h \nabla^2 \rho$. There is intense upwelling in the western boundary current, which is in balance with lateral density mixing (primarily) and lateral advection. As Cummins *et al.* point out, the basin-wide pattern of w seems to be controlled by this anomalous western-boundary numerics. The dominant sinking region is the northeast corner.

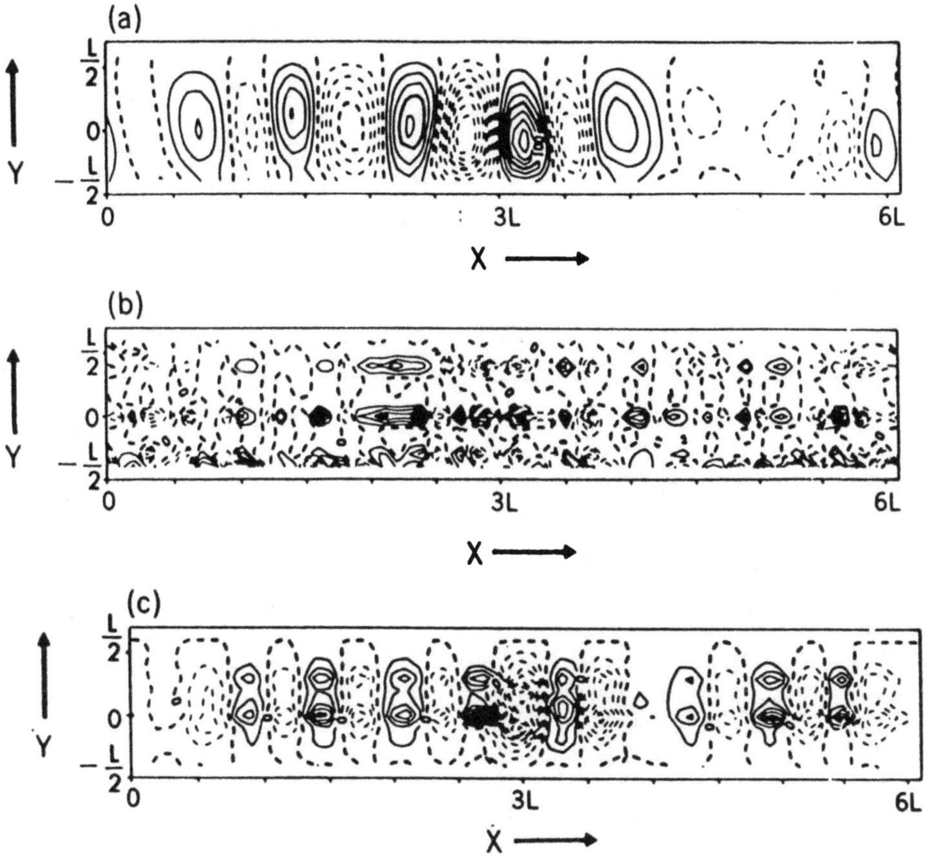

Figure 9: Numerical simulation of bottom-trapped topographic waves, Gerdes, (1991). Perturbation density at the deepest grid level at day 120, for wind-forced bottom-trapped topographic waves (sinusoidal wind pattern close to resonance). (a), sigma-coordinate (bottom following) model, which is close to the analytical solution; (b), z-coordinate Cox-Bryan model (staggered B-grid) with 15 levels; (c), z-coordinate model with 32 levels (Δz = 50m). L=200 km, Δx = Δy=20 km.

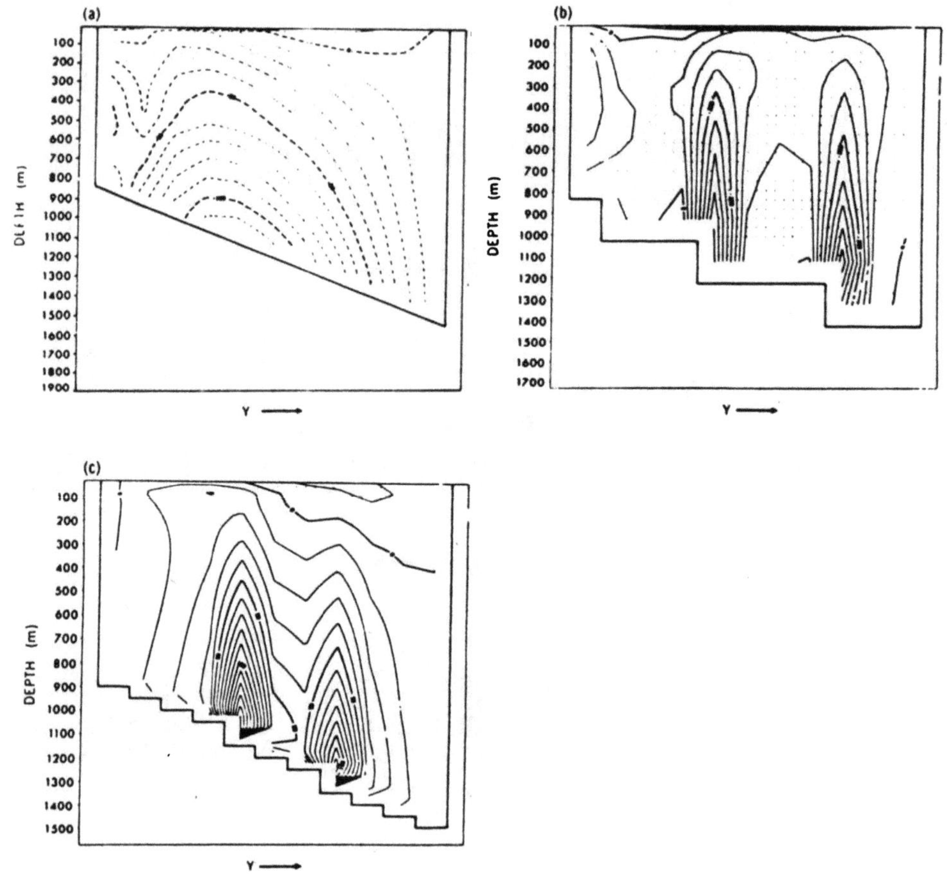

Figure 10: Cross-section of the bottom-trapped wave solution, as in figure 9, at mid-channel. The analytic solution is exponential in the vertical, and a half-sine wave in y, very close to (a).

able 'plotting surface'. But at the sea surface the fairly-well known air-sea transfer of buoyancy converts fluid from one density to another. If the ocean is close enough to a statistically steady state, this conversion must be reversed somewhere else. It is thus of utmost importance to analyze the water-mass conversion across potential-density space: in nature and in models. Walin (1982) initiated the idea, which is simply that if one has maps of air-sea buoyancy flux and maps of sea-surface density, then an annually averaged model suggests that where cooling and salination occur due to evaporation, the heavier fluid must be replaced by flow across isopycnals of warmer or fresher water, in order that a steady state persist. McCartney and Talley (1982) make a qualitative analysis of the subpolar North Atlantic, following the cyclonic surface circulation as it cools and densifies, moving finally into the Labrador Sea sinking region. Tziperman (1986) set up a model calculation driven this way, and Speer and Tziperman (1992) describe the observed North Atlantic water mass conversion using 3 representations of the air-sea fluxes. The flux $F(\rho)$ in Sverdrups across density space (figure 11a) is a crucial descriptor which should be a target for numerical models. Locally it expresses the rate of movement of water across isopycnal space, or 'transformation' owing to surface processes; its integral is the net conversion of water by the entire surface of the North Atlantic; its derivative $dF/d\rho$ (figure 11b) is the 'formation' rate of water masses (i.e., the rate of accumulation between adjacent ρ surfaces). The dominant feature in the North Atlantic might be called the 'Bunker bullet' of heat loss in the Sargasso Sea, in and seaward of the Gulf Stream. It corresponds to the formation of water near 18°C, $\sigma_\theta \cong 26.5$, in the tight inertial recirculation centered at 35°N, 70°W. The volume flux across density space peaks at this point, at about 32 Sverdrups. The second maximum in 'formation' is a broad range of the subpolar mode water, $\sigma_\theta \cong 27.6$ corresponding to cooling the surface of the entire subpolar gyre. What is remarkable about this analytical approach is that one finds a convincing flux rate, in Sverdrups, relating to complex events of formation and meridional flux. For example, 14 Sv. of formation occur in the 18° water peak, and 16.7 Sv in the subpolar peak. About 10.5 Sv of this total flux reaches densities of the Labrador Sea surface and greater. Estimates from geostrophic calculations of the observed circulation are quite similar to these numbers, although there is continual revision in progress (e.g., Schmitz and McCartney 1992). The separate contributions F_S and F_T to F from salinity and heat are plotted in figure 11a. The evaporation - precipitation component is large at low latitude, but the heat component dominates elsewhere.

The production rate of Labrador Sea Water is an example of the uncertainties that face us. Estimates range from 2 Sverdrups by L.V. Worthington to 8.5 Sverdrups by McCartney and Talley (op.cit.); this latter number seems difficult to support. As for the forcing function, neither the average nor the seasonal maps of air-sea buoyancy flux capture the intensity of cooling that actually occurs. The average maps of Schmitt *et al.* (1989) show a zero of buoyancy flux passing through the middle of the Labrador Sea, with nearby values of a few 10's of W/m^2. Yet, violent winter winds cause fluxes locally in excess of 10^3 W/m^2. Our current averaging schemes can obscure what amounts to a globally dominant mode of water mass production.

Interpretation of these observations is otherwise fascinatingly complex. For example, neither the meridional flux in the Gulf Stream system nor the production of lower North Atlantic Deep Water are apparent in the Speer *et al.* calculation. In the former case the 18°-water mass (or more generally, subtropical mode water), must be acting as the 'radiator' of this heat engine, venting off the heat gained from mixing with the nearby boundary current. The crucial role of western boundary currents in the meridional global heat flux requires this link, which represents a subsurface connection not initially apparent (more gradual warming of the 18° water in mid-ocean seems far less important). For every water mass formed at the sea surface, there must be counter-balancing interior diapycnal mixing and transport if anything like a steady climatological state is present. The geomorphology of isopycnal surfaces deserves much attention, for example the comparitive volume and vertical thickness (the 'throat') of fluid between two isopycnal surfaces and the sea-surface area (the 'mouth') between the same two surfaces where they outcrop, has been plotted by Rhines (1986a), and plays an important role in Marshall and Nurser's (1992) attempt to track the volume flux and potential vorticity flux in fluid subducting from the sea-surface to the interior. The constraints of the observed outcrop lines and air-sea buoyancy flux in the Southern Ocean are interesting to contemplate, for the Ekman induced upwelling and northward surface flow would seem to violate the continuing densification of surface waters.

5.5 Vorticity Dynamics.

Despite the slowness of the motion, it is likely that

$$\beta v = f w_z \tag{17}$$

still holds, and describes the control of planetary stiffness over the flow. Given the possibly wide dominance of this Sverdrup dynamics, it is essential to develop estimates of the size of neglected terms. Terms missing

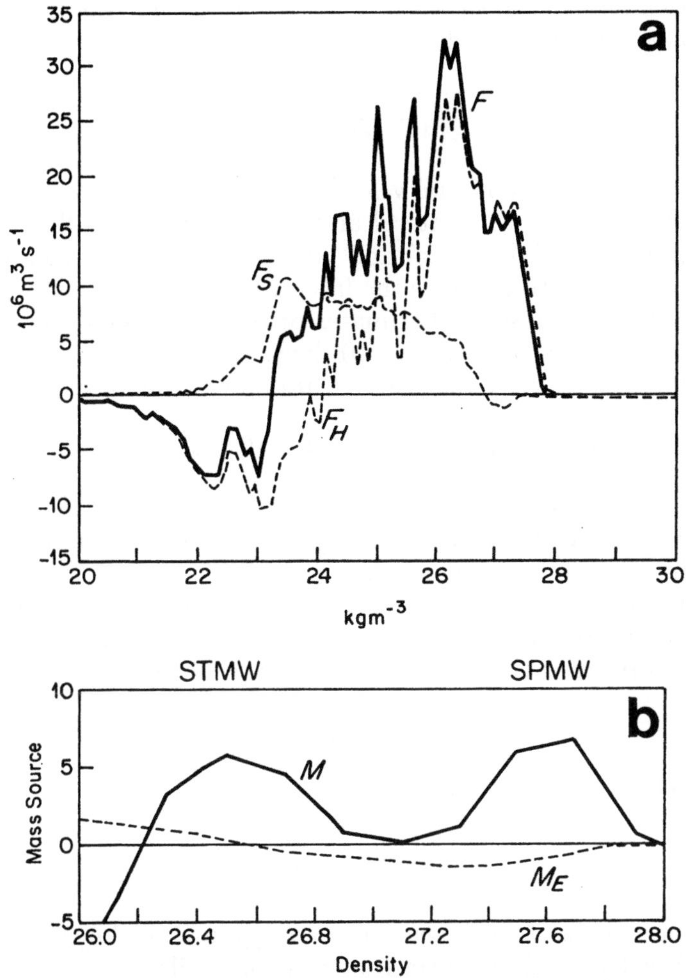

Figure 11: (a), Transformation $F(\sigma_\theta)$ (ordinate, in Sv) of water masses across density space in the North Atlantic, calculated from air-sea buoyancy flux and observed surface density field (abscissa, in $^\circ/_{00}$) (Speer and Tziperman, 1991). About 32 Sv. moves to greater density, at σ_θ=26.0. (b), the derivative $dF/d\rho$, which is the 'formation' rate of water masses at the surface. The two peaks correspond to 18° water and a broad range of subpolar mode waters, the heaviest of which is Labrador Sea Water.

from (17) are advection $\mathbf{u}\cdot\nabla\zeta$, diffusion $(\kappa_h\nabla^2\zeta+\kappa_v\partial^2\zeta/\partial z^2)$, and $\partial\zeta/\partial t$. These are, respectively, of order

$$\frac{U'^2}{U\beta L^2},\ \frac{\kappa_h}{\beta L^3}+\frac{\kappa_v}{\beta H^2 L},\ \text{and}\frac{1}{\beta LT},$$

where U is a typical mean lateral velocity, U' the mesoscale eddy velocity, κ_h and κ_v are the respective lateral and diapycnal eddy diffusion coefficients. H and L are the vertical and lateral scales of the mean motion (with L assumed also to measure the lateral scale of the average eddy intensity) and T the timescale. Estimates of these neglected effects based on L = 500 km, H = 500 m, T = 10 days, U' =.1 m s^{-1}, U = .001 m s^{-1}, $\kappa_h = 10^3$ m^2 s^{-1}, $\kappa_v = 10^{-5}$ m^2 s^{-1} are all extremely small:

$$0.4\times10^{-5}, 0.4\times10^{-3}+0.4\times10^{-5}, \text{and } 10^{-3}$$

respectively. A crucial parameter is the aspect ratio H/L of the water mass: here the estimate is 1:1000, yet it will be larger in boundary currents, while smaller in mid-ocean.

Equation (17), as described earlier, governs not only steady circulations, but long baroclinic Rossby waves. It can also encompass the mean response to eddy potential vorticity flux; of the two components

$$\langle\mathbf{u}'\zeta'\rangle,$$

the flux of relative vorticity and

$$\left(\frac{f}{\rho_0}\right)\langle\mathbf{u}'\frac{\partial\rho'}{\partial z}\rangle,$$

which is proportional to the the flux of isopycnal thickness variations, the former breaks the literal Sverdrup relation while the latter does not (as it enters from the density equation). The generalized Rossby-wave drag, or 'form' drag, exerted through eddies from one layer to the next represents a lateral potential vorticity flux, yet a vertical flux of horizontal momentum. Models elsewhere in this volume, and Rhines and Holland (1979), verify that it is the major momentum transfer mechanism in the ACC, permitting the wind-stress to appear as an inviscid pressure force on sloping bottom topography, and that form drag is also the source of abyssal gyres found near surface jets (see also Hogg, 1983).

In a simpler world, equation (17) might apply immediately to observations, connecting a measureable meridional velocity with an unobservably weak vertical velocity. However, there are few examples in which the Sverdrup balance is unambigously clear, other than the general sense of the wind-driven gyres driven by an apparent difference in w from top to the base of the thermocline. The spatial-temporal averaging necessary to 'pull out' the Sverdrup balance from both harmless and dynamically potent eddy effects has been explored by Holland and Rhines (1980) in the context of a circulation model, and it is a more significant problem than the above scale analysis would admit. The components of the vertical profile of w that arise from density mixing and deep convection have been mentioned above. These have been used to drive layered models of the general circulation (Rhines, Holland and Chow 1985, Luyten and Stommel 1986, Dewar 1991) and diagnostic use of the models for the North Atlantic (Spall 1991).

It is welcome (and somewhat surprising) to have such a robust dynamical statement, with planetary 'stiffness' holding even at such long timescales. There are notable exceptions to Sverdrup dynamics in boundary currents and near-surface formation regions, and in the benthic boundary layer. Diffusion and nonlinear advection, not apparent here, are dominant effects in the density and tracer conservation equations. For a typical tracer, or potential temperature, or salinity, obeying

$$\frac{DC}{Dt} = \kappa_h \nabla^2 C + \kappa_v \frac{\partial^2 C}{\partial z^2}$$

the relative measure of advection and diffusion is the Peclet number, UL/κ_h and $UH^2/\kappa_v L$. Warren (1976), for example, exploits the greater effect of mixing on the density field to explain the great width of observed deep western boundary currents. Now, using the same scale estimates, we have Pe $\cong$ 0.5 based on lateral eddy mixing, and $\cong$ 50 based on cross-isopycnal mixing. Note that this apparent victory of lateral mixing over vertical is sensitive to estimates of the length scale of the water mass: the ratio of vertical to lateral Peclet numbers is $(\kappa_v/\kappa_h)L^2/H^2$ (so that for L = 5000 km and H = 500 m the vertical and lateral mixing processes are comparable.) Estimates of κ_v are still uncertain, with direct microstructure measurements reporting values typically $\kappa_v \cong 10^{-5} - 10^{-6}$ in the thermocline, yet global estimates of $\kappa_v \cong 10^{-4}$ m s^{-1}. If in addition to mean advection there is an eddy component $\nabla \cdot (\mathbf{u}'C')$ one has to estimate flux of the tracer C by mesoscale eddies, small-scale turbulent mixing and even by mixing at the boundary.

On the face of it, lateral eddy mixing would appear to be more potent and active in diluting water masses than diapycnal mixing. This is a different view than one had in the 1960's, when the complexity of the full thermocline equations led to adoption of a simple one-dimensional vertical model of the thermocline, with vertical diffusion in balance with vertical advection. Of course, to close the global pattern of T, S and tracer fluxes there must be regions like the deep North Pacific where fluid inevitably rises across isopycnal surfaces, and there the water mass adjusts its property gradients so as to make vertical diffusion effective. The famous global-balance estimates of κ_v, e.g., Munk (1966), seem to overpredict the vertical diffusivity by 1 or 2 orders of magnitude, perhaps owing to an incorrect reliance on vertical diffusion: the average large-scale w required to return fluid to the surface may occur as fluid glides more along isopycnal surfaces than across. There are signs of this balance in the study of Cummins *et al* (1990).

The elementary impression one gets from 'cruising' through the ocean hydrographic database, is of long-lived water masses which carry their properties for 10's of thousands of km before losing identity (figure 12). Neither lateral nor vertical mixing is potent enough to prevent this wide-reaching advection. Yet episodes of violent mixing, with inevitable diapycal component, are clearly visible. CTD $\theta - S$ profiles from the Brazil basin, in the South Atlantic Ventilation Experiment (figure 13) show massive interleaving at the top of the North Atlantic Deep Water, which is flowing generally southward beneath fresher, and often cooler Antarctic Intermediate Water. The effect of this fine structure is to cut off the 'knee' in $\theta - S$ (figure 12) which gives the water mass a sloping top, as one proceeds southward [although the circulation is surely not confined to the plane of the figure, the degradation of the knee is visible in virtually all the 3D region].

Numerical general circulation models are hard-pressed in reproducing evolution of this kind. The detailed, possibly double-diffusive, interleaving is here taking place in a rather quiet part of the ocean, far from any sharp fronts. It seems to be associated with the temperature inversion, an unusual feature in large-scale water masses, and will be difficult to parameterize. It clearly will not be accurately represented in models with smooth or uniform mixing coefficients. Further dynamical study of tracers, q, θ, and S, in gcm's needs to be carried out, for we are entering a period of great interest in climate change over 1000-year scales. Exploration of the effect of the diffusivity ratio κ_v/κ_h on deep circulation would be of particular interest, if resolution can be improved enough to reach the oceanic scaling.

Laboratory simulations on the advection and diffusion of deep ocean water masses can be carried out, although without full spherical-Earth physics: at present, one has to accept the limited dynamics (no Sverdrup relation). 3-dimensional flows at the scale of 1 meter easily exceed the resolution possible in numerical gcms. Isotropic diffusion of a laboratory laminar flow in a container with aspect ratio between 1 and 10 mimicks an oceanic water mass with $\kappa_h/\kappa_v = 10^8$ and aspect ratio ranging from

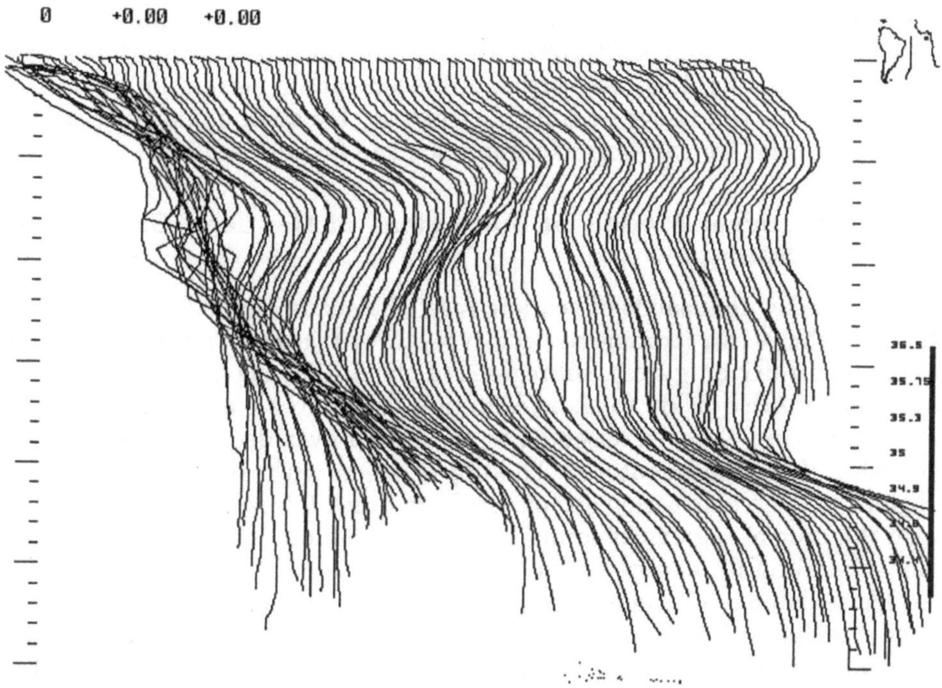

Figure 12: Profiles of dissolved silicate along 25°W longitude in the South Atlantic, from the SAVE/Hyros program. The North Atlantic deep water minimum (actually two minima) narrows southward. In just that region the top of the water mass is eroded by fine structure intrusions.

10^{-4} to 10^{-5}. The aspect ratio scales as $(\kappa_v/\kappa_h)^{\frac{1}{2}}$ using a similarity rule in which $\kappa_h \nabla^2 C$ and $\kappa_v \partial^2 C/\partial z^2$ are kept in the same ratio. Thus the laboratory flows will access vertical diffusion rather more than we would like, based on the above oceanic parameters. By adding eddy stirring in the laboratory, we can begin to approach the favored scaling.

5.6 Dynamical Models of Subregions of the deep Circulation and Spinup.

While it takes thousands of years to reset the entire global θ, S and tracer fields of the ocean (compared with typically 10 years to reset circulation of the wind-driven gyres), subregions of the 3-dimensional ocean can react much more quickly to changes in buoyancy forcing and winds. There are interesting changes observed in the ocean over decades, which must be using these 'propagators'. The upper ocean with its mixed layer and regions of deep convection are particularly quick to respond, and they are also the crucial regions affecting biological productivity and new production of carbon for export to the deep.

The existence of many topographic basins and barriers in the deep ocean tends to break the global thermohaline circulation into distinct subunits, with sometimes major transition between them. The vorticity forcing and recirculation contributed by topography seem to create generally stronger deep circulation; for example Reid's (1988) maps of the deep circulation in the Brazil Basin show water masses moving at speeds in excess of 1 cm $\sec^{-1}$, recirculating (anticyclonically!) about the basin, rather than dividing neatly into narrow boundary currents and weak poleward interior drift.

Development of layered 'non-Doppler' models carries over also to the deep circulation with non-trivial bottom topography. In its full generality, the deep circulation is more complex and difficult to model than the wind-driven flow of the upper kilometer of ocean. The slower deep flows must navigate through complex topography, which defines deep basins and severely limits the travel of abyssal water. In addition, topography can alter water masses through boundary mixing, and propel them through mesoscale Rossby-wave drag. Water masses mix along and across potential density surfaces at rates that are not readily understood. They have intense western boundary currents, but the interior circulation may take centuries to transit a major ocean. Small gyres are induced dynamically by upper ocean eddy activity and potential vorticity mixing, and the 'Stokes drift', the difference between current-meter mean and fluid particle-mean circulation, is likely to be locally important.

If we take as a representative number the transport that sinks in the far northern North Atlantic (of order 13 Sverdrups) and that sinking in the Weddell Sea (of order 5 Sverdrups), and assume that the fluid rises quasi-uniformly through the 1.25 x 10^{18} m^3 of the ocean's volume, this gives an average turnover time of about 2200 years. Yet there is much quicker response along the faster 'waveguides' of the system. The stark contrast with the O(10 year) spinup time for the wind-driven circulation gyres suggests that the dynamics, and heat- salt- ice- and moisture-dynamics of the thermohaline circulation may be intricate, and slight effects like geothermal heating may be significant.

In addition to their role in ocean biology and the carbon cycle, chemical tracer measurements are available to diagnose deep flows, and they provide a well-integrated measure of the circulation: if only we could know their boundary conditions better. Transient tracers (^{14}C, tritium, CFC's, carbon tetrachloride) injected by man's activities give an exciting look at a fresh 'dye' washing through the system.

Next we give four examples of simple solutions (both linear and non-linear) of deep ocean circulation: pieces of the circulation rather than the great global machine envisaged by Stommel and Arons.

5.7 Examples of Deep Ocean Circulation

(i) β-plumes. Perhaps the simplest prototype of buoyancy driven circulation on a β-plane is the 'β-plume'. A version of this (Rhines, 1977) is the single-layer equation

$$R\nabla^2\psi + \beta\frac{\partial\psi}{\partial x} = F(x,y) \tag{18}$$

where R is the bottom friction (analogous to Ekman-layer suction) and $F(x,y)$ is either a mechanical stress-curl or H/f times the volume flux from a source of fluid injected at the upper boundary. If F is simplified as a delta function, $\delta(x,y)$, at the origin, then the solution is

$$\psi(x,y) = (2\pi R)^{-1}e^{-Bx}K_0(Br) \tag{19}$$

where K_0 is a modified Bessel function, and $B = \beta/2R$, figure 14. The simplicity comes from the close relation between (18) and the Helmoltz equation of the form $\nabla^2\phi - \phi = \delta$, where $\psi = exp(-Bx)\phi$. This is the Green function for the self-adjoint equation, and hence the general solution is its convolution with F. Near the mass source the flow simply circulates about the origin: $\psi \cong (2\pi R\ell n(r))(Br << 1)$. In the far field ($Br >> 1$) circulation is

$$\psi(x,y) \cong (8\pi R^2 Br)^{\frac{1}{2}}exp(-Br(1+\cos\theta))$$

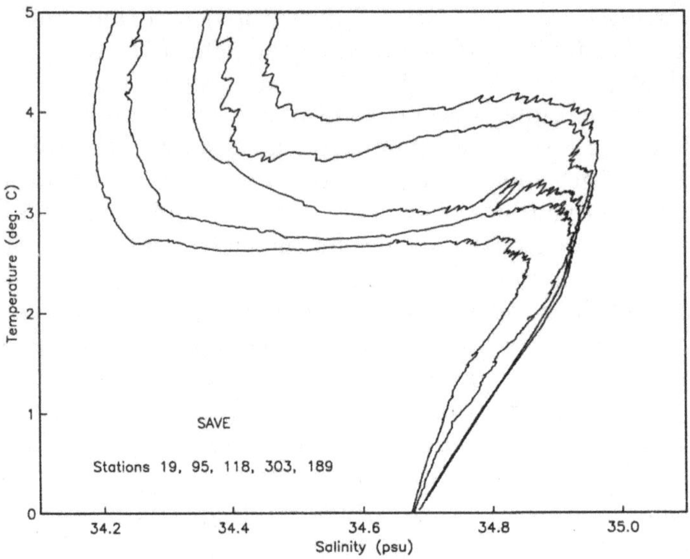

Figure 13: $\theta - S$ profiles in the Brazil Basin, showing intrusive fine structure, and the disappearance of the 'knee' of water upper NADW. The profiles begin at latitude 10°S,29°W (the saltiest/warmest 'knee') and then 19S,31W; 26S,34W; 36S,41W; 47S,42W. This destruction of the upper NADW by fine structure is visible as the top of the low-silicate tongue slopes downward to the south, figure 12. Such specialized mixing will be very difficult to incorporate in a gcm.

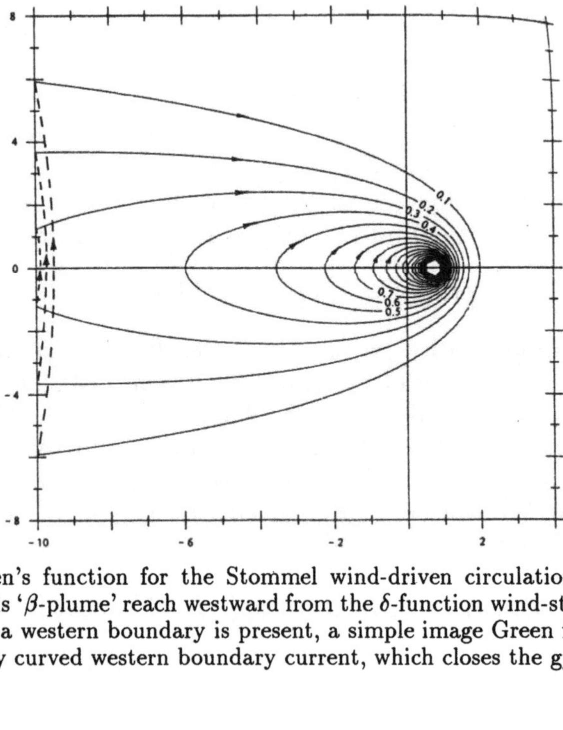

Figure 14: Green's function for the Stommel wind-driven circulation equation. The streamlines of this 'β-plume' reach westward from the δ-function wind-stress curl, crossing latitude lines. If a western boundary is present, a simple image Green function produces a narrow, slightly curved western boundary current, which closes the gyre.

where r, θ are polar coordinates. The streamfunction has an exponential decrease with distance, except that as $\theta \to \pi$ it extends much farther out, the flow reaching westward in a plume. This solution bears an important relationship with Rossby-wave propagation. The linear damping $R\nabla^2\psi$ resembles $i\omega\nabla^2\psi$, which would be the vorticity acceleration in undamped waves of frequency ω. Rossby waves of low frequency (yet finite wavelength) have group velocity, c_g, given by

$$c_g \cong (-\frac{\beta}{\ell^2}, 0)$$

where ℓ is the north-south wavenumber. As $\omega \to 0$ these waves remain potent in carrying signals westward from a source. Friction, with a charateristic damping rate R, will 'arrest' the waves after they have propagated a distance $\cong c_g/R = \beta L^2/R$, where L is now the typical north-south scale ℓ^{-1} of the forcing. The delta function forcing above is 'scale-free', yet a natural scale is set by the near-field, purely frictional, response. A quick way to visualize the ψ-field is to realize that the exact same equation describes the temperature field in a plate which is steadily translating westward (speed $= U$) past a δ-function heat source (in the analogy R/β is identified with κ/U, where κ is the heat diffusion coefficient in the metal plate).

Equation (18) is of parabolic type, and of course gives the Green function for Stommel's formulation of the wind-driven circulation of a single-layer ocean. Gill (1968) has used it to discuss the coupling of a model circumpolar ocean with a model subtropical ocean dominated by Sverdrup response. But the Green-function solution above emphasizes that free flow on a β-plane is not purely zonal, so that sources and sinks are not strictly connected by 'pipes' lying along latitude circles. Mass sources and sinks of finite dimension force the fluid vorticity very strongly, creating gyres like expression (19) while the mass flow threads almost invisibly from source to sink.

A complementary problem for the stratified ocean was articulated by Gill and Smith (1970). Stommel (1982), Speer and Rona (1989) and Hautala and Riser (1991) have developed the idea for oceanic water masses, possibly driven by hydrothermal heating at the East Pacific Rise, in the South Pacific. If the damping of the vorticity equation is through vertical diffusion rather than lateral mixing (though our scale analysis above does suggest that lateral eddy stirring will remain a potent competitor), the equation is found as follows: for baroclinic, large-scale flow the fluid reponds 'stiffly', $\beta v = f w_z$, and w is found from mass-conservation plus a equations for heat and salt conservation, abbreviated as

$$\frac{D\rho}{Dt} = \kappa_h \nabla^2 \rho + \kappa_v \rho_{zz} + \frac{\mathcal{H}}{g}$$

where $\mathcal{H}$ is prescribed buoyancy generation by heating, cooling or mixing. The dominant balance for slow, steady circulation is

$$w\rho_z = \kappa_h \nabla^2 \rho + \kappa_v \rho_{zz} + \frac{\mathcal{H}}{g}$$

If vertical diffusion is dominant, this combines with Sverdrup balance to give

$$\beta \rho_x = \frac{f^2 \rho}{g} \left[\left(\frac{\kappa_v \rho_{zz}}{\rho_z}\right) + \frac{\mathcal{H}}{g\rho_z} \right]_{zz} .$$

If κ_v is uniform and if we linearize about a large mean stratification,

$$\rho_{zzzz} + \gamma \rho_x = -\left(\frac{1}{g\kappa_v}\right)\mathcal{H}_{zz}$$

is the resulting equation for stratified β-plumes, where $\gamma = \beta N^2 / f^2 \kappa_v$. One can either analyze this using vertical mode decomposition, or for a small buoyancy forcing region in a deep ocean, in terms of a dispersive (in x and z) similarity solution

$$\rho = \left(\frac{-\gamma}{x}\right)^{\frac{1}{4}} F\left(\frac{-\gamma z^4}{x}\right) .$$

For a heat-source, the differentiation in $\mathcal{H}_z$ yields a pair of forced gyres, an anticyclone above a cyclone, with plume-like extension westward. The exact same ideas of arrested waves, used to discuss the barotropic problem above, apply now to baroclinic Rossby waves in the 'stiffly' responding fluid. Their group velocity is just $\beta(NH/f)^2 \equiv \beta L_\rho^2$, where H and L_ρ refer to motions with vertical scale H. They reach westward of the forcing a distance c_g times the damping time $= \beta H^4 N^2 / \kappa_v f^2 \equiv \gamma H^4$. On a spherical Earth or equatorial β-plane (rather than mid-latitude β-plane on which f is considered constant) there is an important new effect: plumes are asymmetric north and south, the signal nearest the Equator reaching farther westward owing to faster Rossby-wave propagation there (recall that the Rossby-wave speed for Equatorial modes scales scales as $c_g \cong \beta L_\rho^2$ where now $L_\rho = (NH/\beta)^{\frac{1}{2}}$ is the Equatorial Rossby radius). They also are dispersive in x and z, such that layered 'water masses' can appear, with fine vertical structure near the forcing region, growing coarser at a distance.

(ii) Kelvin- and Rossby waves and the spinup of the Stommel-Arons circulation. With the inclusion of coastal boundaries, an elaborate pattern of arrested waves describes linear circulation problems. The faster baroclinic Rossby-wave velocity at the Equator (reaching several m $\sec^{-1}$) means that the ocean response to broader baroclinic forcing will involve plumes reaching west along the Equator. But, in addition, Kelvin waves give rapid eastward signal propagation there: both rotation-dominated and gravity-dominated waves arising from a single forcing event, with similar propagation rates yet differing directions. Kawase (1987) has given important synthesis of these events for forced flows in β-plane basins. Kawase forces the flow with a mass-source, say at $y = y_0$, on the western boundary of a trans-equatorial basin. It is just the Stommel-Arons problem, but with time-dependence and diffusive damping.

The arrested-wave family now includes the internal Kelvin wave. This coastal trapped wave is one of a family of edge waves that arise when there are two 'orthogonal' restoring effects in a fluid: this translates mathematically to mixed boundary conditions at a solid boundary. In the ocean it is in fact the bottom-trapped quasigeostrophic wave that plays the role of the baroclinic Kelvin wave, in situations where the continental rise is wider than one Rossby radius, L_ρ. But in either case the group velocity scales with N/f, the width and height scaling as N/f.

The steady, damped momentum equations

$$\begin{aligned} Ru - fv &= -gh_x \\ Rv + fu &= -gh_y \\ \lambda h + H(u_x + v_y) &= 0 \end{aligned}$$

and corresponding vorticity equation

$$R(v_x - u_y) + \beta v - (\frac{f\lambda}{H})h = 0$$

descibe a single layer of fluid with a free surface, and depth h, on an equatorial β-plane ($f \equiv \beta y$) with linear, 'Rayleigh' damping of momentum at rate R and interface height h at rate λ. The model is meant to represent the deep ocean, with a much thicker, dynamically inactive layer above (a '$1\frac{1}{2}$ layer' model). Such linearly damped dynamics is closely related to a wave problem in which R and λ are pure-imaginary, and hence are frequencies rather than damping rates. Rossby waves, equatorial and coastal Kelvin waves all become 'plumes' of steady circulation. The key parameters are $R/\beta L$, the ratio of Stommel frictional boundary current width to basin width, L, and $gH/\lambda\beta L^3$, the ratio of interface-damping time to transit time for Rossby waves across the basin. The dominant terms in western or eastern boundary layers give for the vorticity equation

$$v_{xx} + (\frac{\beta}{\kappa})v_x - \frac{\lambda f^2}{RgH}v = 0$$

The resulting western coastal boundary layer is a combination of arrested Kelvin waves (boundary current width $= (gHR/f^2\lambda)^{\frac{1}{2}}$ and Stommel western boundary current (width $= R/\beta$). The asymmetry of Kelvin waves carries over, with a plume extending in $y < y_0$ from a forcing region with the coast on its right, in the northern hemisphere; it now emerges naturally from the Fourier-transform solution of the forced problem. Simple poles determining the solution are shifted into or outside of the complex integration contour, according to whether $y - y_0$ is positive or negative. This essentially utilizes the group velocity of the complex dispersion relation. β-induced vorticity collaborates with interface-stretching-induced vorticity to thin the western boundary current, while an eastern boundary current is thickened (i.e., weakened) by β. The agreement with Kelvin-wave group velocity ideas is typical of arrested-wave problems, and has much mathematical underpinning (see Lighthill, 1968), based on the complex dispersion relation that occurs with frictional damping.

It is amusing to think about the flow due to a source and sink pair, located on the western boundary. Both send transient Kelvin pulses equatorward, which eventually tend to cancel, leaving a simple source-to-sink boundary jet with a vestigal 'tail' of recirculation extending equatorward from the southern member of the pair.

Time-dependent numerical solutions proceed first with Kelvin dynamics carrying the interface elevation equatorward along the western boundary, thence east along the equator, poleward at the eastern boundary, finally proceeding slowly westward into the interior as Rossby waves (figure 15). Initially the western boundary current has the Kelvin thickness, L_ρ; if the Stommel layer is thinner than this, the added vorticity input due to β thins the layer accordingly. This convoluted pathway of influence is truncated before completing the Stommel-Arons interior solution, if the damping terms do their work before the group-velocity determined time has elapsed (figure 16). With moderate or small damping, the eastern boundary layer acts as a source for baroclinic Rossby waves over longer time. The solutions provide important points of comparison for both numerical models and observations. Here and in McCreary (1981) the diffusive, steady equatorial circulation has a similar relationship with linear wave propagation. Tracer observations in the Atlantic, beginning with Georg Wüst's maps from the Meteor expedition and continuing to the transient tracers of the SAVE program, have shown tongues of tracer departing the western boundary and extending zonally eastward, on or near the Equator.

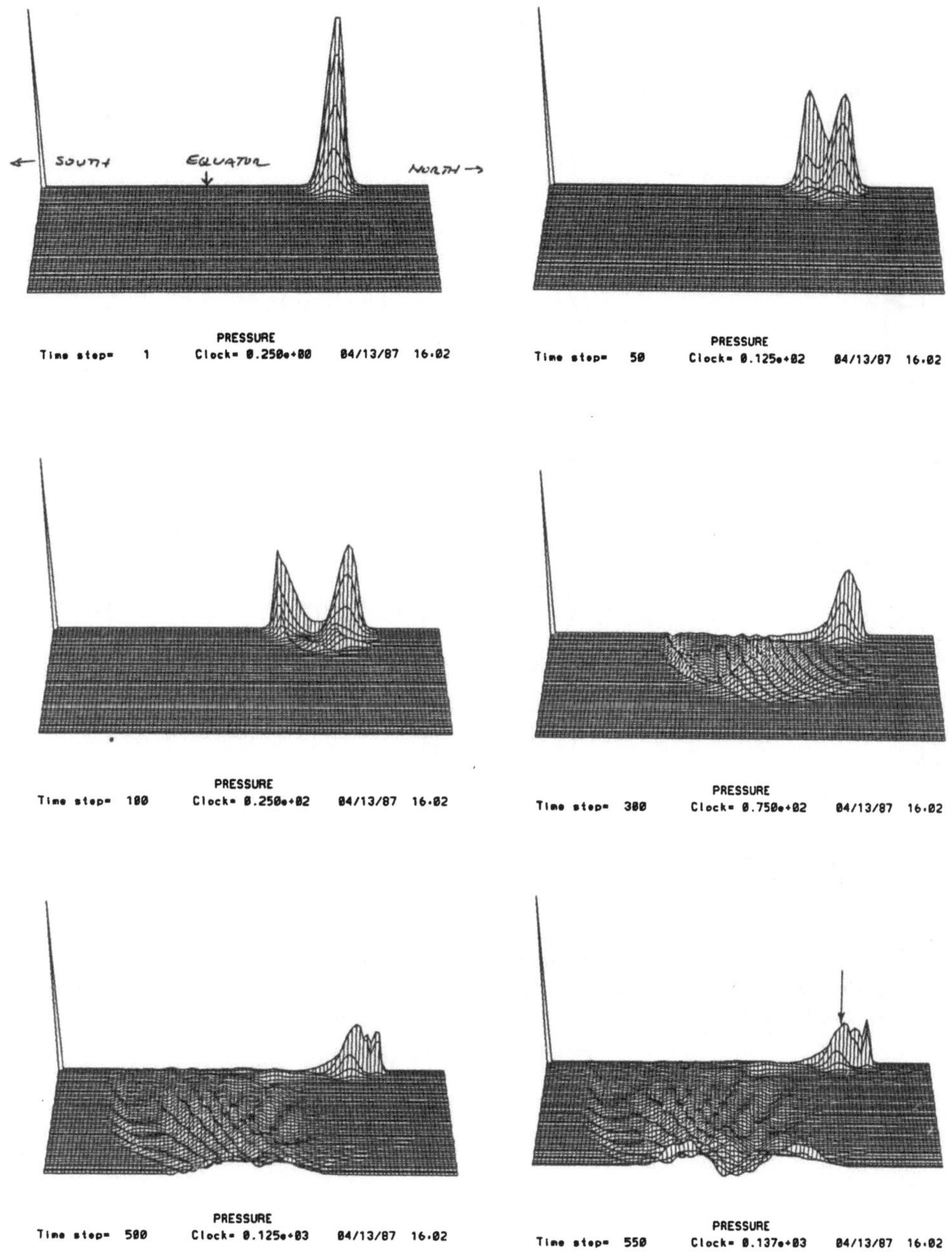

Figure 15: Time history of a geostrophic adjustment in an equatorial β-plane, shallow water primitive equation model. The initial uplift in the interface, in the northwest, sends a steep-faced Kelvin wave Equatorward, while free gravity waves move into the interior. The Kelvin wave crosses the Equator, splits, and heads poleward. Meanwhile the geostrophically adjusted eddy advects northward in the 'anti-Kelvin wave' direction. The arrow marks the initial position of the uplift.

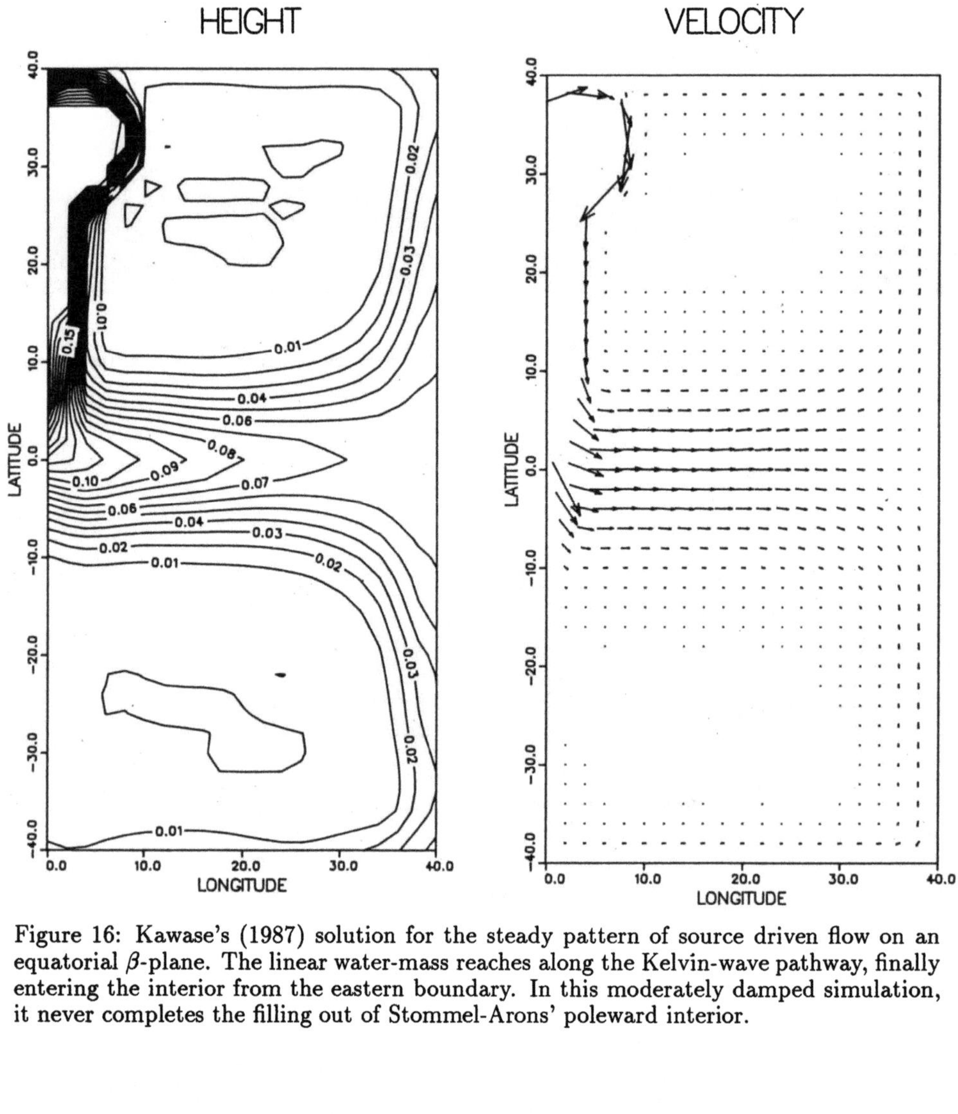

Figure 16: Kawase's (1987) solution for the steady pattern of source driven flow on an equatorial β-plane. The linear water-mass reaches along the Kelvin-wave pathway, finally entering the interior from the eastern boundary. In this moderately damped simulation, it never completes the filling out of Stommel-Arons' poleward interior.

Kawase's model is complementary to the classic Stommel-Arons model in another way: the sink of fluid is a part of the solution rather than being prescribed as spatially uniform. Interface height perturbations h here are damped in proportion to their amplitude. The idea of 'self-damped' water masses is an extreme parameterization, but perhaps no more so than uniform upwelling into the thermocline.

There are many contending effects, beyond the one-layer linear dynamics portrayed here. It remains difficult, for example, to explain the lateral structure of observed deep western boundary currents, which seem to contain narrow (O(50km wide) jets of high velocity, yet tracer plumes and lesser flow that may be as much as 500 km wide. Nonlinear vorticity dynamics of the jet may alter its width greatly (Hermann, Rhines and Johnson, 1989); jet instability and mixing with the interior has been seen very clearly in laboratory experiments on deep circulations (Condie and Kawase, 1992); small, deep, eddy-induced gyres can carry tracers offshore; or, boundary currents can be broadened by bottom slopes. Stommel and Arons (1972) observe that a broad continental rise can broaden a deep current, by allowing it to have both tilted isopycnals (and hence flow) and also quasi-uniform thickness (hence quasi-uniform potential vorticity). There do not seem to be simple bottom-trapped wave modes commensurate with this great width, and hence the rapid setup of broad, deep western boundary currents cannot be carried out by linear propagation. There are some rather weak modes dependent on β, which may take hold at longer time.

The diffusive nature of global circulation models suggests that a mixture of β-plume and arrested boundary-wave dynamics is at work in them. Diagnosis of models using these solutions is an important, as yet unfulfilled activity. The presence of such influence paths in coarse-resolution gcms (e.g., Bryan 1986), provides an important testing ground for both theory and numerics. In the ocean itself, the many scattering and damping effects acting on water masses might be judged by the result; that is, by the degree to which plumes of arrested waves are seen in the response to relatively well-known forcing effects. The study of Hautala and Riser (1991,2) addresses the observed β-plume owing to geothermal heating of the mid-depth South Pacific, in this spirit. Using β-spiral diagnostic procedure, they map lateral and diapycnal mixing as well as reaching a first objective estimate of the plume's circulation.

(iii) Rossby 'hydraulics'; topography; eddy forcing. Large-scale gcm's emphasize planetary-scale modes of deep circulation. Yet deep gyre recirculations are seen at many smaller scales. Rhines and Holland (1979) emphasize the driving by mesoscale eddies, which stir and mix potential vorticity; thus the upper level flow drags the deep water into motion. The result is moderate-scale deep gyres beneath active surface flows. In the North Atlantic the deep recirculations and their form-drag eddy driving has been observed by Hogg (e.g. 1983).

The topography of the sea-bed is an important control over deep circulation. Because of the static stability, fluid cannot generally make large vertical excursions to pass over obstacles, and must either find gaps to pass through, or otherwise enter into a slow, advective-diffusive life-cycle to transport trace properties vertically. Beyond the absolute barriers like the spreading center ridge system, lesser topography can guide and block the flow by vortex stretching. With the exception of deep boundary current cores, the speed of deep ocean mean circulation is usually less than 1 cm sec^{-1}, and the measure of topographic influence, $Ro^{-1}\delta/H$, is often large; δ is the topographic height and H the vertical scale of the current. Below we show how a 300m high topographic feature, less than 500 km in extent, can totally reshape the circulation in a layered 2000x2000km model.

The lateral scale of the general circulation exceeds the Rossby radius, which descends to less than 10 km at high latitude (expressing the strength of rotational stiffness, and weakness of stratification there). If it were not for β, much of the circulation would be close to barotropic. Essentially baroclinic events like deep sinking must be confined to very small lateral scale. Yet, the slowness of flow in the deep interior gives β time to work. Just as the wind-driven gyres shoal with time (since the winds switched on), concentrating into the upper ocean until being arrested by advective effects, Rossby waves in the deep ocean can set the vertical scales and lateral influence paths there.

Layered models have been developed to focus on these questions, both in the context of modification of the Stommel-Arons pattern of deep, poleward interior flow, and in the context of an idealized zonal current like the ACC. Welander (1969) discussed the likely importance of closed geostrophic contours on such circulations. Rhines (1989) developed nonlinear, time-dependent models in one or two layers. The equation for the interface height in a $1\frac{1}{2}$ layer model (a deep active layer with a thick, inert layer above) is yet another example of the 'non-Doppler effect':

$$\eta_t + a\eta_x + b\eta_y = \mathbf{U}_b \cdot \nabla d + \kappa\nabla^2\eta \qquad (20)$$

where (U_b, V_b) is a prescribed barotropic flow, $d(x, y)$ is the topographic height, the layer thickness $h = H_0 - d + \eta$, and c_0 is the speed of the single Rossby-wave mode, $c_0 = \beta g' h/f^2$, with $f = f_0 + \beta y$, and $\kappa\nabla^2\eta$ is a model diffusion of interface height. The coefficients

$$
\begin{aligned}
a &= [U_b - g'd_y/f - c_0]r \\
b &= [V_b + g'd_x/f]
\end{aligned}
$$

combine the advection, Rossby wave propagation, and modification of wave paths by topography. Yet, the baroclinic flow does not advect itself, leaving the problem soluble by characteristics (thus the non-Doppler effect holds for self-interaction with active advection by the barotropic flow alone). The right-most term is a diffusive damping of the interface field. The term $U_b \cdot \nabla d$ is a known topographic forcing for the wave equation (it may be absorbed by a change in dependent variable to leave a homogeneous equation). A complementary problem may readily be written down for a QG model with two active layers. Similar equations without topographic effects have been developed by Williams (1985) for planetary circulations. The time-dependent equation (20) may be solved by integrating along characteristic curves

$$\frac{dy}{dx} = \frac{b}{a}$$

which is readily carried out on a lap-top microcomputer. A great range of physical effects and patterns emerge. The key question initially is whether U_b is greater or less than the topographic + planetary westward group velocity, $g'd_y/f + c_0$. Fast eastward barotropic flows sweep the waves downstream, while westward or slow eastward barotropic flow has wave information reaching westward across the domain. But this implication of hydraulic structure, which would invoke conservative steady flow along streamlines, is incomplete.

Broad, gentle slopes can block an initially uniform flow, expelling the circulation to an outer rim (defined by the hyperbolic characteristics)) where it forms into jets. Planetary-scale flow encountering a mid-ocean ridge develops transient waves that move upstream or downstream, leaving a bound vortex over the topography. The transient and 'bound' vortices are connected by long ridge or trough of dynamic height. At small amplitude the theory resembles classic open-channel hydraulics, but with topography of height only, say, 300m and mean flow speeds of order 1 cm $\sec^{-1}$, we are in a large-amplitude, nonlinear regime. Wave steepening is prominent. Above the topography the characteristics form closed curves, preventing upstream information from entering the region. The density interface winds into a tight spiral, with energy growing like t, limited only by diffusion. This is a form of shear dispersion, but with the sheared field being the group velocity rather than flow velocity. Extensive blocking occurs, in which the topographic disturbance reaches far up- or downstream, permanently modifying the flow there. This is particularly strong when the incident barotropic flow, U_b, has north-south shear. The core of a jet, for example is supercritical yet its edges subcritical, causing the characteristics to 'bounce' off a ridge and create an upstream block.

Despite all this complex dynamics, there is useful insight in the sensitivity to the large-scale incident flow: flows faster than the wavespeed of baroclinic Rossby waves in the appropriate vertical mode tend to develop isopycnal distortions in phase with the topography, while slow (subcritical) flows tend to the opposite, developing troughs above a topographic ridge. But the special dynamics of the two kinds of blocking, and the closed region of characteristics above the topography, are not simple to summarize.

The time-dependent development of the deep interface is shown in figure 17, for the case of a broad jet of incident flow on a 300m-high ridge, which extends about 1000 km north and south, in a 2000 km wide channel. This is for supercritical (strong) eastward flow, which carries a 'starting' vortex eastward, while the 'bound' topographic vortex winds the interface and pv into a tight spiral. Because the oncoming zonal flow is sheared, a plume of blocking reaches westward, upstream, in the region of slower (subcritical) flow. There are intense jets over the topography, and downstream in the trough of isopycnal height, and upstream circumnavigating the blocking plume (which is eventually arrested and held steady by dissipation). Despite its complexity, the final flow closely follows the simple characteristics describing the sum of Rossby propagation and barotropic advection.

Basin solutions. The complement to these 'Southern Ocean' models is a mid-latitude basin with a source-driven deep circulation, Straub and Rhines (1990), Kawase and Straub (1991), Hautala and Riser (1991). As in all such models, the thermodynamics must be handled either by prescribing the upwelling distribution or by developing the full advection-diffusion dynamics for the density field. If we prescribe uniform upwelling at the top of this model, with a source in the northwest corner of a 2000km x 2000km northern hemisphere basin, the circulation (figure 18) fills out the Stommel-Arons poleward flow. Yet, with a rather modest topographic rise in the center, the response is greatly altered. Fast circulation round closed contours distorts the Stommel-Arons pattern above and to the west of the topography. Characteristics that split apart to round the topography meet again on the western side, but there suffer discontinuous behavior. The resulting flow involves an intense, zonal, double jet west of the topography. In this model the topography is confined to the lowest layer. Baroclinic spin-up is rapid, owing to local forcing by topography. When the flow becomes steady, it is cyclonic above the topography,

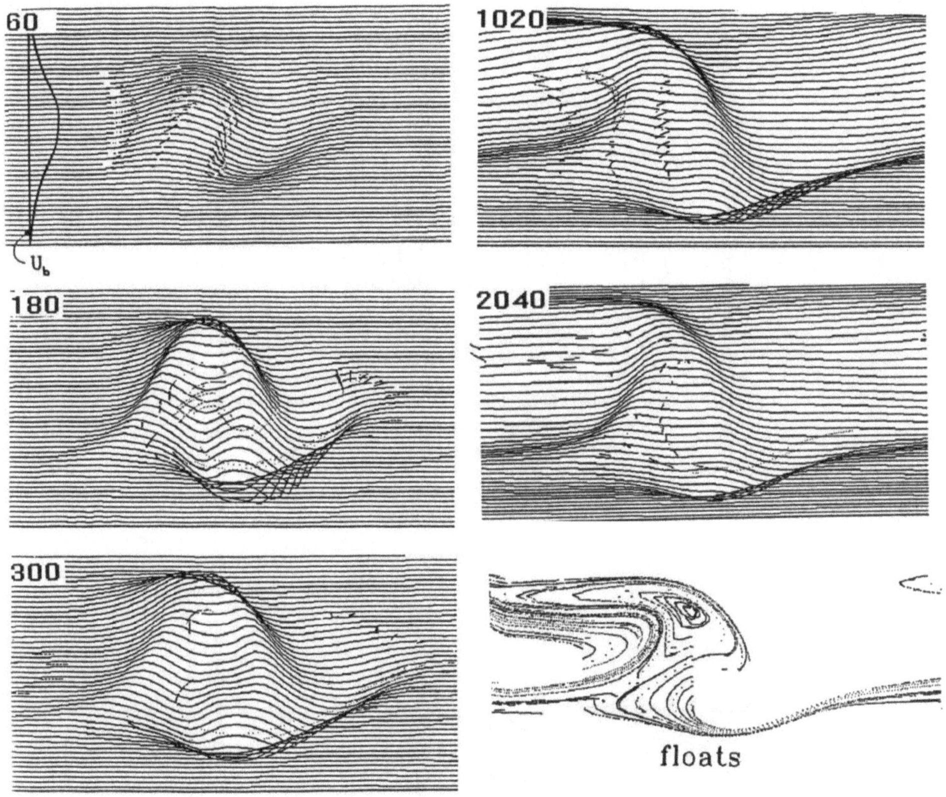

Figure 17: Time sequence of zonal flow past a topographic ridge in mid-basin, with planetary-geostrophic dynamics, $1\frac{1}{2}$ layer model, 2000km x 2000km (Rhines, 1989). The barotropic flow (see upper left panel) is faster than the Rossby-wave speed in mid-channel, yet slower at the sides. The up-down interface deflection propagates both upstream and downstream as a result, leaving a pattern of blocking ridges, troughs, and jets. Particle paths (lower right) slow the eastward zonal flow deflecting far back upstream to round the block, before finally crossing the ridge.

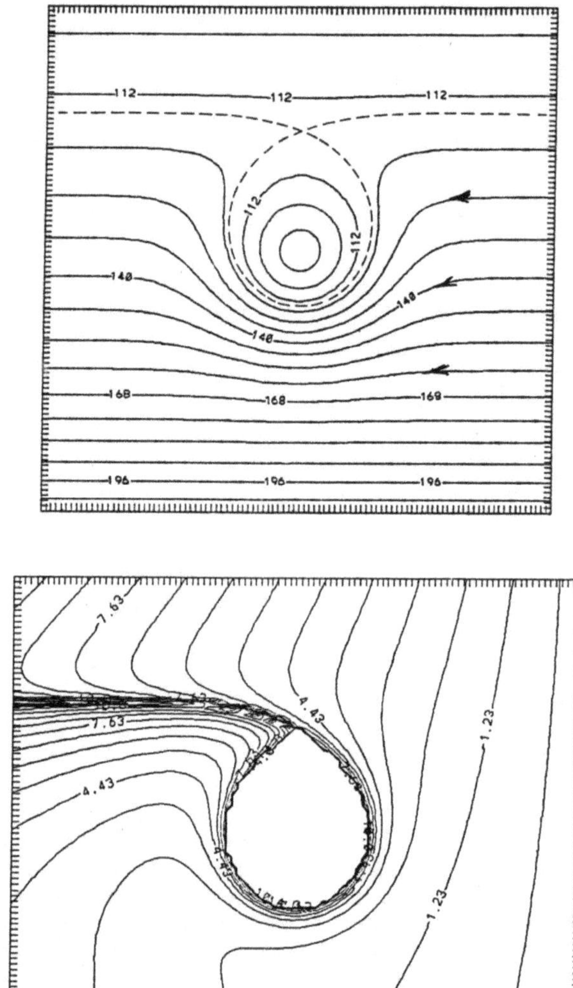

Figure 18: Source driven deep circulation in a layered model, with assumed uniform upwelling (Straub *et al.* 1989). A topographic rise in mid basin greatly alters the poleward Stommel-Arons poleward flow. A double jet communicates with the western boundary and a cyclone sits over the ridge. Upper panel: characteristic field; lower panel: interface-height.

whether it is positive (a rise) or negative (a basin). The papers referenced above show that the planetary-geostrophic solutions resemble closely the full solutions for a primitive equation layered model, although barotropic Rossby waves and baroclinic instability play some role in the latter.

(iv) Hypsometry and the case for anticyclonic deep circulation. When a branch of the deep circulation enters an ocean basin like the North Pacific, and subsequently rises toward the thermocline, it is supposed in the Stommel-Arons theory that the vertical velocity must increase upward, supporting the downward diffusion of the warm waters. The result is horizontal convergence of the flow, which would cause cyclonic vorticity on an f-plane. In the Stommel-Arons flows, Rossby-waves carry the western half of this cyclone to the western boundary, leaving flow that is poleward throughout mid-ocean. The idea would seem to be consistent with a laboratory experiment with a rotating cylinder fed by a source at its rim; the free surface rises in response, stretching planetary vortex lines, and hence creating cyclonic vorticity. If the source were instead at the basin center, its outward flow would generate negative (anticyclonic) circulation, yet the same rising surface forces the vorticity to be positive. This apparent paradox is resolved by the presence of an anticyclonic swirl ($u_\theta \propto r^{-1}$) with zero vorticity outside the origin, plus a solid-body cyclonic swirl ($u_\theta \propto r$).

Consider, however, the effect of basin-shaped topography. This encourages the heavy plume to move toward the basin-center before rising. If the fluid thus displaced in the thermohaline circulation encounters rapidly increasing horizontal area as it rises, it can diverge horizontally rather than converging; indeed, this seems the more natural tendency. But horizontal divergence induces anticyclonic vorticity and circulation on an f-plane, or Equatorward interior flow on a sphere (Rhines and MacCready 1989). Suppose that the rising branch of the circulation has horizontally uniform vertical velocity, and virtually fills an ocean basin. If the horizontal area of the basin is $A(z)$, mass conservation in the rising branch requires that

$$wA = S_0, \quad \text{a constant}$$

or

$$w_z = -w\frac{A_z}{A} = -S_0\frac{A_z}{A^2} < 0$$

where w is the area-averaged vertical velocity. If we take w to be uniform, then Kelvin's theorem for a rising disk of fluid (figure 19) tells us that, on an f-plane

$$(\frac{d}{dt} + R)\int \mathbf{u}\cdot d\boldsymbol{\ell} = -f\frac{DA}{Dt} = -S_0\frac{A_z}{A^2}$$

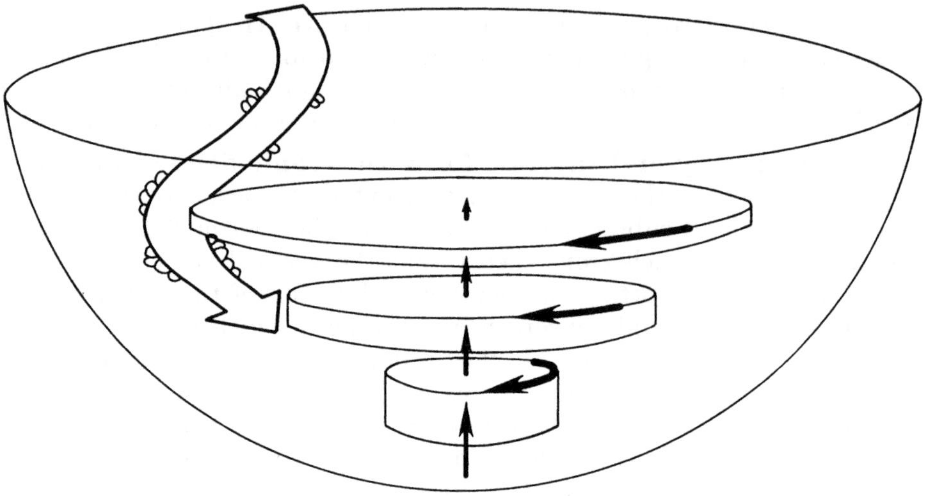

Figure 19: Schematic of the hypsometric effect (the decrease in cross-sectional area with depth), by which a source-driven deep basin develops horizontally divergent, hence anticyclonic, vorticity.

where R is a coefficient of linear drag. Kelvin's circulation about the rim of the disk is thus anticyclonic, and grows strongly until being limited by damping. If w is not uniform, this calculation 'sees' the vertical velocity at the boundary, where the circulation path lies. This modifies the result to

$$(\frac{d}{dt} + R) \int \mathbf{u} \cdot d\boldsymbol{\ell} = f \int (\frac{w}{\nabla h \cdot \hat{\mathbf{n}}}) d\ell$$

again, with R modelling the viscous effects. $\hat{\mathbf{n}}$ is a unit vector normal to the contour of integration, and h is depth. This circulation will also tend to be anticyclonic, though the detailed distribution of upwelling determines its strength. For the β-plane, once again with w taken to be uniform,

$$\beta v = -f S_0 \frac{A_z}{A^2}.$$

Now, the hypsometric control over w forces the deep ocean to move equatorward, opposite to the Stommel-Arons circulation (figure 20). It may be that the deep source orginates as an overflow running down the sloping bottom, which will involve entrainment or detrainment of interior fluid at a rate $S_1(z)$ m^3 sec^{-1} m^{-1}. Or, there may be a broader deep circulation entering the basin as well. In either case, with the upwelling source distributed rather than concentrated at the bottom, we have:

$$\begin{aligned} w_z &= \frac{S_1}{A} - wA_z/A \\ &= \frac{S_1}{A} - (A_z/A^2 \int_{-h}^{z} S_1 dz) \end{aligned}$$

which drives either βv or swirl. When the source enters the basin over a broad range of depths, we can recover the familiar Stommel-Arons case; for example, with $A_z = 0$, S = constant, $w_z = S/A > 0$. Speer and Tziperman (1990) have developed models of the entrainment process. Notice in figure 20 the suggestion that the familiar cyclonic path of the sinking plume persists, even though a great body of displaced fluid circulates in the opposite sense. There is a tendency, perhaps instilled in us by tracer data, to focus on the pathway of the sinking branch and its deep western boundary current. The 'old' fluid driven into motion can be a major component of the circulation.

We have carried out laboratory experiments in a large, rotating bowl which have investigated the stratified, yet f-plane case. The expanding disks of fluid gain anticyclonic circulation until limited by friction. There, the weakness of stratified spin-down allows the circulation to become very strong before being arrested by friction. Observations of the North Pacific (Reid, 1981, Johnson and Toole, 1992) support the idea. The subtropical anticyclone, wind-driven near the surface, seems to expand in size as one moves below 1000m. *Pv* maps (e.g., Holland *et al.* 1984) suggest remarkably deep penetration of circulation, not giving in to βy until great depth. The hypsometric curve for the Pacific (between 55°S and the Aleutians), figure 21, shows persistent area change below 3000m. Johnson and Toole (op.cit.) outline a scheme of abyssal circulation involving a dominant deep anticyclonic flow, with entry at the Samoan Passage and Equatorward flow along the rising topography to the east (figure 22). There is likely involvement with the observation that the deep North Pacific waters rise only to mid-depth before exiting to the south (Fiadeiro 1982, Toggweiler and Samuels 1992).

Ishizaki (1993) has recently carried out a series of primitive-equation gcm (1.25° x 1° x 20 levels) experiments for the Pacific, which are very relevant to the question. He finds a single, dominant anticyclonic deep circulation at depths below 2500m. His numerical solutions give strong support to hypsometric control.

It must be stressed that there are many effects working to determine the nature and sense of the deep circulation; hypsometry is just one of these. Hypsometry is also working to increase the general strength of

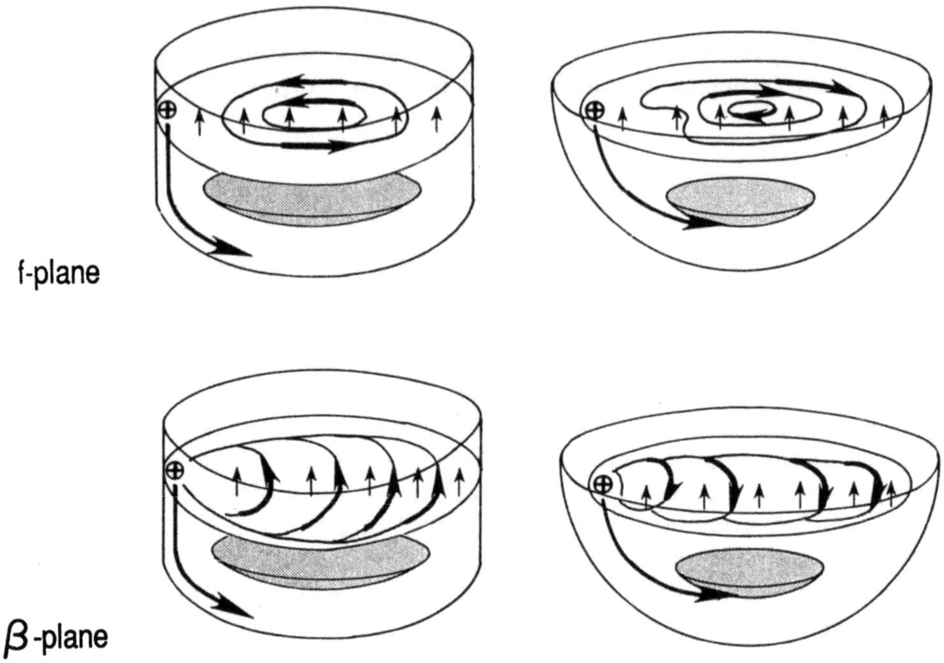

Figure 20: Sketches of f-plane and β-plane response to the source flow, with assumed uniform upwelling. The β-plane interior motion is poleward with uniform depth, yet equatorward in a bowl-shaped basin.

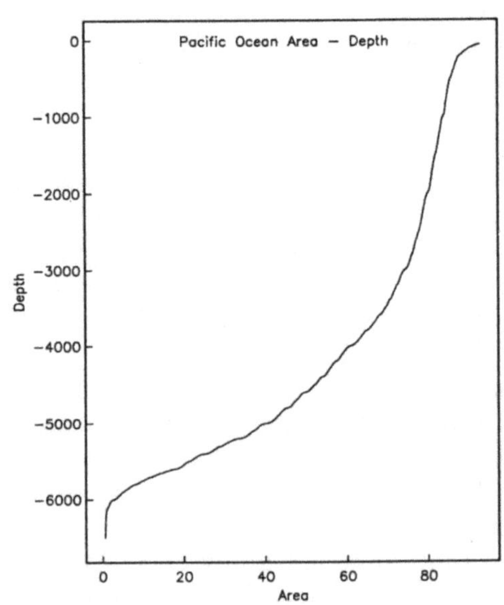

Figure 21: Variation in horizontal area with depth ('hypsometric curve') for the Pacific north of 55°S, calculated from USGS DB5 (5′-resolution) topography.

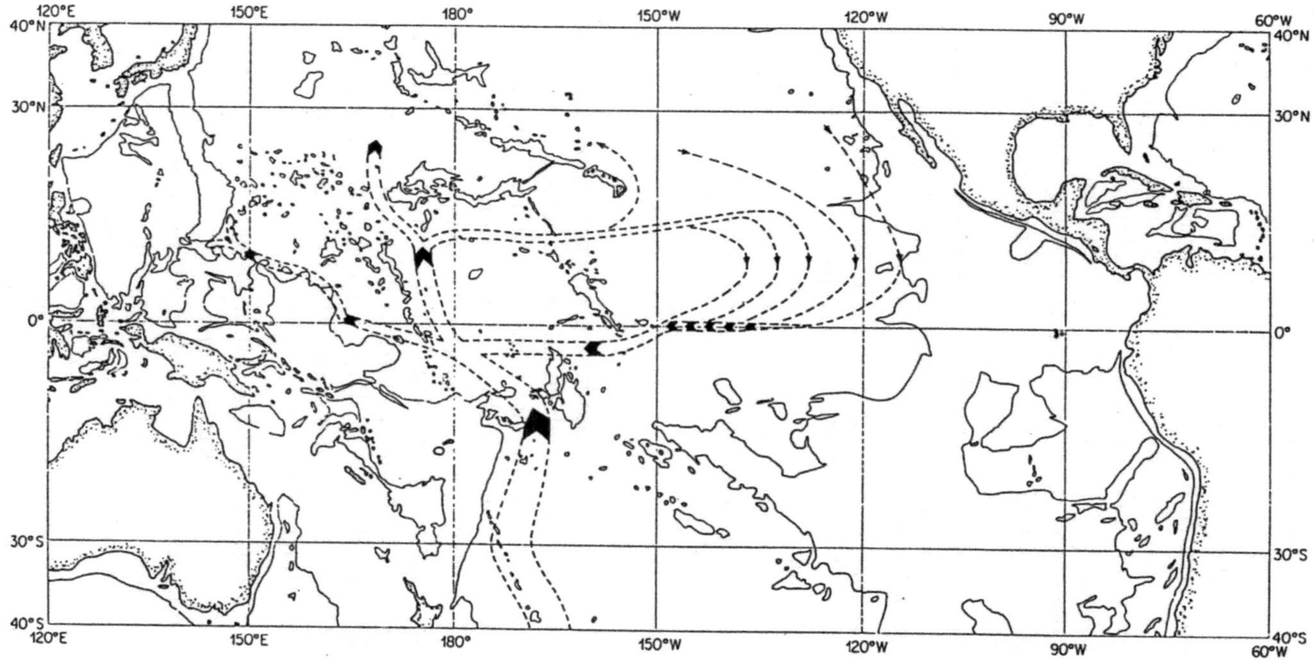

Figure 22: Schematic of the flow of lower central Pacific water, with the 4000m isobath plotted. This is characteristically water colder than 1.2°C, lying about 1000m to 1500m thick on the East Pacific Rise, thus involving depths between 2500m and 4500m. Thickness of arrows indicates transport; Johnson and Toole (1992). Water enters the North Pacific through the Samoa Passage Equatorial jet. The deep flow is predominantly anticyclonic, as the hypsometric model would suggest.

deep upwelling, simply by confining it to a smaller area of the planet.

It is quite ironic that the circulation predicted is opposite in sense to that calculated for the *same* topography, using layered models, (iii) above. These models always develop cylonic vorticity above closed contour regions of either sign (hills or basins), owing to the confinement of the topography to the lowest layer. The heavy source water entering the region does not 'underflow' the layer containing the topography, as it does with continuous stratification. The hypsometric effect thus relies on effects absent from many models: the intersection of isopycnals with the sloping bottom, and stratification that can represent the invasion of a deep plume fully.

Modelling of deep circulation, by theory, numerics or laboratory experiment is faced with the difficult effect of sloping topography encountering isopycnal surfaces (which may also slope). This intersection of density and topography is is just as difficult to handle as is outcropping at the surface of the ocean. It represents a δ-function of potential vorticity, and makes 'arrested' wave modes, benthic boundary layer, spindown, and abyssal boundary currents very different from models with vertical sidewalls. Pointed study of 'in-cropping' is called for.

6 Conclusion

Many current ocean theories and models, predictive and diagnostic, involve internal boxes and layers in which all the structure is concentrated into a few interfaces. The continuous fluid equations, however, contain much more. There are patterns of turbulent distortion, jets, fronts, cusps, shear dispersion and nearly fractal material surfaces and mixing lines and mixed tongues. This is the area in which true fluid dynamics can contribute. We close with an anonymous poem from the newsletter of the first (along with the Russian Polygon experiment) mesoscale eddy program, called MODE (some would say the beginning of the end of small-and-friendly oceanography):

I tossed a pebble into a pond
It went splash
Then someone gave me a large computer
After great effort and ingenuity
I learned to toss a numerical pebble
Into a numerical pond
It went numerical splash
So now I understand splash

- MODE Hot-Line News, 1973

Acknowledgment

My work has been generously supported by the National Science Foundation (current grant OCE-89-16009) and the Office of Naval Research (grant N00014-92-J-1405). I want particularly to thank Jürgen Willebrand and David Anderson for organizing and leading this most enjoyable and productive workshop. I want to thank my students Bob Hallberg, Parker MacCready, David Pierce, Matter Trunnell, and David Straub for many discussions about this material.

References

ANDERSON, D.L.T. AND A.E.GILL 1975, Spin-up of a stratified ocean, with applications to upwelling, *Deep-Sea Res.* **22**, 583-596.

BAUER E. AND G. SIEDLER 1988, The relative contributions of advection and isopycnal and diapycnal mixing below the subtropical salinity maximum, *Deep-Sea Res.* **35**, 811-837.

BARRINGER, M.O. AND J.F.PRICE 1990, A simple model of the descent of the Mediterranean outflow plume in *The Physical Oceaongraphy of Sea Straits, L.J.Pratt ed, Kluwer Acad. Pub., Neth.*, .

BEHRINGER, D. 1972, Investigations of large-scale oceanic circulation using hydrographic data, *Ph.D. dissertation, University of Calif. San Diego.*

BONING, C.W. AND M.D.COX 1988, Particle dispersion and mixing of conservative properties in an eddy-resolving circulation model, *J.Phys.Oceanogr.* **18**, 320-338.

BRETHERTON F.P. AND D.P.HAIDVOGEL 1976, Two-dimensional turbulence above topography, *J.Fluid Mech.* **78**, 129-154.

BREWER, P., W.S. BROECKER, W.J. JENKINS, P.B. RHINES, C.G. ROOTH,J.H.SWIFT, T. TAKAHASHI AND R.T. WILLIAMS, 1983, A climatic freshening of the deep North Atlantic over the past 20 years, *Science* **222**,1237-1239.

BRYAN, F. 1986, High-latitude salinity effects and interhemispheric thermohaline circulations. *Nature* **323**, 301-304.

BUDYKO, M.I. 1963, Atlas of the heat balance of the earth sphere, *Joint Geophysical Committee, Academy of Sciences, USSR, 5pp.*

BUNKER, A.F. 1976, Computations of surface energy flux and annual air-sea interaction cycles of the North Atlantic Ocean, *Monthly Weather Rev.* **104**, 1122-1140.

CHANG, P. AND S.G.H.PHILANDER 1989, Rossby wave-packets in baroclinic mean circulations, *Deep-Sea Res.*, **36**, 17-37.

CHERESKIN, T. AND D.ROEMMICH 1991, A comparison of measured and wind-derived Ekman transport at 11°N in the Atlantic Ocean, *J.Phys.Oceanogr.* **21**, 869-878.

CONDIE, S. AND M.KAWASE 1992, Models of abyssal flow in basins separated by a mid-ocean ridge, *J.Marine Res.* **50**, in press.

COX, M.D., 1985, An eddy-resolving numerical model of the ventilated thermocline. *J.Phys.Oceanogr.* **15**, 1312-1324.

COX, M.D., 1989, An idealized model of the world ocean. Part I: the global-scale water masses, *J.Phys.Oceanogr.* **19**, 1730-1752.

CUMMINS, P.F., G. HOLLOWAY AND A.E. GARGETT 1990, Sensitivity of the GFDL ocean general circulation model to a parameterization of vertical diffusion, *J.Phys.Oceanogr.* **20,** 817-830.

DANIELSEN, E.F. 1990, In defense of Ertel's potential vorticity and its general applicability as a meteorological tracer, *J.Atmos.Sci.* **47,** 2013-2020

DAVEY, M.K., W.W. HSIEH AND R.C. WAJSOWICZ 1983, The free Kelvin wave with lateral and vertical viscosity, *J.Phys.Oceanogr.* **13,** 2182-2191.

DEWAR, W.K., P.B. RHINES AND W.R. YOUNG 1984, The nonlinear spin-up of a stratified ocean, *Geophys. Astrophys. Fluid Dyn.* **30,** 169-197.

DEWAR, W.K. 1991, Arrested fronts, *J.Mar.Res.* **49,** 21-52.

DICKSON, R.R., J. MEINCKE, S.A. MALMBERG AND A.J. LEE 1988, The great salinity anomaly in the northern North Atlantic, 1968-82, *Prog. Oceaongr.* **20,** 103-151.

FIADEIRO, M.E. 1982, Three-dimensional modeling of tracers in the deep Pacific Ocean II. Radiocarbon and the circulation, *J.Marine Res.* **40,** 537-550.

GERDES, R. 1991, A primitive equation ocean circulation model using a general vertical coordinate transformation. *Parts I and II, preprint.*

GILL, A.E. 1968, A linear model of the Antarctic circumpolar current, *J.Fluid Mech.* **32,** 465-488.

GILL, A.E. 1982, Atmosphere-ocean dynamics, *Academic Press.*

GILL, A.E. AND R.K. SMITH 1970, On similarity solutions of the differential equation $\psi_{zzzz} + \psi_x = 0$, *Proc.Camb.Phil.Soc.* **67,** 163-171.

HAIDVOGEL, D. AND P.B. RHINES, 1983, Waves and circulation driven by oscillatory winds in an idealized ocean basin, *Geophys. Astrophys. Fluid Dyn.* **25,** 1-65.

HAUTALA, S. AND S.C. RISER 1991, A simple model of abyssal circulation, including effects of wind, buoyancy and topography, *J.Phys.Oceanogr.* **19,** 596-611.

HAUTALA, S. AND S.C. RISER 1992, A nonconservative β-spiral determination of the deep circulation in the eastern South *Pacific, J.Phys.Oceanogr.,* submitted

HAYNES, P.H. AND M.E. MCINTYRE 1987, On the evolution of vorticity and potential vorticity in the presence of diabatic heating and frictional or other forces, *J.Atmos.Sci.* **44,** 828-841.

HAYNES, P.H. AND M.E. MCINTYRE 1990, On the conservation and impermeability theorems for potential vorticity, *J.Atmos.Sci.* **47** 2021-2031.

HELD, I.M. 1983, Stationary and quasi-stationary eddies in the extratropical troposphere: *theory. in Large-scale dynamical processes in the atmosphere, B.Hoskins and R.Pearce eds., Academic Press, London.*

HERMANN, A.L., P.B. RHINES AND E.R. JOHNSON, 1989, Nonlinear Rossby adjustment in a channel: beyond Kelvin waves, *J.Fluid Mech.* **205,** 460-502.

HOGG N.G. 1983, A note on the deep circulation of the western North Atlantic: its nature and causes, *Deep-Sea Res.* **30,** 945-961.

HOGG, N.G. 1987, A least-squares fit of the advective-diffusive equations to Levitus Atlas data, *J.Marine Res* **45,** 347-375.

HOLLAND, W.R. AND P.B. RHINES, 1980, An example of eddy-induced ocean circulation, *J. Phys. Oceanogr.* **10(7)**, 1010-1031.

HOLLAND, W.R., T. KEFFER AND P.B.RHINES 1984, The general circulation of the oceans: the potential vorticity field, *Nature* **308**, 698-705.

HOLLOWAY, G. AND P.B. RHINES, 1990, Angular momenta of modeled ocean gyres, *J. Geophys. Res,* **96**, 843-846.

HOSKINS, B.J., M.E.MCINTYRE, AND A.W. ROBERTSON 1985, On the use and significance of isentropic potential vorticity maps, *Q.Jour.Royal Met.Soc.* **111**, 877-946.

HUANG, R.X. 1986, Solutions of the ideal fluid thermocline with continuous stratification, *J.Phys.Oceaongr.* **16**, 39-59.

HUANG, R.X. 1990, On the three-dimensional structure of the wind-driven circulation in the North Atlantic, *Dyn.Atmos.Oceans,* **15**, 117-159.

IMBRIE,J., E.A.BOYLE, S.C.CLEMENS, A.DUFFY, W.R.HOWARD, G.KUKLA, J.KUTZBACH, D.G.MARTINSON, A.MCINTYRE, A.C.MIX, B.MOLFINO, J.J.MORLEY, L.C.PETERSON, N.G.PISIAS, W.L.PRELL, M.E.RAYMO, N.J.SHACKLETON AND J.R.TOGGWEILER 1992, On the structure and origin of major glaciation cycles. 1. Linear responses to Milankovitch forcing,*Paleoceanography, submitted.*

ISAYEV, G. 1991, On the absolute velocity of the geostrophic circulation, *J.Fluid Mech.,* submitted.

ISHIZAKI, H. 1993, A simulation of the abyssal circulation in the North Pacific Ocean, *J.Phys.Oceaongr.* submitted

JAMOWITZ, G.S. 1986, A surface density and wind-driven model of the thermocline, *J.Geophys.Res.* **91**, 5111-5118.

JOHNSON, G.C. AND J.M.TOOLE 1992, Flow of deep and bottom waters in the Pacific at 10°N, *Deep-Sea Res. in press.*

JOHNSON, G.C. AND D.R.OLSEN 1992, Frictional rotating channel exchange and ocean outflows, *J.Phys.Oceanogr.* submitted

KAWASE, M. 1987, Establishment of deep ocean circulation driven by deep-water production, *J.Phys.Oceaongr.* **17**, 2294-2317.

KAWASE, M. AND D.STRAUB 1991, Spinup of source-driven circulation in an abyssal basin in the presence of bottom topography, *J.Phys.Oceanogr.* **21**, 1501-1514.

KEFFER, T. 1985, The ventilation of the world's oceans: maps of the potential vorticity field, *J.Phys.Oceanogr* **15**, 509-523.

KESSLER, W. 1990, Observations of long Rossby waves in the North Tropical Pacific, *J.Geophys.Res* **95**, 5183-5217.

KILLWORTH, P.D. 1983, Deep convection in the world ocean, *Revs.Geophys. Sp.Phys.* **21**, 1-26.

KILLWORTH, P.D. 1987, A continuously stratified nonlinear ventilated thermocline, *J.Phys.Oceanogr.* **17**, 1925-1943.

KUSHNIR, Y. 1992, Interdecadal variations in North Atlantic sea surface temperature and associated atmospheric conditions, *J.Climate, submitted*

LAZIER, J. 1992, The salinity decrease in the Labrador Sea over the past 30 years, *preprint*

LEHMAN, S.J. AND L.D.KEIGWIN 1992, Sudden changes in North Atlantic circulation during the last deglaciation, *Nature* **356,** 757-762

LEVITUS, S. 1989, Interpentadal variability of temperature and salinity in the deep North Atlantic, 1970-74 versus 1955-59, *J.Geophys.Res.* **94,** 16125-16131.

LIGHTHILL, M.J. 1978, Waves in fluids, *Cambridge Univ. Press.*

LIGHTHILL, M.J. 1986, An informal introduction to theoretical fluid mechanics, *Oxford Univ. Press.*

LIU, Z, 1992, Interannual planetary wave breaking in the presence of Ekman pumping and mean flow, *J.Fluid Mech., submitted*

LOZIER,S. AND S.RISER 1989, Potential vorticity dynamics of boundary currents in a quasi-geostrophic ocean, *J.Phys.Oceanogr.* **19,** 1373-1396.

LUYTEN, J.R., J.PEDLOSKY AND H.M.STOMMEL 1983, The ventilated thermocline, *J. Phys. Oceanogr.* **13,** 292-309.

LUYTEN, J. AND H.STOMMEL 1986, Gyres driven by combined wind and buoyancy flux, *J.Phys.Oceanogr.* **16,** 1551-1560.

MACCREADY, P. 1992, Frictional decay of the deep western boundary current in the northern North Atlantic, *J.Mar.Res.* submitted.

MACCREADY, P. AND P.B. RHINES 1991, Buoyant inhibition of Ekman transport on a slope and its effect on stratified spin-up, *J.Fluid Mech.* **223,** 631-661.

MACCREADY, P.M. AND P.B. RHINES, 1993, Arrested Ekman layers on a slope, *J.Phys.Oceanogr.* Jàn. in press.

MARSHALL, J.C. AND A.J.G.NURSER 1992, Fluid dynamics of oceanic thermocline ventilation, *J.Phys.Oceanogr.* **22,** 583-595.

MAROTZKE, J., P.WELANDER AND J.WILLEBRAND 1988, Instability and multiple equilibria in a meridional-plane model of the thermohaline circulation. *Tellus* **40a,** 162-172.

MCCARTNEY, M.S. 1982, The subtropical recirculation of mode waters, *J.Marine.Res.* **40,** supp. 427-464.

MCCARTNEY, M.S. AND L.D.TALLEY 1982, The subpolar mode water of the North Atlantic Ocean, *J.Phys.Oceanogr.* **12,** 1169-1205.

MCCREARY, J.P. 1981, A linear stratified model of the Equatorial Undercurrent, *Phil Trans.Roy.Soc.Lon. A,* **298,** 603-635.

MCDOUGAL, T.J. 1988, Neutral-surface potential vorticity, *Prog.Oceanog.* **20,** 185-221

MCDOUGAL, T.J. 1992, Ocean mixing's influence on the absolute velocity vector and on the level of no motion, *J.Phys.Oceanogr.* submitted.

MCDOWELL, S., P.B. RHINES AND T. KEFFER, 1982, North Atlantic potential vorticity and its relation to the general circulation, *J.Phys.Oceanogr.* **12,** 1417-1436.

MCINTYRE, M.E. AND T.N.PALMER 1983, Breaking planetary waves in the stratosphere, *Nature* **305,** 593-600.

MUNK, W. 1966, Abyssal recipes, *Deep-Sea Res.* **13,** 707-730.

NURSER, A.J.G. AND J.C.MARSHALL 1991, On the relationship between subduction rates and diabatic forcing of the mixed layer, *J.Phys.Oceanogr.* **21,** 1793-1802.

OHLSEN, D.R. AND J.E.HART 1989, The transition to baroclinic chaos on the β-plane, *J.Fluid Mech.* **203,** 23-50.

PEDLOSKY, J. 1986, The buoyancy and wind-driven ventilated thermocline, *J.Phys.Oceanogr.*, **16** 1077-1087.

PHILLIPS,N.A. 1963, Geostrophic motion, *Revs. Geophys.* **1,** 123-176.

REID, J.L. AND R.J.LYNN 1971, On the influence of the Norwegian-Greenland and Weddell seas upon the the bottom waters of the Indian and Pacific oceans. *Deep-Sea Res.* **18,** 1063-1088.

REID, J.L. 1981, On the mid-depth circulation of the world ocean, *in Evolution of physical oceanography, B.Warren and C.Wunsch eds., MIT Press, 70-111.*

REID, J.L. 1988, On the total geostrophic circulation of the South Atlantic Ocean: *flow patterns, tracers and transports, Prog. Oceanogr.*

RHINES, P.B., 1977. The dynamics of unsteady currents, *In: The Sea, Vol. VI, E.D. Goldberg (ed.), John Wiley and Sons, Inc., NY, 189-318.*

RHINES, P.B., 1986, Vorticity dynamics of the ocean general circulation, *Ann. Rev. Fluid Mech.* **18,** 433-497.

RHINES,P.B., 1986, Lectures on ocean circulation dynamics, *in Large-Scale Transport Processes in Oceans and Atmospheres, D. Anderson and J.Willebrand Eds. Reidel, Dortrecht. 105-161*

RHINES, P.B., 1989, Deep planetary circulation over topography: simple models of mid-ocean flows. *J. Phys. Oceanogr* **19,** 1449-1470.

RHINES, P.B. AND W.R. HOLLAND, 1979, A theoretical discussion of eddy-driven mean flows, *Dyn. Atmos. Ocean* **3,** 289-325.

RHINES, P.B. AND W.R. YOUNG, 1982A, A theory of the wind-driven circulation I. Mid-ocean gyres, *J. Mar. Res.* suppl. to **40(3),** 559-596.

RHINES, P.B. AND W.R. YOUNG, 1982B, Homogenization of potential vorticity in planetary gyres, *J. Fluid Mech.* **122,** 347-367.

RHINES, P.B., W.R. HOLLAND AND J.C. CHOW, 1985, Experiments with buoyancy-driven ocean circulation, *NCAR Technical Note 260+STR.*

RHINES, P.B. AND R. SCHOPP, 1991, Wind-driven circulation: theory and quasigeostrophic simulations for non-symmetric winds, *J.Phys.Oceanogr.* **21,** 1438-1469. (Cox Memorial vol.)

ROEMMICH, D. AND T.MCALLISTER 1989, Large-scale circulation of the North Pacific Ocean, *Prog.Oceanog.* **22,** 171-204.

ROBINSON, A.R. AND H.M. STOMMEL 1959, The oceanic thermocline and associate thermohaline circulation, *Tellus* **11,** 295-308.

ROOTH, C. 1982, Hydrology and ocean circulation, *Prog. Oceanogr.* **11,** 131-149.

ROSSBY, C.G. 1940, Planetary flows in the atmosphere, *Quart.J.Roy.Met.Soc.* **66** Suppl., 68-87.

SALMON, R. 1986, A simplified linear ocean circulation theory, *J.Marine.Res.* **44,** 695-711.

SCHMITZ,W.J., AND M.MCCARTNEY 1992, On the North Atlantic circulation, *submitted to Revs.Geophys.Sp.Phys.*

SCHMITZ, R.W., P.S.BOGDEN AND C.E.DORMAN 1989, Evaporation minus precipitation and density fluxes for the North Atlantic, *J. Phys. Oceanogr.* **19,** 1208-1221

SPALL, M.A. 1991A, A diagnostic study of the wind- and buoyancy-driven North Atlantic circulation, *J.Geophys.Res.* **96,** 18509-18518.

SPEER, K.G. AND P..RONA 1989, A model of an Atlantic and Pacific hydrothermal plume, *J.Geophys.Res* **94,** 6213-6220.

SPEER, K.G. AND E.TZIPERMAN 1990, Convection from a source in an ocean basin, *Deep-Sea Res.* **37,** 431-446.

SPEER, K.G. AND E.TZIPERMAN 1992, Rates of water mass formation in the North Atlantic Ocean, *J. Phys. Oceanogr.* **22,** 83-92

STOMMEL, H.M. 1961, Thermohaline convection with two stable regimes of flow, *Tellus* **13,** 224-230.

STOMMEL, H.M. 1962, On the smallness of sinking regions in the ocean, *Proc. Nat. Acad. Sci., U.S.A.* **48,** 766-772.

STOMMEL, H.M. 1979A, Oceanic warming of western Europe, *Proc.Nat.Acad.Sci.U.S.A.* **76,** 2518-2521.

STOMMEL, H.M. 1979B, Determination of water mass properties of water pumped down from the Ekman layer to the geostrophic flow below, *Proc.Nat.Acad.Sci.U.S.A.* 3051-3055.

STOMMEL, H.M. 1982, Is the South Pacific helium-3 plume dynamically active? *Earth Planet.Sci.Lett.* **61,** 63-67.

STOMMEL, H.M. AND A.B.ARONS 1958, On the abyssal circulation of the world ocean, I. Stationary planetary flow patterns on a sphere, *Deep-Sea Res.* **6,** 217-233.

STOMMEL, H.M., A.ARONS AND A.J.FALLER 1958, Some examples of stationary planetary flow patterns in bounded basins, *Tellus* **10,** 179-187.

STOMMEL, H.M. AND A.B.ARONS 1972, On the abyssal circulation of the world ocean-V. The influence of bottom slope on the broadening of inertial boundary currents, *Deep-Sea Res.* **19,** 707-718.

STRAUB, D.N. 1992, On the transport and angular momentum balance of channel models of the ACC, *J.Phys.Oceanogr.* In press.

STRAUB, D.N. AND P.B. RHINES 1989, Effects of large-scale topography on abyssal circulation, *J.Marine Res.* **48,** 223-253

SUGINOHARA, N. AND S.AOKI, 1991, Thermohaline circulation as horizontal convection on a β-plane, *J.Marine Res.* **49,** 295-320.

TABATA, S.B.THOMAS AND D.RAMSDEN, 1986, Annual and interannual variability of steric sea level along line P in the North Pacific Ocean, *J.Phys.Oceanog.* **16,** 1378-1398.

TOGGWEILER,J.R., AND B.SAMUELS 1992, New radiocarbon constraints on the upwelling of abyssal water to the ocean surface, *in The Global Carbon Cycle, M.Heimann ed, NATO ASI Series, Reidel.*

TZIPERMAN, E. 1986, On the role of interior mixing and air-sea fluxes in determining the stratification and circulation of the oceans. *J.Phys.Oceanogr.* **16,** 680-693.

VERONIS, G. 1975, The role of models in tracer studies, in Numerical models of the ocean circulation, *Nat.Acad.Sci USA,* 133-146.

WALLACE, J.M. AND Y.ZHANG 1992, Structure and seasonality of interannual and interdecadal variability of the geopotential height and temperature fields in the Northern Hemisphere troposphere, *J.Climate, submitted.*

WALIN, G. 1982, On the relation between sea-surface heat flow and thermal circulation in the ocean, *Tellus* **34,** 187-195.

WARNER, M.J. AND R.F.WEISS 1992, Chlorfluoromethanes in the South Atlantic Antarctic Intermediate Water, *Deep-Sea Res.* in press.

WARREN, B.A. 1976, Structure of deep western boundary currents, *Deep-Sea Res* **23,** 129-142.

WARREN, B.A. 1981, Deep circulation of the world ocean, *in Evolution of Physical Oceanography, B.Warren and C.Wunsch eds., MIT Press, Cambridge.*

WEAVER, A.J. AND E.S.SARACHIK 1990, On the importance of vertical resolution in certain ocean general circulation models, *J.Phys.Oceanogr.* **20,** 600-609.

WEAVER, A.J. AND E.S.SARACHIK 1991, Evidence for decadal variability in an ocean general circulation model: *an advective mechanism, Atmosphere-Ocean, submitted.*

WELANDER, P. 1971, The thermocline problem. *Phil.Trans.Roy.Soc. A.,* **270,** 69-73.

WELANDER, P. 1986, Thermohaline effects in the ocean circulation and related simple models. *in: Large-Scale Transport Processes in the Oceans and Atmosphere, D.L.T. Anderson and J.Willebrand eds, NATO ASI series, Reiderl.*

WELANDER, P. 1989, A new type of double-diffusive instability? *Tellus* **41A,** 66-72.

WIFFELS, S.E., R.W.SCHMITT, H.L.BRYDEN AND A.STIGEBRANDT 1992, On the transport of freshwater by the oceans, *J.Phys.Oceanogr.* **22,** 155-162.

WILLIAMS, R.G., 1989, The influence of air-sea interaction on the ventilated thermocline, *J.Phys.Oceanogr.* **19,** 1255-1267.

WILLIAMS, R.B., 1991, The role of the mixed layer in setting the potential vorticity of the main thermocline, *J.Phys.Oceanogr.* **21,** 1803-1814.

WORTHINGTON, L.V. 1954, A preliminary note on the time scale in North Atlantic circulation. *Deep-Sea Res.* **1,** 244-251.

WORTHINGTON, L.V. 1977, The case for near-zero production of Antarctic Bottom Water, *Geochim.Cosmochim.Acta* **41,** 1001-1006.

WRIGHT, D.G. AND T.F.STOCKER 1991, A zonally averaged ocean model for the thermohaline circulation. Part I: model development and flow dynamics. *J.Phys.Oceanogr.* **21,** 1713-1724.

WINTON M. AND E.S.SARACHIK 1992, Thermohaline oscillations induced by strong steady salinity forcing of ocean general circulation models, *J.Phys.Oceanogr.* accepted

WOODS, J.D. AND W.BARKMANN 1986, A Lagrangian mixed layer model of Atlantic 18°C water formation, *Nature* **319,** 574-576.

YOUNG, W.R. AND G.R.IERLEY 1986, Eastern boundary conditions and weak solutions of the ideal thermocline equations. *J. Phys. Oceanogr.* **16,** 1884-1900.

ON THE OCEANIC THERMOHALINE CIRCULATION

A. COLIN DE VERDIÈRE
Laboratoire de Physique des Océans
UMR127 CNRS, Ifremer, France

1 Introduction

Forced by heat flux, evaporation and precipitation at the air sea interface that vary in latitude, the ocean adjusts to a circulation that feeds upon the temperature and salinity differences through the development of pressure gradients. Under steady state conditions, the role of that circulation is to transport the excess heat absorbed in low latitudes to high latitudes, where cooling predominates, through a poleward warm upper branch and a return equatorward cold lower branch. Under present day conditions, the most conspicuous evidence for such a circulation is the existence of the North Atlantic Deep Water (NADW), a water mass which is formed through a combination of deep convection and lateral mixing in the Norwegian Sea and moves southward to the Antarctic Circumpolar Ocean, and then to the Pacific Ocean. On the other hand there is no such direct deep water mass formation in the Pacific, a difference that may be not too surprising now that the study of paleoclimates has revealed that the thermohaline circulation has also changed temporally and was considerably different during the glaciations period. Broecker *et al.* (1985) suggested the important role of deep ocean circulation in setting the climate of the quaternary and indicated that changes could occur rapidly through modifications of the global hydrological cycle triggering migration from one state of equilibrium to another. This gave fuel to the old idea of Stommel (1961) who showed that a two component fluid such as seawater could exhibit several equilibrium states if the two components temperature and salinity evolve according to different mixing coefficients or surface boundary conditions.

The emphasis of these lecture notes is to provide basic ideas on the underlying dynamics of the thermohaline circulation, a problem that presents itself essentially as one of large-scale convection. Because the ocean circulation, that one tries to explain, has been observed for the last century or so and is being observed now with an unprecedented coverage, it may seem that physical oceanographers have an enormous advantage over paleoclimatologists or over those geophysicists studying

NATO ASI Series, Vol. I 11
Modelling Oceanic Climate Interactions
Edited by J. Willebrand and D. L. T. Anderson

the dynamics of the earth core who have to rely on very indirect methods to reconstruct the convection patterns. This advantage is mitigated, however, by the fact that the large-scale ocean circulation is unstable, and that the energy containing eddies generated by the instabilities with horizontal scales of 100 km or so obscure considerably the large-scale pattern. The observations of temperature and salinity have been shown in the past on $T - S$ diagrams (figure 1) where quite remarkably the observations within ocean basins do not fill the diagram but are crowded along a few curves. The extrema of the curves can be traced back to the regions of formation of specific water masses, where negative buoyancy flux at the sea surface allows convective processes to renew them on an annual basis. These $T - S$ diagrams tell immediately that the places where this may happen are not very numerous, and this is a very fortunate situation that allows the large scale fluid motions to be painted by a small number of dyes of different color injected at the sea surface. The Meteor sections in the Atlantic (1925) were the first to show the layered cake structure of the major water masses that could be followed many thousands of kilometer away from their sources. Shown on such a historical meridional section (figure 2) these structures bear witness that large-scale convection is operating in the ocean. On the other hand, a blown up view of a small piece of ocean as can now be provided through modern water tagging techniques reveals the existence of fully turbulent motions (figure 3) which mix the fluid laterally at lateral diffusivity rates of order 10^3 m^2 s^{-1}. This provides a mixing agent for the water masses that competes with the mean advection to shape up the oceanic properties distribution. Measured by a Peclet number UL/K of order 10 (where U is an horizontal mean velocity of say 1 cm s^{-1}, L a meridional length scale of 1000 km and K the above diffusivity coefficient), the competition in the interior of the ocean between both processes is severe. What is important, however, to account for the huge meridional oceanic heat transport that the ocean must carry to temper the climate of the earth, is that a net zonally averaged mass transport distributed between the warm and cold layers of the ocean persists in the face of much stronger eddying motions.

Although the thermohaline circulation has been the focus of many oceanographic expeditions and its general character is being deciphered, the theoretical understanding is comparatively low. The past emphasis on the wind driven circulation has produced an impressive body of theoretical concepts because the problem reduces to the deductions of velocities given an applied stress at the surface and a stratification. In the case of the circulation driven by surface buoyancy flux, it is necessary to determine the density in addition to the velocities and their nonlinear

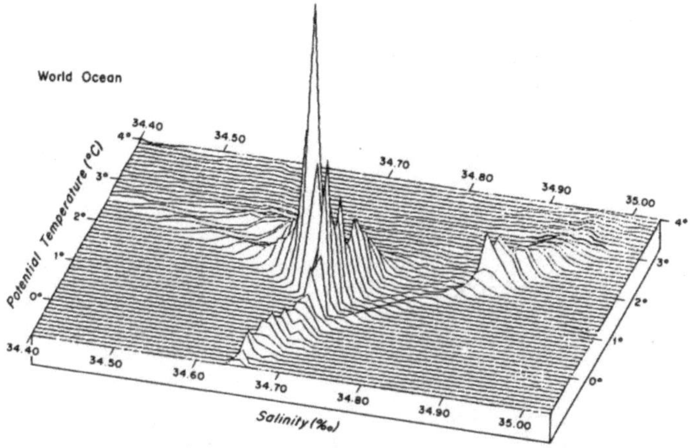

Figure 1: The $T-S$ diagram of the deep water masses of the world ocean with elevation proportional to volume (from Worthington, 1981).

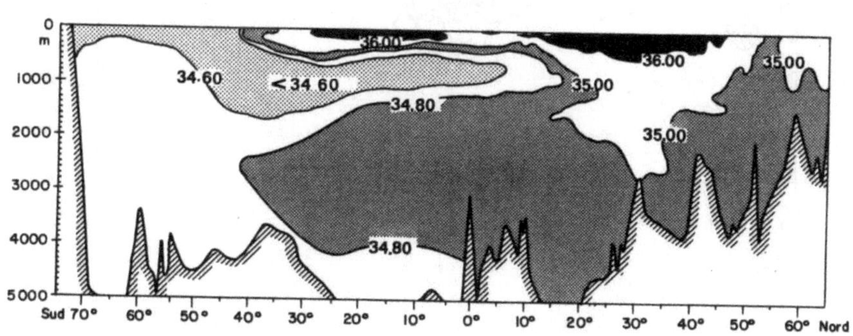

Figure 2: The central Atlantic salinity section of the Meteor expedition (1925-1927). Note the high salinity waters of the thermocline, the injection of low salinity Antarctic Intermediate waters in the Southern ocean and the prominent North Atlantic deep water originating from the Norwegian Sea.

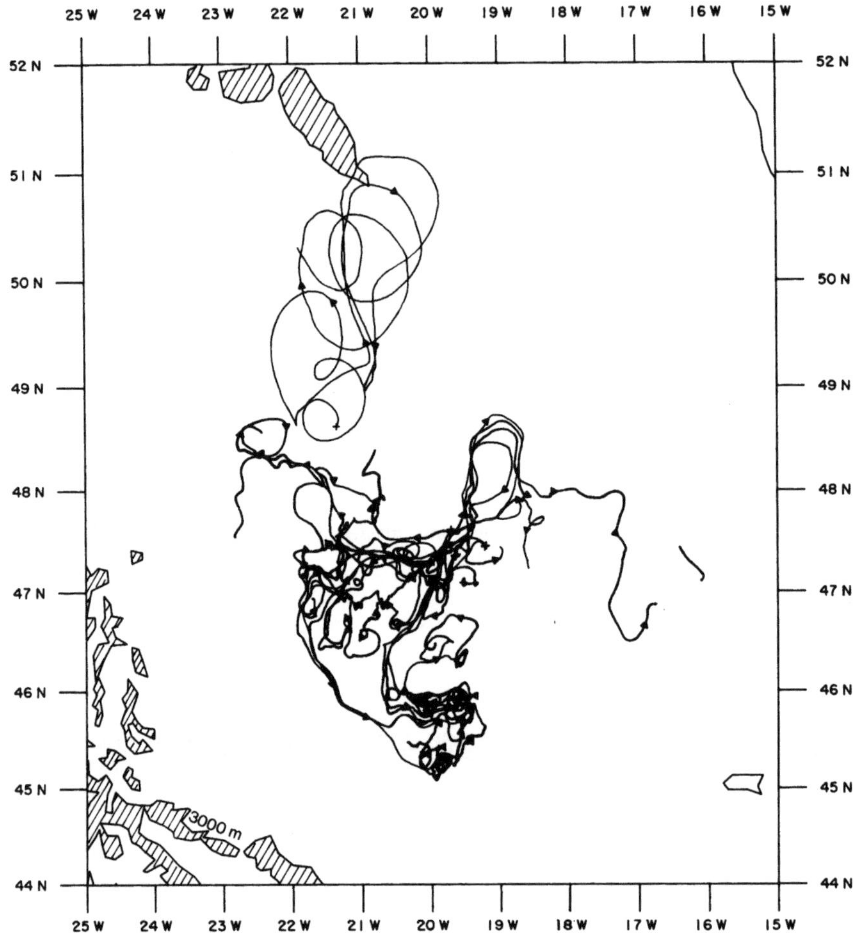

Figure 3: Trajectories of deep (4000 m) neutrally buoyant floats in the North East Atlantic during the NOAMP experiment. There is an arrow every 50 days. The floats reveal the presence of an active topographic turbulence. Note the different trajectory of the float in the upper part of the figure that drifts at 800 m (Ollitrault, 1992).

coupling places convection type problems at the highest level of difficulties in fluid mechanics. Furthermore, while considerable advance on the dynamics of wind driven gyres was made possible by the local character of the mechanical response, another level of complexity is brought about by the *global* character of the thermohaline circulation. On an earth without continents, the thermohaline circulation would respond to a meridional temperature gradient with a zonal thermal wind, and the meridional heat transport, if needed, would be carried out by the instabilities of the zonal mean flow - i.e. an essentially atmospheric description of the ocean circulation. With continental barriers such zonal circulations are forbidden in the ocean. The symmetry about the equator is broken, zonal pressure gradients and hence mean meridional geostrophic flows can be established. As is already well known from the wind driven circulation problem, boundary layers can be expected to close the convective circulations.

Rather than going into the details of the machinery of numerical models that calculate the thermohaline circulation, I wish to present first the important physical elements that must be borne in mind when attacking the problem for the first time and second to show the respective advantages of primitive equations and geostrophic models for simulations of different spatial resolution.

2 Thermodynamics

If, to simplify matters, we assume that the ocean is operating in a closed cycle between two heat sources, a warm source which delivers a quantity Q_H of heat to the ocean at the absolute temperature T_H and a cold sink that removes a quantity Q_L at the absolute temperature T_L, then the Clausius inequality, embodying the Second law of thermodynamics tells us that:

$$\frac{Q_H}{T_H} + \frac{Q_L}{T_L} \leq 0$$

In the above inequality the equal sign is only applicable for reversible processes which provide the maximum efficiency (work produced over heat input) of Carnot heat engines operating in such closed cycles:

$$\eta_{Carnot} = \frac{W}{Q_H} = 1 - \frac{T_L}{T_H}$$

These considerations show

1. that to function as a heat engine, the ocean must receive heat at a higher temperature than the temperature at which heat is given back to the atmosphere,

2. that the Carnot efficiency is very low ($< 10\%$) because the observed temperature difference $T_H - T_L$ is of the order of 20°C at most in the ocean.

Moreover because of irreversibilities caused by various frictional and diffusion processes organized by turbulent cascades to small scales, the actual efficiency is certainly much less than the Carnot efficiency and the ocean can therefore be judged a very inefficient heat engine. As a consequence, the First Law reduces in the steady state to a conservation law for heat, $Q_H + Q_L \sim 0$ to a close approximation. Although vastly inefficient, the quantities of heat exchanged with the atmosphere are so enormous, that a sizable motion occurs which once started will be difficult to stop on account of the high thermal capacity of the working fluid. In terms of the entropy state variable, the heat exchange at the ocean surface decreases the entropy of the ocean while internal irreversibilities tend to increase it, both effects being in balance when the steady state is reached.

We must now evaluate the constraints set by the dynamics on the conditions for which mechanical work can be produced when the ocean operates under such a steady state cycle. Indeed, it had been conjectured in the past that the ocean being heated and cooled at the top should have no significant circulation. Kelvin's theorem (see Gill, 1982) can be used to show that along a steady streamline, the following balance holds:

$$\oint \alpha \, dp = \oint \boldsymbol{F} \cdot \boldsymbol{dl}$$

where α is the specific volume and $\boldsymbol{F}$ denotes the frictional forces.

While there is no term contributed by rotation since the Coriolis force acts at right angles to the particle path, there is of course a hidden effect of the Coriolis force to shape up the path of integration itself which is here envisioned to follow the streamlines of a large scale convection cell. In steady state, the pressure work must be positive to balance the negative work of frictional forces $\boldsymbol{F}$. When integration proceeds along a closed cycle it can be rewritten as:

$$\oint p \, d\alpha = - \oint \boldsymbol{F} \cdot \boldsymbol{dl}$$

a result known as Sandström's theorem (see Dutton, 1986): to oppose the work of decelerating frictional forces ($\boldsymbol{F} \cdot \boldsymbol{dl} < 0$), the work of pressure forces must be such that expansion (compression) happens at high pressure (low pressure) respectively. As reminded by Welander, 1986, within the Boussinesq approximation, the pressure work is zero and work pears instead as that of buoyancy forces, so that we have instead:

$$\oint \rho/\rho_o \boldsymbol{g} \cdot d\boldsymbol{l} = -\oint \boldsymbol{F} \, . \, d\boldsymbol{l}$$

By introducing the geopotential ϕ such that $\boldsymbol{g} = \boldsymbol{\nabla}\phi$, this can be rewritten as:

$$\oint \phi/\rho_o d\rho = \oint \boldsymbol{F}.d\boldsymbol{l}$$

and the above requirement translates that to insure motion, expansion (compression) must occur at high (low) values of the geopotential. When regions of heating coincide with regions of expansion (where $d\rho$ is negative) and conversely for cooling, the internal motions must organise themselves so that heating within the closed fluid loop occurs at a greater depth (higher pressure, higher geopotential) than cooling. That is, although both heat and cold sources are ultimately at the sea surface, the heating must penetrate deeper than the cooling to sustain a global motion. Although there is no difficulty for convective plumes to reach large depths when winter cooling occurs, one of the major obstacle that the thermohaline circulation has to surmount is associated with the necessity for heating to penetrate downwards to sustain accelerating buoyancy torques. In the mean time the depth of penetration of free convection must be limited.

This particular way of forcing the ocean through the upper boundary only (the geothermal flux can be safely neglected in global problems) has other consequences as well. One of the most ubiquitous aspects of the thermohaline circulation is the very restricted number and size of regions where deep water formation occurs. Rather early Rossby (1965) carried out laboratory experiments of non rotating convection with this observation in mind. The bottom boundary is exposed to a constant temperature gradient while all others are kept insulated. This results in a highly asymmetrical convection cell with hot fluid rising in a very limited region and subsiding over a much broader area before being captured into a thermal boundary layer at the bottom of the tank. At such large Rayleigh number ($\sim 10^8$), the free convection is in fact so efficient to transport heat upwards that the much weaker process of conduction needs a much larger area to operate. The same experiments with rotation exhibit much of the same asymmetry (Hignett et al., 1981 ; Speer and Whitehead, 1988) and it is now believed that a similar argument applies in the large scale ocean to reduce the area where free convection and deep water renewal may occur. This is not the only constraint however, for the earth's rotation also has a powerful effect (see Section 3).

It is useful to have order of magnitude knowledge of the buoyancy forcing at the surface of the ocean. Heating of the ocean occurs through solar short wave radiation, 40% of which returns as long wave radiation. The remaining losses to the atmosphere are divided between sensible heat flux (one third) and latent heat flux (two thirds). Outstanding on the maps of North Atlantic net oceanic heat flux (Isemer-Hasse, 1985) is the strongly evaporative cooling in the Gulf Stream region which indicates that the buoyancy forcing depends very much on the oceanic circulation itself. Quite often, the oceanic models are not forced by the observed surface heat fluxes and instead, the surface temperature is restored to a given air temperature over a time scale that depends upon the heat fluxes, the mixed layer depth etc... The important point is that temperature anomalies are controlled by strong negative feedbacks that may eliminate them rapidly. This is not so for the salinity anomalies which are driven by the differences between evaporation and precipitation, two fields independent of the surface salinity. As a consequence, the salinity flux is more naturally held constant in ocean models. It is this difference in boundary conditions for heat and salt that is the cause of the multiple steady states and oscillations of the thermohaline circulation (Stommel, 1961; Marotzke-Willebrand, 1991).

Knowing the temperature and salinity input at the sea surface one may compute the buoyancy input and find out what is the more important, the heat or salinity driving. The temperature change caused by a net heat flux Q distributed over a mixed layer of depth h is $Q/(\rho\ C_p h)$ and the salinity change is approximately $S\ (E-P)/h$ (where $E-P$ stands for the difference between evaporation and precipitation) so that the ratio R of contributions to buoyancy becomes $R\ =\ -\alpha Q/[\rho\ C_p \beta S(E-P)]$ where α and β are respectively the thermal and haline expansion coefficients. While β does not vary much, α decreases by one order of magnitude from 20°C to 0°C. In spite of this, the ratio R is about 10 in polar regions decreasing to about 0.1 in the subtropics, and increasing again in the tropics. The result stems from the strong poleward decrease of $E-P$ itself and thus in general thermal driving is more important north of 30-40°N in the Atlantic (Schmitt et al., 1989 ; Speer, 1992). Interestingly, Tziperman and Speer (1991) have sorted out the buoyancy flux in density classes, a procedure which allows the net production of each water mass formed at the surface to be obtained. In a steady state, the same quantity must of course be exchanged inside the fluid with other water masses.

3 Rotation

The slow velocities and the large scale structure of the thermohaline circulation imply an important control exerted by the rotation of the earth. Diagnostically the large scale flows must be close to geostrophic balance:

$$2\ \boldsymbol{\Omega} \times \boldsymbol{u} = -\nabla p/\rho \tag{1}$$

Now if heating occurs near the tropics and cooling near the poles, one may expect high pressure over the tropics and low pressure near the poles at the surface and the converse at large depths. The geostrophic relation (1) associates an eastward surface current at right angles to the pressure gradient and a westward one at depths which have therefore an awkward orientation to transport heat polewards. While in the atmosphere, the instabilities of these zonal flows allow a release of potential energy and a heat transport by the eddies, the lateral boundaries of the ocean, however, are known sites where the above geostrophic constraint can be broken (in at least one direction), allowing the meridional "heat transporting" flows to occur.

Casual observations show that convection cells in a differentially heated fluid have a fully three dimensional character and the linear instability patterns found in Rayleigh-Bénard convection testify to that as well. Now one of the major effects of rotation (see also the lectures by Rhines in this volume) is to try to make the flow independent of the coordinate along the rotation axis, bidimensional that is, so that competing effects occur when rotation and convection appear together. The transition between a 3D regime of convective plumes leading to intermediate or deep water formation and the rotating regime is necessarily controlled by the size of the Rossby number $U/2\Omega L$ of the flow (where U is a velocity scale, L a length scale) which must be smaller than unity for rotational effects to dominate. The predicted transition scale L hinges on the appropriate value of the velocity. Consider a freely convective plume of density $\rho+\Delta\rho$ reaching a depth h within an environment of smaller density ρ. The only velocity scale is the free fall velocity $(hg\ \Delta\rho/\rho)^{1/2}$, so that a Rossby number of order one occurs for a lateral scale $L \sim (hg\ \Delta\rho/\rho)^{1/2}/2\Omega$, a value which is none other than the internal Rossby radius of deformation based on the penetration depth h of the plume. With oceanic values of h and $\Delta\rho/\rho$ the resultant scale is of the order of a few kilometer and this simple argument shows that the earth rotation is bound to control the thermohaline circulation over a large spectral range and that to free themselves from its influence, the direct convective cells must evolve on very small spatial scales. Such convective filaments being so restricted in size are of course hard to observe, analyse and finally of course to parametrize in large scale models.

At much larger scales, the rotation dominates and the vorticity equation provides a nice way to show the consequences. Under the Boussinesq approximation in an incompressible fluid (two valuable approximations in ocean dynamics) it reads:

$$\frac{D}{Dt}\boldsymbol{\xi} = (\boldsymbol{\xi} + 2\,\boldsymbol{\Omega}).\nabla\boldsymbol{u} - \boldsymbol{g} \times \frac{\boldsymbol{\nabla}\rho}{\rho_o} \tag{2}$$

where $\boldsymbol{\xi} = \nabla \times \boldsymbol{u}$ is the relative vorticity.

Relative vorticity changes come about through 1) the stretching or squashing of vortex lines and 2) buoyancy torques. The above relation simplifies for slow flows in rapidly rotating systems for which the relative vorticity is small compared to the background angular velocity of the system. However care must be excercised because in geophysical situations of interest, the gravity and rotation vector are not parallel. If at mid latitudes the vertical component of relative vorticity at large scale is indeed much smaller than the vertical component of the earth rotation, the same is not true in the horizontal direction since with density stratification vertical shears of 10 cm s^{-1} over 1000 m in the vertical are not uncommon and produce horizontal relative vorticity comparable to the planetary component. Supposing for a moment that there is no such stratification, such shears do not appear however, and the large scale, small Rossby number flows obey:

$$(2\,\boldsymbol{\Omega}.\boldsymbol{\nabla})\boldsymbol{u} = 0$$

The fluid cannot support elongation or contraction along the rotation axis and the flow is effectively bidimensional: this is the so called Taylor-Proudman theorem (to which Poincaré's (1910) name could also be associated), and which implies pure zonal motion on a rotating sphere. When the fluid is stratified, however, buoyancy torques modify the result by creating large relative vorticity at right angles to the gravity vector. In a quasi steady, small Rossby number situation this production is equilibrated by the first term on the LHS side of (2), and the generalized thermal wind equation yields:

$$(2\,\boldsymbol{\Omega}.\boldsymbol{\nabla})\boldsymbol{u}_h = +\boldsymbol{g} \times \boldsymbol{\nabla}\rho/\rho_o \tag{3}$$

where $\boldsymbol{u}_h$ is the velocity vector in the horizontal direction (When carrying out the derivation from (2) to (3), note the existence of metric terms that are safely neglected in the horizontal components of the vorticity equation). (3) reduces further to the familiar formula used in practice at sea when the assumption of *small* aspect ratio (vertical scale $\ll$ horizontal scale) is made. This has the effect of singling out the vertical component of the earth rotation as the dominant one (away from the equator) such that $2\boldsymbol{\Omega}.\boldsymbol{\nabla}$ becomes nearly $2\Omega \sin\theta \frac{\partial}{\partial z}$ (θ is latitude). There are no buoyancy torques in the vertical direction (parallel to g), and the small

Rossby number approximation of (2) gives:

$$(2\,\mathbf{\Omega}.\mathbf{\nabla})w \;-\; \frac{2\Omega\cos\theta}{r}v = 0$$

Again for small aspect ratio, the LHS may be simplified analogously away from the equator and the equation leads to the well known planetary vorticity equation which emphasizes the vertical component of the earth rotation:

$$f w_z = \beta v \tag{4}$$

with $f = 2\,\Omega\sin\theta$, $\beta = 2\Omega\cos\theta/r$, and r the earth radius.

Because of the shape of the earth, a single and important metric term appears in the vertical component of the vorticity equation (4), the βv term. The small Rossby number approximation does not imply a bidimensional flow anymore and elongation or contraction of fluid columns in the direction of the vertical component of the earth vorticity is possible provided that it is accompanied by meridional motions (the v component). From continuity these elongations or contractions are made possible through convergences in the horizontal plane, themselves brought about by readjustment of the density field. (4) is recognised as the corner stone of wind or buoyancy driven ocean circulation theory but note that while the vertical velocity is constrained externally by the Ekman pumping in studies of the wind driven response, it is totally free when one deals with the buoyancy driven response. It is important to note that only the turbulent vorticity fluxes are eliminated in (4) and that no assumption has been made regarding the strength of the turbulent buoyancy fluxes. For this reason it can therefore be used to infer the vorticity dynamics of the large scale thermohaline circulation. Although the equation (4) is widely accepted, it is fair to say that it has not yet received all the experimental support that it deserves. Large scale meridional velocity can be measured with some confidence given the resources to average out the mesoscale turbulence, but there are no instruments that can measure the tiny vertical velocities of order $h/r\;U$ implied by (4) (about 1.510^{-6} m s^{-1} for $h\;\sim 1000$ m, $U \sim 1$ cm s^{-1}, $r\;\sim 6400$ km) and the best that we can hope for is to compare the consequences of (4) with reality.

4 Theories

4.1 The M Equation

The theories aimed at understanding the interior portion of the large scale thermohaline circulation are presently built upon the planetary vorticity equation (4). At the momentum level, it implies strict (not quasi) geostrophy, the relative accelerations being neglected altogether. Known in the litterature as the thermocline equations, the planetary geostrophic equations or the inertia-less equations (Salmon, 1990), the equations of motions that embody (4) are:

$$
\begin{aligned}
f\,\mathbf{k} \times \mathbf{u} &= -\nabla p/\rho_o && \text{(a)} \\
o &= -p_z - g\rho && \text{(b)} \\
\nabla . \mathbf{u} &= 0 && \text{(c)} \\
\tfrac{DT}{Dt} &= K_v T_{zz} + K_h \nabla_h^2 T && \text{(d)} \qquad (5) \\
\tfrac{DS}{Dt} &= K_v S_{zz} + K_h \nabla_h^2 S && \text{(e)} \\
\rho &= \rho_o (1 - \alpha T + \beta S) && \text{(f)}
\end{aligned}
$$

The temperature equation (5d) is a remnant of the internal energy equation in which compressibility effects and dissipation are neglected and when simple diffusion laws have been assumed for the turbulent property fluxes. A linearized equation of state (5f) is used and when diffusivity coefficients and boundary conditions for temperature and salinity are also the same a single equation for density may be derived from 5(d), (e), (f). In this case Welander (1959) has shown that the set (5) can be transformed to a single equation by setting $p/\rho_o = M_z$, since all dependent variables can then be expressed in terms of M:

$$
\begin{aligned}
\mathbf{u}_H &= 1/f\ \mathbf{k} \times \nabla M_z \\
w &= \beta\, \mathbf{i}\, .\, \nabla M/f^2 \\
\rho/\rho_o &= -M_{zz}/g
\end{aligned}
$$

so that the density equation can also be expressed in terms of M only:

$$
M_{zzt} + \frac{1}{f} J(M_{zz}, M_z) + \frac{\beta}{f^2}\, \mathbf{i}.\nabla M\ M_{zzz} = K_v M_{zzzz} + K_h \nabla_h^2 M_{zz} \qquad (6)
$$

where $\mathbf{i}$ and $\mathbf{k}$ are respectively zonal and vertical unit vectors. Written this way, (6) can be used either with cartesian βplane or spherical coordinates.

When the stratification is given, perturbation solutions can be found from the linearized version of (6) (see Gill, 1985 ; Salmon, 1986, for examples). Note that the only remaining wave mode in (6) is the baroclinic Rossby wave which appears as a balance between the first and third terms on the LHS of (6). The wave is non dispersive because of the neglect of inertia but can be deformed non linearly by stratification.This simplification of the large scale dynamics should be accompanied by a similar task for the boundary conditions and here lies one of the major difficulties that thermocline theorists have struggled with and we come back to that point in Sections 4 and 5. Interest in the M equation focussed initially upon the search for similarity solutions that could describe a thermocline away from solid boundaries and the relevant studies have been summarized by Veronis, 1969. (More recently the complete set of similarity solutions has been published by Salmon and Hollerbach, 1991). The main difficulty with such analytical solutions lies in the impossibility of satisfying top and bottom boundary conditions. A new path was taken by Salmon (1990) who analysed numerical steady state solutions of (6) in very interesting situations that apply to the thermocline in the subtropical gyre away from lateral boundaries and from subpolar density outcrops. Salmon first explores the particular solution $M = xw(z)$, in which case the equation satisfied by w derived from (6) is:

$$w w_{zzz} = \mu \; w_{zzzz} \tag{7}$$

where $\mu = f^2 \; K_v/\beta$.

In this simple case the zonal velocity is zero and the density varies as xw_{zz}. The solution of this one dimensional case is obtained for prescribed vertical velocities and temperatures at upper and lower boundaries. What can be mentioned briefly here is that for subtropical gyre conditions (a downward Ekman pumping, a surface temperature that increases westwards and zero values for temperature and vertical velocity at the bottom), an *internal front* (i.e. a thermocline) of thickness proportional to $\mu^{1/2}$ shows up to separate the warm from the cold abyssal layer. The diffusion, the RHS term in (7), plays a fundamental role to patch together the two outer solutions above and below the front where it is unimportant. Salmon's discovery of the conditions under which the vertical density structure evolves spontaneously toward such a two layer ocean gives weight to the many studies of the wind driven circulation that have been made with this model of the stratification. In more complicated two and three dimensional calculations, Salmon observes that an internal thermocline is still obtained in the steady state, along with a decrease of the horizontal density advection with the vertical mixing coefficient. This mode of generation of a thermocline in the limit of small (but non zero) diffusivity is a very valid alternative to the ideal treatment described in 4.4. It points out already the key role played by turbulent vertical mixing.

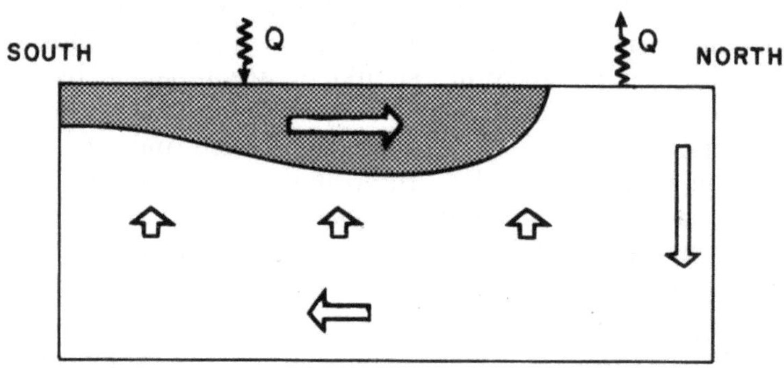

Figure 4: The direct, meridional, assymetric convective cell that transports the excess heat from tropical to polar regions. The shape of the thermocline is shaded.

4.2 Numbers

In view of the difficulties of calculating the "thermocline" that the reader cannot fail to notice in (6), it may be appropriate to pause for a moment and give numbers that reflect the observed thermocline (figure 4) and should guide further thoughts about the physical problem.

After all, we know that there is a thermocline of approximately 1000 m depth in the subtropical gyre that separates the warm and cold water masses. Furthermore the heat fluxes at the surface are known (albeit approximately) and therefore the velocity scales of the convective circulation that must exist to maintain this thermocline may be sought after. In the warm upper layer, the dominant heat balance should be one between meridional advection and surface fluxes giving:

$$\rho C_p \, v \, \frac{\Delta T}{L} \sim \frac{Q}{h}$$

while from continuity $w \sim v\, h/L$.

With $\Delta T = 10°\text{C}$, $h = 1000\,\text{m}$, $L = 4000\,\text{km}$ and $Q = 50\ \textit{watts}\ \text{m}^{-2}$ one obtains $v = 0.6\ \text{cm s}^{-1}$ and $w = 1.2\ 10^{-4}\ \text{cm s}^{-1}$. The meridional velocity is small and represents properly a zonally averaged velocity across the basin. The vertical upward velocity that is required by continuity is fully comparable with the Ekman pumping value indicating that the buoyancy driven part of the circulation is likely to be as important as the wind driven part. If we assume that this upwelling occurs uniformly across the thermocline and must be returned to the deeper layers at a polar boundary, $w \times L^2$ gives the rate of deep water formation or about 20 Sv here (as a way of comparison the formation rate of NADW is believed to be around 10 Sv). Having this velocity estimate is it possible to compute the meridional heat transport carried out by this meridional cell ? Upper and lower branch contribution to the heat flux H coming respectively from above and below the thermocline can be written as:

$$H = \rho\ C_p L\ (v_1 h_1 T_1 + v_2 h_2 T_2)$$

But since when the steady state is reached there can be no mass flux in a closed basin:

$$h_1 v_1 + h_2 v_2 = 0$$

therefore :

$$H = \rho\ C_p\ L v_1 h_1 \Delta T$$

With the previous numbers, H is about 10^{15} watts. This is the order of magnitude of the meridional heat flux, that is computed in the North Atlantic with the help of complete zonal sections of temperature and salinity (Roemmich and Wunsch, 1985). More than anything else, this calculation is shown here to emphasize that it is the circulation in the vertical plane that transports the heat ; horizontal recirculations, that have faster upper layer flows but smaller lateral temperature contrast, are not as important in this respect. However they seem to operate in the Pacific ocean which lacks deep water formation (Bryan, 1962).

The key physical balance that is underlying this global scenario is the one between the interior upwelling and the diffusive turbulent fluxes carrying heat from the surface. It is with this balance in mind that Munk (1966) sought to determine the diffusivity coefficient K_v from observations of vertical profiles of properties (temperature, salinity, carbon 13). If for instance the balance $wT_z = K_v T_{zz}$ becomes valid in a certain depth range, knowledge of T constrains the ratio w/K_v. Estimates of deep water production constrain the spatially averaged w, and the so called "abyssal recipes" give therefore K_v of the order of 1 cm^2s^{-1}. We will see later that this balance holds reasonably well at the base of the thermocline in rather complete numerical calculations of the circulation in regions of heat gain (i.e. away from free convection regions) and the determination of K_v is today one of the major priorities in physical oceanography (see the lectures by A. Gargett in this volume).

The simple reasoning has been essentially two dimensional in a vertical-latitudinal plane and it is worth at this point to examine the consequences of the planetary vorticity balance (4) thought to hold in the interior. Given a generalized upwelling at the base of the main thermocline and zero vertical velocity at the top and bottom, (4) predicts equatorward (poleward) flow respectively above and below the main thermocline. This is the sense of an *indirect cell* oriented against the one that has just been

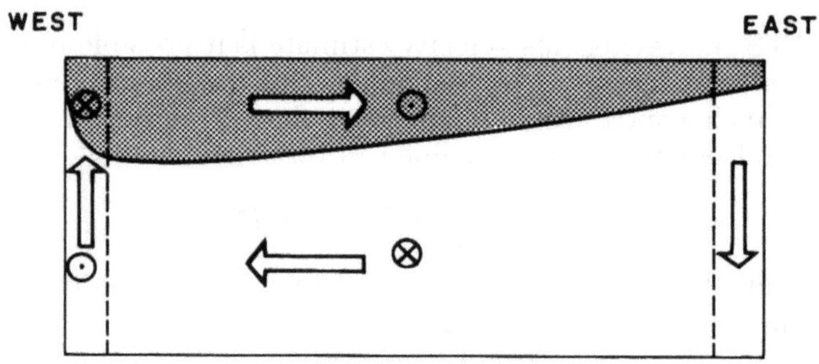

Figure 5: The circulation of the fluid in the zonal plane which adds to Sverdrup type circulations in the interior and closes along the boundaries with the help of enhanced mixing. (Note the shape of the main thermocline that adapts geostrophically to the meridional interior Sverdrup flows).

shown responsible of the heat transport. If we continue to adhere to our assumptions, the elimination of this paradox should come from the fact that the previous meridional velocities in the sense of the direct cell must be interpreted as zonally averaged values and must arise from the boundary layers where the geostrophic constraint (4) can be broken. Furthermore the meridional mass transport in these boundary layers must *exceed* the interior transport at the same depth. By continuity, we see a need for a vertical mass exchange in the boundary layers and a zonal cell in the vertical-longitudinal plane appears quite naturally to close the mass balance (figure 5). It is important to note in the picture the vertical exchange that occurs also at the eastern boundary which now plays a rather active role by accepting and injecting fluid in the interior to be compared with the more passive role that has been assigned to it in the wind driven problem. Note that the upwelling or downwelling layers result from a different local balance of terms of the density equation permitted by enhanced lateral mixing which happens when the flow evolves on a smaller horizontal scale (see Section 5.2.).

It is finally useful to also have ideas about the time scales of such convective large scale cells. The very existence of a main thermocline points out that the heat fluxes at the surface are much larger than those below the main thermocline and that the diffusive heat divergence within the thermocline is balanced out by the poleward advective upper layer flow. The time scale T_u for the advective upper layer flow can therefore be deduced from the given surface fluxes Q as $\rho\, C_p\, \Delta T h/Q$ or about 25 years with the standard values given before. This is relatively rapid compared to the spin up of the deep thermal structure which is carried out by the smaller diffusive fluxes at the base of the main thermocline. The time scale for the abyss is given by the diffusion time scale $T_l = H^2/K_v$ where the standard value of K_v of the order $10^{-4}\mathrm{m}^2\mathrm{s}^{-1}$ indeed results in small deep heat fluxes of order $K_v \rho C_p \Delta T/H$ or about 1 W m^{-2} when an abyssal depth of 4000 m is chosen. The spin up of the deep circulation T_l now reaches 5000 years! The obvious question arises immediately whether the present thermohaline circulation is in steady state. In view of the limited period of observation (the last century) it is impossible to answer such a question with any degree of certainty and while the upper layer thermal structure is probably well adjusted to present day air-sea heat fluxes, the deep temperature and salinity could still be recovering from the Little Ice Age!

4.3 The Abyssal Circulation

It is helpful at this point to present the abyssal circulation schemes of Stommel-Arons (1960) that have had some much influence in shaping up ideas about the flow of deep waters. Given the necessity for a diffusive-advective property balance in the vertical to maintain the thermocline, Stommel and Arons *impose* a vertical upwelling pattern W_s at a *given* depth below the main thermocline. The determination of the deep flow, now becomes completely similar with the determination of the upper layer flow below the wind driven Ekman layer. Assuming a deep ocean of uniform density and constant depth H, the planetary vorticity balance (4) gives: $v = fW_s/(\beta H)$. In cartesian coordinates the continuity equation $u_x + v_y + W_s/H = 0$, allows u to be obtained by zonal integration that naturally proceeds from the east, in which case:

$$Hu(x) \;=\; Hu(x_\epsilon) + \int_x^{x_\epsilon} \left[W_s + \frac{f}{\beta} W_{sy} \right] dx \qquad (8)$$

Therefore the problem is solved entirely provided that the eastern boundary conditions $u(x_\epsilon)$ are known. Stommel-Arons assumed a zero velocity there but this is not necessarily the case. Baroclinic boundary layers can exist at the eastern boundary to supply fluid to the interior as in the sketch of figure 5, and this is a point to which we will have to come back later. Since Stommel-Arons did not consider such vertical fluid exchange along lateral walls, the quantity of water upwelled in the interior should reappear as modified deep water *specified* at some location along the poleward boundary of the model basin. As a consequence of these assumptions a mass imbalance that occurs at the western boundary (if Hu in (8) is not zero at the western wall) must be treated by the appearance of western boundary currents (WBC) that break (4) and redistribute waters *horizontally* across different latitudes. It is then a simple water budget calculation to find the WBC transport at each latitude. The simplest abyssal Stommel-Arons picture consists therefore of equatorward WBC with poleward interior recirculation directed *towards* the source of deep water. While Stommel-Arons's predictions of several deep western boundary currents before their discoveries is a sufficiently rare event in oceanography to be noted (Warren, 1981), it is more difficult to be certain of the existence of their poleward interior flow of abyssal waters. The difficulty lies partly with the number of deep water masses that move meridionally at various density range with sources in the Northern or Southern hemisphere (see figure 2) and partly with the complicated topography of the ocean floor that introduces many new effects besides acting as a noise generator at small scales. However indirect evidence from geochemical tracers now exists for several deep basins (see for instance Warren and Speer, 1991).

The Stommel-Arons calculation remains of limited predictive quantitative value because 1) the horizontal structure of the upwelling function W_s is not known observationally and cannot be deduced simply from the surface heat fluxes and 2) the possible role of the baroclinic side wall layers should be considered.

4.4 The Ventilated Thermocline

Moving up from the abyss to the warm upper layer, the upper branches of the thermohaline circulation interact nonlinearly with the wind driven flow. While the latter has been studied in isolation in the past, one important building block of circulation theory that must be mentioned is the so called problem of the ventilation of the thermocline initiated by Luyten, Pedlosky and Stommel (1983) and followed by rather numerous other calculations reviewed recently by Huang (1991). Although originally presented as a study of the wind driven response, it is clear that implicit buoyancy fluxes are present and that the calculation has much broader scope than thought originally. The ocean envisioned by LPS is divided into a number of immiscible layers of constant density (so that diffusive fluxes vanish), and the geostrophic response to steady Ekman pumping is found to cure some problems of the linear wind driven flow theory. The linear stratified wind driven spin-up problem studied by Anderson and Gill (1975) had shown that each westward propagating Rossby mode sets up in its wake its own Sverdrup flow and that, if one waits long enough, the wind driven interior flow would vanish but in the uppermost directly forced layer. In the absence of diffusive fluxes or lateral injection from a boundary layer, the layers which are not exposed to any direct forcing, must become motionless in the steady state. This can be seen readily by vertical integration of (4) between two density interfaces separated by a thickness h to obtain:

$$\boldsymbol{u} \, . \, \boldsymbol{\nabla}(f/h) = 0 \tag{9}$$

Given that h remains finite and that the normal velocity component vanish at the eastern boundary, (9) can be integrated along the characteristics (the potential vorticity contours f/h) that emanate from that boundary to yield zero horizontal velocity everywhere. Faced with this difficulty of generating motions at depths, LPS proposed to relax the assumptions of small layer thickness variations to allow h in (9) to vanish at specified locations called outcrops of a given density surface, a situation that mimics the observation of the large scale density field in the winter season. To obtain a tractable analytic solution, they assumed furthermore the ocean to be made of only two or three active layers above one at rest, decoupling thereby the upper thermocline from the deep circulation. When a given outcrop lies south of the line of zero wind stress curl, fluid is now pumped from the Ekman layer into the geostrophic layer below which moves southward and subducts below the upper exposed layer when it reaches the next outcrop of a shallower density surface. The mechanism provides for direct injection of young water masses of low potential vorticity that "ventilates" the upper thermocline. The major result of the theory is that the ventilation cannot occur right next to the eastern boundary. Again (9) provides the reason. As in the abyssal circulation theory, the vorticity equation is naturally integrated from the east. Now if a southward motion of subducted fluid occurs along the boundary, the variation of f in (9) implies a variation of h associated necessarily under such geostrophic dynamics with a non zero normal flow at the boundary. To alleviate this difficulty, LPS showed 1) that the subducted fluid moves southwards and westwards (i.e. away from the coast) leaving an unventilated "shadow zone" along the boundary and 2) that to satisfy the zero normal velocity, each layer below the top one must surface along the *eastern boundary.* This latter consequence is highly artificial of course, and shows that in such baroclinic problems the geostrophic constraint needs to be broken not only at western but also at eastern boundaries. The interior geostrophic regime has to give room to a coastal ageostrophic regime in which transport between layers through boundary mixing is permitted. Clearly possibilities exist for such lateral boundary ventilation of the interior as already alluded to in the previous section. This possibility is observed in numerical models (Section 5), but is not easily formulated analytically due to the non linear coupling between the geostrophic interior and the coastal boundary layers. Observations can play a role here to provide these "coastal" rates of water mass transformation associated with boundary flow. Given the observed values of layer thicknesses at the eastern boundary, the LPS type theory is easily modified to obtain the geostrophic interior consistent with the non-zero velocities just outside such coastal boundary layers.

Cast originally in a wind driven context, the reference to air-sea surface forcing is implicit in the LPS theory. Specification of the density outcrops at the outset implies in fact strong air-sea forcings. Furthermore, the constant injection of the upper layer fluid into exposed layers requires some diffusive sinks (neglected by LPS) somewhere to compensate and allow water masses to reach constant volume in a steady state. The ventilation in the LPS theory proceeds at the rate of the Ekman pumping and the observation of the oceanic distribution of tracers such as tritium-Helium 3 reveals much higher values of ventilation. Quite clearly the heat flux has something to do with this and a number of authors have included the *lateral* injection of fluid following deep winter convection events from deep mixed layers as a more efficient ventilation process (see Nurser and Marshall, 1991; Williams, 1989).

After the presentation of these major ideas on the dynamics of upper layer ventilation and abyssal flow, it remains to look at the role that numerical models may have to test them and cast them in a more realistic geophysical context.

5 Numerical Models

Ultimately the need for prediction of the future ocean state becomes more pressing and one must turn to numerical models for answers. Testing the stability of the present ocean climate to various changes in the forcing requires first that the present circulation be reproduced to a certain degree of accuracy. Because the surface forcing has only been known (albeit with large errors in the Southern Ocean for instance) for a few decades, which represent a small fraction of the adjustment time of the thermohaline circulation, it becomes obvious that the search for the present ocean state (interior and boundary conditions) must supplement the forcing data with those of the ocean interior whose $T-S$ properties are known. It is the important task of inverse modelling to find out the best circulation compatible with these $T-S$ properties and dynamical constraints that look plausible. This is not reviewed here and the priority is given instead to the description and use of models for long term predictions, the so-called prognostic models. The most widely used are those based upon the primitive equations (PE) (Cox-Bryan, 1984) whose major assumptions are hydrostatics, incompressibility, Boussinesq approximation and rigid lid. Incompressibility filters out the sound waves and the rigid lid suppresses the external gravity (or tidal) waves. The wave modes that remain include both the fast internal gravity modes and the slow Rossby modes (see figure 6). The former have maximum phase speeds NH/π which amounts to 3-5 m s^{-1} when realistic values for the (uniform) Brunt-Väisälä frequency are used. This is about an order of magnitude in excess of the flow speed observed at scales above the mesoscale (100 km), indicating that the internal gravity waves will dominate the numerical stability criteria of PE models, severely limiting the time steps of integration to "hours" at a 100 km resolution. Because the thermohaline circulation problems are of a global nature and require centuries long integration, this short time step limits the spatial resolution that can be used and one of the central issues in thermohaline circulation modelling becomes whether the mesoscale energy-containing eddies should be resolved. In the atmosphere, the answer is definitely positive because processes such as baroclinic instability have powerful controls upon the mean heat balance for instance. However the larger scale of the atmospheric eddies and the shorter equilibration time scales due to the smaller thermal inertia of the air indicate a lighter computational burden than for the ocean. There are regions in the ocean which are dynamically analogous to the atmosphere, such as the Antarctic circumpolar current, where very similar processes have been shown to occur (see the lectures by D. Olbers in this volume). The ocean eddies are also known to play a crucial role in driving the inertial recirculations of the Gulf Stream which are also regions where the major cooling of the North Atlantic gyre occurs. The problem is that to run a 10 km eddy resolving PE ocean model of 4000 $\times$ 4000 km for 100 years requires say 4 $\times$ 10^5 steps over 400 $\times$ 400 $\times$ 20 grid points which at 10 μsec per grid point on a Cray demands 5 months of present computer time. Such computational efforts are being undertaken at a number of laboratories and there is no doubt that the gradual increase of the speed of computing machines will make this approach more feasible.

In the mean time the long natural time scales of the thermohaline circulation point out filtering alternatives in the equations of motion to suppress high-frequency internal gravity waves. One way to do this, as in the set of equations (5), is to neglect entirely the relative accelerations in the momentum equations so that the dynamics reduce to geostrophy and planetary vorticity balance, suppressing internal gravity waves and part of the Rossby waves (see figure 6).

When the density surfaces have large vertical excursions h over the horizontal scale L, the thermal wind equation scales horizontal velocities as N^2h^2/fL so that the small Rossby number implied by the geostrophic approximation requires a small Burger number N^2h^2/f^2L^2 as well. That is the errors involved after such a filtering are likely to be small if the geostrophic models using these equations are run at coarse resolution well in excess of the internal Rossby radius deformation scale. The various models using this planetary geostrophic approximation differ by the choice of their frictional operators. Classical Laplacian friction was used by Colin de Verdiére (1989), while Rayleigh friction, used initially in such modelling by Killworth (1985), was shown later to be incompatible with hydrostatics by Salmon (1986) who proposed a better method of solution (see 5.2.). It is important to note that this filtering approximation does not suppress the important baroclinic instability process that remains in these equations as the long wave Green modes of the full problem (Colin de Verdière, 1986). Because the growth rate of the inviscid modes is unbounded at high wavenumbers the instability calculation shows that the addition of friction is a necessary condition for using these type of equations in ocean models. There are other less drastic choices, such as balance models that retain time derivatives in the vertical vorticity equation but not in the horizontal divergence equation that have been proposed and are being explored for the study of the thermohaline circulation at mesoscale resolution (Gent and McWilliams, 1983).

5.1 Eddy Parametrization

Run at coarse resolution, (coarse meaning here at scales above the internal Rossby radius deformation scale), the thermohaline circulation models must parametrize the mesoscale eddy terms. The need for conduction terms in the heat equation is of course vital to the existence of a large scale convection cell that could not be sustained without a turbulent field transporting temperature anomalies down to smaller scales where molecular diffusion occurs.

In regions of stable stratification the parametrization is frequently one of classical diffusion for temperature and salinity with constant diffusion coefficients K_v of order $10^{-4}\mathrm{m}^2\mathrm{s}^{-1}$, K_h of order $10^3\mathrm{m}^2\mathrm{s}^{-1}$ in the vertical and horizontal directions respectively, values which are inferred from the turbulent fields responsible for the mixing. It has been suggested for a long time, however, that the mixing of temperature and salinity anomalies happens mostly along mean density surfaces rather than across it. This has led to new parametrizations of thermohaline mixing along isopycnals (see Gent and McWilliams, 1989 and references there in) which must be supplemented by prescriptions of the diapycnal mixing across isopycnals that take over the role of the vertical mixing in producing the buoyancy circulation.

When the density profile becomes gravitationally unstable the situation is different: free convection occurs and the way to represent it in large-scale models is very ad-hoc. At a given time step temperature and salinity are mixed vertically and instantaneously until neutral conditions are restored while conserving the heat and salt present initially in the water column.

The difficulty of fitting lateral boundary conditions to the geostrophic thermocline equations has already been mentioned in Section 4. Suppose that you insist on pure geostrophy to describe the dynamics up to solid boundaries. With the vanishing of normal velocities on solid boundaries, the pressure must then be uniform along the periphery of the domain, but when this is so, the net meridional velocity zonally averaged across the basin vanishes and therefore no large scale convection cell can be supported! The same conclusion follows from the energetic argument of Section 2 that indicates that buoyancy torques require frictional eddy torques for equilibrium under steady state conditions and the neighborhoods of solid boundaries are likely sites where the frictional eddy terms can break the geostrophic balance.

To show that at the suspected rate of mixing of momentum the large scale interior is rather inviscid, write out the steady vertical vorticity balance including the advection of relative vorticity and eliminate the vertical velocity with the help of the steady density equation to obtain equation 10:

$$\begin{array}{ccccccccc} \mathbf{u}.\nabla q & = & - \; f\mathbf{u}_{hz}.\nabla\rho & - & \rho_z\mathbf{u}.\nabla\xi & + \; f\frac{\partial}{\partial z} & (K_h\nabla_h^2\rho & + \; K_v\rho_{zz}) & - \; A\rho_z\nabla_h^2\xi \\ & & \frac{U}{fL} & & \left\{\frac{U}{fL}, \frac{U}{\beta L^2}\right\} & & \frac{K_h}{UL} & \frac{K_v}{wh} & \left\{\frac{A}{\beta L^3}, \frac{A}{fL^2}\right\} \\ 0\,(1) & & 10^{-4} & & 10^{-3} & & 10^{-1} & 10^{-1} & 10^{-4} \end{array}$$

where ξ is the vertical component of relative vorticity and A the lateral friction coefficient, q represents the large scale potential vorticity $f\ \rho_z$ and the RHS terms represent different sources and sinks. Indicated below (10) are the orders of magnitude of these terms when the following interior scales are chosen: $U \sim 2$ cm s^{-1}, $L \sim 500$ km, $h \sim 1000$ m, $w \sim 10^{-4}$cm s^{-1}, $K_v = 1$ cm^2 s^{-1}, $A = K_h = 10^7$cm^2 s^{-1}. It appears that the potential vorticity PV q is indeed approximately conserved, the limiting factors being the diffusion terms measured by Peclet numbers UL/K_h, wh/K_v which are of order 10. This mid-latitude scaling argument points out that the eddy stirring of temperature and salinity is potentially the most important factor to consider in large scale models to provide for sources or sinks of large scale PV. Friction on the other hand is shown to have a smaller influence at these scales. A side benefit of the calculation indicates that the relative advection of vorticity may easily reach order one for such semi-permanent jets as the offshore branches of the Gulf Stream, which have scales comparable to the mesoscale turbulence. They may be seen as mesoscale *internal boundary layers* and taking them into account is not possible in planetary geostrophic models but requires full mesoscale resolution.

5.2 Boundary Layers

It is the size of the boundary layers that dictates the minimum value of frictional coefficients to use in coarse resolution models. There is a choice to be made between classical Laplacian or Rayleigh friction. The former involves an operator that is scale dependent that may keep the larger scales relatively inviscid while the latter imposes the same friction at all scales. Referring to Salmon's (1986) paper for Rayleigh friction boundary layers, the case of Laplacian friction used in both PE and some geostrophic models is considered here.

The barotropic boundary layer width δ_M for the vertically integrated transport of horizontal momentum is that of Munk (1949) and is obtained from (10) by requiring the last term on the RHS to be order one. Non dimensionally:

$$\delta_M = E^{1/3}(R_E/L)^{1/3}$$

where E is the horizontal Ekman Number $A/f\ L^2$ and R_E is the earth's radius. This $E^{1/3}$ layer imposes a value of A of 10^9 cm^2 s^{-1} given an horizontal resolution of 200 km and using, 2 or 3 grid points to resolve the barotropic boundary current. Such a coarse resolution requires large frictional coefficients well above the suspected geophysical value at the mesoscale.

The constraints in the baroclinic case (for Laplacian friction) are less well known in large scale modelling, although Pedlosky (1969) gives a relevant discussion in the presentation of his linear ocean calculations. Note first that for the horizontal Ekman flow (tangential and normal velocities) to vanish along vertical side walls requires friction to be important at the level of the momentum equation implying a boundary layer width of order $E^{1/2}$ thinner than the Munk layer since E is much smaller than unity. The resolution of these $E^{1/2}$ layers would require $A \sim 4\ 10^{10}$ cm^2 s^{-1} for the same 200 km grid size and with this high value of A the conservation of the large scale interior PV in equation (10) would be in doubt. One solution is to reduce the boundary requirements and to find out first the width of the boundary layer over which the normal velocity only is forced to vanish. In the inviscid Euler equations of classical fluid mechanics, no friction is required to impose the kinematic requirement of no normal

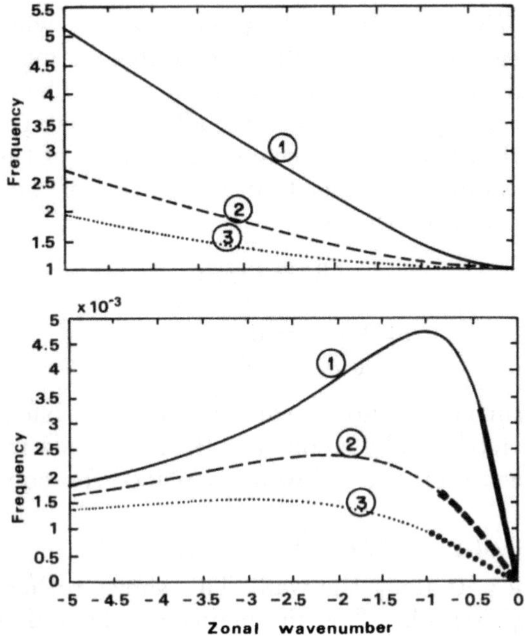

Figure 6: The dispersion diagram for baroclinic oceanic waves: The top panel represents fast internal gravity waves and the bottom panel the slow Rossby waves. Frequency is scaled by the Coriolis parameter at 30°N and wavenumber by the first mode internal Rossby radius. The first three baroclinic modes are shown in each case. Note the three orders of magnitude difference in frequency. The thick parts of the curves in the lower panel indicate the remaining wave modes in planetary geostrophic or thermocline modelling.

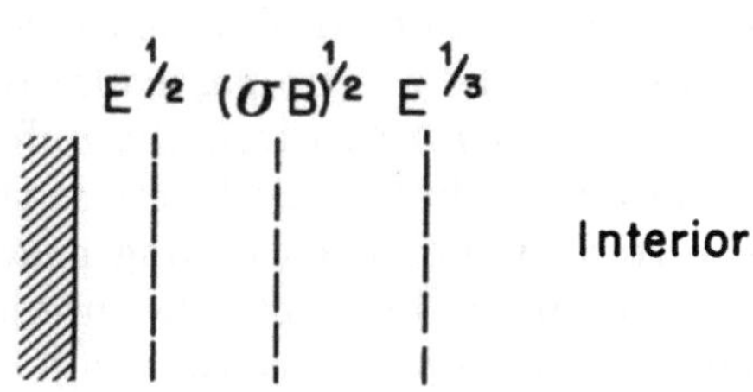

Figure 7: The different boundary layer widths which occur in the study of the baroclinic oceanic circulation.

velocity at a wall. The situation is different, however, when the inertial terms have been neglected as the following simple example will demonstrate. Assume a solid boundary along the y axis. The along-the-wall velocity v is kept geostrophic while friction allows to balance the Coriolis force associated with u velocities. Because the x scale of the flow normal to the boundary is so much smaller than the y scale of the flow along it, the equations of motion reduce to:

$$\begin{aligned} fv &= p_x/\rho_o \\ fu &= A\ v_{xx} \\ u_x + w_z &= 0 \\ w\ \rho_z &= K_h\ \rho_{xx} \end{aligned}$$

When density perturbations are linearized around a background mean density profile characterized by a constant Brunt Väisälä frequency N (f is also assumed constant), the following Laplace-type equation for u is obtained:

$$u_{zz}\ +\ A/K_h\ N^2/f^2\ \ u_{xx}\ =\ 0$$

to be solved with $u = 0$ along the wall. This shows rather directly that given an interior geostrophic flow impinging on the wall the condition of no normal flow gives rise to a baroclinic boundary layer width given non-dimensionally by:

$$\delta_{bc}\ =\ (\sigma\ B)^{1/2}$$

where the Burger number B is N^2h^2/f^2L^2 and σ is the horizontal Prandtl number A/K_h. In coarse resolution situations where the Burger number is small, the resolution of this baroclinic layer imposes a large horizontal Prandtl number. An added degree of freedom is found here that allows the satisfaction of the zero normal velocity requirement by adapting the value of σ, without having to resolve the $E^{1/2}$ layer that remains the width over which the tangential velocity vanishes. There is no guarantee, however, that the solution so obtained is a good description of the geophysical situation for there is no particular observational support for a turbulent Prandtl number σ substantially different from one. The nice feature of these baroclinic boundary layers is the possibility of vertical circulations permitted by enhanced mixing near the boundaries. These can in turn inject the mixed fluid into the interior and provide for other causes of ventilation of deep water recirculation (see Section 4). The existence of these boundary layers recycling waters in the vertical reveals the degree of arbitrariness of analytical solutions of the abyssal circulation or thermocline ventilation. The method for coping with this extra fluid pathway in theories is not clear at the moment as is borne out by the fact that it is difficult to couple the stratification in such regions to the interior.

Finally there are also important computational requirements to follow when choosing an eddy parametrization that depends on the numerical scheme which is used to solve the advection-diffusion equation (5d, 5e). As an example, centered differencing in the vertical has often caused a problem when the Peclet number $w\ \Delta Z/K_v$ based on the largest grid spacing and largest vertical velocity is allowed to become larger than one.

5.3 Results

Only the numerical simulations that have been carried out in closed square basins subject to idealized buoyancy and wind forcings will be described here. The models used have been the classical PE model and the less expensive planetary geostrophic models. The objective here is not to describe the details of such solutions (see Cox-Bryan, 1984 ; Colin de Verdière, 1989) but to comment on some of their salient features that relate to the general ideas presented above. After a few thousand years a typical model response to a two gyre wind pattern and buoyancy forcing (described briefly in figure 8) shows among other things:

1. The asymmetry between the subtropical and the subpolar gyre. The buoyancy forcing intensifies the subtropical gyre and replaces the subpolar gyre by a broad drift to the north-east.
2. The abyssal flow pattern with its strong southward deep WBC and the weaker poleward interior flow.
3. The presence of an interior vertical upwelling albeit non uniform at the base of the main thermocline.
4. The existence of zonal cells closing in baroclinic boundary layers in addition to the direct meridional cell.
5. The presence of a main thermocline in the subtropical gyre shallowing towards low latitude.

At coarse resolution it is possible to run many sensitivity simulations varying parameters one at a time and the greatest sensitivity is, not surprisingly, that due to the vertical mixing coefficient (see K. Bryan (1991) for a review of these sensitivity studies). When it increases, the heat gain can penetrate further downwards, the thermocline deepens and the buoyancy driven part of the circulation intensifies while the upper wind driven response becomes distributed over a wider depth range. Simple scaling arguments help in understanding what is happening. Consider first the

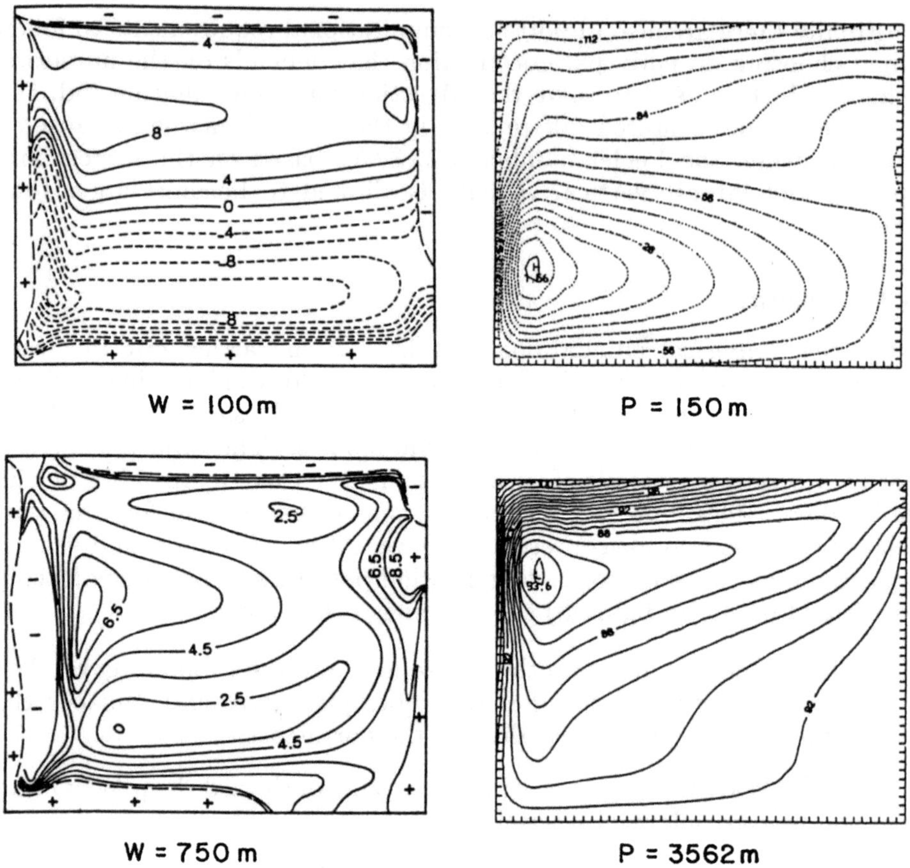

Figure 8: The surface and deep geostrophic pressure fields in a combined wind and buoyancy driven oceanic numerical simulation (from Colin de Verdière, 1989). In spite of the two gyres wind forcing, only the subtropical gyre shows up in the response. The vertical velocity shows the familliar Ekman pumping near the surface and an interior upwelling at the base of the thermocline whose net effect is to increase (decrease) the squashing of upper layer fluid columns in the southern (northern) regions respectively. In the abyss weak poleward interior flows and fast equatorward deep WBC stand out.

wind driven problem for an adiabatic situation. Given the vertical Ekman pumping velocity, planetary vorticity (4) and density conservation allow to derive a velocity scale U and a depth scale δ_w as:

$$U \sim f\ w_E/\delta_w\beta$$

and

$$w_e\ N^2 \sim \boldsymbol{u}_H \,.\, \boldsymbol{\nabla}\ g\rho'/\rho_o \sim f\ U^2/\delta_w$$

so that

$$U = w_E^{2/3}\left(\frac{N^2}{\beta}\right)^{1/3} \quad \text{and} \quad \delta_w = \left(\frac{U}{\beta}\right)^{1/2}\frac{f}{N}$$

This shows readily that when N is lowered, the wind driven response penetrates over a larger depth range as measured by δ_w. Determination of the vertical structure of the wind driven circulation has been the subject of intense studies (Rhines and Holland, 1979) in which this vertical scale has played an important role.

Appropriate scaling arguments have been provided for the buoyancy circulation by Welander (1971). Suppose that what is given instead, is an external temperature gradient at the sea surface. The new velocity scale U and depth δ_b may be related as before, using thermal wind and planetary vorticity balance:

$$\beta U \sim f\ \frac{w}{\delta_b\beta}$$

$$f\ \frac{U}{\delta_b} \sim \frac{g}{\rho_o}\frac{\Delta\rho}{L}$$

However an additional constraint is needed since w is unknown. It must come from the energy equation that relates rates of heating and mass fluxes. In regions of heat gain, it has been argued (section 4.2) that the balance that controls this is one between vertical advection and vertical diffusion holding at the base of the thermocline so that:

$$\delta_b \sim K_v/w$$

Then :

$$\delta_b = \left(\frac{K_v}{\beta}\right)^{1/3} B^{-1/3}$$

$$U = f\left(\frac{K_v}{\beta}\right)^{1/3} B^{2/3}$$

where $(K_v/\beta)^{1/3}$ is a vertical length scale while $B = (g\ \Delta\rho)/(\rho_o\ Lf^2)$ is a modified Burger number. Such a scaling explains well the sensitivity

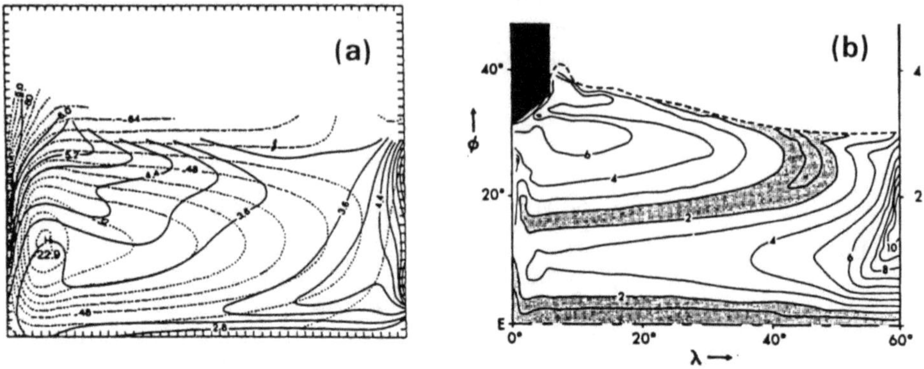

Figure 9: The potential vorticity contours along a given density surface within the main thermocline from planetary geostrophic (a) and primitive equations (b), two simulations which compare very favorably, Colin de Verdière, 1989 and Cox-Bryan, 1984, respectively. (Dashed contour in (a) indicate the Bernouilli function which gives the flow fields along density surface).

of the buoyancy circulation runs to the vertical mixing coefficient. It shows well the problem that thermohaline circulation modelers are facing since the value of the vertical mixing coefficient appropriate for large scale flows suffers from major uncertainties. If a constant coefficient is chosen for simplicity the results of the simulations produce reasonable heat fluxes and deep water production when K_v is around 1 cm^2s^{-1}. With such values, it was found in planetary geostrophic simulations (Colin de Verdère, 1989) that the vertical advection-diffusion balance held well in the interior providing a maximum upward velocity at the base of the main thermocline. In view of the strong control exerted by this balance upon the thermohaline circulation, it is surprising that such a balance has not yet appeared from the more complete PE models results.

It is finally appropriate to present a comparison of the ventilated thermocline as seen from planetary geostrophic and PE models. Figure 9 shows large similarities between the PV mapped on a given isopycnal in the main thermocline. Three sources of PV may be readily identified: the low PV tongue coming from northern density outcrops, the high PV tongue injected by the WBC and finally a high PV tongue that emanates from the eastern boundary on the southern side of the shadow zone of

LPS. This last injection is allowed by the non trivial structure of the eastern baroclinic layer described previously. Such comparisons are encouraging to point out the relevance of PG models at large scales. Their economy and simplified dynamics are certainly an asset for thermohaline circulation studies. However to exploit their full potential in global problems, it remains to be seen how well they perform at the equator. While in principle there is no dynamical difficulty to run them in basins straddling the equator, the coarse resolution of such models may yield an unrealistic exchange pattern across the equator, by doing too much harm to the small meridional scales of equatorial flows.

The more natural avenue of future research for PE models is the explicit consideration of mesoscale eddies. Models of North Atlantic basin size have already been run at 1/3° resolution for comparison with the coarser resolution models (Cox, 1985). The important result is that there is little difference in the global heat transport whether the eddies are present or not. Most of the heat transport is still being carried by the large-scale advective meridional cell. While eddies influence local patterns of the mean circulation through rectification mechanisms, the rectified flows oppose the explicit eddy transports leaving a picture that is not at this time very different from the coarse resolution simulation. However the great benefit of finer and finer resolution models will be the possibility of a much better parametrization of water mass transformations in regions of cooling and along coastal boundary layers, transformations that happen necessarily at small scales where the constraint of the earth rotation weakens.

Acknowledgments

I wish to thank K. Speer for many interesting comments and suggestions made on an earlier draft of this paper and J. Le Gall for the organisation and the typing of this manuscript.

References

ANDERSON D.L.T., GILL A.E., 1975. Spin up of a stratified ocean with applications to upwelling. *Deep Sea Res.*, **22**, 583-596.

BROECKER W.S., DENTON G.H., 1989. The role of ocean-atmosphere reorganisations in glacial cycles. *Geochim. et Cosmochim. Acta*, **53**, 2465-2501.

BRYAN K., 1962. Measurements of meridional heat transport by ocean currents. *J. Geophys. Res.*, **67**, 3403-3414.

BRYAN K., 1991. Poleward heat transport in the Ocean. A review of a hierarchy of models of increasing resolution. *Tellus*, **43**(AB), 103-115.

COLIN DE VERDIÈRE A., 1986. On mean flow instabilities within the planetary geostrophic equations. *J. Phys. Oceanogr.*, **16**, 1981-1984.

COLIN DE VERDIÈRE A., 1988. Buoyancy driven planetary flows. *J. Marine Res.*, **46**, 215-265.

COLIN DE VERDIÈRE A., 1989. On the interaction of wind driven and buoyancy driven gyres. *J. Mar. Res.*, **47**, 595-633.

COX M.D., 1985. An eddy resolving numerical model of the ventilated thermocline. *J. Phys. Oceanogr.*, **15**, 1322-1324.

COX M.D., BRYAN K., 1984. A numerical model of the ventilated thermocline. *J. Phys. Oceanogr.*, **14**, 674-687.

DUTTON J.A., 1986. The ceaseless wind. An introduction to the theory of atmospheric motion. *Dover Publ. Inc. New-York.*

GENT P.R., MCWILLIAMS J.C., 1983. Regimes of validity for balanced models. *Dyn. Atmos. Oceans,* **7**, 167-183.

GENT P.R., MCWILLIAMS J.C., 1989. Isopycnal mixing in ocean circulation models. *J. Phys. Oceanogr.*, **20**, 150-155.

GILL A.E., 1982. Atmosphere-Ocean dynamics. *Academic Press, 662 p..*

GILL A.E., 1985. An explicit solution of the linear thermocline equations. *Tellus,* **37**A, 276-285.

ISEMER H.J., HASSE J.L., 1985. The bunker climate atlas of the North Atlantic ocean. *Springer Verlag.*

HIGNETT P., IBBETSON A., KILLWORTH P.D., 1981. On rotating thermal convection driven by non uniform heating from below. *J. Fluid. Mech.,* **109**, 161-187.

HUANG R.X., 1991. The three - dimensional structure of wind-driven gyres: ventilation and subduction. *Rev. Geophysics. Suppl.,* 590-609.

KILLWORTH P.D., 1985. A two-level wind and buoyancy driven thermocline model. *J. Phys. Oceanogr.*, **15**, 1414-1432.

LUYTEN J.R., PEDLOSKY J., STOMMEL H., 1983. The ventilated thermocline. *J. Phys. Oceanogr.*, **13**, 259-309.

MAROTZKE J., WILLEBRAND J., 1991. Multiple equilibria of the global thermohaline circulation. *J. Phys. Oceanogr.*, **21**, 1372-1385.

MUNK W.H., 1950. On the wind driven ocean circulation. *J. Meteor.*, **7**, 79-93.

MUNK W.H., 1966. Abyssal recipes. *Deep Sea Res.*, **13**, 707-730.

NURSER A.J., MARSHALL J.C., 1991. On the relationship between subduction rates and diabatic forcing of the mixed layer. *J. Phys. Oceanogr.*, 1793-1802.

OLLITRAULT M., 1992. Deep float dispersion in the North East Atlantic. *To be submitted to Deep Sea Research.*

PEDLOSKY J., 1969. The linear theory of the circulation of a stratified ocean. *J. Fluid. Mech.*, **25**, 185-205.

POINCARÉ H, 1910. Théorie des marées. *Leçon de mécanique céleste.* **3.** Gauthiers Vilars Paris.

RHINES P.B., HOLLAND W.R., 1979. A theoretical discussion of eddy driven mean flows. *Dyn. Atm. and Oceans,* **3**, 289-325.

ROEMMICH D., WUNSCH C., 1985. Two transatlantic sections: meridional circulation and heat flux in the subtropical North Atlantic ocean. *Deep Sea Res.,* **32,** 619-664.

ROSSBY H.T., 1965. On the thermal convection driven by non uniform heating from below. An experimental study. *Deep Sea Res.,* **12,** 9-16.

SALMON R., 1986. A simplified linear ocean circulation theory. *J. Mar. Res.,* **44,** 695-711.

SALMON R., 1990. The thermocline as an internal boundary layer. *J. Mar. Res.,* **48,** 437-469.

SALMON R., HOLLERBACH R., 1991. Similarity solutions of the thermocline equations. *J. Mar. Res.,* **49,** 249-280.

SCHMITT R.W., BOGDEN P.S., PORMAN C.E., 1989. Evaporation minus precipitation and density fluxes for the North Atlantic. *J. Phys. Oceanogr.,* **19,** 1208-1221.

SPEER K., WHITEHEAD J.A., 1988. A gyre in a non uniformly heated rotating fluid. *Deep Sea Res.,* **35,** 1069-1077.

SPEER K., TZIPERMAN E., 1992. Rates of water mass transformation in the North Atlantic Ocean. *J. Phys. Oceanogr.,* **22,** 93-104.

SPEER K., 1992. Personal communication.

STOMMEL H., ARONS A.B., 1960. On the abyssal circulation of the world ocean. Part 2. *Deep Sea Res.,* **6,** 217-233.

STOMMEL H., 1961. Thermohaline convection with two stables regime of flow. *Tellus,* **13,** 224-230.

VERONIS G., 1969. On theoretical models of the thermocline circulation. *Deep Sea Res.,* **16** (suppl.), 301-323.

WARREN B.A., 1981. Deep circulation of the world ocean. In: Evolution of Physical Oceanography edited by B.A. Warren and C. Wunsch. *The MIT Press, Cambridge, Mass.*

WARREN B., SPEER K.G., 1991. Deep circulations in the eastern South Atlantic. *Deep Sea Res.,* **32,** suppl. n° 1A 281-322.

WELANDER P., 1959. An advective model of the ocean thermocline. *Tellus,* **11,** 309-318.

WELANDER P., 1971. The thermocline problem. *Phil. Trans. R. Soc. Lond.,* **A 270,** 415-421.

WELANDER P., 1986. Thermohaline effects in the ocean circulation and related simple models. In "Large scale transport processes in oceans and atmosphere" *(eds. J. Willebrand and D.L.T. Anderson) D. Reidel Publ. Co., pp. 163-200.*

WILLIAMS R.G., 1989. Influence of air sea interaction on the ventilated thermocline. *J. Phys. Oceanogr.,* **9,** 1255-1267.

WORTHINGTON L.V., 1981. The water masses of the world ocean : some results of a fine scale census. In "Evolution of Physical Oceanography" Eds. B.A. Warren and C. Wunsch. *The MIT Press, Cambridge Mass.*

PARAMETERIZING THE EFFECTS OF SMALL-SCALE MIXING IN LARGE-SCALE NUMERICAL MODELS

ANN E. GARGETT
Institute of Ocean Sciences
Sidney, B.C., Canada

Abstract

This paper first reviews some results from the field of small-scale turbulence measurement. Features which are of particular importance to the ways in which the effects of turbulent mixing are incorporated in large-scale numerical ocean models are the following: the relation between mixing and "mean" shear, the potential for variable mixing efficiency, and the possibility that diffusivities differ for heat (temperature) and salt. The next section reviews the process(es) of deriving values for the constant diffusivity coefficients which are presently used in most numerical ocean models, while a final section explores effects of using instead variable diffusivities, modelling these roughly on results of the past decade of work on ocean microscales.

1 Introduction

It is rare to find an account of efforts to build numerical models of the ocean for climate prediction purposes without some mention of the importance of small-scale mixing of mass, represented as a vertical "eddy" diffusivity K_ρ. Major model features such as water mass properties, thermocline depth, and thermohaline transport are extremely sensitive to the value chosen for K_ρ. Moreover, while increased computer power may soon lessen the need for the large horizontal diffusivities which presently represent the effects of the geostrophic eddy scales, we will never make global computations down to the ocean microscale. The observed sensitivity of models to parameters which represent the effects of these microscales, combined with the lack of a brute strength computational solution, suggests that we must continue to work on improving parameterizations of the effects of ocean microscales in large-scale numerical models. To do so, we must continually attempt new syntheses of the results of ocean observations, laboratory studies, and associated theory, assessing the results in numerical test cases whose results under standard assumptions are well known.

NATO ASI Series, Vol. I 11
Modelling Oceanic Climate Interactions
Edited by J. Willebrand and D. L. T. Anderson

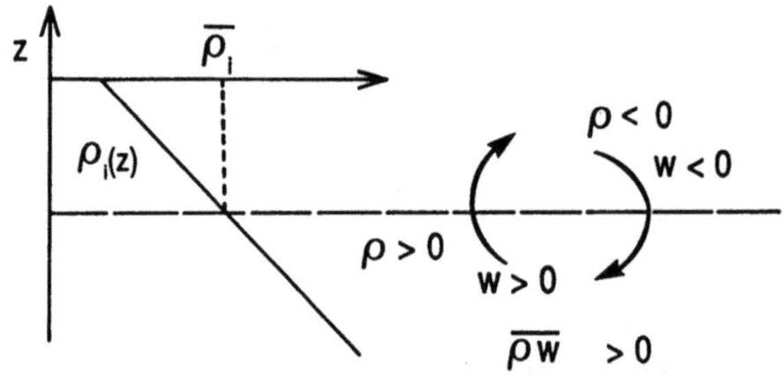

Figure 1: Schematic of the vertical transport of species, here fluid density, by small-scale turbulence in the stably stratified interior of the ocean. $\rho = \rho_i - \overline{\rho_i}$, where ρ_i is instantaneous and $\overline{\rho_i}$ time-average *in situ* density.

2 Microscale Observations and "Conclusions"

A distinguishing characteristic of turbulence is its ability to transport properties irreversibly from regions of high property concentration to regions of low property concentration. Schematically, Fig. 1 shows transport of density anomaly $\rho = \rho_i - \bar{\rho}_i$, where ρ_i is instantaneous and $\bar{\rho}_i$ time-average *in situ* density, in a fluctuating (zero-mean) vertical velocity field w. If upward vertical velocity $w > 0$ is on average associated with heavier than average fluid ($\rho > 0$), and downward $w < 0$ with lighter than average fluid ($\rho < 0$), the time-average density flux $\overline{\rho w} > 0$, acting in the down-gradient direction. By analogy with molecular diffusion, the turbulent flux is modelled as the product of a turbulent diffusivity K_ρ and the mean gradient, here $-\overline{\rho_z} > 0$, so that

$$\overline{\rho w} = -K_\rho \overline{\rho_z} \tag{1}$$

defines K_ρ, the vertical turbulent eddy diffusivity for mass.

For a number of reasons, direct measurement of turbulent fluxes in the ocean is extremely difficult and to date estimates of $\overline{\rho w}$ have been derived indirectly from microscale quantities we *are* able to measure, namely ϵ, the dissipation rate of turbulent kinetic energy, and χ_T, the dissipation rate of (fluctuation) temperature. To go from the dissipation rates which we measure to the fluxes we wish to know requires conceptual models: it is necessary to understand the assumptions inherent in these models in order to assess the reliability of the results and whether (how) they may be applied within a particular numerical model.

2.1 K_ρ from ϵ:

To interpret ϵ, the dissipation rate of turbulent kinetic energy, we start with the fluid momentum equation under the Boussinesq approximation (Phillips 1977):

$$\frac{du'_i}{dt} = \frac{\partial u'_i}{\partial t} + u'_j \frac{\partial u'_i}{\partial x_j} = -{\rho_o}^{-1} \frac{\partial p'}{\partial x_i} - \rho' {\rho_o}^{-1} g \delta_{i3} + \nu \frac{\partial^2 u'_i}{\partial {x_j}^2} \tag{2}$$

where $\rho' = \rho_i - \rho_o$, ρ_o is a constant reference density, and ν the molecular viscosity. We then make what is known as the Reynolds decomposition, writing each primed variable as the sum of a zero-mean fluctuating field (unprimed) and a time-mean (overbar) field, the latter assumed to be a function of z only, with $\bar{U}_3 = 0$. Making these substitutions in eqn. (2), taking the inner product with fluctuation velocity u_i, time-averaging, and subtracting the equation governing mean kinetic energy finally yields the turbulent kinetic energy (TKE) equation:

$$\frac{d}{dt}\left(\frac{\overline{u_i^2}}{2}\right) + \left(\overline{u_i u_j \frac{\partial u_i}{\partial x_j} + \rho_0^{-1} \frac{\partial (u_i p)}{\partial x_i}}\right) + \overline{u_i w} \frac{\partial \bar{U}_i}{\partial z} = -g \rho_o^{-1} \overline{\rho w} - \epsilon \tag{3}$$

where $d/dt = \partial/\partial t + \bar{U}_j \partial/\partial x_j$. We then make the further assumption that TKE is steady-state and horizontally homogeneous (so the first term is zero), and drop both the triple correlation term (because we can't measure it) and the pressure/velocity correlation term (because we can't measure it either), so that the second term is also zero. This leaves:

$$\overline{u_i w} \frac{\partial \bar{U}_i}{\partial z} = -g \rho_o^{-1} \overline{\rho w} - \epsilon \tag{4}$$

a balance between a source term (lhs) which is the rate at which the turbulent stresses $\overline{u_i w}$ remove energy from the "mean" shear $d\bar{U}_i/dz$ and two sink terms (rhs) representing the rate at which work is done against buoyancy and the rate at which molecular viscosity dissipates turbulent kinetic energy. Now we define a flux Richardson number R_f

$$R_f \equiv \frac{-g \rho_o^{-1} \overline{\rho w}}{\overline{u_i w} \frac{\partial \bar{U}_i}{\partial z}} \tag{5}$$

as the ratio of buoyancy sink to shear source terms. Using eqn. (1) for K_ρ, and the definition of buoyancy frequency, $N = (-g {\rho_o}^{-1} \overline{\rho_z})^{\frac{1}{2}}$, eqn. (4) can be rewritten as:

$$K_\rho = \frac{R_f}{1-R_f}\frac{\epsilon}{N^2} \quad or \quad K_\rho \simeq 0.25\frac{\epsilon}{N^2} \tag{6}$$

if we make the final assumption that $R_f \simeq$ constant $\simeq 0.2$.

2.2 K_ρ from χ_T:

To interpret χ_T, the rate of dissipation of fluctuation temperature, we start with the equation for conservation of fluctuation density (cf. Gargett and Holloway, 1984). Making the Reynolds decomposition of the velocity field, multiplying by fluctuation density and time- averaging yields the fluctuation density equation:

$$\frac{d}{dt}\left(\overline{\frac{\rho^2}{2}}\right) + \overline{\rho u_i \frac{\partial \rho}{\partial x_i}} + \overline{\rho w}\,\overline{\rho_z} = -\kappa_\rho \overline{\left(\frac{\partial \rho}{\partial x_i}\right)^2} = -\chi_\rho \tag{7}$$

After assuming that the system is steady-state and *isotropic* (not just horizontally homogeneous), the first two terms vanish, leaving:

$$\overline{\rho w}\,\overline{\rho_z} = -\chi_\rho$$

a balance between generation of small-scale density variance by the action of the turbulent buoyancy flux on the mean density gradient (lhs) and dissipation of that variance by molecular forces (rhs). Using eqn. (1) for K_ρ, this yields:

$$K_\rho = \frac{\chi_\rho}{(\overline{\rho}_z)^2} \quad or \quad K_\rho = \frac{\chi_T}{(\overline{T}_z)^2}$$

since in fact we don't measure χ_ρ, but $\chi_T = \kappa_T\overline{(\partial T/\partial x_i)^2}$, the dissipation rate for fluctuation temperature. Note that this last step involves a final assumption, that $K_\rho = K_T$.

2.3 What may be wrong?

Having gone through the above exercises, it's clear that 1) ocean "measurements" of K_ρ are really indirect *estimates* of K_ρ and 2) these estimates have significant uncertainties which are based in the assumptions required to derive them. What are likely to be the most important sources of uncertainty?

Perhaps *the* fundamental conceptual problem, particularly for application to numerical models, is the Reynolds decomposition of the velocity

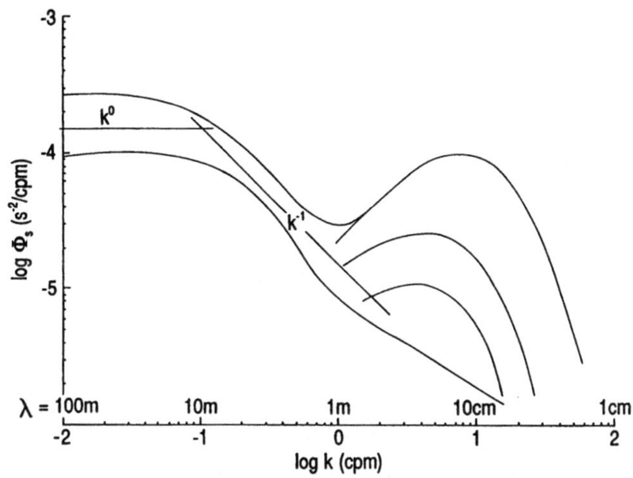

Figure 2: A composite spectrum ϕ_s of the vertical shear of horizontal velocity as a function of vertical wavenumber k, originally due to Gargett *et al.* (1981). In the ocean interior, shear approaches critical values (leading to shear instability and turbulence) only at scales near 10m, far smaller than scales resolved by ocean models.

field. In most of the stratified interior of the ocean, the shear $d\bar{U}_i/dz$ associated with the time-mean velocity field is actually very small, so small that the mean field Richardson number

$$\overline{Ri} = N^2/(d\bar{U}_i/dz)^2 \gg 1$$

Since $Ri < 1/4$ is a generally accepted criterion for stratified flow instability leading to turbulence, this means that turbulence is not associated with the large-scale "mean" shear. Instead, oceanic measurements of $\phi_s(k)$, the spectrum of vertical shear of horizontal velocity as a function of vertical wavenumber, have found that shears sufficient to cause instability are associated with the (time-dependent) internal wave field.

The average shear spectrum is flat over wavenumbers corresponding to scales from 100m down to 10m, then falls off before rising again to a variable level "bump" associated with turbulence. If we define a "wave" Richardson number

$$Ri(k) \equiv \frac{N^2}{\overline{\left(\frac{\partial u_i}{\partial z}\right)^2}} = \frac{N^2}{\int_o^k \phi_s(k')dk'}$$

which includes the shear from progressively smaller vertical wavelengths, we find that "the" Richardson number is actually a function of the vertical scale over which it is calculated. Moreover, if shear spectral level scales as Garrett and Munk (1979, henceforth GM), then $Ri(k) \sim 1$ when $k \sim (10m)^{-1}$: that is, the wave field Richardson number reaches a statistically constant value near that associated with shear instability when we add up all the wave shear down to a vertical scale of order 10m. It appears that in the stratified interior of the ocean, the internal wave field hovers on the brink of shear instability, so that extra energy can be easily and rapidly transferred from the instability scale (order 10m or smaller) to dissipation scales.

Because shear instability leading to turbulence in the ocean interior is associated with time scales and vertical spatial scales which are much smaller than those presently used in ocean numerical models, it is generally not possible to turn on dissipation and diffusion based on a Richardson number criterion as calculated by the model. Except in extremely limited regions of the upper equatorial oceans, use of a Richardson number dependent diffusivity, like that of Pacanowski and Philander (1981), is a waste of computational time. Moreover it is clear that to interpret ocean measurements of ϵ, we need to replace the conceptual model of Sec. 2.1 with a new framework in which 1) we do *not* make the Reynolds decomposition of the velocity field and 2) we *do* incorporate the concept of a statistically constant near-critical Richardson number maintained by the internal wave field.

There are presently two theories which incorporate both of these important features (I refer to these as "theories" rather loosely, as both involve significant assumptions, some no more (or less) justifiable than those present in the original "theory": the reader is urged to refer to the original papers for the details which limited space forces me to omit here). Gargett and Holloway (1984) argue for the importance of the triple correlation term in eqn. (3) and predict that

$$\epsilon \sim EN^{1.0} \quad or \quad \epsilon \sim EN^{1.5},$$

where E is internal wave energy level (Gargett, 1990). The difference in N-dependence is due to different internal wave vertical velocity spectra: $N^{1.0}$ results if the vertical velocity spectrum is narrow-band, as suggested by the observations of Pinkel (1981), $N^{1.5}$ if it is GM broadband. Henyey *et al.* (1986) use theory of linear wave interaction together with features of the observed internal wave shear spectrum to predict that

$$\epsilon \sim E^2N^2.$$

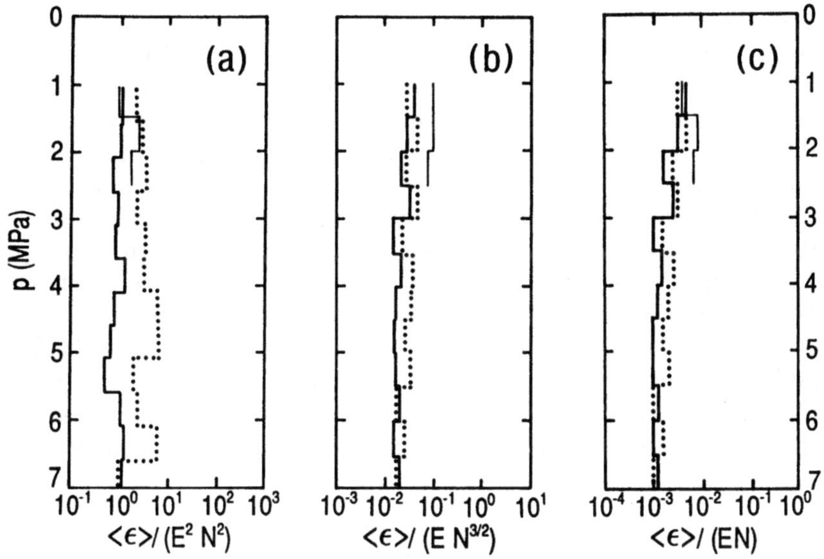

Figure 3: A comparison of ocean data with existing theories [(a) Henyey *et al.* (1986) (b) and (c) Gargett and Holloway (1984)] for the dependence of ϵ, the dissipation rate of turbulent kinetic energy, on buoyancy frequency N and internal wave energy level E. Individual curves are the PATCHEX (heavy solid), CSALT (dotted), and RING82I (light solid) data sets of Gregg (1989). The existing data base is not sufficient to favour one theory over the other, hence we are unsure whether the associated estimate of K_ρ is constant or stratification-dependent.

If we then use eqn. (6), it is clear that these two theories lead to quite different predictions for K_ρ, respectively $K_\rho \sim EN^{-0.5}$ or $K_\rho \sim E^2N^0$ if both are based on an underlying GM spectrum, $K_\rho \sim EN^{-1}$ if the spectrum is narrow band. Unfortunately, available data does not yet clearly distinguish among the various predictions for the functional form of ϵ, hence K_ρ. Fig. 3 shows three sets of oceanic measurements of $\epsilon(z)$, scaled by the three predicted dependences on $E(z)$ (if this can be estimated from measurement of shear at 10m vertical scale, as suggested by Gregg 1989) and $N(z)$. Successful scaling would collapse the scaled data to a constant value, independent of z: all three proposals leave residual variability of about the same magnitude, with little reason to prefer one over the other. Thus, even if the internal wave field were assumed to have on average the GM energy level, we are presently left with the question of whether K_ρ depends upon stability (N) or is independent of it.

This is not the only question we are left with, however, since both "theories" require the assumption of constant R_f to predict K_ρ via eqn. (6). From energetics alone, $R_f < 1$, while Stewart (1959) convinces us that R_f is considerably less than 1; so apparently we needn't worry about $R_f/(1 - R_f)$ being very large. We can't so easily dismiss the possibility that R_f may become very small. An early compilation of laboratory-scale measurements of R_f by Linden (1979) suggested that R_f was a function of Ri, reaching a maximum value of $R_f \simeq 0.25$ for $Ri \simeq 0.1$, and falling rapidly to near zero by $Ri \simeq 0.4$. In subsequent work, McEwan (1983) pointed out the importance of boundary effects in laboratory results and claimed that when these are removed, mixing by random internal wave breaking operates at constant $R_f \simeq 0.2$. While this result (and expediency) is the basis for estimates of oceanic K_ρ, there is fragmentary evidence that R_f is *not* constant in the ocean. Fig. 4, redrawn from Dillon and Park (1987), plots the values of a mixing efficiency

$$\eta \equiv \frac{N^2 \chi_T}{\epsilon(\overline{T}_z)^2} = \frac{R_f}{1 - R_f} \frac{(K_\rho)_\chi}{(K_\rho)_\epsilon}.$$

K_ρ should be the same whether determined from ϵ or from χ_T, i.e. $(K_\rho)_\chi = (K_\rho)_\epsilon$ and hence $\eta = R_f/(1 - R_f)$. These data suggest that η, hence R_f, is *not* constant, but varies as a function of a ratio of length scales (which need not concern us here).

The data of Fig. 4 come from a region in which water column T and S gradients resemble those in region (i) of Fig. 5. The final major problem in parameterizing the effects of small-scale processes comes from those large volumes of the world ocean where water property distributions resemble those of regions (ii) or (iii) of Fig. 5, in which double-diffusive instabilities are possible. The reader should consult Chapter 8 of Turner (1973) for details of the double-diffusive layering (DDL) instability which can occur in region (ii), and the salt-fingering (SF) instability which can arise in region (iii). For present purposes, the important point is that while double-diffusive instabilities cause normal down-gradient T and S fluxes, *the associated flux of density is counter-gradient.* Thus any model which carries (and diffuses) density as a single state variable cannot incorporate the effects of double-diffusive processes. A more appropriate procedure is to carry T and S as separate state variables, *and* to allow them to diffuse at different rates. Because the ocean density field is so closely coupled to the velocity field, omission of processes which affect the model density field through diffusion may also have secondary effects through consequent modification of the advective fields of the model.

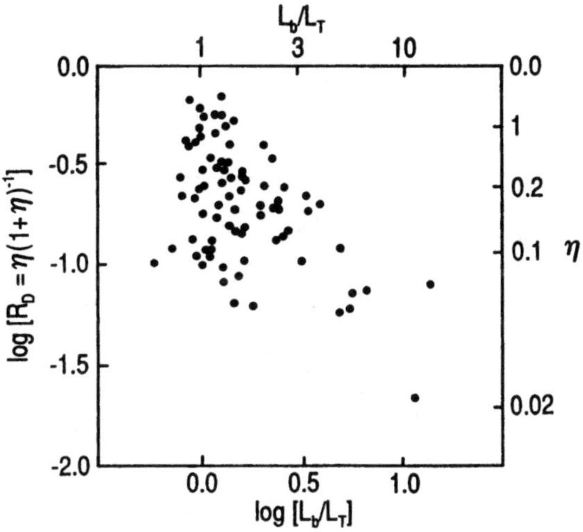

Figure 4: Mixing efficiency η may not be the constant value required if we are to derive K_ρ from measurements of ϵ. Figure redrawn with permission from Dillon and Park (1987).

3 Effects of Various Choices for K_ρ in Numerical Models

At present, diapycnal diffusion by sub-grid scale processes is handled in most numerical models by choosing a sufficiently low constant value for $K_\rho = K_T = K_S$. A value for "sufficiently low" has been determined in two ways. First, the results of microscale "measurements" of the kind detailed in the previous section suggest that, at least in the upper kilometer of the ocean, K_ρ may be of order 0.1 cm^{2}s^{-1} or smaller (Gregg, 1987). The second technique uses results from the first systematic study of the effects of the various eddy diffusivities (both horizontal and vertical, for both mass and momentum) used in coarse resolution ocean models. Bryan (1987) showed that major features of these models were most sensitive to K_ρ, the vertical turbulent eddy diffusivity for mass. The effects of varying K_ρ from 0.1 to 2.5 cm^{2}s^{-1} are illustrated in Fig. 6 for the meridional overturning streamfunction. Smaller (larger) values of K_ρ yield smaller (larger) values of mass transport, as well as pycnocline depth and meridional heat transport (not shown). The only one of these properties which is observationally well-determined is the pycnocline depth: Bryan found it necessary to use fairly small values of K_ρ, of order 0.1-0.3 cm^{2}s^{-1}, to obtain an appropriate depth, and most subsequent studies have used values

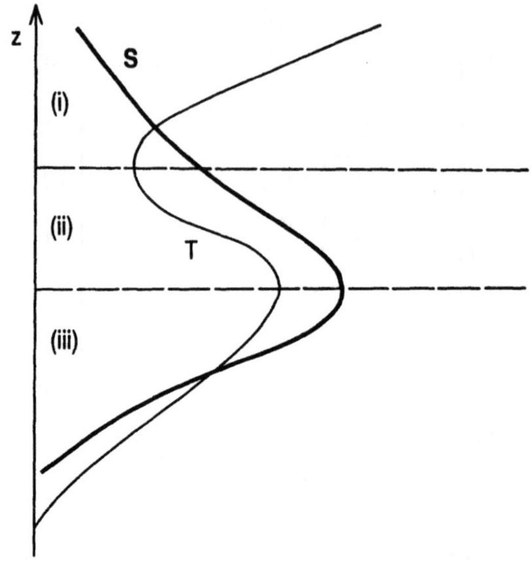

Figure 5: A schematic of vertical distributions of temperature T and salinity S, assumed to be statically stable. Because of the differing molecular diffusivities of heat and salt, region (ii) may be unstable to the diffusive layering instability and region (iii) of the salt fingering instability. Both instabilities transport T and S down-gradient at rates greater than molecular, but the resulting density flux is counter-gradient. Even in region (i), where double diffusive processes cannot occur, "normal" turbulent diffusivities may possibly be different for T and S (Gargett, 1988).

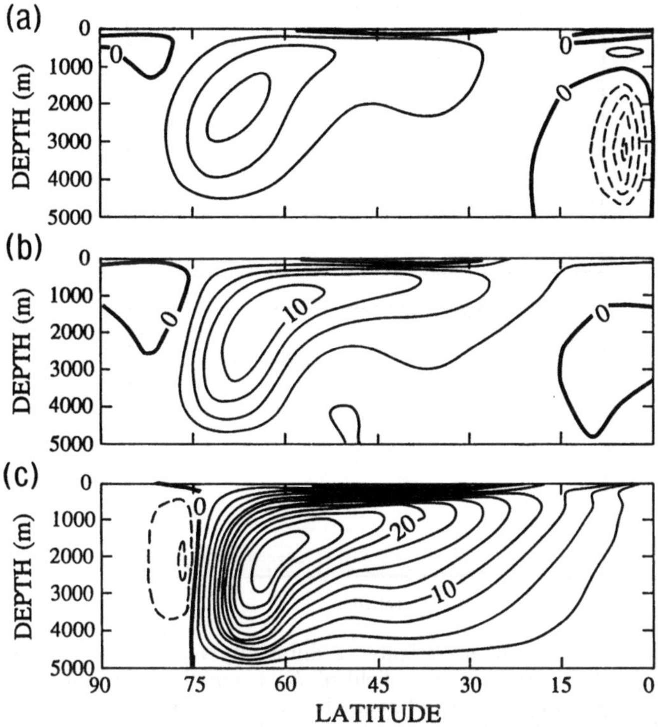

Figure 6: Variation of meridional overturning streamfunction with values of (constant) K_ρ: (a) 0.1 cm^{2}s^{-1} (b) 0.5 cm^{2}s^{-1} (c) 2.5 cm^{2}s^{-1}, redrawn with permission from Bryan (1987). Contour interval is 2.5 Sv (1Sv = 10^6 m^3 s).

in this range. In light of the discussion in Section 2, however, this sensitivity study can only be considered as a first step. Since major features of models are apparently very sensitive to the magnitude of K_ρ, we are left wondering whether they are also sensitive to non-constancy of K_ρ, or to differences between K_T and K_S which are possible when double-diffusion or differential diffusion (Gargett, 1988) contribute to diapycnal fluxes in the ocean.

The first of these questions was considered by Cummins *et al.* (1990, henceforth CHG), who used the GFDL ocean model to investigate the effects of a parameterization $K_\rho = a_o N^{-1}$, with $a_o = 10^{-3}$ cm^2 s^{-1} (Gargett 1984) in which the field of K_ρ is determined by the model density field. With model configuration and forcing very nearly that of Bryan (1987), the resultant steady-state field of K_ρ (Fig. 7) exhibits upper ocean values which are comparable to those estimated from microscale measurements and from Bryan's study. Values rise rapidly with increasing depth how-

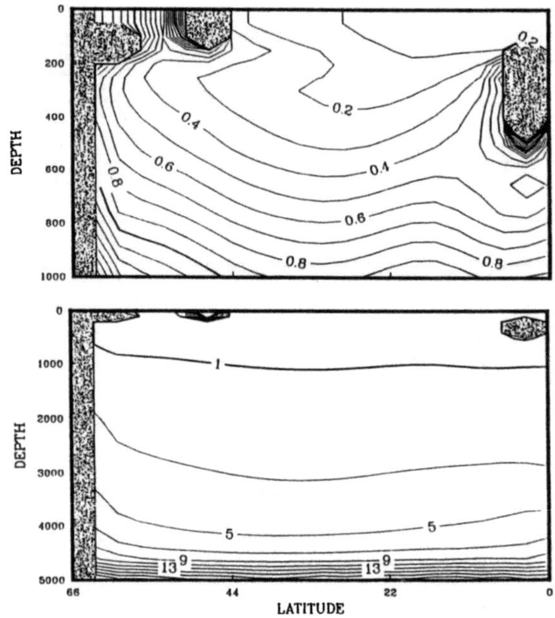

Figure 7: Cummins *et al.* (1990) investigated a diffusivity $K_\rho = a_o N^{-1}$ which depends on the model density field: the resulting field of K_ρ is small in the upper ocean, and increases with depth (shading denotes regions of convective instability, where K_ρ is set to 10^4 cm^{2}s^{-1}). Most model fields were relatively unchanged from those obtained with constant $K_\rho \simeq 0.3$cm^{2}s^{-1}.

ever, reaching 1 cm^{2}s^{-1} by 1 km depth and considerably larger values by 3-4 km depth. Surprisingly, despite considerable differences between this stratification-dependent diffusivity and the small constant values which produce Bryan's "best fit", CHG find relatively minor changes in model features. Although deep ocean stratification increases and the deep western boundary undercurrent becomes more vigorous, such major features as pycnocline depth, and meridional mass and heat transports remain effectively unchanged. Part of the reason for this behaviour is found in the fact that what you put into a numerical model is not necessarily what you get. To illustrate this, CHG use a diagnostic which they call the diapycnal diffusivity ratio DDR (Fig. 8). Because isopycnal surfaces generally have non-zero slope m, where

$$m^2 = \frac{\nabla_h \rho \cdot \nabla_h \rho}{\rho_z^2},$$

the action of the horizontal diffusivity for mass K_h which must be included for stability of the numerical integration scheme also produces diapycnal transport in primitive equation models framed in geographic coordinates. DDR is defined as the ratio of the implicit diffusivity $K_i = K_h$ m^2 caused by this effect to the explicit diffusivity K_ρ whose effects are supposedly under investigation. We are truly carrying out a sensitivity study on K_ρ only if $DDR << 1$ ($log(DDR) << 0$) throughout the model domain. Fig. 8(a) shows the field of log(DDR) for a GFDL model run with $K_\rho = 0.1$ cm^{2}s^{-1}: the total diapycnal diffusivity is equal to this value only in the limited regions of the upper ocean where $log(DDR) < 0$. Elsewhere K_i dominates, leading to a *total* diapycnal diffusivity field which increases with depth. When the same experiment is run with $K_\rho = a_o N^{-1}$, explicit diffusion dominates over most of the volume. Since the *total* diapycnal diffusivity in both cases is similar, small in the upper ocean and increasing in magnitude with depth, it is less surprising that the results should be so similar. This discussion implies that we must be very careful about implicit diffusion when attempting sensitivity studies. Ideally, *if* we believe that diffusive processes tend to act in and across isopycnal planes rather than geographic planes, it is best to rotate the diffusivity tensor to the (local) density surface (Salomon, 1981; Redi, 1982). However, when $K_\rho = a_o N^{-1}$, the resulting improvement in DDR is not striking (Fig. 8(c)), so it seems safe to use this form for K_ρ without tensor rotation.

Finding that models aren't enormously sensitive to fairly significant (defined by the fact that microscalers are still arguing about them) differences in K_ρ is encouraging for numerical modellers. Less cheering are the results of Gargett and Holloway (1992, henceforth GH92) who considered the question: are models sensitive to reasonable differences in diffusivity between heat (temperature) and salt? The answer is an unequivocal "yes". GH92 argue that, given what we know of double diffusive processes and/or differential diffusion, we may reasonably expect a variation of at least a factor of two in the diffusivity ratio

$$d \equiv \frac{K_s}{K_T},$$

where K_S and K_T are respectively the diffusivities for S and T. With the same model configuration and forcing used by CHG, GH92 use diffusivities for T and S which differ by the factor d (with underlying $a_o N^{-1}$ form). The resulting steady-state meridional and zonal streamfunctions are shown in Fig. 9 for three values of d. Compared to the standard $d = 1$ case of CHG, d = 0.5 (typical of diffusive layering or differential diffusion: for discussion see GH92) causes a 40% increase in volume flux, while $d = 2$ (typical of salt-fingering) is associated with near collapse of

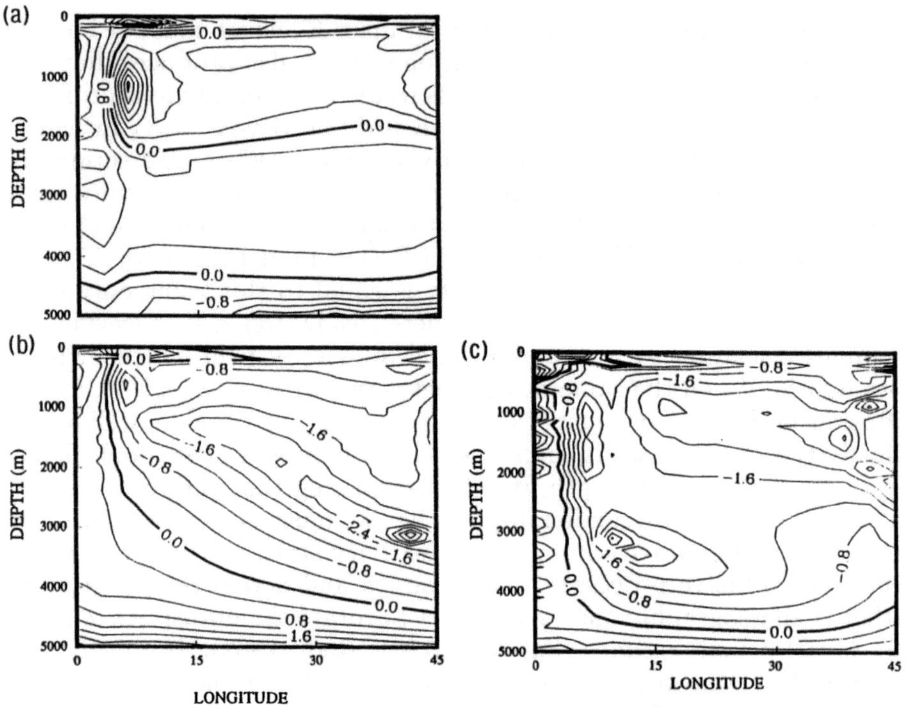

Figure 8: The total diapycnal diffusivity present in a numerical model run is the sum of the explicit diffusivity K_ρ and an implicit diffusivity K_i caused by horizontal diffusion acting across sloping isopycnals. Contours of $log(K_i/K_\rho)$ are shown for runs of a GFDL model with (a) $K_\rho = 0.1\text{cm}^2\text{s}^{-1}$ (b) $K_\rho = a_o N^{-1}$ (c) $K_\rho = a_o N^{-1}$ *and* rotation of diffusivity tensor to local isopycnals. From Cummins *et al.* (1990).

the normal meridional cell and a reversal of flow direction in the deep subtropical gyre. Such significant changes in the thermohaline flow field are associated with large changes in deep ocean water mass properties, as seen in the N/S salinity sections of Fig. 10. For $d = 0.5$, the deep ocean is even more homogeneous than the standard case, while $d = 2$ is associated with enhanced N/S salinity (and temperature) gradients, and with a mid-depth salinity minimum S_{min} extending southwards from the subpolar gyre beneath the subtropical gyre. GH92 show that the strength and location of this S_{min} are particularly sensitive functions of how one implements different diffusivities for T and S. At first glance, model meridional heat and salt fluxes (Fig. 11) are not extremely sensitive to d. Note however that d does affect the *divergence* of these fluxes (the slopes of the curves) at high latitudes. Since the divergence of oceanic fluxes force the atmosphere, this more subtle change may be important for coupled ocean/atmosphere models.

4 Summary

The last decade of oceanic microscale measurements has produced results which are important to those who seek appropriate ways to include the effects of small-scale mixing in numerical models which cannot explicitly resolve these scales. It is now clear that in most of the ocean volume, shear-driven instabilities occur at vertical length scales which are far smaller than those resolved by large-scale models, so that mixing cannot be parameterized in terms of model-generated Richardson numbers. In seeking more realistic parameterizations, we find that observations are not yet sufficient to differentiate between constant and stratification-dependent diffusivities. From both laboratory and ocean observations, it seems clear that processes leading to different diffusivities for T and S are likely operative in the sea, although proportions are yet unknown. Many outstanding questions about the reliability of our present methods of determining diapycnal fluxes await direct flux measurements.

Preliminary basin-scale experiments indicate that coarse resolution numerical models are relatively insensitive to a reasonable variation of K_ρ with stability, but quite sensitive to relatively minor (factor of 2) difference between K_T and K_S. If these results still hold in global simulations, there are serious implications for both the modelling and microscale communities. There is, for example, the very real possibility that we are deluding ourselves as to the usefulness of our numerical oceans over the long time scales of climate change. Ocean models run with vertical eddy diffusivity K_ρ chosen to best simulate the present ocean circulation may have no predictive value if K_ρ is in reality dependent upon water mass proper-

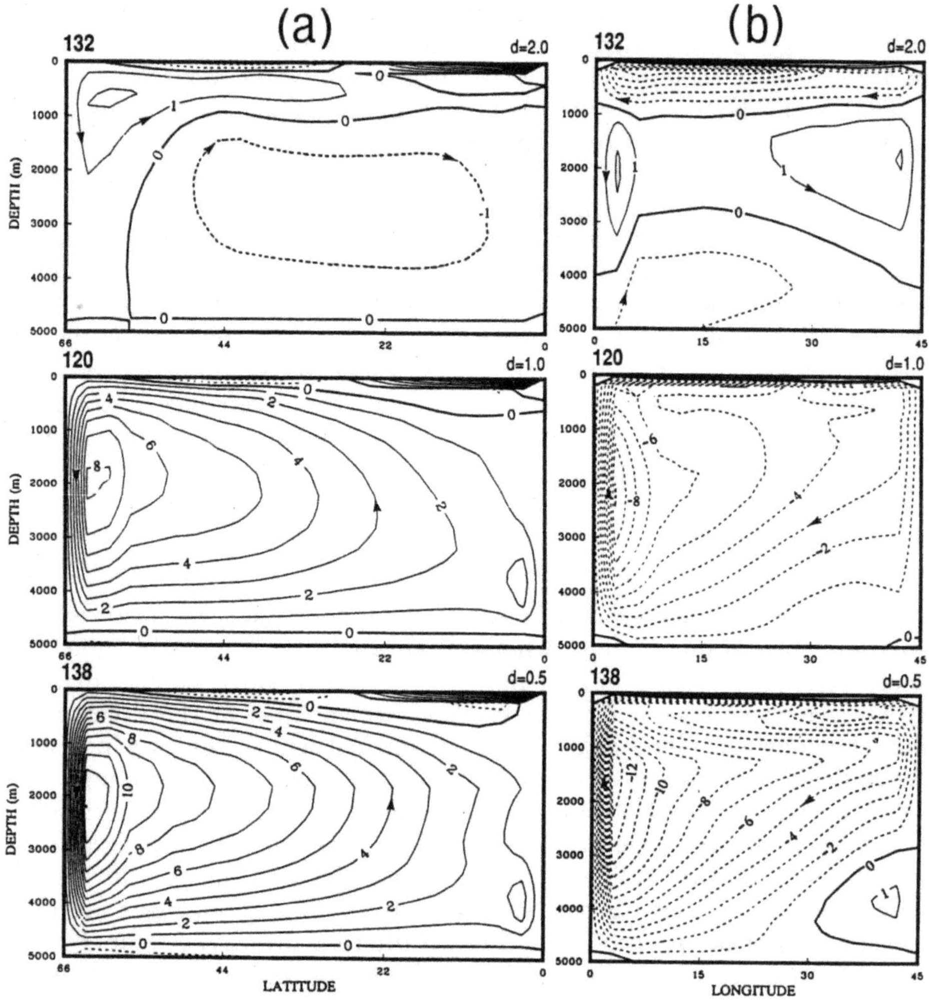

Figure 9: Both (a) meridional and (b) zonal overturning cells vary substantially with the value of $d = K_S/K_T$. The center panels illustrate the standard use of $K_S = K_T$, hence $d = 1$. When $D = 0.5$ (bottom panels) transport increases by 40%: when $d = 2$, the normal meridional cell nearly collapses, and flow in the deep subtropical gyre is actually reversed. From Gargett and Holloway (1992)

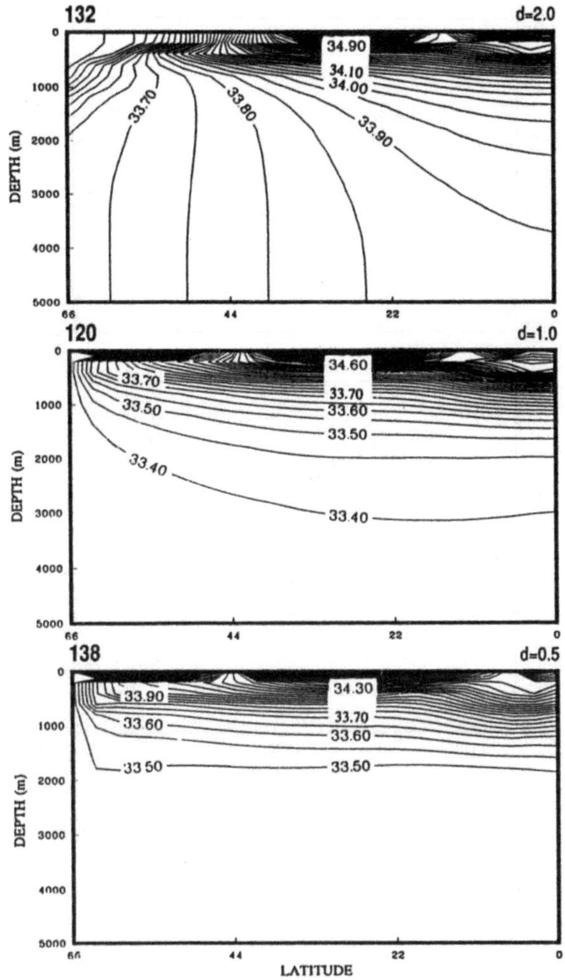

Figure 10: Relaxing the assumption that $K_S = K_T$, hence $d = 1$ (center panel), also brings changes in deep water properties, here illustrated with sections of S with latitude. With $d = 2$, note the development of a salinity minimum originating in the subpolar gyre and extending beneath the subtropical gyre at intermediate depths.

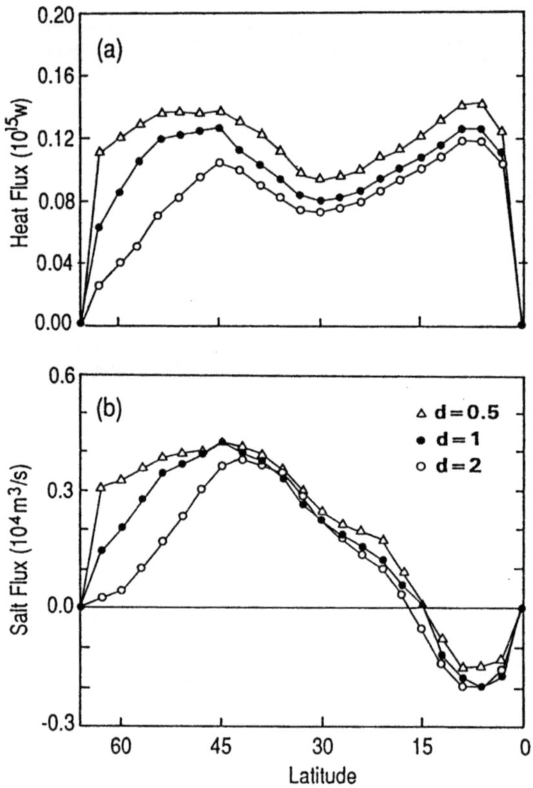

Figure 11: Magnitudes of meridional heat and salt fluxes are relatively insensitive to the value of d: note however that d does affect *divergences* of heat and salt fluxes at high latitudes, with consequent changes in the fluxes which drive a coupled atmosphere.

ties which will change under changing surface forcing. As well, diffusive properties which evolve as functions of the model properties themselves present a whole new range of feedback mechanisms which we must begin to consider. The sensitivity of model ocean properties to $K_T \neq K_S$ also indicates an important direction for the next decade of microscale research. The task will not be easy, since we are not presently able to measure the turbulent flux of anything, much less of T and S separately. Nevertheless, the importance of the answer suggests that numerical modellers will follow with interest as the microscale community grapples with the problem of direct measurement of possibly differential fluxes.

References

BRYAN F. (1987). On the parameter sensitivity of primitive equation ocean general circulation models. *J. Phys. Oceanogr.* **17,** 970-985.

CUMMINS P.G., HOLLOWAY G. AND GARGETT A.E. (1990). Sensitivity of the GFDL ocean general circulation model to a parameterization of vertical diffusion. *J. Phys. Oceanogr.* **20**(6), 817-830.

DILLON T.M. AND PARK M.M. (1987). The available potential energy of overturns as an indicator of mixing in the seasonal thermocline. *J. Geophys. Res.* **92**(C5), 5345-5353.

GARRETT C. AND MUNK W. (1979). Internal waves in the ocean. *Ann. Rev. Fluid Mech.* **11,** 339-369.

GARGETT A.E. (1990). Do we really know how to scale the turbulent kinetic energy dissipation rate ϵ due to breaking of oceanic internal waves? *J. Geophys. Res.* **95**(C9), 15971-15974.

GARGETT A.E. (1988). Reynolds number effects on turbulence in the presence of stable stratification. Small-Scale Turbulence and Mixing in the Ocean. Ed. JCJ Nihoul and BM Jamart. *Elsevier Science Publishers 517-528.*

GARGETT A.E., HENDRICKS P.J., SANFORD T.B., OSBORN T.R., AND WILLIAMS III, A.J. (1981). A composite spectrum of vertical shear in the upper ocean. *J Phys Oceanogr* **11**(9), 1258-1271.

GARGETT A.E. AND HOLLOWAY G. (1984). Dissipation and diffusion by internal wave breaking. *J. Mar. Res.* **42,** 15-27.

GARGETT A.E. AND HOLLOWAY G. (1992). Sensitivity of the GFDL ocean model to different diffusivities for heat and salt. *J. Phys. Oceanogr.* **22**(10), 1158-1177.

GREGG M.C. (1989). Scaling turbulent dissipation in the thermocline. *J. Geophys. Res.* **94**(C7), 9686-9698.

GREGG M.C. (1987). Diapycnal mixing in the thermocline: a review. *J. Geophys. Res.* **92**(C5), 5249-5286.

HENYEY F.S., WRIGHT J. AND FLATTÉ S.M. (1986). Energy and action flow through the internal wave field: an eikonal approach. *J. Geophys. Res.* **91,** 8487-8495.

LINDEN P.F. (1979). Mixing in stratified fluids. *Geophys. Astrophys. Fluid Dyn.* **13,** 3-23.

MCEWAN A.D. (1983). Internal mixing in stratified fluids. *J. Fluid Mech.* **128,** 59-80.

PACANOWSKI R.C. AND PHILANDER S.G.H. (1981). Parameterization of vertical mixing in numerical models of tropical oceans. *J. Phys. Oceanogr.* **11**, 1443-1451.

PHILLIPS O.M. (1977). The dynamics of the upper ocean. *Second edition Cambridge Univ Press 336 pp.*

PINKEL R. (1981). Observations of the near-surface internal wavefield. *J. Phys. Oceanogr.* **11**, 1248-1257.

REDI M.H. (1982). Oceanic isopycnal mixing by coordinate rotation. *J. Phys. Oceanogr.* **12**, 1154-1158.

SOLOMON H. (1971). On the representation of isentropic mixing in ocean circulation models. *J. Phys. Oceangr.* **1**, 233-234.

STEWART R.W. (1959). The problem of diffusion in a stratified fluid. *Adv Geophys.* **6**, 303-311.

TURNER J.S. (1973). Buoyancy effects in fluids. *Cambridge Univ Press UK 367 pp.*

LINKS OF THE SOUTHERN OCEAN TO THE GLOBAL CLIMATE

DIRK OLBERS
Alfred-Wegener-Institute for Polar and Marine Research, Bremerhaven, Federal Republic of Germany.

1 Introduction

Roughly 30% of the world ocean volume has temperatures below $2°C$. Paleoceanographic data have revealed that this was not always the case. Before the Drake Passage opened due to continental drift about 30 Myr BP the climate of the ocean was considerably warmer. In the course of the establishment of the Southern Ocean in its present shape the difference between surface and bottom temperatures in equatorial regions changed from about $7°C$ to its present value of about $26°C$ (Berger 1981). The polar climate of the southern hemisphere got increasingly colder by the growth of glacial ice on the Antarctic continent and the gradual development of the sea ice cover around it. Today the seasonal cycle changes the sea ice extent in the Southern Ocean from about $3 \cdot 10^6 km^2$ in austral summer to $20 \cdot 10^6 km^2$ in austral winter, an area which is roughly as big as the North Atlantic Ocean.

The general cooling of the southern high latitudes of our planet was accompanied by a corresponding development of the present cold water sphere in the entire world ocean. The production of the cold water takes place in the polar regions. The coldest and heaviest water masses - combined under the name Antarctic Bottom Water (*AABW*) - are formed in the Southern Ocean all around Antarctica, predominantly adjacent to the continent in the subpolar seas - the Weddell and Ross Seas. The formation is a consequence of a complex chain of interactions between the ocean, sea and shelf ice and the atmosphere. Ventilation processes originating from heat transfer across the air-ocean interface and brine release due to sea ice production generate meridional overturning of the world ocean which carries passive tracers such as oxygen and carbon dioxide into the deep ocean. These deep ventilation processes are believed to be responsible for the anomalously cold state of the modern ocean.

The opening of the Drake Passage also established the strongest current system in the world ocean, the Antarctic Circumpolar Current (*ACC*), with mass transport of roughly $130 Sv$ ($1 Sv = 10^6 m^3 s^{-1}$).

NATO ASI Series, Vol. I 11
Modelling Oceanic Climate Interactions
Edited by J. Willebrand and D. L. T. Anderson

Though primarily driven by the wind the *ACC* contains a strong thermohaline driven component originating from the deep-reaching convection under the sea ice. As the most important link between the ocean basins of the Atlantic, Pacific and Indian Oceans, it serves as a conduit of all active and passive tracers which affect the climate, notably heat and salt which strongly influence the oceanic mass stratification and consequently the ocean circulation, and the greenhouse gas carbon dioxide.

The strong circulation of the *ACC* acts as barrier to the meridional exchange of properties between the Southern Ocean and the oceans to the north. Unlike other oceans where the poleward transfer of heat is mainly carried by the mean geostrophic currents, the Southern Ocean circulation has to establish the heat transfer in frictional boundary layers and by transient motions. The vigorous meso-scale eddy field originating from internal instabilities of the *ACC* is the main carrier of heat poleward to the south. This meridional heat transport is intimately related to the dynamical balance of the *ACC* which can only establish the observed size of the current by downward flux of momentum due to the eddy action and subsequent loss by pressure action at submarine topography.

The remote location and the hostile climate of the Southern Ocean is mostly responsible for our lack of knowledge of many important processes which act in the area of the Southern Ocean and influence other ocean regions and ultimately our climate. Models of the sea ice cover, water mass formation and ocean circulation thus constitute an important tool to investigate and interrelate these processes. In the following we will discuss the processes involved in the heat transport across the Polar Front, the dynamical balance of the *ACC*, and formation of bottom water in the Weddell Sea. A broader view of the subject is given by Gordon (1988), Nowlin and Klinck (1986) and Whitworth III (1988).

2 The heat link: the transport of heat across the polar front

The Southern Ocean is bounded to the north by a circumpolar system of fronts, the Polar Front (*PF*) and the Subantarctic Front (*SAF*) which separate the cold Antarctic water from the warmer waters to the north. The area south of this frontal boundary is exposed to the cold polar air masses which leads to substantial cooling of the surface waters. Gordon and Owens (1987) have estimated that on average about $0.3PW$ ($1PW$ = 1Petawatt= $10^{15}W$) leave the ocean to the atmosphere in the region south of the Polar Front. This loss of heat must be compensated by a meridional net southward heat transport across the front. It is customary to separate the total heat transport into different components which are associated with different components of the velocity field such as the time mean flow or the eddy field (see e.g. Bryan 1982). The aim of course is to identify different physical mechanisms which have an effect on the mean heat transport and its possible variations in space and time.

2.1 The components of heat transport

For this purpose we consider the heat balance of the piece of ocean bounded to the south by the Antarctic continent (with coast line Γ_S) and to the north by a circumpolar belt (a surface Γ with circumpolar outcrops Γ_N at the sea surface and Γ_B at the bottom). Different shapes of the surface Γ will be specified later. The mass and heat balance of this piece of ocean may be expressed as

$$\begin{aligned} MF &= \int_\Gamma da\rho\omega = (MF)_{ICE} - \frac{\partial}{\partial t}MS \\ HF &= \int_\Gamma da\rho c_p T\omega = Q_{SURF} - L(MF)_{ICE} - \frac{\partial}{\partial t}HS \end{aligned} \tag{1}$$

where ω is the velocity component normal to Γ, $(MF)_{ICE}$ the mass flux of ice across Γ_S, L the latent heat of fusion of ice, da the area element on Γ, and Q_{SURF} the net heat flux across the ocean surface to the atmosphere (about $0.3PW$ if Γ_N is identified with the Polar Front). The storage terms for mass and heat are denoted by MS and HS, respectively. The density ρ and specific heat c_p will be treated as constant for the sake of simplicity.

The mass flux of icebergs $(MF)_{ICE}$ is in fact a fairly small contribution to any mass flux (and hence $L(MF)_{ICE}$ to any heat flux) considered in this problem. According to estimates of the mass balance of the Antarctic ice shield (e.g. Warrick and Oerlemans 1990) we find $(MF)_{ICE} \approx 10^{-7}kgs^{-1}$ corresponding to $10^{-4}Sv$ and hence $L(MF)_{ICE} \approx 10^{-3}PW$. We therefore neglect the ice flux across Γ_S. Furthermore, we split the circulation into its time mean (denote by an overbar) and deviation (denoted by a dash), i.e

$$q = \bar{q} + q' \tag{2}$$

for any quantity q. The part q' is identified with the transient eddies. The time mean mass and heat balance then is written in the simple form

$$\begin{aligned} \overline{MF} &= 0 \\ \overline{HF} &= \overline{Q}_{SURF}. \end{aligned} \tag{3}$$

Notice that the heat flux $\overline{HF}$ across the circumpolar belt Γ only depends on the shape of the surface outcrop Γ_N, all possible subsurface shapes must yield the identical normal heat transport equal to $\overline{Q}_{SURF}$. As a consequence of the vanishing of the net mass flux $\overline{MF}$ we further learn that $\overline{HF}$ is independent of the zero of the temperature scale, i.e.

$$\overline{HF} = \int_\Gamma da \rho c_p \overline{(T - T_0)\omega} \tag{4}$$

for any constant T_0. In fact this property enables us to identify $\overline{HF}$ as the flux of heat.

Our aim is now to identify the role of different physical mechanisms contributing to $\overline{HF}$ such as the heat transport by the time-mean flow and the transient eddies, by geostrophic and ageostrophic motions, by the barotropic and baroclinic flow or standing and transient eddies. It is quite obvious that any such separation must sum up to the same value $\overline{Q}_{SURF}$ for any given Γ_N. We will learn further that the importance of almost any of the mentioned components of the flow in carrying heat across Γ_N is highly dependent on the shape of Γ. In fact there is no sense in attributing specific magnitudes to any heat flux component in general.

The time mean heat flux may be expressed as sum of the heat flux HF^m due to the time mean flow and the contribution HF^e by the transient eddies

$$\overline{HF} = \int_\Gamma da \rho c_p \left(\bar{\omega}(\bar{T} - T_0) + \overline{\omega' T'}\right) = HF^m + HF^e. \tag{5}$$

Further separation concerns the time mean current $\bar{\omega}$. It may again be split into the geostrophic and ageostrophic parts

$$\bar{\omega} = \bar{\omega}_g + \bar{\omega}_{ag} \tag{6}$$

implying

$$\overline{HF} = HF_g^m + HF_{ag}^m + HF^e. \tag{7}$$

The ageostrophic motion is associated with the Ekman transport confined to the planetary boundary layers at the ocean surface and possibly frictionally controlled flow near the ocean bottom. Since vertical variations of the temperature are small in each of these boundary layers the heat flux by ageostrophic motion is fairly well approximated by

$$HF_{ag}^m = \oint_{\Gamma_N} ds \rho c_p (\bar{T}_{SURF} - T_0) V_{ag} \tag{8}$$

for the surface layer and correspondingly for the bottom part. Here

$$V_{ag} = \frac{\tau_s}{f} \tag{9}$$

is the local Ekman transport established by the wind component τ_s which is parallel to Γ_N. In contrast to $\overline{HF}$ or HF^m the Ekman contribution HF_{ag}^m is not independent of the temperature constant T_0 (since the net of V_{ag} does not vanish). We may thus obtain any desired value by suitably choosing T_0, e.g. $HF_{ag}^m = 0$ for a circumpolar path along $\bar{T}_{SURF} = T_0$.

A corresponding arbitrariness is attached to the geostrophic part

$$HF_g^m = \int_\Gamma da \rho c_p (\bar{T} - T_0) \bar{\omega}_g \tag{10}$$

by different choices of T_0 but also for different circumpolar belts Γ. Splitting $\bar{\omega}_g$ into the horizontal and vertical components we have

$$\begin{aligned} \bar{\mathbf{u}}_g &= (u_g, v_g) = \frac{1}{f}\hat{\nabla}\bar{p} \\ \bar{w}_g &= -\frac{\beta}{f}\int_z^0 dz' \bar{v}_g \end{aligned} \tag{11}$$

where u and v are zonal and meridional velocities, respectively, and $\hat{\mathbf{a}}$ is the vector resulting from couterclockwise rotation of $\mathbf{a}$ by $\pi/2$. We may rewrite (10) as

$$HF_g^m = \int_\Gamma da \rho c_p(\bar{T} - T_0)(\bar{\mathbf{u}}_g, \bar{w}_g) \cdot (\mathbf{n}_h, n_v) \tag{12}$$

where $(\mathbf{n}_h, n_v)$ is the unit vector which is locally perpendicular to Γ. The magnitude of the horizontal part is roughly estimated to be of order $1PW$ (with $\Delta T \sim 1K, \bar{v}_g \sim 0.01ms^{-1}$, ocean depth $h \sim 4000m$ and a circumpolar path of length $20000km$) while the contribution of the vertical part is usually much smaller (by a factor $\beta B/f \sim 0.05$ where B is the horizontal width covered by the belt Γ). However we easily find a belt for which the horizonal part vanishes identically and HF_g^m thus becomes very small of order $10^{-2}PW$. We may choose Γ as defined by the relation

$$\bar{p}(\mathbf{x}, z) - \bar{p}(\mathbf{x}_0, z) = 0 \tag{13}$$

for any point $\mathbf{x}_0$ on a circumpolar surface isobar. Then $\mathbf{n}_h \cdot \bar{\mathbf{u}}_g = 0$ and only the small vertical part of (12) remains,

$$HF_g^m = \int_\Gamma d^2x \rho c_p(\bar{T} - T_0)\bar{w}_g \tag{14}$$

Here $\bar{T}$ and $\bar{w}_g$ are evaluated on Γ but the integration extends over the horizontal projection of it. Notice that Γ is vertical at $\mathbf{x} = \mathbf{x}_0$ but is generally tilted at other locations. Obviously, transient eddies and frictional flow must carry the heat (at the required rate $\overline{Q}_{SURF}$) across the corresponding circumpolar path.

In quasigeostrophic scaling the above vertical contribution would not appear in the lowest (geostrophic) order of the theory. A further point to make in this context concerns the separation of the time mean fields into the zonal average and the deviation, called the standing eddy part (following atmospheric notation). Then the question of relative importance of the standing versus transient eddies in carrying heat across latitudes arises. However answers cannot be general if one extends the concept of standing eddies to other circumpolar paths. Obviously, there are no standing geostrophic eddies for the geostrophic streamline stack considered above.

Another peculiar belt was chosen by DeSzoeke and Levine (1981) in the only attempt (known to the author) made so far to estimate the heat flux

HF_g^m from oceanic observations. To circumvent the well-known problem of determining the absolute geostrophic current from hydrography they took Γ defined by the vertical belt of constant vertically integrated potential temperature

$$\Gamma : \int_{-h(\mathbf{x})}^{0} dz\bar{T}(\mathbf{x}, z) = T_1 = constant. \tag{15}$$

Separating the geostrophic velocity (which is now horizontal) into the barotropic and baroclinic components

$$\begin{aligned} \bar{u}_g &= \bar{u}_{bt}(s) + \bar{u}_{bc}(s,z) \\ \frac{\partial \bar{u}_{bc}}{\partial z} &= -\frac{g}{f}\frac{\partial \rho}{\partial s} \\ \int_{-h(\mathbf{x})}^{0} dz\bar{u}_{bc} &= 0 \end{aligned} \tag{16}$$

it is found that the unknown barotropic contribution cancels from the heat flux

$$\begin{aligned} HF_g^m &= \int_\Gamma da(\bar{u}_{bt} + \bar{u}_{bc})(\bar{T} - T_0) \\ &= \oint_\Gamma ds \int_{-h(\mathbf{x})}^{0} dz\bar{u}_{bc}(\bar{T} - T_0) \end{aligned} \tag{17}$$

if $T_0 = T_1$.

2.2 An estimate from hydrographic data

DeSzoeke and Levine (1981) applied this concept to a circumpolar path of constant vertically integrated potential temperature $T_1 = 1.3°C$ which is close to the Polar Front. It was found that the Ekman transport carries $28Sv$ and $0.15PW$ (with respect to $1.3°C$) across this path to the north. They further showed that the geostrophic heat transport is negligibly small, $HF_g^m \simeq 0$ with a minimum error of $0.05PW$ (assuming that data at different levels are totally uncorrelated) and a worst error of $0.23PW$ (assuming perfect correlation which is highly unlikely). The heat balance then reads

$$\begin{array}{ccccccccc} HF_g^m & + & HF_{ag}^m & + & HF^e & + & HF_{miss} & = & \overline{Q}_{SURF} \\ 0 & + & 0.15 & + & ? & + & ? & = & -0.30PW \end{array}$$

where HF_{miss} stands for possibly overlooked contributions, perhaps associated with frictional flow near the bottom, inertial flows in deep canyons or unresolved geostrophic motion. This latter problem may particularly concern Antarctic Bottom Water ($AABW$) escaping the course station spacing close to the bottom. The transport of the cold $AABW$ is not well known, values between 2 and $25Sv$ are given by various authors (see e.g. Fahrbach et al. 1992). The lower range of these values applies to more southerly latitudes close to the formation regions around Antarctica. Further north the definition of $AABW$ is widened to include waters entrained by mixing processes. We will discuss one of the most important formation processes of $AABW$ in chapter 4 in more detail and suggest that $10Sv$ is a realistic upper bound for the formation rate all around Antarctica. With a volume transport of $10Sv$, at $-0.5°C$, the $AABW$ would transport $0.07PW$ (with respect to $1.3°C$) to the south.

The above balance thus requires a substantial contribution from the turbulent eddy field which is known from satellite altimetry and other in-situ observations to be particularly vigorous in this part of the world ocean. To close the heat balance of the Southern Ocean south of the Polar Front the eddies must carry about $0.4PW$ to the south (the uncertainty of this estimate is quite large, possibly $\pm 0.2PW$). Bryden (1983) has estimated the heat of an individual eddy in the Drake Passage area as $0.013PW$ so that about 20 to 30 eddies of this size are required at any time around the circumpolar belt to accomplish the $0.4PW$. This appears to be a reasonable number. Nowlin and Klinck (1986) support this magnitude of the heat flux due to single eddies by other observations.

We can also support this magnitude by results from consideration of the momentum balance of the ACC. We will show below in chapter 3 that lateral transport of heat and the vertical transfer of momentum through the water column are mutually dependent. It is for this reason that the momentum balance of the flow along the Polar Front has strong implications for the heat transport across it.

3 The circumpolar link: the strength of the Antarctic Circumpolar Current

The wind-driven branch of the circulation in the world ocean is structured in cells which are approximately confined to separate ocean basins. In the thermohaline branch the basins are however connected. The deep ventilation is triggered by production of cold dense water masses in high latitudes of the North Atlantic and the Southern Ocean and penetrate from there into the other basins. The deep water formed in the North Atlantic moves south across the equator into the circumpolar region and is carried by the *ACC* into the Indian and Pacific Oceans. It spreads in these basins even further north with gradual upwelling all along its path and concentrated upwelling in the equatorial regions. The loop is closed by the current system in the surface layers and transport (of salinity and passive tracers) induced by transient motions, forming a huge conveyor belt (e.g. Gordon 1986) which connects all basins of the world ocean. The water masses formed in the Southern Ocean (see chapter 4) spread with and across the *ACC* and fill the bottom layer of all ocean basins (e.g. Emery and Meincke 1986).

The *ACC* is thus a major conduit in the thermohaline ventilation system. Its strength is set by the reaction of the fluid to the forcing functions – wind and buoyancy flux at the surface – and the geometry and topography of the circumpolar domain. Measuring the transport and internal structure of the current and understanding its dynamical balance has long been a major task of research (see e.g. the review of Nowlin and Klinck 1986). The strong eastward surface wind stress in the circumpolar belt must obviously be considered as principle driving agent. The problem of early concepts of the balance was to find an effective sink of momentum. In the latitude band of the Drake Passage there are no continental barriers which could support a net zonal pressure gradient as in the other basins and thereby oppose the acceleration of the current by the wind. Conventional parametrisation of friction – in the form of lateral or vertical diffusion of momentum or bottom friction – gave rather unsatisfactory results (Hidaka and Tsuchiya 1953, Gill 1968 and others). The current amplitudes in these models are proportional to the applied wind stress and inversely proportional to the values of the frictional parameters, e.g. for lateral friction the mass transport scales as $\tau_s D^3/A_h$ where τ_s is the wind stress amplitude, D the current width and A_h the lateral eddy viscosity. Hidaka thus faced the difficulty of explaining the observed transport values of about $100Sv$ with reasonable values for the model friction and the observed wind stress.

In one of the earliest papers on this problem Munk and Palmén (1951) suggested a solution which today still appears as the only acceptable way out of a frictionally controlled balance of the current. Though continental barriers are absent, there are significant submarine ridges to build up net zonal pressure gradients. Munk and Palmén speculate that this could enable a transfer of momentum to the solid earth. A balance with the wind stress can occur only if the flow establishes pressure differences across the ridges with the appropriate sign (higher pressure value on the western side for westerly winds) and magnitude. To balance a stress of $10^{-4}m^2s^{-2}$ a pressure difference of only a few dynamic centimeters is required across a ridge of a few thousand kilometers.

For this balance to operate, the momentum given to the ocean at the surface must be transferred down to the blocked depths where the flux by pressure differences across topography into the solid earth - the so-called topographic form stress or mountain drag - can be effective. This vertical transport of momentum must be extremely large compared to other oceanic areas. To transmit the wind stress of the above magnitude in the observed vertical current shear of about $0.1ms^{-1}$ over 1000m by vertical diffusion, an equivalent value of vertical eddy viscosity of $1m^2s^{-1}$ is needed. This is three or even four orders of magnitude larger than elsewhere (large scale oceanic models generally use $10^{-4}m^2s^{-1}$). However, inverse models of the ACC which are based on hydrographic data, fully support these high values of the vertical transport of momentum (Olbers and Wenzel 1989).

Three-dimensional turbulence of small scale is unlikely to support such a large transport of momentum. However, turbulence of larger scale – meso-scale quasigeostrophic eddies – populate the current. Generated by mixed barotropic-baroclinic instability of the mean flow the mesoscale eddies are crucial to achieve the balance of the current.

3.1 The role of eddies in the dynamical balance

Like other large-scale ocean currents the *ACC* is in geostrophic balance: the horizontal velocity component in the direction of the flow is – to a very good approximation – given by the pressure gradient across it. However, this relation does not give insight into the dynamical balance of the flow, asking for the principle external forces imparting momentum into the fluid, the mechanisms which redistribute it in the interior, and the sinks by which the momentum leaves the fluid, ultimately allowing for a steady state (possibly after averaging over the time scale of transient eddies) and determining the strength of the current. The answer to this question resides in projections of the momentum balance equations which are not dominated by the geostrophic relation, as e.g. the component of the balance along the flow.

Introducing horizontal coordinates s along an arbitrary circumpolar path on the depth level z and n normal to it the s-component of the momentum balance reads

$$[\frac{\partial \bar{u}}{\partial t}] - (f + \bar{\zeta})\bar{v} + \bar{w}\bar{\eta} = -\frac{\partial}{\partial s}\left(\bar{p} + \frac{1}{2}(\bar{u}^2 + \bar{v}^2 + \bar{w}^2)\right) + \mathcal{F} + \mathcal{E} \quad (18)$$

where (u, v, w) is the velocity vector in (s, n, z)-coordinates and η and ζ are the n- and z-components of the relative vorticity, respectively. $\mathcal{F}$ denotes frictional terms and $\mathcal{E}$ the terms arising from transient eddies. The vanishing time rate of change is included here for easier identification. As mentioned above, $\bar{u}$ is in approximate geostrophic balance

$$f\bar{u} = -\frac{\partial \bar{p}}{\partial n} \quad (19)$$

and $\bar{v} = 0$ if the path is oriented along the streamline of the time mean flow (s, n then define the natural coordinates of the mean flow).

The external forcing by wind stress couples to the friction term $\mathcal{F}$ if (18) is integrated vertically from the surface to any depth. Thermohaline forcing, however, enters only indirectly via its effect on the baroclinic part of the pressure field.

The geostrophic terms almost entirely cancel by averaging (18) around the circumpolar path,

$$\oint ds\,\left(-(f + \bar{\zeta})\bar{v} + \bar{w}\bar{\eta}\right) = \oint ds(\mathcal{E} + \mathcal{F}) - \sum_{ridges} \Delta(\bar{p} + \frac{1}{2}\bar{v}^2). \quad (20)$$

The only pressure forces remaining are those acting on submarine ridges which intersect the path at the level z.

Terms involving vertical advection (time mean and eddies) are small, in quasigeostrophic (QG) scaling they do not appear in the lowest order approximation of (18). Furthermore, since the circumpolar average of the geostrophic velocity vanishes, the Coriolis term is dominated by the ageostrophic velocity. Abbreviating the circumpolar integration by cornered brackets we have

$$< f\bar{v} > \approx f_0 < \bar{v}_{ag} > + \sum_{ridges} \Delta\bar{p}_g \quad (21)$$

and in QG scaling (20) then reads

$$-f_0 < \bar{v}_{ag} > - < \bar{v}_g \bar{\zeta}_g >=< \mathcal{F} > + < \mathcal{E} > - \sum_{ridges} \Delta(\bar{p}_{ag} + \frac{1}{2}\bar{v}_g^2). \quad (22)$$

Considering a geostrophic streamline as a circumpolar contour (so that $\bar{v}_g$ vanishes) it becomes clear at this stage that eddy and frictional effects must be important at levels where the path is not interrupted by topography. Below such levels the pressure force on submarine mountains may come into play. We will substantiate the importance of these terms below.

In the circumpolar average the term $\mathcal{E}$ representing the direct effect of transient eddies becomes

$$< \mathcal{E} >=< \overline{v'\zeta'} > - \sum_{ridges} \Delta(\frac{1}{2}\overline{v'^2}). \quad (23)$$

There are however indirect effects of transient eddies buried in the ageostrophic Coriolis force. Integrating the mass conservation equation over the horizontal area enclosed by the circumpolar path and the coastline of Antarctica and utilizing the buoyancy balance in the adiabatic form, it is found that the ageostrophic flow across the path is related to the lateral (mean and eddy) flux of buoyancy in the form

$$f_0 < \bar{v}_{ag} >= \frac{\partial}{\partial z}\frac{f_0}{N^2} < \bar{v}_g \bar{b} + \overline{v'b'} > \quad (24)$$

where b is the buoyancy and N the Brunt-Väisälä frequency. The averaged momentum balance (22) may then conveniently be rewritten in terms of the lateral (mean and eddy) flux of the QG potential vorticity

$$Q = f + \zeta_g + \frac{\partial}{\partial z}\frac{f_0}{N^2} b. \quad (25)$$

The flux due to transient eddies is given by

$$< \overline{v'Q'} >=< \overline{v'\zeta'} > + \frac{\partial}{\partial z}\frac{f_0}{N^2} < \overline{v'b'} > \quad (26)$$

The thermal wind relation allows to commute v with the vertical derivative. A corresponding expression holds for $< \bar{v}_g \bar{Q} >$. The balance (22) then becomes

$$-(< \bar{v}_g \bar{Q} > + < \overline{v'Q'} >) = < \mathcal{F} > - \sum_{ridges} \Delta \left(\bar{p}_{ag} + \frac{1}{2}(\bar{v}_g^2 + \overline{v'^2}) \right). \quad (27)$$

This equation expresses the dynamical balance of the ACC as the interplay of three distinct physical processes: those arising from eddies, friction and the time-mean pressure field. The frictional force $\mathcal{F}$ stems from the transfer of momentum due to mechanically driven three-dimensional small scale turbulence limited to the near-surface layer which is directly stirred by the wind. The terms on the lhs represent the lateral fluxes of potential vorticity. Notice that the first term describes the flux due to stationary standing eddies (evaluated with respect to the given arbitrary circumpolar contour). The second term is the flux of potential vorticity due to the transient eddies in the current. This flux is thus achieved by meso-scale turbulent motion which is quasi-two-dimensional and geostrophically balanced.

The eddy flux of potential vorticity consists of two parts. The lateral eddy relative vorticity flux $< \overline{v'\zeta'} >$ represents the convergence of eddy momentum fluxes. We show below that this is just the familiar (negative) meridional divergence of the Reynolds stress $< \overline{u'v'} >$ if the path is oriented zonally. The physics of the second part may be more easily understood if the buoyancy b' is replaced by the vertical displacement $z' = -b'/N^2$ of the mean isopycnals by the eddies. The eddy buoyancy flux

$$< \overline{v'b'} > = -N^2 < \overline{v'z'} > = -\frac{N^2}{f_0} < \overline{z'\frac{\partial p'}{\partial s}} > = \frac{N^2}{f_0} < \overline{p'\frac{\partial z'}{\partial s}} > \quad (28)$$

is then related to the eddy interfacial form stress $< \overline{z'\partial p'/\partial s} >$. It describes the acceleration of fluid by horizontal pressure forces $\partial p/\partial s$ on the eddy scale across the tilted interface $z' = const$ of a material fluid volume thus establishing a vertical transport of horizontal momentum (see e.g Rhines and Holland (1977) or Johnson and Bryden (1989)). In this Lagrangian framework the eddy buoyancy flux represents a vertical transfer of momentum through the water column. A similar reasoning applies to the contribution due to standing eddies.

The third process – the pressure forces on submarine mountains – acts only when the circumpolar contour hits submarine ridges. In the *ACC* belt this occurs at levels below about $2000m$ depth. This pressure force is a property of the time mean flow; it results by adjustment of the mass field to wind and thermohaline forcing and to the geometry and topography of the ocean basin (see e.g. Olbers et al. 1993).

The momentum balance for any layer of fluid is obtained by vertical integration. If the layer extends from a level $z = z_0$ to the bottom $z = -h(s)$ on the contour, the ageostrophic pressure force converts to a stress

$$\int_{-h_{max}}^{z_0} dz \sum_{ridges} \Delta\bar{p}_{ag} = - < \bar{p}_{ag}\frac{\partial h}{\partial s} >=< h\frac{\partial \bar{p}_{ag}}{\partial s} > \tag{29}$$

The corresponding geostrophic part of this stress results from the integration of the stretching term of the potential vorticity flux of the standing eddies

$$\int_{-h_{max}}^{z_0} dz < \bar{v}_g\bar{Q} >= \int_{-h_{max}}^{z_0} dz < \bar{v}_g\bar{\zeta}_g > - < h\frac{\partial \bar{p}_g}{\partial s} > \tag{30}$$

The stress $< h\partial\bar{p}/\partial s >$ is known as the bottom form stress or mountain drag. It transfers momentum from the fluid to the earth (or vice versa, in this case it is not a drag) by the time mean horizontal pressure force at the bottom. Apparently, this can work only if the ocean depth and the bottom pressure are out of phase. To lowest order in geostrophic scaling the bottom form stress may be evaluated using the geostrophic pressure field at the bottom.

The standing eddy contribution vanishes identically if the contour is a geostrophic streamline. In this case the flux of potential vorticity due to transient eddies remains to balance the frictional stress divergence and the pressure force. We see again – as discussed above for the heat flux – that the importance of contour-averaged flux terms depends very much on the shape of the contour. At levels which are not constrained by topography we then get the dynamical balance in the simple form

$$- < \overline{v'Q'} >=< \mathcal{F} >= \frac{\partial < \bar{\tau} >}{\partial z} \tag{31}$$

where τ is the frictional stress component parallel to the contour at level z (lateral friction may be neglected if eddies are resolved). Below the wind-mixed layer where the frictional stress vanishes the lateral flux of potential vorticity must vanish identically. This does not mean that eddies are absent or have no effect on the mean flow, rather it is indicating that different contributions to the eddy flux $< \overline{v'Q'} >$ compensate. As we know from chapter 2 the eddies must be quite vigorous to carry heat across circumpolar latitudes. The connection to the flux of potential vorticity is drawn below in section 3.3. We should realize here that the vanishing of $< \overline{v'Q'} >$ in the ocean interior requires the absence of a large-scale gradient of potential vorticity. Overwelming evidence of nearly constant potential vorticity on isopycnals in the Southern Ocean has been presented by Marshall et al. (1993).

Equation (27) gives a convenient framework to analyse the dynamical balance of a circumpolar flow. In the traditional view, s is taken along the zonal direction. Alternatively, natural coordinates may be taken, in which case s is along the streamline of the time mean geostrophic flow (e.g. Marshall et al. 1993). This approach highlights the underlying physics and role of the transient eddies, since standing eddies are absent. The relation between the normal eddy flux across an arbitrary path - established by standing and transient components - and the flux across a mean streamline - established by transients alone - is readily obtained by integrating the time-average potential vorticity balance over the enclosed area, resulting in

$$\oint_{any\ path} ds(\bar{v}_g\bar{Q} + \overline{v'Q'}) - \oint_{streamline} \overline{v'Q'} = \\ = \int_{enclosed\ area} dA\, \mathbf{k} \cdot \nabla \times \frac{\partial \tau}{\partial z}. \tag{32}$$

If there is no applied stress or friction – as in the interior where $\tau = 0$ – we see that the sum of steady and transient eddies across any contour is exactly equal to the transient eddy flux across a mean streamline. This points out the eminent role of the transient motions in the dynamical balance.

Other forms of the potential vorticity flux may be more familiar to the reader. Considering for example a zonal contour and for simplicity quasi-Cartesian coordinates we find

$$< \overline{v'Q'} > - \sum_{ridges} \frac{1}{2} < \overline{v'^2} > = -\frac{\partial}{\partial y} < \overline{u'v'} > + \frac{\partial}{\partial z} < \overline{p' \frac{\partial z'}{\partial x}} > \tag{33}$$

$$= (\frac{\partial}{\partial y}, \frac{\partial}{\partial z}) \cdot (- < \overline{u'v'} >, < \overline{p' \frac{\partial z'}{\partial x}} >). \tag{34}$$

As outlined above the first term in (33) describes the horizontal convergence of momentum, it arises directly from eddy terms in the momentum balance. The second term decribes vertical convergence of momentum and arises from the mean ageostrophic Coriolis force. The second identity (34) involves the divergence of the Eliassen-Palm vector, here written for the transient eddies (Eliassen and Palm 1961, see also Stevens and Ivchenko (1993) for the context of the *ACC*).

We should emphasize that the balances have been derived in this chapter on the basis of quasigeostrophic dynamics. Their essential form still applies to more complete physics. If we abandon the *QG* scaling the terms associated with vertical advection (mean and eddy) of momentum and buoyancy reappear in the balance (27), in particular the vertical eddy-induced flux $< \overline{u'w'} >$ of momentum. These terms are small. Including the thermodynamics considered in primitive equation models, however, adds another term in (27). In particular south of the *ACC* diabatic sources of buoyancy must be considered which force ageostrophic flow across the circumpolar path. A source S (the divergence of a flux of buoyancy due to convection or diffusion) in the buoyancy balance at any vertical level z converts to an integral over the area enclosed by the coastline and the circumpolar path in the form

$$f_0 V_0(z) = \frac{\partial}{\partial z} \int dA \frac{gS}{N^2} \tag{35}$$

in (24) and appears in addition to the standing and transient eddy flux divergences of buoyancy in the dynamical balance (27) which then takes the more general form

$$\begin{aligned} &-(< \bar{v}_g \bar{Q} > + < \overline{v'Q'} >) = \\ &= f_0 V_0(z) + < \mathcal{F} > - \sum_{ridges} \Delta \left(\bar{p}_{ag} + \frac{1}{2}(\bar{v}_g^2 + \overline{v'^2}) \right). \end{aligned} \tag{36}$$

Notice that the diabatic buoyancy sources stand on equal footing with the viscous terms, i.e. they appear in the form of vertical fluxes.

3.2 The dynamical balance in eddy resolving models of the Southern Ocean

It has only recently become possible to run full primitive equation models with eddy resolution in large ocean domains. The Southern Ocean presents a particular problem because of its vast area and its weak stratification (the baroclinic Rossby radius is between $5km$ and $15km$ (Houry et al. 1987)). If full thermodynamics are included and the stratification is realistic as e.g. in the Fine Resolution Antarctic Model (FRAM), eddies tend to be only marginally resolved (see FRAM 1991). On the other hand *QG* models of the *ACC* can choose the Rossby radii in relation to the given resolution of the model. These models therefore generally run with eddy resolution but too strong stratification.

The dynamical balance of a zonally-periodic current has been investigated in many eddy resolving multilayer channel models with *QG* dynamics and idealised topography (e.g. McWilliams et al. 1978, Wolff and Olbers 1989, Treguier and McWilliams 1990, Wolff et al. 1991, Marshall et al. 1993). These investigations have promoted the scenario first proposed by Munk and Palmén (1951): the momentum imparted by the wind to the fluid at the surface is carried down through the water column by the eddies and leaves the fluid via bottom form drag (this mechanism deccelerates the earth). These processes have been discussed above. Below the surface layer the momentum balance is thus essentially frictionless. Recent analysis of a *QG* model of the Southern Ocean with realistic geometry and topography (Wolf-Gladrow et al. 1993) and the *FRAM* model (Stevens and Ivchenko 1993) only marginally altered the basic concept. The latter authors showed in particular that the ageostrophic flow induced by diabatic physics cannot be neglected, i.e. the term $f_0 V_0$ is not small in (36).

The realistic *QG* model of the *ACC* is an extension of the classical β–plane channels used as an idealization of the circumpolar flow around the Antarctic continent. It has three layers of depth $1000m, 1500m$ and $2500m$ and a horizontal resolution of $1/3°$ zonally and $1/6°$ meridionally which the resolves the wave scale of the Rossby radius quite well (the baroclinic Rossby radii are $18km$ and $9km$). The domain extends from $40°S$ to the Antarctic continent. The coast line is the $500m$ depth contour around the continents, while at $40°S$ the model domain is closed by a solid wall. The deep ocean topography is a linear unsmoothed interpolation the 12′–data of the *DBDB5* data (Heirtzler 1985) on the model grid. The topographic height above $5000m$ is then scaled down to 20% following quasi-geostrophic requirements.

An instantaneous view of the circulation in the upper layer is shown in Figure 1. It exhibits the *ACC* (visible as the thin continuous bands associated with the fronts) imbedded in an intense mesoscale eddy variability. The latter is mostly found in areas with strong topographic features - namely the area behind the Drake Passage, the Kerguelen in the sector of the Indian Ocean and the Maquarie Ridge area south of New Zealand. The flow has a strong barotropic component so that the currents at deeper levels appear rather similar.

Figure 2 displays the terms of the zonally averaged balance of the zonal momentum of the three layers of the model. The dynamical balance (27) is thus considered with averaging along latitudes and integration is performed over the individual layers. The vertical divergence of the interfacial form stress then appears as the difference between the values at the top and bottom of each layer and the pressure force on the bottom converts to the bottom form stress as given by (29) and (30). Zonal averaging has also been applied to latitudes that are blocked by continents (South America north of $55°S$ and the Antarctic Peninsula south of $62°S$) so that in addition to the terms considered for unconstrained latitudes, a pressure force appears in the balance by which the flow acts on the continental barriers. In contrast to the bottom form stress which to lowest non-vanishing order can be evaluated geostrophically, this pressure force on continents contains the ageostrophic pressure (the geostrophic pressure cancels since the continental boundary is a geostrophic streamline). It is shown in Wolf-Gladrow et al. (1993) how the ageostrophic pressure is calculated on solid lateral boundaries.

It is seen in the figures that the wind input in the top layer is compensated almost entirely by the eddy interfacial form stress. This mechanism hands the momentum down through the water column to the bottom layer. Here it leaves via the bottom form stress. All explicit friction terms are negligible, also the horizontal Reynolds stresses caused by the eddies and the pressure force on continents have only a minor role. They are not shown on the figures.

The terms of the balance are combined in Figure 3 to the lateral flux of potential vorticity by the eddies (transient and standing) as described in the preceding chapter. Apparently, in the middle layer, which has no explicit friction, the eddy flux of potential vorticity is fairly small. This is also valid in the bottom layer since bottom friction is small. In the top layer, however, the eddy flux is balanced by the prescribed wind stress. It should be noted that in the interior where vorticity sources are

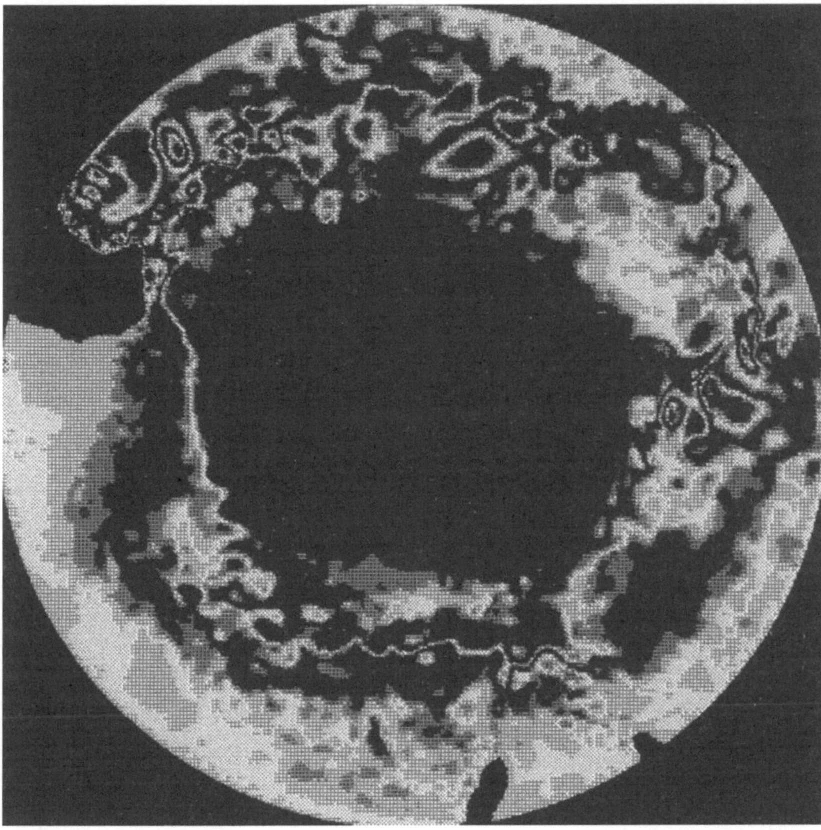

Figure 1: An instananeous view of the streamfunction in the top layer of the quasigeostrophic simulation of the *ACC*. The streamfunction is displayed in gray shading (the scale is repeated three times, each step corresponds to approximately $2Sv$. The transport in this layer is $74Sv$, the two deeper layers carry $105Sv$.

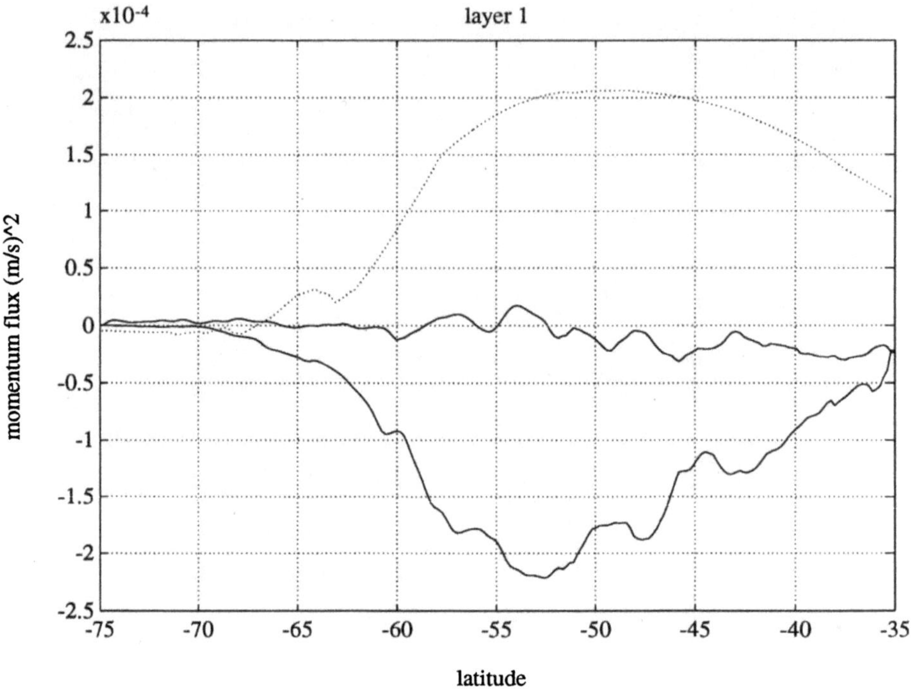
x10-4
layer 1
2.5
2
1.5
1
0.5
0
-0.5
-1
-1.5
-2
-2.5
momentum flux (m/s)^2
-75
-70
-65
-60
-55
-50
-45
-40
-35
latitude

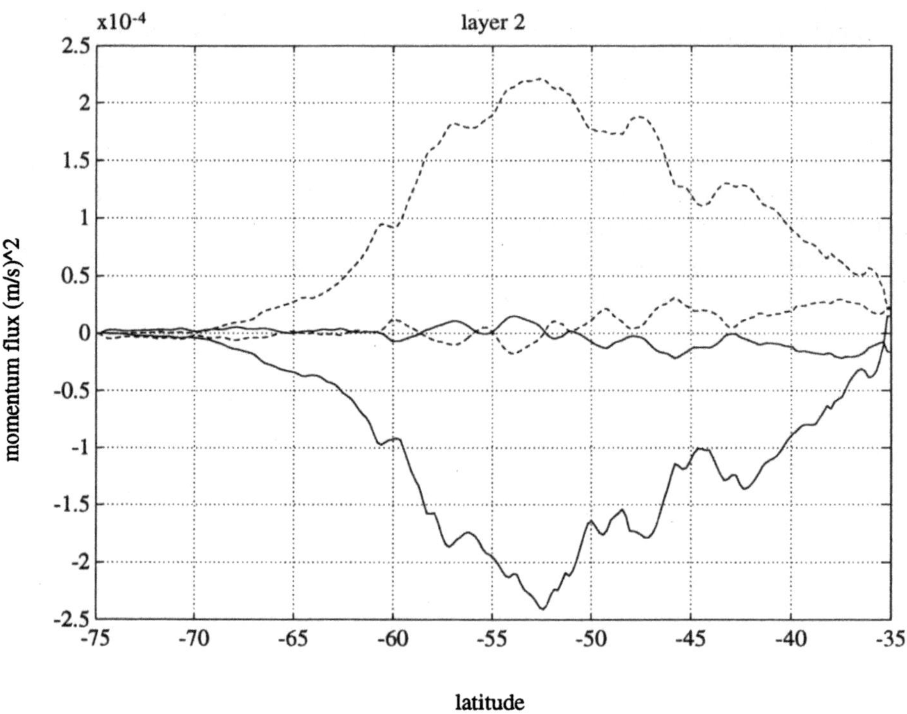
x10-4
layer 2
2.5
2
1.5
1
0.5
0
-0.5
-1
-1.5
-2
-2.5
momentum flux (m/s)^2
-75
-70
-65
-60
-55
-50
-45
-40
-35
latitude

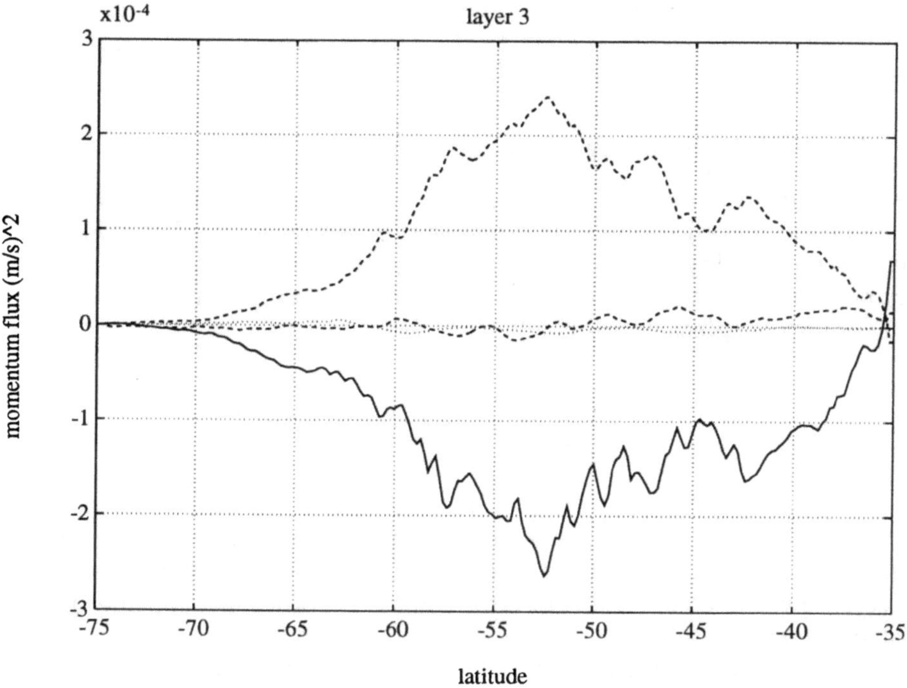

Figure 2: The time and zonally averaged zonal momentum balance for the three layers of the quasigeostrophic simulation of the *ACC*. The three panels display the wind input (in the top layer 1), the bottom friction and the bottom form stress (in the bottom layer 3), and the standing and transient interfacial form stresses (in all three layers). The frictional fluxes are plotted in dotted lines, the form stresses in dashed lines if they are sources of momentum for the particular layer, or solid lines if they are sinks of momentum. In each case the transient eddies term is the smaller contribution.

absent, the potential vorticity becomes homogeneous as a consequence of the mixing by the vigorous eddy field. This is in good agreement with observation of this quantity (Marshall et al. 1993).

3.3 The relation to the eddy heat flux

The momentum balance of the mean flow can be used to infer the magnitude of the meridional heat flux across the *ACC* (i.e. the Polar Front). The interfacial eddy form stress can be expressed in terms of the lateral eddy heat transport by relating the buoyancy and temperature variations of the eddies (neglecting salinity)

$$\frac{< \overline{v'b'} >}{N^2} = \frac{< \overline{v'T'} >}{\frac{dT_m}{dz}} \tag{37}$$

where $T_m(z)$ is the profile of the mean stratification in the area. A corresponding relation holds for the contribution due to standing eddies. We can however concentrate on the transient eddies by considering a geostrophic streamline.

We have seen above that the redistribution of momentum by lateral Reynolds stresses is small. Below the wind-driven layer, frictional effects are small so that by (26) and (31) the eddy heat flux in the water column is a constant. Its magnitude is set by the surface wind stress τ_s (Marshall et al. 1993). Integration of (31) yields for these conditions

$$[< \overline{v'T'} >)]_{z=0}^{z} = \frac{\tau_s}{f_0}\frac{dT_m}{dz}. \tag{38}$$

This implies that the eddy-driven northward heat flux is given approximately by

$$HF^e = hL\rho c_p \frac{\tau_s}{f_0}\frac{dT_m}{dz} \tag{39}$$

where h is the ocean depth and L is the circumference of the path in question. The total meridional heat flux by the eddies in the circumpolar belt is thus set via the dynamical balance of the *ACC* by the surface wind stress. With reasonable numbers for the stratification at the Polar Front ($dT_m/dz \sim 3°C/4000m, \tau_s \sim 10^{-4}m^2s^{-2}$) we find an eddy heat flux

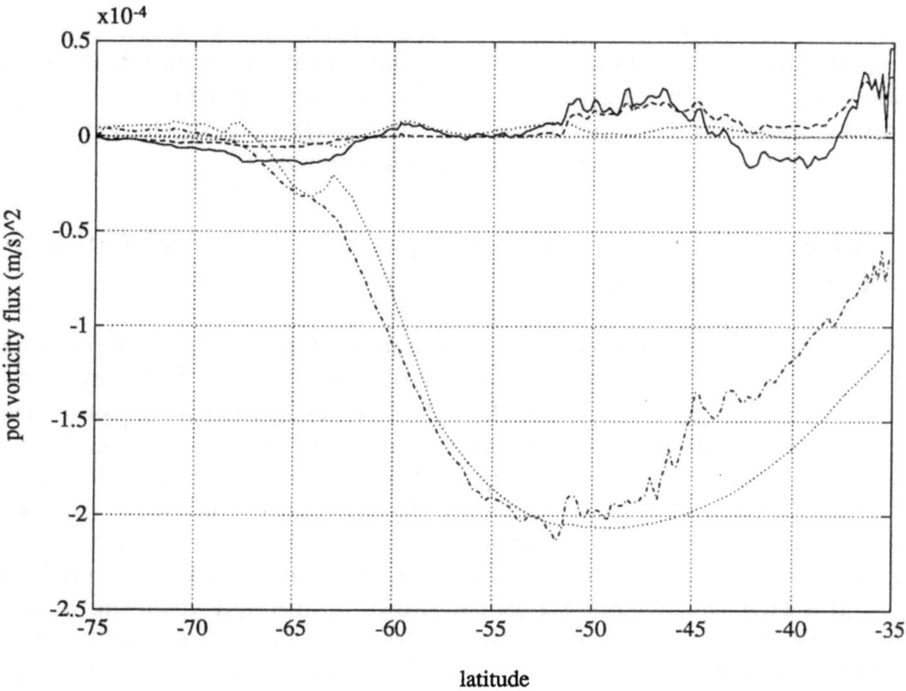

Figure 3: The time and zonally averaged potential vorticity flux (the sum of standing and transient eddies) and the friction terms for the three layers of the QG model. The friction terms are dotted lines, potential vorticity flux in the top layer is dashdot, the flux in the intermediate layer is the dashed line, and the flux in the bottom layer is the solid line.

of magnitude $0.24PW$ to the south as required by the rough analysis given in chapter 2. Though this result has been supported here by a model based on reduced physics it should also apply in a full ocean eddy resolving model and in the real ocean. It emerges out of the generally accepted scenario of a dynamical balance for the *ACC*.

4 The water mass connection: formation of Antarctic Bottom Water

Oceanic motions are approximately adiabatic. Once a water parcel has got its imprint of temperature and salinity by heat and freshwater exchange with the air at the sea surface it conserves potential temperature and salinity thus moving on isopycnals. The distribution of potential density in the world ocean – as exemplified by the zonally averaged meridional distribution (Figure 4) – shows that the deep cold water masses must originate from the polar seas. There is a drastic mismatch between the volume of the cold water sphere and its window at the sea surface through which the density structure allows interaction with the atmosphere. As an example, water with temperatures below $4°C$ occupy 75% of the ocean volume but are in contact with atmosphere at only 4% of the sea surface. These polar gateways of the cold water sphere are thought to define the key role of the polar oceans in global climate.

Formation of deep water takes place in northern and southern high latitudes but the deepest and coldest water of the world ocean originates entirely in the south. Antarctic Bottom Water (*AABW*) spreads far into the northern hemisphere filling the deep layers and influencing directly about 30% of the ocean volume (Worthington 1981).

AABW is produced by a range of delicate formation processes which involve deep reaching convection within cells of very small diameter, and intricate mixing processes of water masses modified by sea ice freezing and shelf ice melting on the continental shelf and slope. The formation occurs south of the Polar Front either in the open ocean or along the continental margin, the latter predominantly in the southern Weddell Sea where according to Carmack (1977) 70% of the total *AABW* is formed. The basic mechanisms of the formation are understood (see e.g. Carmack 1986). Only little, however, is known of the interplay of all the ingredients of *AABW* production and transport - convection, mixing and propagation - in particular the rates of formation and transport are subject to wild speculations. We can be sure that all these processes are not adequately included in any of the coarse ocean circulation models used for climate studies.

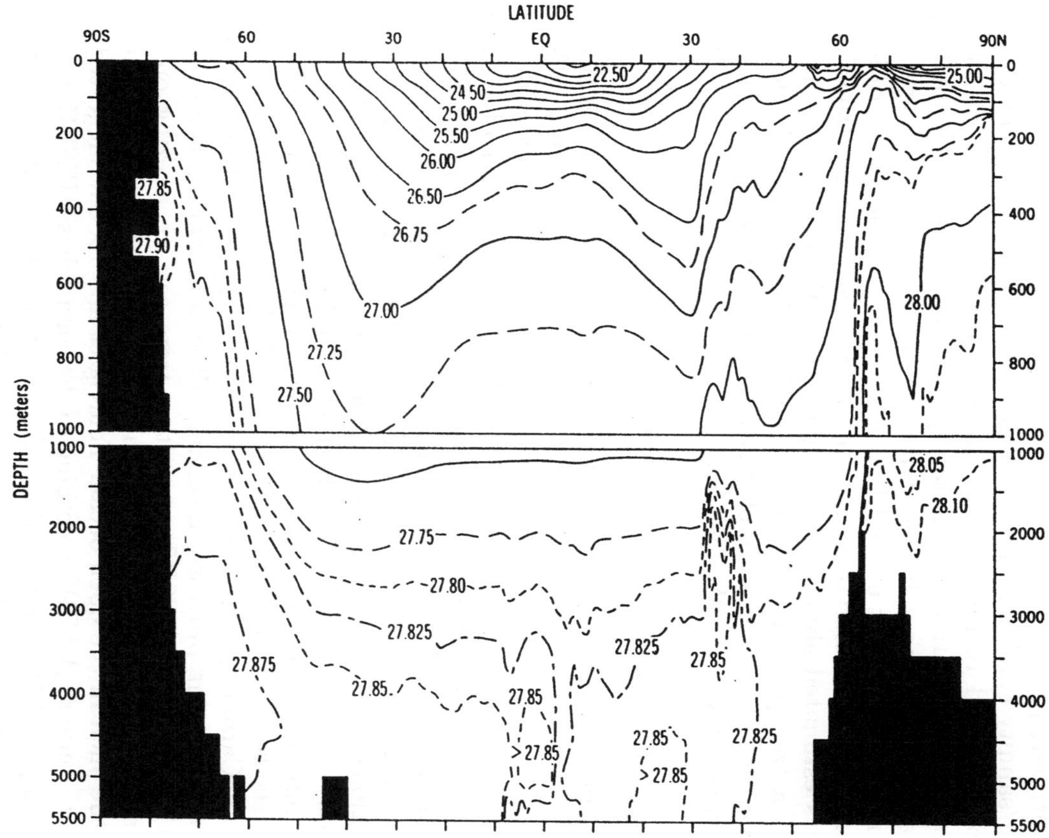

Figure 4: A section of the zonally averaged potential density in the world ocean (units $10^{-3}gcm^{-3}$), (from Levitus 1982).

The ultimate product *AABW* - in various different characteristic temperature/salinity combinations as a consequence of the different formation processes and regions - is assumed to leave the Southern Ocean via deep boundary currents along topographic barriers into the northern ocean basins. Any change in the environmental conditions leading to the production and transport of *AABW* will have an effect on the meridional heat transport. The *AABW* component of the heat flux across the Polar Front was estimated above to be roughly $0.07PW$ to the south. It was based on a transport of $10Sv$ across the Polar Front. We will demonstrate the range of uncertainty of this value by modelling the dominant formation process in the Weddell Sea.

4.1 Formation of *AABW* in the Weddell Sea

The interior of this subpolar gyre is filled with three water masses sitting on top of each other: the Warm Deep Water (*WDW*, roughly 30% of volume, below the surface water masses down to roughly $1000m$ depth) with temperatures in the range $0°C$ to $0.8°C$, a broad layer of *AABW* (about 57%, between $1000m$ and $4500m$ depth) in the range $-0.8°C$ to $0°C$ and a thin layer of cold Weddell Sea Bottom Water (*WSBW*, roughly 3%) in the range $-1.4°C$ to $-0.8°C$. Figure 5 shows these water masses for a section across the Weddell Gyre made during the Winter Weddell Gyre Study in October 1989 (Fahrbach et al. 1991).

Traditionally *AABW* is considered as the mixing product of *WDW* and *WSBW* (see the T-S-diagrams in Figure 6). Following the hypothesis of Foster and Carmack (1976) *WSBW* is formed on the continental shelf in a series of mixing processes involving *WDW* and two further species: Western Shelf Water (*WSW*) - resulting from sea ice freezing on the shelf - and Winter Water (*WW*) which is a remnant of sea ice formation outside the shelf in the open ocean. Foldvig et al. (1985a) suggested an alternative route to *AABW*: according to their hypothesis *WSBW* results from mixing of *WDW* and Ice Shelf Water (*ISW*). This latter is a very cold species (-2.4 to $-2.0°C$) which is formed by thermohaline processes under the huge Filchner and Ronne Ice Shelves flowing from the Antarctic continent out on top of the ocean. Potential temperature sections parallel to the ice shelf edges, representing typical Antarctic summer conditions, reveal distinct cores of *ISW* with temperatures as low as $-2.2°C$ which is well below the freezing point at the surface (about $-1.87°C$ at salinity 35.0).

Sverdrup (1940) assumed that *ISW* results from the modification of shelf water masses due to heat loss at the base of deep-drafting ice shelves. A more detailed description concerning the whole sub-ice shelf system

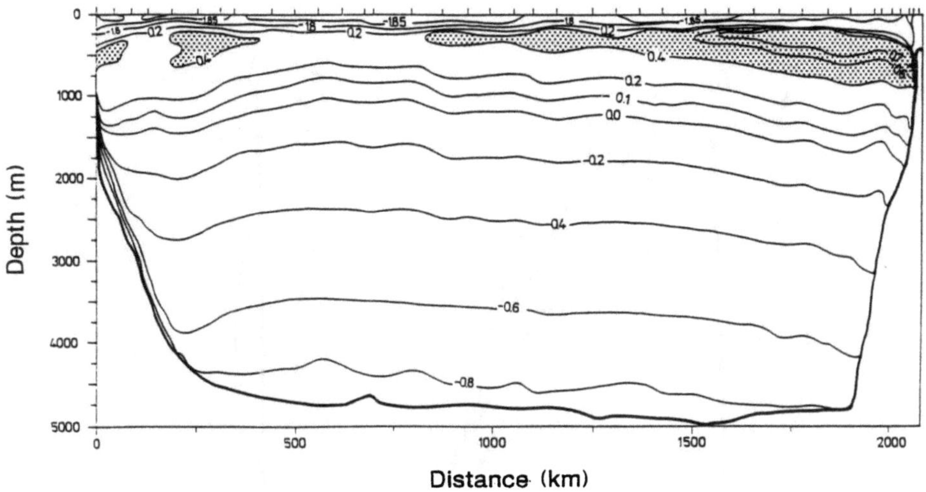

Figure 5: The section of temperature through the Weddell Sea in austral winter of 1989 (from the Winter Weddell Gyre Study 1989), at left the Antarctic Peninsula and at right Cape Norwegia on Antarctica. There is a break in the contour interval at $0.2°C$ to display the temperatures in the surface layer which are close to the freezing point (from Fahrbach et al. 1991).

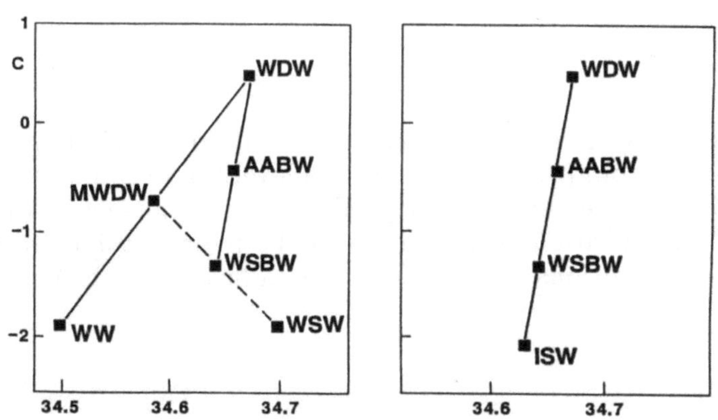

Figure 6: The mixing processes for production of $AABW$ in the T, S-diagram (from Fahrbach et al. 1992). Water masses are indicated by the squares, mixing path ways by lines connecting the squares. The left panel shows formations of $AABW$ through the Modified Warm Deep Water ($MWDW$) route (Foster and Carmack 1976), while the right panel shows the formation though the Ice Shelf Water (ISW) route (Foldvig et al. 1985a).

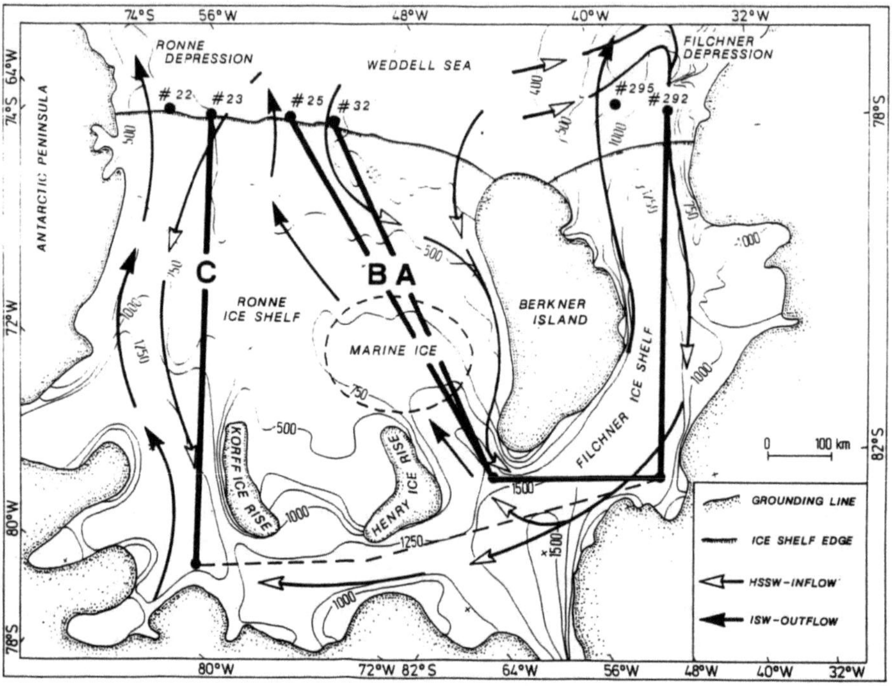

Figure 7: Map of the Filchner-Ronne Ice Shelf area, including the topography of ice bottom (in meters below sea level) and the possible paths of Western Shelf Water and Ice Shelf Water along the sections labeled A, B, and C. The location and numbers of the hydrographic stations used as boundary conditions for the model are indicated at the coast line.

was given by Robin (1979). He proposed that melting occurs at the grounding line and accumulation of ice near the ice shelf edge and related these processes to a single, essentially two-dimensional circulation regime. Hellmer and Olbers (1989) were the first to consider the full dynamical-thermodynamical implications of such a circulation cell in a two-dimensional model and revealed the essential interconnection of thermohaline processes at the ice shelf base and the sub-ice shelf circulation. Melting and freezing at the base of an ice shelf not only result from the sub-ice shelf circulation but also drive it and, more specifically, lead to the modification of the shelf water masses.

With reference to Hellmer and Olbers (1991) we discuss some results on the formation of *ISW* under the Filchner (*FIS*) and Ronne (*RIS*) Ice Shelves. *ISW* is observed in front of the *FIS* and can be traced from the ice shelf edge until it flows over a sill at the continental shelf break (Foldvig et al. 1985a,b) where it starts its mixing processes to produce *AABW* as desribed above. Two hypotheses are being discussed concerning the path of Western Shelf Water (*WSW*) which is the source water mass of *ISW*. This water mass originates from the freezing of sea ice along the coast of the western Weddell Sea. Its temperature is therefore at the freezing point (about $-1.86°C$) corresponding to surface pressure and salinity. *WSW* is observed to enter the *FIS* in the Filchner Depression east of Berkner Island (Robin et al. 1983). As an alternative path Foldvig et al. (1985a) have postulated that *WSW* enter on the western side of the island and flows under the *RIS* around to the Filchner side.

In any case the flow is driven under the ice by the thermohaline torque exerted by the temperature difference between the *WSW* at the front and the local freezing temperature at the shelf ice grounding line. The water comes in contact with the ice at the grounding line and is cooled and diluted by melting of basal ice. It creepes upward under the bottom of the shelf ice towards the front where it is expelled as the very cold *ISW*. If the water achieves local freezing temperature before the front is reached further pressure reduction will lead to super-cooling beyond the in-situ freezing point. This thermodynamic disequilibrium is compensated by the spontaneous formation of ice crystals in the water column and the accumulation of this marine ice under the ice shelf. Freezing rates of $1m/yr$ or higher may occur.

Hellmer and Olbers (1991) simulated the thermohaline circulation around Berkner Island with a 2-dimensional model for different vertical sections indicated on Figure 7. At the shelf ice front, temperature and salinity profiles are prescribed from observations (stations 22, 25, and 32 from Foldvig et al. 1985b, station 292 from Carmack and Foster 1975).

Though the paths under the RIS are only a few hundred km apart the circulation along them differs drastically (Figure 8). Along path A (Figure 8a) the *FIS* is dominated by an anticlockwise cell transporting warm and salty water near the bottom into the sub-ice cavity. It diverges at the deepest point of the section at the tip of Berkner Island, the major part flows back to the *FIS* side initiating the thermohaline scenario described above. A small branch supports a flow through the sub-ice channel towards the front on the *RIS*. Under the *RIS*, however, an internal cell arises which is essentially driven by an ice pump mechanism (Lewis and Perkin 1986): in the deeper part the ice melts and the diluted plume rises to lower pressures where freezing takes over, and the ice is indirectly pumped upwards. The circulation cell is driven by the re-supply of basal ice by the ice flow. Notice that water does not enter at the *RIS* front for path A. The meltwater plume which leaves there is modified *WSW* from the *FIS* side. The corresponding temperatures are displayed in Figure 8b. Path B gives only a slight variation of A.

The basic difference for path C (Figure 8c) is a slightly more saline bottom water at the *RIS* front (34.68 for C compared to 34.65 for A and 34.67 for B). Both regimes are dominated by cells of the same orientation but in C the transport of the *FIS*-cell is decreased by a factor of two and the *RIS*-cell is increased by a factor of four. The latter now transports bottom water from the *RIS* side into the interior and feeds the meltwater plumes of *ISW* on both sides of the Berkner Island.

Further investigation revealed that a critical salinity (for the present model version this is near 34.67) exists for the bottom layer at the *RIS* front which distinguishes between the two flow regimes shown here for the paths A (achieved for lower salinities) and C (achieved for higher salinities). They differ considerably with respect to their strength, their melting rates and of course also in their production rates and heat content of the resulting *ISW*. It is indeed conceivable that the seasonal cycle of ice freezing in the western shelf area of the Weddell Sea switches between the conditions of path A and path C.

4.2 An estimate of the formation rate of *AABW*

In view of the complex picture of *AABW* formation, we must realize that the route of reasoning for connections between the poleward heat transport and thermohaline processes in the remote areas around Antartica is full of side ways (we have not even considered open ocean convection) and paved with our lack of knowledge on many singles steps. Nevertheless, let us attempt to infer the formation rate of *AABW* from the model and mixing concept described above.

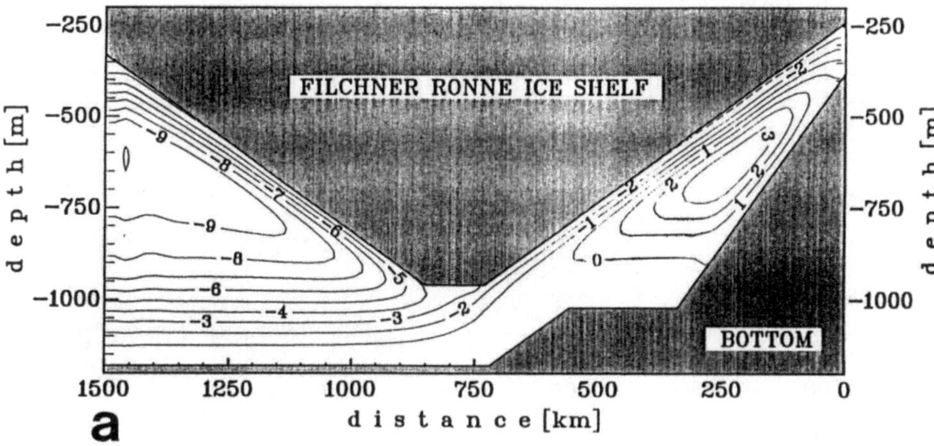
FILCHNER RONNE ICE SHELF
BOTTOM
depth [m]
distance [km]

a

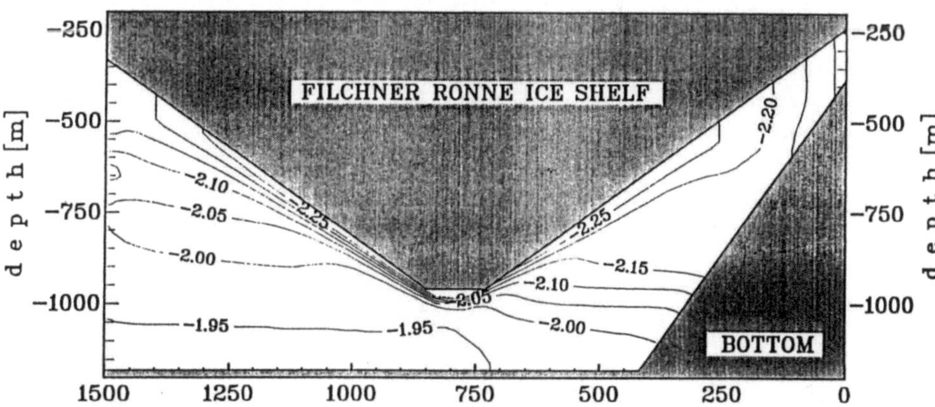
FILCHNER RONNE ICE SHELF
BOTTOM
depth [m]
distance [km]

b

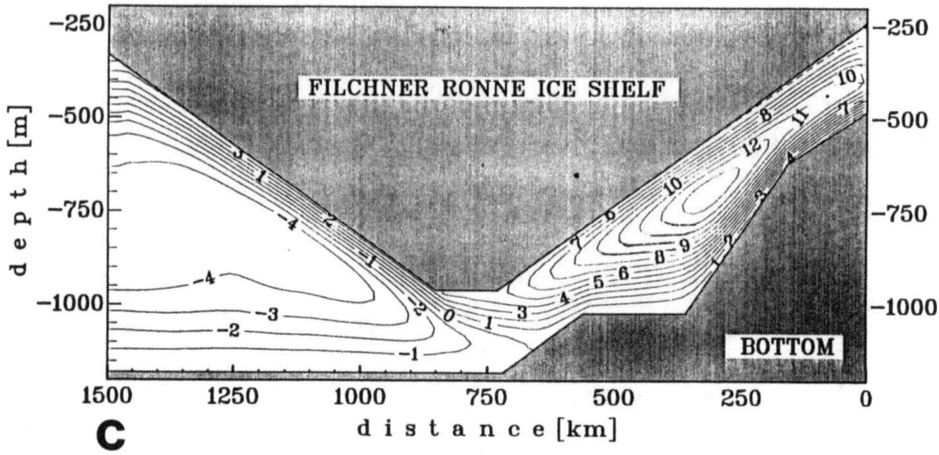

C

Figure 8: The flow beneath the Filcher-Ronne Ice Shelf for scenario A (top panel) with corresponding potential temperatures (middle panel) and the flow for scenario C (bottom panel). Contour interval of the streamfunction is $1m^2s^{-1}$, for temperature $0.05°C$.

In both circulation patterns under the shelf cavity the two-dimensional mass transport of *ISW* out of the *FIS* is $\approx 10 m^2 s^{-1}$, and, if multiplied by $120 km$, the observed width of the *ISW* core on the western flank of the Filchner depression (Foldvig et al. 1985b) we arrive at about $1 Sv$ for the rate of production of *ISW*. This agrees favourably with Foldvig et al. (1985a) who inferred the transport of *ISW* in the Filchner depression from current meter messurements.

Following now Foldvig's route *ISW* + *WDW* → *WSBW* and *WSBW* + *WDW* → *AABW* it is an easy matter to estimate the fraction of *ISW* in each parcel of *AABW* using either heat or salt conservation in this mixing scheme. Thus we find from the heat conservation equations

$$
\begin{aligned}
(1-\alpha)T_{ISW} + \alpha T_{WDW} &= T_{WSBW} \\
(1-\beta)T_{WSBW} + \beta T_{WDW} &= T_{AABW}
\end{aligned}
\tag{40}
$$

that the basic members of the mixing process satisfy

$$
(1-\alpha)(1-\beta)T_{ISW} + (\alpha(1-\beta)+\beta)T_{WDW} = T_{AABW} \tag{41}
$$

With $T_{ISW} = -2.2°C, T_{WDW} = 0.5°C, T_{WSBW} = -1.2°C$ and $T_{AABW} = -0.4°C$ (mean values of the range in the Weddell Sea given by Foldvig et al. (1985a) and Carmack (1977)) we get $\alpha \approx 0.3$, $\beta \approx 0.5$ so that the fraction of *ISW* in *AABW* is given by $(1-\alpha)(1-\beta) \approx 0.35$.

Hence the formation rate of *AABW* in the Weddell Sea is approximately 3 times that of *ISW*. Following Carmack's (1977) estimate that only 70% of *AABW* is formed in this area we arrive, with a brave extrapolation, at a production rate of $5 Sv$ for *AABW* from the *ISW*-route all around Antarctica. The formation process (Figure 6, left panel) involving Modified Warm Deep Water suggested by Foster and Carmack (1976) would add to this number. However, measurement of the level of the stable oxygen isotope O^{18} in *AABW* (Schlosser et al. 1990) indicates that the bottom water receives a significant fraction of meteoric melt water resulting from the melting of shelf ice. It is therefore believed that the *ISW*-route dominates the production of *AABW*. The value of $10 Sv$ used for the transport of *AABW* across the Polar Front in chapter 2 can thus be seen as an upper limit.

5 Conclusions

The Southern Ocean is responsible for the cold climate of the present state of the world ocean. South of the Polar Front relatively warm water masses entering from northern oceans - mainly the Atlantic Ocean - are transformed to cold and heavy waters - notably the Antarctic Bottom Water - which spread northward and fill the bottom layers of the world ocean. The processes producing the cold deep water masses occur on small scales such as deep convection cells originating as a consequence of intricate thermohaline interaction of the ocean surface layer with the air and sea ice or with the the glacial ice at the bottom of the huge ice shelves around the Antarctic continent. The modification involves the mixing of various locally-produced water masses and others induced from remote areas. This process occurs on scales associated with the topography of the continental shelves and slopes. Also these scales are small compared to the broad ocean circulation.

The replenishment of heat for the Antarctic zone south of the Polar Front is mainly accomplished by eddy activity arising from instability processes of the Antarctic Circumpolar Current. Hence this important process is again operating on scales much smaller than the mean circulation itself. We have also shown that the heat flux is intimately related to the dynamical balance of the mean flow in which mesoscale eddies play a dominant role.

We have demonstrated that most of the above-mentioned processes can be simulated by suitably designed numerical models. In particular we have discussed results from an eddy resolving model of the circumpolar circulation in the Southern Ocean and from a model of the thermohaline induced circulation and water mass modification under the Filchner-Ronne Shelf Ice. Evidently a favourable comparison of model results with reality is mainly due to the adequate high resolution of each model, which was necessary to simulate the relevant processes on their intrinsic scales.

It also became evident that none of the processes in question could be resolved in any respect in the ocean components of the coarse climate models used today for investigating natural climate variabilty and man-made climate trends. The resolution of present coupled ocean-atmosphere models have a resolution of 4° to 6° in latitude and longitude (see e.g. Cubasch et al. 1992, Manabe et al. 1990) which is totally inadequate to implement eddy dynamics, convection, shelf and slope processes and even the thermohaline processes of air-ocean-ice interaction, all of which have a particular importance in the immediate region of the Southern Ocean as well as substantial influence on the rest of the world ocean.

References

BERGER, W. H., 1981: Paleoceanography: The deep-sea record. *In: The oceanic lithosphere The Sea Vol.* **7** (Ed. C. Emiliani), Wiley-Interscience, NY, pp. 1437-1519.

BRYAN, K., 1982: Poleward heat transport by the ocean. *Ann. Rev. Earth Planet. Sci.* , **10,** 15-38.

BRYDEN, H. L., 1983: The Southern Ocean. Eddies in Marine Science, A. R. Robinson, Ed., Springer, *265-277.*

CARMACK, E.C. AND T.D. FOSTER, 1975: Circulation and distribution of oceanographic properties near the Filchner Ice Shelf. *Deep-Sea Research,* **22,** 77-90.

CARMACK, E.C., 1986: Circulation and mixing in ice-covered waters. *In: Untersteiner, N. (Ed.), The geophysics of sea ice. NATO ASI Series B. Physics,* **Vol 146,** Plenum Publishing Corp., New York, 641-712.

CARMACK, E.C., 1977: Water characteristics of the Southern Ocean south of the Polar Front. In: *A voyage of Discovery (Ed. M. V. Angel), Pergamon Press, Oxford, 15-41.*

CUBASCH, U., HASSELMANN, K., HÖCK, H., MAIER-REIMER, E., MIKOLAJEWICZ, U., SANTER, B.D., AND R. SAUSEN 1991: Time-dependent Greenhouse warming computations with a coupled ocean-atmosphere model, 1992. *Climate Dyn.,***8,** 55-69.

DESZOEKE, R. A., AND M. D. LEVINE, 1981: The advective flux of heat by mean geostrophic motions in the southern ocean. *Deep-Sea Res.,* **28,** 1057-1085.

ELIASSEN, A. AND E. PALM, 1961: On the transfer of energy in stationary mountain waves. *Geofys. Publ,* **22,** 1-23.

EMERY, W.L. AND J. MEINCKE, 1986: Global water masses: summary and review. *Oceanologica Acta,* **9,** 383-391.

FAHRBACH, E., M. KNOCHE AND G. ROHARDT, 1991: An estimate of water mass transformation in the southern Weddell Sea. *Marine Chemistry,* **35,** 25-44.

FAHRBACH, E., AUGSTEIN, E., AND D. OLBERS, 1992: Impact of shelf and sea ice on water mass modifications and large scale oceanic circulation in the Weddell Sea. *Subm. to Antartic Science–Global Concern, Conference Volume.*

FOLDVIK, A., T. GAMMELSRÖD AND T. TÖRRESEN, 1985A: Circulation and water masses on the southern Weddell Sea Shelf. *Antarctic Research Series,* **43,** 5-20.

FOLDVIK, A., T. GAMMELSRÖD, N. SLOTSVIK AND T. TÖRRESEN, 1985B: Oceanographic conditions on the Weddell Sea Shelf during the German Antarctic Expedition 1979/80. *Polar Research,* **3,** 209-226.

FOSTER, T. D., AND E. C. CARMACK, 1976: Frontal zone mixing and Antarctic Bottom Water formation in the southern Weddell Sea. *Deep-Sea-Res.* **23,** 301-317.

FRAM GROUP, 1991. An eddy-resolving model of the Southern Ocean. *EOS, 72,* **15,** 169-175.

GILL, A. E., 1968: A linear model of the Antarctic Circumpolar Current. *J. Fluid Mech.,* **32,** 465-488

GORDON, A.L. 1986: Interocean exchange of thermocline water. *J. Geophys. Res.,* **91,** 5037-5046.

GORDON, A. L., 1988: The Southern Ocean and global climate. *Oceanus* **31,** 39-46.

GORDON, A.L. AND W.B. OWENS, 1987: Polar Oceans. *Review of Geophysics,* **25,** 227-233.

HEIRTZLER, J.R. (ED.), 1985: Relief of the surface of the earth. *Rep. MGG-2, National Geophysical Data Center, Boulder, Colorado.*

HELLMER, H.H., AND D. OLBERS, 1989: A two-dimensional model for the thermohaline circulation under an ice shelf , *Antarctic Science* **1** , 325-336.

HELLMER, H.H., AND D.J. OLBERS, 1991: On the thermohaline circulation under the Filchner-Ronne Ice Shelf, *Antarctic Science,* **3,** 433-442.

HIDAKA, K., AND M. TSUCHIYA, 1953: On the Antarctic Circumpolar Current. *J. Mar. Res.,* **12,** 214-222

HOURY, S., E. DOMBROWSKY, P. DE MEY, AND J.-F. MINSTER, 1987: Brunt Väisälä frequency and Rossby radii in the Southern Ocean, *J. Phys. Oceanogr.,* **17,** 1619-1626.

JOHNSON, G. C. AND H. L. BRYDEN, 1989: On the size of the Antarctic Circumpolar Current. *Deep-Sea Res.,* **36,** 39-53.

LEVITUS, S., 1982 Climatological atlas of the world ocean. *NOAA Tech. Pap.* **3** , 173.

LEWIS, E. L., AND R. G. PERKIN, 1986: Ice pumps and their rates. *J. Geophys. Res.* **91,** 11756-11762.

MANABE, S., K. BRYAN AND M. J. SPELMAN, 1990: Transient response of a global ocean-atmosphere model to a doubling of atmopheric carbon dioxide. *J. Phys. Oceanogr.* **20,** 722-749.

MARSHALL, J., OLBERS, D. AND H. ROSS, 1993, Potential vorticity constraints on the dynamics and hydrography in the Southern Ocean. *J. Phys. Oceanogr., to appear.*

MCWILLIAMS, J. C., HOLLAND, W. R., CHOW, J. H. S., 1978: A description of numerical Antarctic Circumpolar Currents. *Dyn. Atmos. Oceans,* **2,** 213-291.

MUNK, W. H., AND E. PALMÉN, 1951: Note on the dynamics of the Antarctic Circumpolar Current . *Tellus,* **3,** 53-55.

NOWLIN JR., W.D. AND J.M. KLINCK, 1986 'The Physics of the Antarctic Circumpolar Current' *Reviews of Geophysics* **24,** 469-491.

OLBERS, D. J., AND M. WENZEL, 1989: Determining diffusivities from hydrographic data by inverse methods with applications to the circumpolar current. *Modelling the ocean general circulation and geochemical tracer transport. J. Willebrand and D. L. T. Anderson, Eds., NATO ASI Series C, Kluwer Acad. Publ.,* **284,** 95-122.

OLBERS, D., WÜBBER, CHR., AND J.-O. WOLF, 1993: The dynamical balance of wind and buoyancy driven circumpolar currents. *Submitted.*

RHINES, P.B., AND W.R. HOLLAND, 1979: A theoretical discussion of eddy-driven mean flows. *Dyn. Atmosph. Oceans* , **3,** 289-325.

ROBIN, G. DE Q., 1979: Formation, flow, and disintegration of ice shelves. *Journal of Glaciology,* **24** , 259-271.

ROBIN, G. DE Q., C.S.M. DOAKE, H. KOHNEN, R.D. CRABTREE, S.R. JORDAN AND D. MÖLLER, 1983: Regime of the Filchner-Ronne ice shelves, Antarctica.*Nature,***302,** 582-586.

STEVENS, D.P., AND V.O.IVCHENKO, 1993: The zonal momentum balance in a realistic eddy resolving general circulation model of the Southern Ocean. *In preparation.*

SVERDRUP, H.U., 1940: Hydrology, Section II: Discussion. B.A.N.Z. Antarctic Research Expedition 1929-31, Reports-Ser. A, III, Oceanography, *Part 2, 88-126.*

TREGUIER, A. M., AND J. C. MCWILLIAMS, 1990: Topographic influences on wind-driven, stratified flow in β-plane channel: an idealized model for the Antarctic Circumpolar Current, *J. Phys. Oceanogr.*, **20**, 321-343.

WARRICK, R.A., AND H. OERLEMANS, 1990: Sea level rise. *In: Houghton, J.T. et al.(Eds), Climate change, the IPCC scientific assessment. University Press, Cambridge, 256-281.*

WHITWORTH III, T., 1988: The Antarctic Circumpolar Current. *Oceanus* **31**, 53-58.

WOLF-GLADROW, D., A. HENSE, J. SCHRÖTER AND D. OLBERS, 1993: A quasigeostrophic eddy-resolving model of the Antarctic Circumpolar Current, *In preparation.*

WOLFF, J.-O. AND D. OLBERS, 1989: The dynamical balance of the Antarctic Circumpolar Current studied with an eddy resolving quasigeostrophic model, *in: Mesoscale/synoptic coherent structures in geophysical turbulence, Ed. J. Nihoul, Elsevier Science Publ., 435-458.*

WOLFF, J.-O., E. MAIER-REIMER UND D.J. OLBERS, 1991: Wind-driven flow over topography in a zonal -plane channel: A quasigeostrophic model of the Antarctic Circumpolar Current, *J. Phys. Oceanogr.* **21**, 236-264

WORTHINGTON, L.V., 1981: The water masses of the world ocean: some results of a fine-scale census. *In: Warren, B,A., and C. Wunsch (Eds.): Evolution in physical oceanography. The MIT Press, Cambridge, 42-69.*

MODELLING SEA ICE - MIXED LAYER INTERACTION

P. Lemke
Alfred-Wegener-Institute for Polar and Marine Research, Bremerhaven, Federal Republic of Germany.

1 Introduction

Sea ice plays an important role in the climate system since it modifies the surface radiation balance due to its high albedo, and it influences the exchange of momentum, heat and matter between atmosphere and ocean because of its insulating behaviour. During cooling periods the freezing of sea ice initiates brine expulsion and subsequent convection with deepening of the surface mixed layer and the formation of deep and bottom water. During the melting period relatively fresh water stratifies the oceanic surface layers, i.e. the mixed layer retreats to shallower depths. In contrast to low latitudes the mixed layer evolution in polar regions is dominated by surface fluxes of salt or freshwater (positive or negative freezing rates).

Because of the sea ice motion the melting generally occurs in places far away from the formation area. Averaged over one seasonal cycle the net freezing rate in a certain area is therefore rarely zero, but rather positive or negative depending on the divergence or convergence of the sea ice flow. The net freezing rates represent strong non-stationary surface buoyancy fluxes which heavily affect the density structure and therefore the baroclinic flow in the ocean. The characteristics of the sea ice motion (especially convergence or divergence and therefore the net freezing rate) depend on the rheological aspects of the sea ice as a plastic material, on the geometry of coastlines and on the atmospheric forcing fields, in particular the wind.

The influence of the ocean on the sea ice acts mainly through the oceanic heat flux into the base of the mixed layer, which then enters the ice cover. Estimates range from $4 - 20W/m^2$ (Gordon, 1981). It was shown that this heat flux has large spatial gradients, with large values in early and mid-winter at the sea ice margin and in divergent drift areas, mainly near coastlines, where new ice is formed and strong convection occurs. Relatively small values were obtained under the central part of the ice pack (Lemke et al., 1990, hereinafter LOH).

In the next section, a detailed description of the sea ice and mixed layer model dynamics is presented. In the following section, a review is

NATO ASI Series, Vol. I 11
Modelling Oceanic Climate Interactions
Edited by J. Willebrand and D. L. T. Anderson

given on recent Weddell Sea ice - mixed layer simulations (LOH, Owens and Lemke, 1990, hereinafter OL). From a series of model simulations it is shown that the pattern and the magnitude of the oceanic surface buoyancy flux is rather sensitive to the rheological properties used in the sea ice model.

A set of experiments investigating the collapse of polynyas after they are formed by anomalous oceanic heating suggests that advective effects coupled with the sea ice thermodynamics and mixed layer dynamics create a strong negative feedback system, damping the response to these short-time perturbations and restoring the model back towards the original seasonal cycle. This system eliminates a polynya much faster than a thermodynamics-only model. Finally, a set of experiments with modified atmospheric and oceanic boundary conditions is discussed. The simulation experiments indicate that dynamic – thermodynamic sea ice models are less sensitive to modifications of atmospheric forcing fluxes and to short time pertubations within ocean and sea ice than simple thermodynamic models. This implies that the large polar response to CO_2 increase of sensitivity experiments with coupled atmosphere-ocean circulation models which generally use thermodynamic-only sea ice models has to be interpreted rather carefully.

2 Description of the Model

In order to investigate the interaction between sea ice and ocean via the buoyancy flux a two-layer dynamic-thermodynamic sea ice model (Hibler, 1979, 1986) is coupled with a one-dimensional, prognostic mixed layer-pycnocline model (Lemke, 1987). The sea ice model consists of a momentum balance coupled to ice thickness and compactness (fraction of area covered by sea ice) equations. In the latter equations the vertical oceanic heat flux is determined by the mixed layer dynamics which also prescribes the depth, temperature, and salinity of the mixed layer and the underlying layer.

2.1 Sea ice

In dynamic sea ice models it is generally assumed that the sea ice can be treated as a two-dimensional continuum, which is characterized by five variables, i.e. the fields of snow and ice thickness, compactness and horizontal velocity. The prognostic equations are derived from conservation equations for snow and ice mass and compactness, and from a momentum balance, including internal ice stress. Dynamic-thermodynamic sea ice models consist of four major components:

1. a surface energy balance which uses the atmospheric forcing (wind, temperature, humidity, incoming solar radiation) and the heat conduction through the ice to compute the surface temperature. This energy balance contains furthermore parameterizations of incoming and outgoing long-wave radiation and sensible and latent heat fluxes.

2. a model which describes the heat conduction through the ice for given surface temperature and oceanic heat flux into the ice. The difference between the conductive and the oceanic heat fluxes determines the freezing (or bottom melting) rate. The surface melting is calculated if the surface temperature is above $0°C$. Then the surface temperature is set to zero and the net energy for this temperature is used to melt snow and ice.

3. a momentum balance including acceleration, Coriolis force, atmospheric and oceanic stress, sea surface tilt and internal ice stress, from which the ice velocity is calculated. The internal ice stress is determined from a particular rheology which relates the internal stress to the deformation. Our model is based on the viscous-plastic rheology first used for sea ice by Hibler (1979).

4. and balance equations for snow and ice mass and ice concentration. These balance equations use the sea ice velocity and the freezing/melting rate determined from the heat conduction model together with the surface energy balance for ice and open water to calculate the new ice thickness and concentration in each grid cell.

Although there are some differences in the thermodynamic part (1. and 2.) of the models, especially in the modification of the sea ice compactness due to atmospheric and oceanic heat fluxes, the basic difference between existing dynamic sea ice models lies in the treatment of the internal ice stress.

2.1.1 Ice Momentum Equation

The momentum equation, written in cartesian coordinates is

$$m\frac{D}{Dt}\mathbf{u} = -mf\mathbf{k} \times \mathbf{u} + \tau_a + \tau_w - mg\nabla H + \mathbf{F} \tag{1}$$

where $\mathbf{u}$ is the sea ice velocity, m is the sea ice mass per unit area, f is the Coriolis parameter and g is the gravitational acceleration. Eq. (1) includes inertial terms, Coriolis force, wind (τ_a) and water (τ_w) stresses , surface tilt H, and internal ice interactions $\mathbf{F}$. The wind and water stresses are determined by integral boundary layers assuming constant turning angles (e.g. McPhee, 1979). The surface tilt is related to the prescribed ocean currents using geostrophy.

According to continuum mechanics the internal ice forces can be written in terms of the divergence of the two-dimensional stress tensor σ , $\mathbf{F} = \nabla \cdot \sigma$, where a specific constitutive law has to be defined which describes σ as a function of the deformation, ϵ, or the deformation rate, $\dot{\epsilon}$, $\sigma = \sigma(\epsilon, \dot{\epsilon}, h_i, A)$. If the stress depends only on the deformation rate (which is given by the horizontal gradients of the velocity field), then the problem can be solved with the usual Eulerian approach.

In the Hibler (1979) model, the sea ice is considered to follow a nonlinear viscous plastic constitutive law

$$\sigma_{ij} = 2\eta\,(\dot{\epsilon}_{ij}, P)\,\dot{\epsilon}_{ij} + [(\zeta(\dot{\epsilon}_{ij}, P) - \eta(\dot{\epsilon}_{ij}, P))\dot{\epsilon}_{kk} - P/2]\,\delta_{ij} \tag{2}$$

where $\dot{\epsilon}_{ij}$ is the strain rate tensor, $P/2$ is a pressure term related to the ice thickness, $\dot{\epsilon}_{kk} = \dot{\epsilon}_{11} + \dot{\epsilon}_{22}$ is the ice divergence, and ζ and η are the nonlinear bulk and shear viscosities, respectively. The viscous-plastic constitutive law prescribes

$$\zeta = \frac{P}{2\Delta} \qquad \text{and} \qquad \eta = \frac{\zeta}{e^2} \tag{3}$$

where

$$\Delta = \left[(\dot{\epsilon}_{11}^2 + \dot{\epsilon}_{22}^2)(1 + \frac{1}{e^2}) + 4\frac{\dot{\epsilon}_{12}^2}{e^2} + 2\dot{\epsilon}_{11}\dot{\epsilon}_{22}(1 - \frac{1}{e^2})\right]^{\frac{1}{2}} \tag{4}$$

and e is the ratio of the principal axes (eccentricity) of the elliptical yield curve. The viscosities are confined to finite values as the strain rates become small by limiting their values to be less than

$$\zeta_{max} = (2.5 \times 10^8 s)P \qquad \text{and} \qquad \eta_{max} = \frac{\zeta_{max}}{e^2}. \tag{5}$$

The ice strength is coupled to ice thickness characteristics according to

$$P = P^* h_i \exp(-C(1 - A)) \tag{6}$$

where h_i is the areal average ice thickness and A is the ice compactness or fraction of the area covered by ice.

In this formulation three model parameters have to be specified a priori which describe the mechanical properties of the sea ice floe field, the ratio of principal axes, e, and the yield strength parameters P^* and C. For this study the standard experiment (described in LOH) uses the same values as Hibler and Ackley (1983), $e = 2, P^* = 2.75 \times 10^4 Nm^{-2}$, and $C = 20$.

2.1.2 Sea ice and snow conservation equations

Local changes of snow and ice thickness (h_s, h_i) and compactness (A) are computed from continuity equations for these variables,

$$\begin{aligned}\dot{h}_i &= -\nabla \cdot (\mathbf{u}h_i) + S_h + Diffusion &(7)\\ \dot{A} &= -\nabla \cdot (\mathbf{u}A) + S_A + Diffusion + S_{op} &(8)\\ \dot{h}_s &= -\nabla \cdot (\mathbf{u}h_s) + S_s + Diffusion &(9)\end{aligned}$$

where $A \leq 1$. The diffusion terms are small and introduced solely for numerical purposes. S_h , S_A and S_s are thermodynamic growth terms given by

$$S_h = A\,f(\frac{h_i^*}{A})\Theta + (1-A)f(0) - \frac{Q_o}{\rho_i L} \tag{10}$$

$$\begin{aligned} S_A = &\begin{cases} (1-A)\,f(0)/h_0 & \text{if} \quad f(0) > 0 \\ 0 & \text{if} \quad f(0) \leq 0 \end{cases} \\ &+ \begin{cases} 0 & \text{if} \quad S_h \geq 0 \\ AS_h/(2h_i) & \text{if} \quad S_h < 0 \end{cases} \end{aligned} \tag{11}$$

$$\begin{aligned} S_s = &\begin{cases} AP_w\rho_w/\rho_s & \text{if} \quad T_a < 0°C \\ 0 & \text{if} \quad T_a \geq 0°C \end{cases} \\ &+ \begin{cases} Af(h_i^*/A)\rho_i/\rho_s & \text{if} \quad f(h_i^*/A) < 0 \quad \text{and} \quad h_s > 0 \\ 0 & \text{otherwise} \end{cases} \end{aligned} \tag{12}$$

with

$$\Theta = \begin{cases} 0 & \text{if} \quad f(h_i^*/A) < 0 \quad \text{and} \quad h_s > 0 \\ 1 & \text{otherwise} \end{cases} \tag{13}$$

where $f(h)$ is the growth rate for ice of thickness h. The growth rates are determined from conductive heat fluxes through the ice which arise from conservation of energy at the upper surface of the snow/ice system (Semtner, 1976; Parkinson and Washington, 1979). The S_h term is the sum of the ice growth over both open water and that part of the grid cell covered by ice. Since the growth rate, $f(h)$, is a highly nonlinear function of ice thickness and, in reality, each cell is covered by a distribution of ice thicknesses which give the average thickness, h_i, the procedure of Hibler (1984) is generally adopted by estimating the growth rates $f(h)$ from an equal, seven level distribution between zero and twice the average thickness.

The inclusion of a snow layer modifies the calculation of the growth rates of ice, $f(h)$. Following Semtner (1976) and Parkinson and Washington (1979) linear temperature gradients through the snow and ice layers and no internal heat sinks in either layer are assumed. In this case the conductive heat fluxes through the two layers must be the same, $Q_i = Q_{sn}$, and one can obtain the following

$$Q_i = Q_{sn} = \kappa_i \frac{T_f - T_s}{h_i + \frac{\kappa_i}{\kappa_s} h_s} \tag{14}$$

where κ_i and κ_s are the thermal conductivities of ice and snow, respectively, and T_s and T_f are the temperatures at the top of the snow or ice layer and at the interface between the ice and water, respectively. It can be seen in this equation that in computing the ice growth rate one needs only replace the ice thickness, h_i, with $h_i^* = h_i + (\kappa_i/\kappa_s)h_s$ and substitute a higher surface albedo (0.85 compared to 0.75 for frozen ice and 0.66 for melting ice) if a snow layer is present. Under atmospheric melting conditions, it is assumed that the snow is completely melted before any melting of ice begins. This is taken care of by Θ in (10). Q_O in (10) denotes the vertical oceanic (entrainment) heat flux which is determined from the mixed layer model, and ρ_i and L are the density of sea ice and the latent heat of fusion, respectively.

The S_A term characterizes how the growth rates of thick and thin ice affect the areal fraction of ice coverage. The first term in equation (11) parameterizes the closing of leads and open water under freezing conditions. It is assumed that the open water is not instantly covered by ice, but rather that the fraction of open water decays exponentially at a rate of $h_0/f(0)$. For example, in the Antarctic frazil ice formation, followed by strong wind events redeposits thin ice under the ice floes (Ackley et al., 1980) which tends to slow the closing of open water. Nevertheless, although h_0 has a physical basis, and we estimate that this decay rate is of the order of 5 days, the exact value of h_0 is not known and is an important model parameter. For this study a standard value of $0.5m$ has been used.

The second term in equation (11) accounts for the decrease in the fraction of the area covered by ice due to the melting of thick ice, assuming its thickness is distributed between 0 and twice h_i/A and all the ice melts at the same rate.

For the S_S term (12) we have assumed a uniform precipitation rate of $P_w = 0.35$ meters of water per year. When the atmospheric temperature, T_a, is below freezing, this precipitation is in the form of snow (weighted with the ice concentration A), otherwise it is rain which we assume runs off through the leads and enters the oceanic mixed layer. The second term in (12) accounts for the melting of snow. ρ_w and ρ_s denote the densities of sea water and snow, respectively.

2.2 Mixed Layer

Models of the oceanic mixed layer are of special importance in climate related problems, since the mixed layer transfers information between the atmosphere and the deep ocean. Because of this interaction it also links the fast time-scales (atmosphere: days) to the longer time-scales (ocean: decades to centuries). There are a variety of models that describe the time evolution of the mixed layer properties due to wind mixing and seasonal varying heating and cooling (Niiler and Kraus, 1977; Martin, 1985).

In most parts of the ocean temperature and salinity near the surface tend to vary more along a vertical distance of $100m$ then along a horizontal distance of $1000km$. This means that vertical exchange processes affect local conditions more than horizontal advection or mixing. Therefore it seems permissible to derive one-dimensional models of the upper ocean by ommitting the horizontal derivatives in the governing equations for momentum, heat and salt conservation. To solve this set of one-dimensional equations for the mean temperature, salinity and velocity one has to find explicit expressions for the turbulent fluxes. This can be achieved with several methods, with turbulence closure models, eddy coefficient or mixing length hypothesis, or with mixed layer models.

In the latter models it is assumed that mean temperature, salinity and horizontal velocity are uniform in the vertical within the mixed layer. This assumption permits vertical integration of the governing equations. The mixed layer properties are now described as functions of the associated fluxes at the bottom and the top of the mixed layer. Since the vertical integration introduces the time-dependent mixed layer depth as a new variable, the system has to be closed by an additional equation for the entrainment velocity, i.e. the deepening of the mixed layer.

This is usually obtained from the turbulence energy equation, which balances the turbulent kinetic energy input by the surface stress minus dissipation with the change of potential energy of the oceanic stratification during the deepening of the mixed layer. This Kraus-Turner approach (see Niiler and Kraus, 1977) will also be used here, although in an extended version in which a second (pycnocline) layer is introduced (Lemke and Manley, 1984; Lemke, 1987)

The vertical structure of the one-dimensional mixed layer-pycnocline model used here to represent the upper ocean is shown in Fig.1. Temperature and salinity are constant within the mixed layer and show an

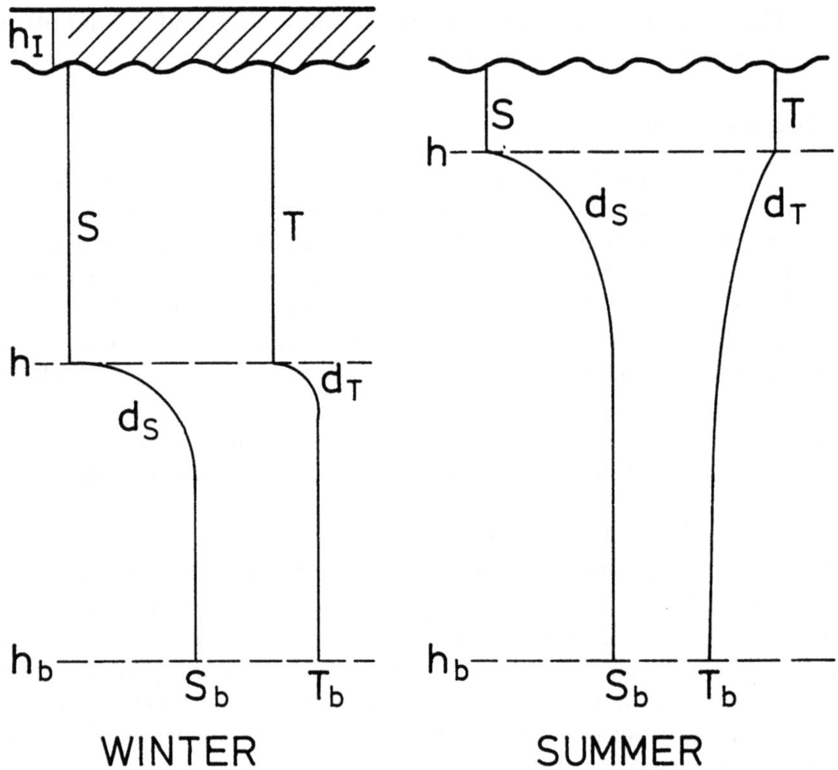

Figure 1: Vertical structure of the mixed layer - pycnocline model.

exponential profile within the pycnocline.

$$\left.\begin{aligned} T(z) &= T \\ S(z) &= S \end{aligned}\right\} \qquad 0 > z > -h$$

$$\left.\begin{aligned} T(z) &= T_\infty + (T - T_\infty)\exp[(z+h)/d_T] \\ S(z) &= S_\infty + (S - S_\infty)\exp[(z+h)/d_S] \end{aligned}\right\} \quad -h > z > -h_b \tag{15}$$

The prognostic variables are the mixed layer temperature T and salinity S, the mixed layer depth h, and the e-folding depths of the thermocline d_T and halocline d_S. The prognostic equations of the mixed layer-pycnocline model are derived from the conservation of heat and salt, potential energy considerations and a parameterization for the entrainment heat and salt fluxes. The lower level of the model, h_b, is set at 3000m. Generally, the scale depths of the thermocline and halocline, $d_S, d_T \ll h_b - h$, so that $T_b = T(-h_b) \approx T_\infty$ and $S_b = S(-h_b) \approx S_\infty$. T_b and S_b are considered as given boundary conditions and taken to be constant in space and time with typical values for the Weddell Sea, $0.5°C$ and $34.8 psu$, respectively.

With the above model profile the salt content H_S of the two layer system is given by

$$H_S = (S - S_b)(h + d_S) + S_b h_b \tag{16}$$

The salt balance states that the change of the salt content is balanced by the salt flux at the sea surface, Q_S, and the upwelling W,

$$\dot{H}_S = \dot{S}(h + d_S) + (S - S_b)(\dot{h} + \dot{d}_S) = Q_S + W(S_b - S) \tag{17}$$

The upwelling term represents the net effect of the oceanic circulation, i.e., the divergence of the horizontal salt transport, required to balance the excess of precipitation over evaporation at high latitudes. For the heat content a similar equation can be derived.

$$\dot{H}_T = \dot{T}(h + d_T) + (T - T_b)(\dot{h} + \dot{d}_T) = Q_T + W(T_b - T) \tag{18}$$

The rate of change of the mixed layer salinity and temperature is determined by the appropriate surface and entrainment fluxes

$$\dot{S} = \frac{Q_S + B_S}{h} \tag{19}$$

$$\dot{T} = \frac{Q_T + B_T}{h} \tag{20}$$

where the surface fluxes of salt, Q_S, and heat, Q_T, are determined from the sea ice model. B_S and B_T are the entrainment fluxes of salt and heat, respectively. These fluxes are parameterized in terms of a turbulent length scale δ and the entrainment velocity W_e. This approach represents a translation of the Kraus-Turner type parameterization for a discontinuous profile to our case of a continuous profile with a sharply discontinuous gradient at $z = -h$.

Consider an entrainment zone of thickness δ which is not resolved by the model. This zone is characterized by a mean temperature T^* and salinity S^* which are given by

$$\begin{aligned} T^* &= \frac{1}{\delta}\int_{-h-\delta}^{-h} T(z)dz = T_b + (T_b - T)\frac{d_T}{\delta}\left(\exp(-\delta/d_T) - 1\right) \\ S^* &= \frac{1}{\delta}\int_{-h-\delta}^{-h} S(z)dz = S_b + (S_b - S)\frac{d_S}{\delta}\left(\exp(-\delta/d_S) - 1\right) \end{aligned} \tag{21}$$

such that

$$T^* - T = a_T(T_b - T) \qquad \text{and} \qquad S^* - S = a_S(S_b - S) \tag{22}$$

where

$$a_{S,T} = 1 + \frac{d_{S,T}}{\delta}\left(\exp(-\delta/d_{S,T}) - 1\right) \tag{23}$$

$a_{S,T}$ are the fractions of the salt or temperature difference between the mixed layer and the base of the second layer used in the calculation of the entrainment fluxes.

$$B_T = (T^* - T)W_e \qquad \text{and} \qquad B_S = (S^* - S)W_e \tag{24}$$

Note that for this mixed layer model $a_{S,T}$ are prognostic variables while they are fixed values (= 1.0) for the original Kraus-Turner model (Niiler and Kraus, 1977).

For mixed layer models the closure is generally taken from potential energy considerations. Wind and ice keel stirring provide the energy $\tilde{K}$ needed to balance the increase of the potential energy due to the surface and entrainment fluxes, Q and B, respectively. In wintertime, convection provides additional energy for the deepening of the mixed layer. Other turbulent kinetic energy sources besides wind stress and convection are neglected. We assume that below the mixed layer there is always enough turbulence in the pycnocline to provide the energy needed to maintain the exponential profile.

After determination of the potential energy of the model profile and taking the time derivative we find for the potential energy balance for the entrainment process (Niiler and Kraus, 1977)

$$\tilde{K} - \varepsilon = \frac{h}{2} g(B - Q) \tag{25}$$

where ε is a dissipation term which is parameterized in terms of the active turbulence generating processes: wind or ice keel stirring, and convection. The surface and entrainment buoyancy fluxes, Q and B, respectively are determined from

$$Q = \beta Q_S - \alpha Q_T \qquad \text{and} \qquad B = \beta B_S - \alpha B_T \tag{26}$$

where α and β represent the expansion coefficients of the density with respect to temperature and salinity.

Inserting the fluxes and solving for the entrainment velocity we find

$$W_e = \frac{2KD_1 + hQD_2}{hE} \tag{27}$$

where $K = \tilde{K}/g, E = \beta(S^* - S) - \alpha(T^* - T)$ and $(1 - D_1)$ and $(1 - D_2)$ represent the depth–dependent dissipation of mechanical and convective energy input at the sea surface. The depth-dependence is assume to be exponential,

$$\begin{aligned} D_1 &= \exp(-h/h_w) \\ D_2 &= \begin{cases} \exp(-h/h_c) & Q > 0 \\ 1 & Q < 0 \end{cases} \end{aligned} \tag{28}$$

The scale depth of dissipation h_w and h_c are determined together with the tubulent length scale δ from a least squares fit similar to Lemke and Manley (1984) of the salinity part of the model to observations obtained during the Arctic Ice Dynamic Joint Experiment (AIDJEX). The optimal values were obtained as $\delta = 8m$, $h_w = 7m$ and $h_c = 50m$. In order to allow for deep convection in case of a weak stratification, the dissipation of convective energy is taken to be constant for $h > h_{max} = 50m$, such that $D_2 = \max(\exp(-h/h_c), \exp(-h_{max}/h_c))$

The model equations are now given by

$$\dot{S} = \frac{1}{h}[Q_S + (S^* - S)W_e] \tag{29}$$

$$\dot{T} = \frac{1}{h}[Q_T + (T^* - T)W_e] \tag{30}$$

$$\dot{d}_S = \frac{d_S}{S_b - S}\dot{S} + \left(\frac{S^* - S}{S_b - S} - 1\right) W_e \tag{31}$$

$$\dot{d}_T = \frac{d_T}{T_b - T}\dot{T} + \left(\frac{T^* - T}{T_b - T} - 1\right) W_e \tag{32}$$

where the change of the mixed layer depth h is determined from the entrainment and the upwelling velocity

$$\dot{h} = W_e - W \tag{33}$$

The above equations apply only for the deepening of the mixed layer (entrainment phase, $W_e > 0$). During the period of increased heating (melting) the stress induced energy at the surface is generally insufficient to overcome the stabilizing effect of the surface buoyancy flux ($W_e < 0$). In this case the mixed layer retreats to an equilibrium depth, $\hat{h}$, given diagnostically by the Monin-Obukhov length which is determined by (27) with $W_e = 0$ or

$$2KD_1 + \hat{h}Q = 0 \tag{34}$$

Changes of the surface salinity and temperature during the retreat phase are determined from (29) and (30), respectively, with $W_e = 0$. The thermocline and halocline variables d_T and d_S, finally are determined from the conservation of heat and salt, respectively.

According to the AIDJEX analysis (McPhee and Smith, 1976) the kinetic energy input, $\tilde{K}$, is computed from the simulated sea ice velocity $\mathbf{u}$ relative to the prescribed ocean currents

$$\tilde{K} = C_W \|\mathbf{u}\|^3 \cos\gamma \tag{35}$$

where γ is the frictional turning angle, found to be 24°, and the drag coefficient C_W was estimated to be 0.005.

2.3 Sea ice - ocean coupling

In polar regions the determination of the surface buoyancy flux Q differs a little from the more straightforward calculation in the open ocean (see Niiler and Kraus, 1977). In contrast to lower latitudes, the pycnocline under polar sea ice is generally warmer than the mixed layer. Therefore in ice covered oceans the interaction between the heat and salt budgets is essential. When entrainment takes place due to enhanced winds or freezing (brine convection), warm water is mixed into the upper layer. This heat is then used to melt sea ice, thereby modifying the surface heat and salt fluxes; i.e. the surface buoyancy flux is partly determined through the entrainment velocity.

The heat flux at the sea ice - ocean interface, Q_T, is given by the entrainment heat flux, which is lost to the melting of ice. We assume that the mixed layer temperature remains at the freezing point as long as ice is present. From (30) with $\dot{T} = 0$ we find

$$Q_T = -B_T = -(T^* - T)W_e \tag{36}$$

The total surface salt flux is accordingly determined by the freezing of ice due to heat loss to the atmosphere, Q_A, and by the melting of ice due to the oceanic heat flux, and by the net surface freshwater flux F.

$$\begin{aligned} Q_S &= (S - S_i)\frac{\rho_i}{\rho_w}\dot{h}_i - FS \\ &= (S - S_i)\frac{\rho_i}{\rho_w}\left(\frac{Q_A}{\rho_i L} - c\frac{T^* - T}{\rho_i L}W_e\right) - FS \end{aligned} \tag{37}$$

where S_i is the salinity of sea ice taken to be $5psu$, and $c = 4.26 \times 10^6 Jm^{-3}K$. Q_A is the atmospheric freezing rate determined from the ice model

$$Q_A = S_h + \frac{Q_o}{\rho_i L} \tag{38}$$

where the oceanic heatflux is given by

$$Q_o = B_T = (T^* - T)W_e \tag{39}$$

The equations above show that the total surface buoyancy flux Q, determined from (26), includes two terms which are proportional to the entrainment velocity W_e. This leads to a reformulation of the entrainment rate,

$$W_e = \frac{2KD_1 + hQ^*D_2}{h(E + E^*)} \tag{40}$$

with

$$E^* = \left(\frac{\beta c(S - S_i)}{\rho_w L} - \alpha\right)(T^* - T) \tag{41}$$

$$Q^* = \frac{\beta(S - S_i)Q_A}{\rho_w L} - \beta FS \tag{42}$$

F is the net surface freshwater flux determined by the excess of precipitation over evaporation, P_w, and the melting of snow.

$$\begin{aligned} F = &\begin{cases} P_w(1 - A) & \text{if} \quad T_a \leq 0°C \\ P_w & \text{if} \quad T_a > 0°C \end{cases} \\ + &\begin{cases} Af(h_i^*/A)\rho_i/\rho_w & \text{if} \quad f(h_i^*/A) < 0 \quad \text{and} \quad h_s > 0 \\ 0 & \quad\quad \text{otherwise} \end{cases} \end{aligned} \tag{43}$$

Q^* contains now only the atmospherically induced freezing rate. Since E^* is positive, the entrainment oceanic heat flux always leads to a reduction of the entrainment rate.

The sea ice motion generally creates regions with annual net melting and freezing. In order to keep the salinities in those areas from decreasing or increasing indefinitely we have applied a Newtonian damping which couples the mixed layer salinities to observations with a time constant of one year. This is a crude attempt to include advective effects in our one-dimensional mixed layer-pycnocline model.

2.4 Model domain

The model equations described above were solved on a nominal 2° grid, Fig.2, which is the same as that used by Hibler and Ackley (1983). The continental boundaries are closed while the hatched grid cells are considered to be open boundaries treated in the same manner as Hibler (1979). The atmospheric forcing fields were also those used by Hibler and Ackley (1983), and consisted of a year long series of daily atmospheric data derived from the Australian analysis of the 1979 FGGE winds and solar radiation estimated from empirical parameterizations similar to those used by Parkinson and Washington (1979). Prescribed geostrophic ocean currents were those calculated from historical hydrographic data given in Gordon, et al. (1978).

Initially the model was integrated with a fixed mixed layer depth for four years, starting with $2m$ of ice everywhere, 100% compactness, and a fixed oceanic heat flux of $2Wm^{-2}$. This produced an ice field for 1 January similar to that described by Hibler and Ackley (1983). These distributions of ice thickness, compactness, and mixed layer temperatures were then used as intial conditions for our model studies. The coupled sea ice - mixed layer model was then integrated for a further four years to allow it to reach a cyclo- stationary state. Analysis of an integration for a final fifth year were used to desribe the results.

3 Results

3.1 Standard experiment

The seasonal cycle of the integrated ice covered area in the Weddell Sea compares sufficiently well with observations (see LOH and OL). The seasonal cycle of the integrated sea ice mass shows an ice mass which is four times larger in winter than in summer.

The geographical distributions of the mean September sea ice velocity, the sea ice thickness and mixed layer depth on October 6, and the net ice production in the Weddell Sea are displayed in Fig.3. The September mean ice velocity shows the well known gyre circulation with sea ice being

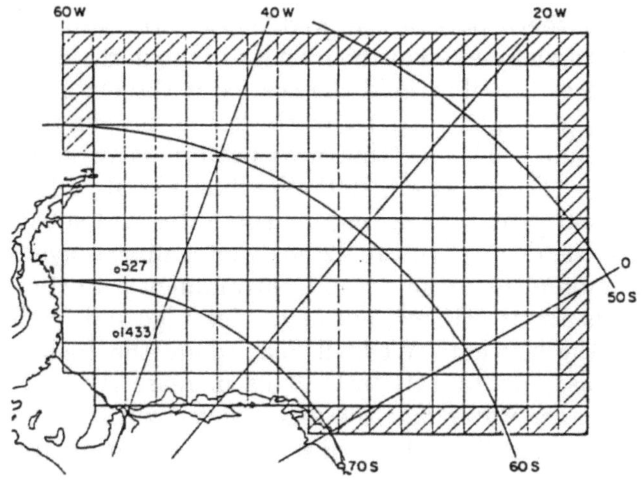

Figure 2: Grid used for numerical simulations (from Hibler and Ackley, 1983).

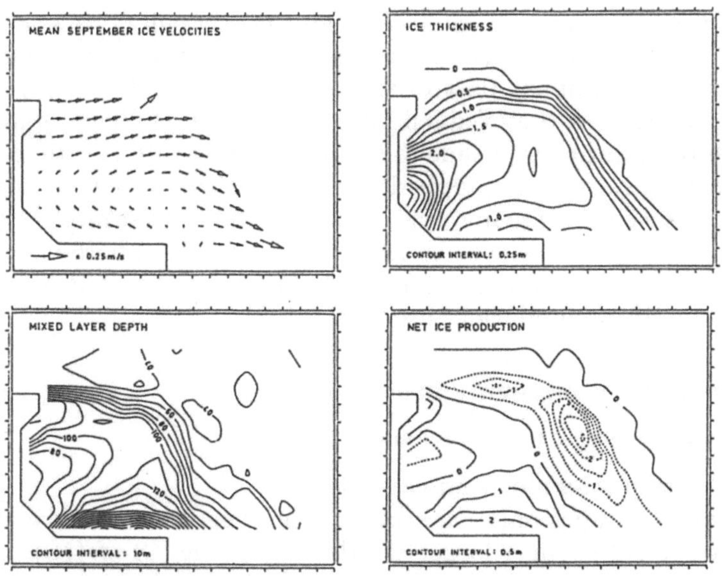

Figure 3: Simulated mean September sea ice velocities, sea ice thickness and mixed layer depth on 6 October, and annual net ice production in the Weddell Sea from the experiment with the prognostic snow cover (from OL).

formed in the southern part of the Weddell Sea, transported northward and then eastward and melted at the ice edge in the north-eastern Weddell Sea. As a result of this flow field the sea ice is thinner in the southern part of the grid (divergent drift) and thicker at the western wall - the Antarctic Peninsula. The ice thickness pattern is in good agreement with observations (Eicken et al., 1991; Wadhams et al., 1987).

In the divergent drift zone strong freezing rates and subsequent brine convection give the deepest (up to $220m$) winter mixed layer. Over most of the model domain the mean oceanic entrainment heat flux in September (see LOH) is about $2W/m^2$, but at the ice edge and in the divergent drift region in the southern Weddell Sea, where the freezing rates are largest, it increases up to $14W/m^2$. This increase is the result of entrainment of warmer water into the mixed layer which is mainly due to convective overturning caused by brine rejection during freezing. The simulated heat fluxes at the ice edge are close to observations by Gordon et al. (1984).

The effects of the velocity field are also apparent in the pattern of the annual net freezing rate. In the southern part of the grid up to $2.5m$ of ice is frozen per year, is transported through the Weddell Gyre and is finally melted at the ice edge. This pronounced freezing and melting pattern represents the dominant part of the oceanic surface buoyancy flux and strongly determines the Weddell Gyre circulation (Olbers et al., 1993). The pattern and the amplitude of the net freezing rate is rather sensitive to the rheological aspects applied in the constitutive law characterizing the sea ice as a solid state material. It will be shown that models with simplified sea ice dynamics used so far in coupled atmosphere-ocean GCMs do not produce a realistic freezing and melting pattern, and consequently obtain a modified circulation in polar regions.

3.2 Sensitivity experiments

3.2.1 Modification of sea ice dynamics

a) no advection (thermodynamics-only)
The thermodynamics-only experiment which neglects ice advection and snow cover resulted in a larger ice thickness (Fig. 8 of OL) compared to the standard run. The ice thickness contours are more or less zonal and reflect the pattern of the surface air temperature. Since the seasonal cycle is strongly reduced (too much ice in summer), the ice thickness is relatively high throughout the year, and therefore the freezing rates are smaller than in the standard run. Accordingly the mixed layer depth is drastically reduced in the southern Weddell Sea and reaches a maximum of only $80m$. As it should be in thermodynamic sea ice models, the annual

net freezing rate is zero. After reaching equilibrium the ice frozen locally in winter is melted at the same spot in summer.

b) Free drift

The seasonal cycle of the ice extent for the free drift experiment ($\mathbf{F} = 0$ in (1) differs only slightly from the standard experiment. The ice volume on the other hand is substantially increased. The reason for this large increase of the total ice volume is the fact that due to the strongly modified ice velocity field (Fig.4) the ice thickness is drastically increased in the south-western Weddell Sea. At the eastern edge of this thick pack ice there is a small band of thin ice. In this region the mixed layer depth has its largest values. Under the heavy ice cover in the south-western corner of the grid the mixed layer is rather shallow due to the smaller freezing rates. The pattern of the net freezing rate is strongly modified in this region. The overall amplitude of net freezing and net melting is similar to the standard run.

c) Free drift modifications

In order to improve the free drift results another experiment was performed in which the sea ice velocity was allowed to follow the free drift as long as the sea ice thickness was less than $2m$. For thicknesses more than $2m$ the ice velocity was set to zero. This is analogous to the often used treatment of sea ice in global coupled atmospheric and oceanic GCMs. The seasonal cycles of extent and volume are improved by this modification as compared to the free drift case (see Fig.9 of OL), but they still differ from the standard experiment.

Because of this modification the sea ice velocity is zero in the southern half of the model domain (Fig.5) in strong contrast to the standard run, and the sea ice thickness in this region is therefore similar to the thermodynamics-only case (compare Fig.8 of OL)). As in the thermodynamics-only experiment the mixed layer depth is also strongly reduced in this area, and there is no pronounced net freezing region which we observe in the standard run.

To further improve the spatial patterns of the sea ice and ocean variables towards the standard experiment, an additional modification was imposed. The sea ice velocity was set to zero only if the sea ice was thicker than $2m$ and if the motion was convergent, i.e. the ice was allowed to follow a divergent free drift motion even if it was thicker than $2m$.

Of all free drift modifications the flow field (Fig.6) for this experiment best agreed with the standard results. The sea ice thickness was improved, and the mixed layer depth and the net freezing rate look very similar to the standard run. The seasonal cycles of extent and volume (Fig.9 of OL) are closer to the standard experiment than the first free drift modification.

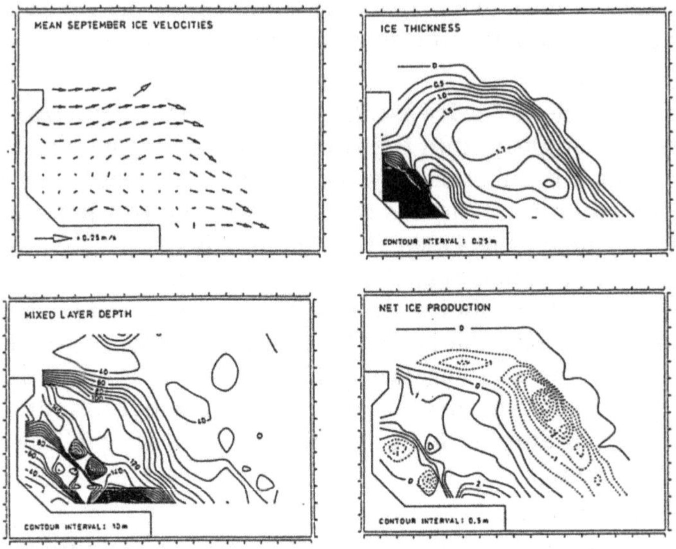

Figure 4: Simulated mean September sea ice velocities, sea ice thickness and mixed layer depth on 6 October, and annual net ice production in the Weddell Sea from the free drift simulation (from OL).

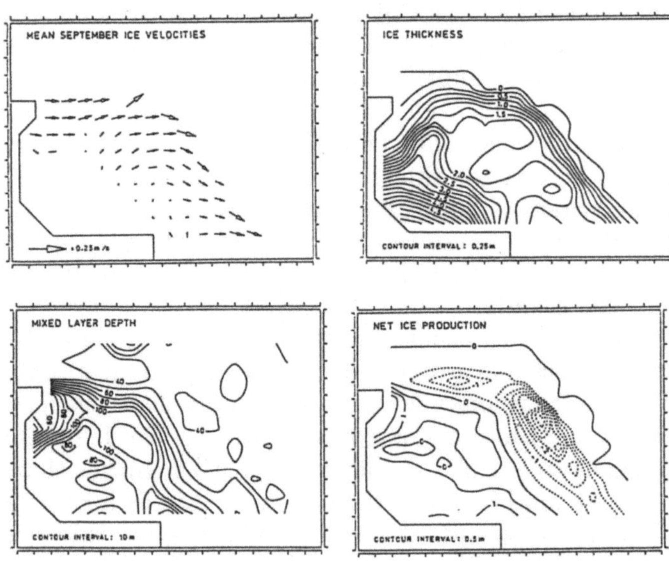

Figure 5: Simulated mean September sea ice velocities, sea ice thickness and mixed layer depth on 6 October, and annual net ice production in the Weddell Sea from the experiment with the first free drift modification (no motion for $h_i > 2m$) (from OL).

In general, the free drift modification, which sets the sea ice velocity to zero if the ice is thicker than $2m$ and if the motion is convergent, yields the best results of all competitive rheologies tested. Nevertheless, the original Hibler rheology shows the most convincing results, but it uses more computer time than the free drift modification.

3.2.2 Polynya experiments

Two mechanisms that destabilize the oceanic stratification, allowing stronger entrainment of warm water and a subsequent reduction of the sea ice thickness, were discussed in Lemke (1987) using a simple thermodynamic sea ice model. In order to investigate the effect of the sea ice dynamics on the formation and termination of a polynya, one of these mechanisms was also tested in LOH with the Hibler type sea ice model.

Gordon and Huber (1984) observed large warm subsurface eddies which travelled from the east into the Weddell Sea. A consequence of these warm cells was a significant lifting of the mixed layer base and an increase of the pycnocline temperature, which generally leads to more intense entrainment of warm and salty water.

In the polynya experiment presented in LOH the warm cells were modelled by resetting the mixed layer depth to $40m$ for 50 days beginning with June $3rd$, at a time when the mixed layer is normally deepening, and by increasing T_b by $1°C$. After 50 days, T_b is again relaxed to the standard value of $0.6°C$. This perturbation was applied to nine gridpoints in the southern Weddell Sea. The initial conditions for this experiment were taken from 31 December of the fifth year of the standard run.

The response of the sea ice thickness on 6 October (75 days after the end of the perturbation) is shown in Fig.7 (upper left panel), where the difference between the polynya experiment and a simulation without perturbation is displayed. It is clear that the warm cell - perturbation leads to a strong reduction of the sea ice thickness of up to $1.5m$. Due to the ice advection the extent of the response pattern is larger than the actual perturbation area (dashed line). The zonal scale of the response is more pronounced than the meridional one since this area is in the southern portion of the Weddell Gyre where the velocities are mainly zonal.

One year later (1 October) the sea ice thickness has nearly returned to normal conditions (Fig.7, upper right panel). This is in contrast to Lemke (1987) where a thermodynamic sea ice model was used. In order to investigate whether the sea ice dynamics reduces the lifetime of a polynya, a similar experiment was performed with the thermodynamics-only version of the Weddell sea ice model. Fig.7 (lower left panel) indicates that

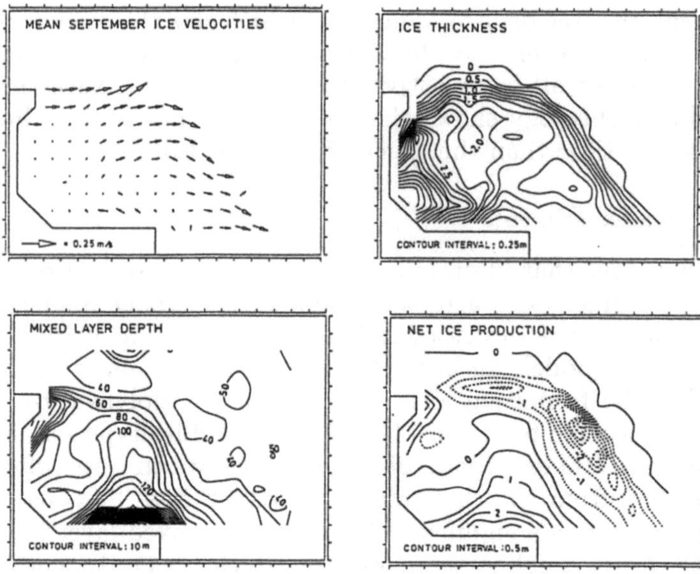

Figure 6: Simulated mean September sea ice velocities, sea ice thickness and mixed layer depth on 6 October, and annual net ice production in the Weddell Sea from the experiment with the second free drift modification (no motion for $h_i > 2m$ and convergent $\mathbf{u}$) (from OL).

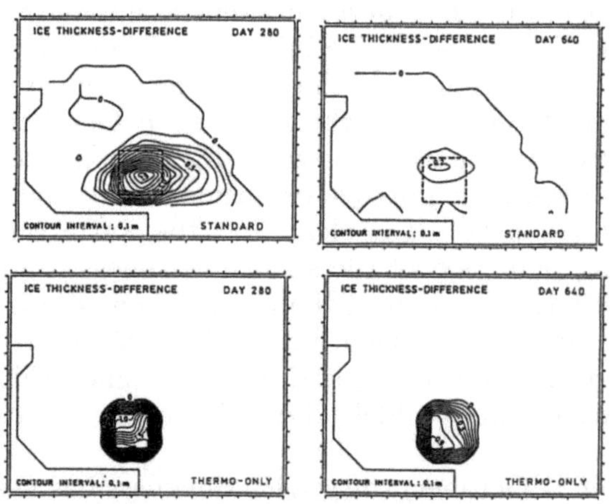

Figure 7: Sea ice thickness difference between the polynya experiment and a simulation without perturbation using the standard dynamic model for Day 280 (6 October), 75 days after the end of the 'warm cell perturbation' (upper left panel) and for Day 640 (1 October), 435 days after the end of the 'warm cell perturbation' (upper right panel). The dashed lines indicate the region where the perturbation was applied. The lower panels shows the corresponding ice thickness differences for the experiment with the thermodynamics-only version of the model (modified from LOH).

the magnitude of the response of the thermodynamics-only model in the winter following the perturbation is similar to that of the dynamic model, but the response is limited to the warm cell area, where the perturbation was applied. From Fig.7 (lower right panel) it is evident that one year later the thermodynamics-only model still shows a pronounced reduction of the ice thickness.

These experiments show that the sea ice dynamics increases the spatial scale of the response, and it reduces the lifetime of the polynya-response by importing ice into the thin ice and low compactness region, where there is less resistance to convergence. In other words, the dynamic sea ice model is less sensitive to short-time perturbations than the purely thermodynamic model. Experiments with a larger warm cell perturbation area (24 gridpoints) showed a larger spatial scale and a slightly increased lifetime of the response.

3.2.3 Paleoclimate experiments

The deep ocean boundary conditions and the atmospheric forcing during earlier paleoclimatic periods are believed to have been quite different from today's values. In this section the response of the standard model to boundary conditions corresponding to different possible paleoclimatic epochs is presented. The results are compared with the standard experiment without snow cover (see LOH). Fig.8 presents the results of four response experiments:

1. The deep ocean temperature was reduced to $0°C$ everywhere.
2. The incoming solar radiation was reduced by 10%.
3. The air temperature was reduced by $2°C$ everywhere.
4. All three boundary modifications were applied simultanaeously.

It is obvious that the seasonal response increases from experiment 1 to 4. A 10% reduction of the solar radiation leads to more ice than the cooling of the deep ocean. And the reduction of the atmospheric temperatures by $2°C$ yields even more sea ice.

The application of all three modifications of the boundary conditions resulted in an expansion of the ice covered area of $0.7 \times 10^6 km^2$, and the volume increased by $2 \times 10^3 km^3$. The velocity pattern for this experiment was similar to the standard one. The winter ice thickness increased by about $0.3m$, and the mixed layer depth was larger by about $20m$. The magnitude of the oceanic heat flux also increased, whereas the pattern remained similar. Due to the larger atmospheric cooling the net freezing rate was enhanced by $0.5m/year$.

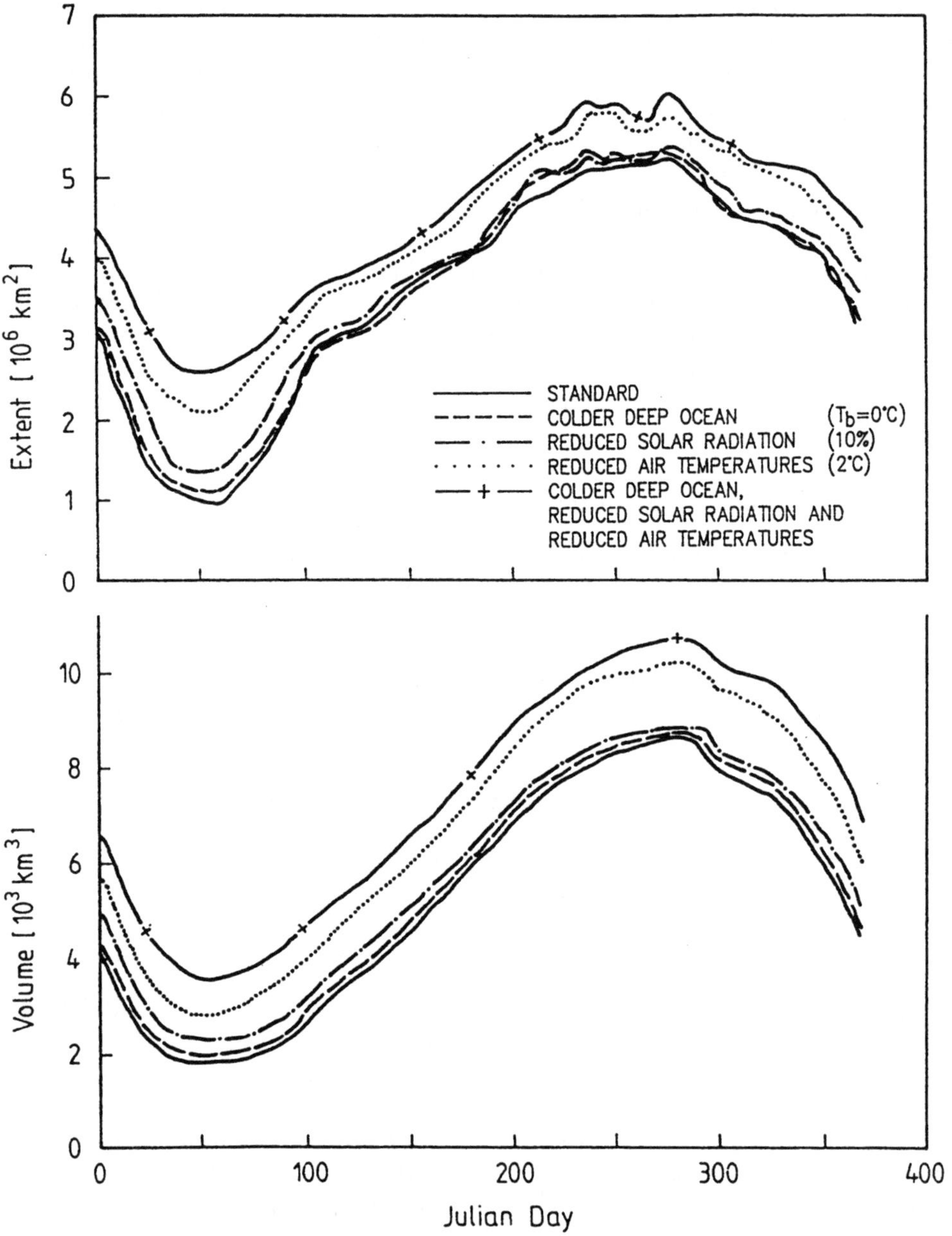

Figure 8: Seasonal cycle of the integrated ice covered area (upper panel) and ice volume (lower panel) for different paleoclimate experiments (from LOH).

Another experiment with a larger surface freshwater flux (melting ice sheets) resulted in a increased seasonal response of ice thickness and extent similar to the cold deep ocean experiment. The velocity and thickness patterns differed only slightly from the standard run, but due to the enhanced stability the mixed layer depth was reduced significantly by about $50m$. Accordingly the oceanic heat flux decreased allowing the sea ice to move to lower latitudes. Due to the larger freshwater flux the surface salinity was reduced by $0.2 - 0.4psu$.

In general the dynamic-thermodynamic sea ice model is less sensitive to modifications of surface or bottom boundary conditions than a simple thermodynamic sea ice model (Lemke, 1987).

3.2.4 Radiative forcing variations

Sensitivity experiments concerning various forcing and internal model parameters are presented in Hibler (1984), Lemke et al. (1990), Owens and Lemke (1990) and Stössel et al. (1990). Here we will discuss results of sensitivity experiments concerning the radiative forcing. The following radiation perturbation experiments have been performed:

1. a 10% increase of the incoming longwave radiation. This represents an increase of cirrus clouds which are more or less transparent to solar radiation but not to the outgoing longwave radiation.
2. a 0.1 increase of the ice albedo (from 0.75 to 0.85)
3. comparison of the model including snow cover (with the dry snow albedo $\alpha_s = 0.85$) with the model without snow (ice albedo $\alpha_i = 0.75$).
4. a 0.1 reduction of the snow albedo (from 0.85 to 0.75). In this experiment also the albedo of melting snow and bare sea ice ($\alpha_{sm} = \alpha_i = 0.65$) and melting sea ice ($\alpha_{im} = 0.56$) have been reduced by 0.1 as compared to the standard experiment with snow.
5. a 10% reduction of the incoming solar radiation (see previous section)
6. a $2°C$ reduction of the air temperature (see previous section)

The results of these sensitivity experiments concerning summer minimum and winter maximum of the ice covered area and the ice volume (thickness) are displayed in Table 1.

The 10% increase of the incoming solar radiation leads to a substantial reduction of ice extent and volume in summer and in winter. The 0.1 increase of the sea ice albedo towards the value of dry snow (there is no

Table 1: Results of sensitivity experiments (LW↓ = incoming long-wave radiation, α = albedo, SW = incoming solar radiation, T_a = air temperature)

Experiment	changes of extent		changes of volume	
	summer	winter	summer	winter
10% increase of LW↓	−30%	−10%	−37%	−17%
0.1 increase of α	+19%	+ 1%	+ 9%	+ 2%
snow - snow	−41%	− 4%	−47%	−24%
0.1 reduction of α	−23%	− 1%	−14%	− 1%
10% reduction od SW	+28%	+ 3%	+10%	+ 4%
T_a minus 2°C	+88%	+10%	+40%	+20%

snow in this experiment) results in an increased summer ice extent and volume. The winter maximum is only slightly affected, because there is no solar radiation.

The inclusion of a snow model (which affects the results via the higher albedo and the lower heat conductivity as compared to the experiment without snow) leads to a substantial reduction of summer and winter sea ice extent and volume. This is basically due to the thermal insulation of the sea ice by the snow cover. A reduction of the albedo by 0.1 (representing dust on snow and ice) results in a significant reduction of the summer extent and volume. The winter maximum is only slightly affected.

As a comparison the last two experiments which have been discussed in the paleoclimate section are also shown here. They indicate that a 2°C reduction of the air temperature has a larger effect on the sea ice than a 10% reduction of the solar radiation. This is presumably due to the fact that a temperature reduction influences the turbulent heat fluxes and the incoming longwave radiation, which is parameterized in terms of the air temperature.

For further sensitivity experiments concerning the use of a Kraus-Turner mixed layer model and other internal model parameters see LOH and OL. An extension to the entire Southern Ocean is presented in Stössel et al. (1990).

4 Conclusions

In polar regions sea ice plays a dominant role in determining the surface buoyancy flux for the ocean. Dynamic-thermodynamic sea ice models have become more realistic in the last decade, but their use in global coupled atmosphere-ocean circulation models is still rather limited. The importance of sea ice dynamics is demonstrated in a series of experiments with a coupled sea ice - mixed layer model for the Weddell Sea. The results show the well known gyre circulation with a divergent sea ice drift, and accordingly thinner ice, larger freezing rates, higher salinities, and deeper mixed layers in the south off the Antarctic continent. These results agree favourably with observations.

The coupled sea ice - mixed layer model has been shown to be sensitive to changes in both its thermodynamics and dynamics. In particular both the inclusion of a snow layer and changes of the lead closing rate parameter h_0 reduce the ice thickness significantly, while only slightly modifying the extent. Changes of the dynamic model parameters had little effect on ice extent. The volume, on the other hand, was modified significantly. Varying the parameters P^* and C to increase the ice strength produced smaller ice volume because of stronger resistance of the sea ice to compression. Changes of the ratio of bulk to shear viscosity have a smaller effect on the model results as compared to modifications of the strength parameters. It appears that the neglection of the shear viscosity has little influence on the performance of the ice - mixed layer model.

On the other hand, ice velocities as well as ice thickness are rather sensitive to these ice rheology parameters. Although we do not see any way in the near future to obtain seasonal coverages of ice thicknesses over a large region such as the Weddell Sea, one can expect to obtain such coverage for the ice velocities estimated by tracking features from one satellite Synthetic Aperture Radar (SAR) image to the next. When this data is available one could consider inverse modelling analyses to better determine the ice rheology parameters.

The applications of simplified ad hoc rheologies indicates that the free drift simulation can be improved substantially by assuming that the free drift motion stops if it is convergent and if the sea ice thickness exceeds a critical value, which in this paper was taken to be $2m$. The simulation using this simplified advection scheme has a close resemblance to the standard experiment with the viscous-plastic rheology.

These experiments demonstrate that a proper modelling of the sea ice rheology is important, not only to realistically describe the sea ice extent, thickness and compactness, but also in order to provide the correct forcing fields for the ocean (salt and heat fluxes, net freezing rate), which strongly depend on the nature of the applied dynamic sea ice model.

The results also indicate that the vertical oceanic heat flux shows a pronounced time and spatial dependence. The largest values are generally found at the ice edge and in the divergent drift region in the southern Weddell Sea, where large entrainment rates occur due to off-shore winds.

In summertime relatively large entrainment heat fluxes are produced by the surface stress induced by wind and sea ice drift, which is generally higher in this season.

Experiments with short-time disturbances, such as a warm eddy, which destabilize the oceanic stratification compared with the normal seasonal conditions, show that such short-lived perturbations are capable of producing an anomalous reduction of the sea ice thickness, but the sea ice dynamics reduces the lifetime of such a polynya-response considerably compared to a thermodynamics-only model. This is due to the fact that under convergent winds sea ice is imported into the thin-ice-low-compactness region, which has only a small ice strength, and is melted, thereby increasing the oceanic stratification and reducing the entraiment. This suggests that the observed long life of the polynya in 1974/75 (Carsey, 1980) must have been sustained by anomalous meteorologic or oceanic conditions.

This reduced response means that the dynamic-thermodynamic sea ice model is less sensitive to short-time perturbations than the thermodynamics-only model, and from the paleoclimate experiments it is evident that this is also true for modifications of the atmospheric and deep ocean boundary conditions. The reduced sensitivity can be explained by the specific interaction between the dynamics and the thermodynamics of the sea ice model. In regions where the thermodynamics reduces the ice thickness, the ice gets weaker and the dynamics (under favourable conditions, i.e. convergence) can readily increase the ice thickness (by importing ice into the region). In regions where the dynamics reduces the ice cover (divergence), the thermodynamics (under favourable conditions, i.e. cooling) can easily increase the sea ice thickness. These interactions produce a negative, i.e. stabilizing feedback, which in previous CO_2 experiments was not included. The reduced sensitivity of a more realistic sea ice model will obviously modify the large polar response that has been observed in the numerical experiments with coupled circulation models.

References

ACKLEY, S.F., A.J.GOW, K.R. BUCK, AND K.M.GOLDEN,1980 Sea ice studies in the Weddell Sea aboard USCGS Polar Sea, *Ant. J. U.S.* **15 (5)**, 84-86.

CARSEY, F.D.,1980 Microwave observation of the Weddell Polynya, *Mon. Wea. Rev.* **108**, 2032-20

EICKEN, H., M.A.LANGE AND G.S.DIECKMANN,1991 Spatial variability of sea ice properties in the northwestern Weddell Sea, *J. Geophys. Res.*, **96**, 10603-10615.

GORDON, A.L.,1981 Seasonality of Southern Ocean sea ice, *J. Geophys. Res.* **85**, 4193-4197.

GORDON, A.L., E.MOLINELLI AND T.BAKER,1978 Large scale relative dynamic topography of the Southern Ocean, *J. Geophys. Res.* **83**, 3023-3032.

GORDON, A.L., AND B.A.HUBER,1984 Thermohaline stratification below the Southern Ocean sea ice, *J. Geophys. Res.* **89**, 641-648.

GORDON, A.L., C.T.A.CHEN, AND W.G.METCALF,1984 Winter mixed layer entrainment of Weddell Deep Water, *J. Geophys. Res.* **89**, 637-640.

HIBLER, W.D.,1979 A dynamic - thermodynamic sea ice model, *J. Phys. Oceanogr.*, **9**, 815-846.

HIBLER, W.D.,1984 The role of sea ice dynamics in modelling CO_2 increases. In *Climate Processes and Climate Sensitivity, Geophys. Monogr. Ser.*, **29**, edited by J.E.Hansen and T.Takahashi, pp. 238-253, AGU, Washington, D.C..

HIBLER, W.D.,1986 Ice dynamics, in The Geophysics of Sea Ice, edited by N.Untersteiner, *Plenum Pub. Corp., New York, pp. 577-640.*

HIBLER, W.D. AND S.F.ACKLEY,1983 Numerical simulation of the Weddell Sea pack ice, *J. Geophys. Res.* **88**, 2873-2887.

LEMKE, P.,1987 A coupled one-dimensional sea ice - ocean model, *J. Geophys. Res.* **92**, 13164-13172.

LEMKE, P. AN T.O. MANLEY,1984 The seasonal variation of the mixed layer and the pycnocline under polar sea ice, *J. Geophys. Res.* **89**, 6494-6504.

LEMKE, P., W.B. OWENS AND W.D. HIBLER III,1990 A coupled sea ice-mixed layer-pycnocline model for the Weddell Sea, *J. Geophys. Res.* **95**, 9513-9525.

MARTIN, P.J.,1985 Simulation of the mixed layer at OWS November and Papa with several models, *J. Geophys. Res.* **90**, 903-916.

MCPHEE, M.G.,1979 The effect of the oceanic boundary layer on the mean drift of pack ice: Application of a simple model, *J. Phys. Oceanogr.* **9**, 388-400.

MCPHEE, M.G., AND J.D.SMITH,1976 Measurement of the turbulent boundary layer under pack ice, *J. Phys. Oceanogr.*, **6**, 696-711.

NIILER, P.P., AND E.B.KRAUS,1977 One-dimensional models of the upper ocean, in Modelling and Prediction of the Upper Layers of the Ocean, edited by E.B.Kraus, pp. 143-172, *Pergamon, New York.*

OLBERS, D., C.WÜBBER AND J.-O. WOLF,1993 The dynamical balance of wind and buoyancy driven circumpolar currents, *Submitted.*

OWENS, W.B. AND P. LEMKE,1990 Sensitivity studies with a sea ice - mixed layer-pycnocline model in the Weddell Sea, *J. Geophys. Res.*, **95**, 9527-9538.

PARKINSON,C.L. AND W.M.WASHINGTON,1979 A large-scale numerical model of sea ice, *J. Geophys. Res.*, **84**, 311-337.

SEMTNER, A.J.,1976 A model of the thermodynamic growth of sea ice in numerical investigations of climate, *J. Phys. Oceanogr.*, **6**, 379-389.

STÖSSEL, A., P. LEMKE AND W.B. OWENS,1990 Coupled sea ice - mixed layer simulations for the Southern Ocean, *J. Geophys. Res.*, **95**, 9539-9555.

WADHAMS, P., M.A.LANGE, AND S.F.ACKLEY,1987 The ice thickness distribution across the Atlantic sector of the Antarctic Ocean in Midwinter, *J. Geophys. Res.*, **92**, 14535-14552.

GORDON, A.L. [illegible] R.A. HUBER, 1984 Thermohaline stratification below the Southern Ocean sea ice. *J. Geophys. Res.*, 89, 641-648.

[illegible] C.T.A. [illegible] W.S. [illegible], 1984 [illegible] hypersaline [illegible] Deep Water. *J. Geophys. Res.*, [illegible]

[illegible] W.D. 1981 A dynamic [illegible] *J. Phys. Oceanogr.*, 11, [illegible]

[illegible] 1984 [illegible] Climate Sensitivity, *Geophys. Monogr.*, 29, edited by J. Hansen and [illegible], pp. 288-[illegible] AGU, Washington, D.C.

[illegible]

[illegible]

[illegible] 29, [illegible]

[illegible]

[illegible] W.D. [illegible] Res., 89, [illegible]

MARTIN, [illegible] 1981 [illegible] *Res.*, [illegible]

[illegible] The effect of [illegible] and the [illegible] application of [illegible] *J. Phys. Oceanogr.*, 14, [illegible]

[illegible] of the [illegible] *J. Phys. Oceanogr.*, [illegible]

[illegible]

[illegible]

[illegible] 1983 [illegible] *J. Geophys. Res.*, 88, [illegible]

[illegible] and W.M. WASHINGTON, 1979 A large-scale numerical model of sea ice. *J. Geophys. Res.*, 84, 311-337.

SEMTNER, A.J., 1976 A model for the thermodynamic growth of sea ice in numerical investigations of climate. *J. Phys. Oceanogr.*, 6, 379-389.

[illegible] and W.D. [illegible] *Res.*, [illegible]

[illegible] 1987 [illegible] distribution [illegible] of the [illegible] Ocean [illegible] *J. Geophys. Res.*, 92, [illegible]

BIOCHEMICAL PROPERTIES OF THE OCEANIC CARBON CYCLE

CATHERINE GOYET
Woods Hole Oceanographic Institution,
Woods Hole, MA 02543

PETER G. BREWER
Monterey Bay Aquarium Research Institute,
160 Central Avenue, Pacific Grove, CA 95930

The earth's climate is regulated by the radiative balance between the solar energy received by the earth-atmosphere system and the thermal radiation sent back to space. While most of the former is absorbed by the earth's surface, most of the latter originates from the atmosphere. As a result, thermal exchanges across the ocean-atmosphere interface directly influence the atmospheric density as well as the surface ocean density, thus forcing oceanic and atmospheric circulations. The continuous exchange of gases and energy across the ocean-atmosphere interface strongly influences the physical and biological processes in the upper ocean. Penetration of heat and light in the surface seawater intensifies not only biological activity, but also ocean stratification and circulation which in turn influences the penetration of greenhouse gases into the ocean, thus modulating the global warming of our planet. CO_2 gas is the major natural and anthropogenic greenhouse gas and therefore has played both a primordial role and a contemporary role in the intensity of the global warming.

The ocean is a huge reservoir of carbon, principally in the form of bicarbonate ion, approximately 60 times larger than that of the atmosphere, and can potentially absorb virtually all the atmospheric CO_2 generated by man over time scales of many hundreds of years. However, CO_2 exchange across the ocean-atmosphere interface is a relatively slow process compared to the input of anthropogenic CO_2 to the atmosphere, and the volume of "fresh" sea water ventilated by the atmosphere each year is a very small fraction of the oceanic volume. As a result, anthropogenic CO_2 gas is quickly accumulating in the atmosphere. Atmospheric CO_2 concentrations at 354 μatm in 1990 (Keeling and Whorf, 1991) were about 25% higher than the presumed pre-industrial (1800) value of 280 μatm. The current atmospheric level is the highest observed to date and is still increasing at a rate of approximately 0.5% per year. The increase presently measured corresponds to approximately 60% of the total

NATO ASI Series, Vol. I 11
Modelling Oceanic Climate Interactions
Edited by J. Willebrand and D. L. T. Anderson

carbon emissions (essentially caused by burning of fossil fuel and heavy deforestation). The remainder is therefore assumed to be taken up by the oceans and the terrestrial biosphere. However, due to the obscuring of secular changes by large natural variations of partial pressure of CO_2 (pCO_2) in the surface waters of the ocean, direct quantification of the relative importance of land and oceans in absorbing atmospheric CO_2 gas remains extremely difficult to achieve.

If we are to measure, understand, and predict the uptake of atmospheric CO_2 gas by the ocean, then two essential steps are necessary; the accurate measurement of the CO_2 properties in the ocean and the accurate description of the biogeochemical processes which affect the CO_2 concentration in the ocean and provide the large background signal now being modulated by man. These must be combined accurately with knowledge of the physical circulation. Although the first CO_2 measurements in the ocean were performed as early as the late 19th century, little reliable information on the carbon cycle was gained. The data were scarce and the techniques of measurements were not appropriately documented. It is only within the past two decades that significant progress has occurred. Our ability to accurately measure the CO_2 properties in sea water with state of the art technology has greatly improved our ability to describe the biogeochemical processes influencing the oceanic carbon cycle. Recently Brewer *et al.*, (1989) suggested a new approach to the determination of CO_2 flux linked to climate change. Since CO_2 gas exchange across the air-sea interface is influenced by surface seawater pCO_2 which in turn is influenced by surface seawater temperature which in turn influences our climate, their approach was to treat both the total CO_2 concentration and the heat content of seawater in a similar manner in order to estimate CO_2 fluxes compatible with estimates of heat fluxes. As an example, they computed the CO_2 fluxes transported by ocean currents in different layers of the water column, across latitude 25°N in the North Atlantic Ocean. This was done by combining measured vertical oceanic profiles of CO_2 with earlier estimates of mean layer transports (Bryden and Hall, 1980; Hall and Bryden, 1982). The sum of these northward and southward total CO_2 fluxes yield the net transport of total CO_2 by ocean currents in a manner identical to estimates of the heat flux at 25°N. This unique approach which takes into account the physical constraints of ocean circulation, as used to determine heat fluxes, in the determination of CO_2 flux could be extended over the global ocean to improve our understanding of the interaction of biogeochemical processes in the context of climate change. Their local (North Atlantic Ocean) result of CO_2 flux is virtually identical to the CO_2 flux as estimated independently by Tans *et al.*, (1990) using a more global approach. They compiled observations of seasurface pCO_2 obtained over many years, and combined these estimates of $DpCO_2$ (pCO_2^{air} - pCO_2^{sea}) with transfer coefficients to estimate uptake by the global ocean. This superficial agreement however does not imply that these estimates solve the problem of recognition of the fossil fuel CO_2 invasion in the ocean, for both estimates leave severe CO_2 mass balance problems for the atmosphere (observed CO_2 concentration in the atmosphere is lower than expected).

This chapter describes and documents the recent advances in CO_2 measurement technology, and presents current observations and understanding of the ocean response to global warming with respect to biogeochemistry. The present discussion will focus on the method of determining the oceanic CO_2 uptake from observations and will not review the literature on CO_2 uptake determinations by means of ocean models. For the sake of teaching course integrity, this chapter will successively describe: 1) the basic CO_2 chemistry in seawater, 2) the determination of CO_2 flux across the ocean-atmosphere interface, 3) the recognition of the anthropogenic CO_2 signal in the ocean, 4) the use of freons as an analogous tracers of anthropogenic CO_2 invasion in the ocean, 5) the potential feedback through temperature and dissolved organic carbon (DOC) oxidation rates.

1 Oceanic CO_2 chemistry

The fundamental chemical equilibria involving dissolved carbon dioxide in seawater can be represented by the following series:

$$CO_2(g) = CO_2(aq) \quad (1)$$
$$CO_2(aq) + H_2O = H_2CO_3 \quad (2)$$
$$H_2CO_3 = H^+ + HCO_3^- \quad (3)$$
$$HCO_3^- = H^+ + CO_3^{2-} \quad (4)$$

In practice, since it is very difficult to distinguish between species CO_2(aq) and H_2CO_3 , the sum of these species is expressed as a hypothetical species CO_2^*(aq). Reactions 1,2, and 3 are then redefined as :

$$CO_2(g) = CO_2^*(aq) \quad (5)$$
$$CO_2^*(aq) + H_2O = H^+ + HCO_3^- \quad (6)$$

The thermodynamic equilibrium constants (Sillen and Martel, 1964,1971) of carbonic acid (for reactions 6 and 4) in seawater are defined as:

$$K_1 = [H^+][HCO_3^-]/[CO_2^*] \quad (7)$$
$$K_2 = [H^+][CO_3^{2-}]/[HCO_3^-] \quad (8)$$

They have been carefully measured (Mehrbach, 1973; Hansson et al., 1973; Goyet and Poisson, 1989) and their pK (pK = $-\log_{10}(K)$) are known to within +/- 0.011 in pK_1 and +/- 0.020 in pK_2.

Currently, each of these carbonate species in seawater is not directly measurable. The measurable parameters of the carbonate system in seawater are defined as follows:

1.1 CO_2 partial pressure (pCO_2)

pCO_2 is defined as the product of CO_2 mole fraction and total pressure, and is expressed in μatm. In theory, due to the non-ideal nature of CO_2 gas, the CO_2 fugacity that can be computed from the solubility coefficient (Weiss, 1974) and a carbonate property pair is different from the measured CO_2 partial pressure. In practice, the correction for the non-ideal nature of CO_2 gas is negligible ($< 1\mu atm$) and CO_2 partial pressure is often treated identically to its fugacity (the fugacity of a real gas is defined as the pressure an ideal gas would have at the same temperature and pressure, to have the same free energy). Currently, pCO_2 is measured at a known temperature in the gas phase in equilibrium with the seawater sample to be measured using either a gas chromatographic technique (Weiss, 1981) or a non-dispersive infra-red analyzer system (Takahashi, 1961; Inoue *et al.*,1987). pCO_2 can be measured either on discrete water samples or on a continuous water flow (surface seawater). The time required for a measurement depends on the system used; it varies from a couple of minutes to approximately fifteen minutes.

1.2 Total CO_2 (C_T)

Total dissolved inorganic carbon is expressed in μmol/kg and is defined as the sum of the carbonate species in seawater:

$$C_T = [CO_2^*] + [HCO_3^-] + [CO_3^{2-}] \tag{9}$$

Currently C_T is measured using a coulometric technique (Johnson *et al.*, 1985; Johnson *et al.*, 1987). A seawater sample is acidified to transform all the carbonate species to CO_2 gas. This latter is stripped with an ultra pure nitrogen gas stream and is absorbed by chemical reaction in ethanolamine dissolved in a dimethylsulfoxide solution containing a small amount of water and thymolphthalein indicator dye. The product is then titrated by electrochemically generated OH^-. The equivalence point is detected photometrically, and the amount of total CO_2 titrated is related to the time-integrated current using Faraday's law. The precision of C_T

measurements using this technique is close to 1.5 μmol/kg, and the desired accuracy is 4 μmol/kg. Once the system is calibrated, the rate of measurement is approximately fifteen minutes per sample.

1.3 Total alkalinity (A_T)

Total alkalinity is expressed in μmol/kg and is rigorously defined (Dickson, 1981) as: "... the number of moles of hydrogen ion equivalent to the excess of proton acceptors (bases formed from weak acids with a dissociation constant $K < 10^{-4.5}$, at 25°C and zero ionic strength) over proton donors (acids with $K > 10^{-4.5}$) in one kilogram of sample."

$$A_T = [HCO_3^-] + 2[CO_3^{2-}] + [B(OH)_4^-] + [OH^-] + [HPO_4^{2-}]$$
$$+2[PO_4^{3-}] + [SiO(OH)_3^-] + [HS^-] + [NH_3] + ...$$
$$- [H^+] - [HSO_4^-] - [HF] - [H_3PO_4] - .... \qquad (10)$$

where the ellipses stand for a variety of minor acid base species. Currently A_T is commonly measured by potentiometric titration (Dyrssen, 1965; Bradshaw et al., 1981). The automated titration is performed in a closed cell maintained at constant temperature (25.0 ± 0.1°C), and the ionic strength of the titrant (0.1N hydrochloric acid solution) is adjusted with NaCl in order to be similar to that of seawater of salinity 35. The precision of AT measurements is close to 0.1%. The kinetics of the reactions involved during the titration impose a minimal titration time of approximately half an hour per sample.

1.4 pH

pH is defined as:

$$pH = -log a_{H^+} \qquad (11)$$

However there are actually three pH scales in use by ocean chemists; the NBS (National Bureau of Standard) pH scale (which refers to non-saline solutions), the "free" hydrogen ion concentration scale (which refers to saline solutions free of fluoride and is expressed in mol H^+/kg H_2O), and the pH sea water scale (which refers to saline solution including fluoride and is expressed in mol.H^+/kg.solution). For measurement in sea water the latter pH scale, which refers to the total hydrogen ion concentration in the solution, is recommended (UNESCO 1987). Currently pH is measured either with the conventional pH electrode (Almgren *et al.*, 1975; Dickson, 1984), or by spectroscopy (Byrne and Breland, 1989) with a precision of 0.001 and a desired accuracy of 0.003. A few minutes are required to perform a pH measurement.

The state of the carbonate system in seawater at a given temperature and salinity is completely defined from the measurement of only two measurable carbonate properties and from the knowledge of the dissociation constants. Of the four measurable carbonate properties, C_T and A_T are the only two temperature independent properties. Since they are mass properties, they are the only ones that can be used in conservation statements.

The interested reader will find additional details on the CO_2 chemistry in seawater and on the measurement techniques mentioned above, in more specialized literature (Almgren *et al.*, 1975; Bradshaw *et al.*, 1981; Brewer *et al.*, 1986; Byrne and Breland, 1989; Dickson, 1981; Dickson, 1984; DOE handbook, 1991; Goyet and Hacker, 1992; Johnson *et al.*, 1987; UNESCO, 1987; Weiss 1974, Weiss, 1981).

2 CO_2 flux across the ocean-atmosphere interface

The net CO_2 flux across the ocean-atmosphere interface can not be easily assessed by direct measurement in the atmosphere (Jones and Smith, 1977; Jones, 1980). Calculation from air-sea partial pressure differences (Broecker *et al.*, 1978; Liss, 1983), is the universally preferred scheme. Currently, most of the CO_2 flux estimates are calculated from the gas exchange flux equations (Liss, 1983a,b) of the form:

$$F = K([CO_2]^{int} - [CO_2]^{sea}) \qquad (12)$$

Where F represents the CO_2 flux, K represents the transfer velocity, $[CO_2]$int represents the concentration of CO_2 gas at the air-sea interface, in the surface film layer which is presumed to be in thermodynamic equilibrium with the atmospheric CO_2 gas concentration. In practice, $[CO_2]^{int}$ is not directly measurable but is derived from the measured atmospheric pCO_2 using Henry's law:

$$[CO_2]^{int} = b_{sl} pCO_2^{air}, \qquad (13)$$

where bsl represents the solubility of CO_2 gas in seawater (Weiss, 1974) at the temperature of the surface skin layer and pCO_2air represents the measured pCO_2 in the water-vapor- saturated air above seawater. $[CO_2]$sea represents the concentration of dissolved CO_2 ($[CO_2^*(aq)]$) in the mixed layer. Similarly, $[CO_2]^{sea}$ is computed from the measured pCO_2 in seawater using the relationship:

$$[CO_2]^{sea} = \beta_{ml} pCO_2^{sea}, \qquad (14)$$

where β_{ml} represents the solubility of CO_2 gas in seawater at the temperature of the mixed layer and pCO_2^{sea} represents the measured pCO_2 in the mixed layer.

2.1 Transfer velocity

The transfer velocity, K, is assumed to be primarily dependent on windspeed and is therefore derived from measurements of wind velocities, measured at 10 m above sea level, using empirical equations (Liss and Merlivat, 1986; Upstill-Goddard *et al.*, 1990; Watson *et al.*, 1991, Peng and Takahashi, 1990). The uncertainty associated with the determination of transfer velocity is currently estimated to be close to a factor of 1.7 (Watson *et al.*, 1991). Additional at-sea experiments are now underway to reduce this significant uncertainty.

2.2 pCO_2^{air}

The spatiotemporal variations of pCO_2^{air} are globally well documented (Komhyr *et al.*, 1985), as pCO_2^{air} is relatively easy to continuously monitor from shore bases and using flask networks (Beardsmore and Pearman, 1987; Conway and Tans, 1990; Keeling *et al.*, 1982; Pearman and Beardsmore, 1984) . CO_2 in air is chemically inert, and the troposphere is relatively well mixed.

2.3 pCO_2^{sea}

pCO_2^{sea} is poorly documented because pCO_2^{sea} has to be measured at sea on ships and consequently is not continuously monitored. The large local changes in pCO_2^{sea} driven by thermal and biological effects are not rapidly erased by gas exchange, or homogenized by mixing. In order to ameliorate this challenging problem, efforts are now being directed to develop instruments for unattended monitoring of pCO_2^{sea} (Goyet *et al.*, 1992; Degrandpre, in press) for future use on ocean moorings. Another source of uncertainty in the estimation of a global annual mean air-sea CO_2 flux remains the uncertainty in the seasonal variation of pCO_2^{sea}. Recent observations (Peng *et al.*, 1987, Murphy *et al.*, 1991; Goyet *et al.*, 1991, Metzl *et al*, 1991; Goyet and Brewer, in press), and models (Bacastow and Maier-Reimer, 1990) indicate that pCO_2^{sea} in surface waters can fluctuate by up to $\pm$ 150 μatm around pCO_2^{air}. These variations depend not only upon the temperature of the seawater but also upon the geographical location, the biological activity, and the dynamic mixing of the water mass considered. As a result of both the present scarcity of the spatiotemporal observations of surface seawater pCO_2^{sea} and the different mathematical approaches of calculations, uncertainties associated with global estimates of net ocean-atmosphere CO_2 fluxes are large; in the latitudinal zone 15°N - 50°N, Keeling *et al.*, (1989)'s estimate of CO_2 uptake by the world ocean (by means of an atmospheric transport model), is more than a factor of 4.5 larger than that of Tans *et al.*, (1990) (by means of air-sea pCO_2 differences), and close to a factor of 10 larger than that of Bacastow and Maier-Reimer (1990) (by means of a 3-D ocean carbon cycle model). The problem is not simply scarcity of data but the ability to resolve small mean changes over large areas in the face of fluctuating, noisy natural variability. Large ocean observing programs such as the Joint Global Ocean Flux Study (JGOFS) and the World Ocean Circulation Experiment (WOCE) are now addressing this question, and one of their main objectives is to determine the controls on oceanic CO_2 budgets, including spatiotemporal variations of surface seawater pCO_2^{sea}. Results from the 1989 JGOFS North Atlantic Bloom Experiment (NABE) at 47°N 20°W illustrate some of the complexity of the temporal variations of the carbonate properties in the upper ocean. Within the upper 150 m, total alkalinity normalized at constant salinity 35 remained approximately constant during this experiment (from April 25 to May 31, 1989), and C_T was highly variable. The variations of pCO_2 in surface seawater were mainly controlled by those of C_T (not by temperature). Thus our ability to model the temporal variations of pCO_2 in the surface ocean strongly depends on our ability to parameterize the variations of C_T (the variations of alkalinity in the surface ocean are generally small). Our recent analysis of this JGOFS NABE carbonate data set demonstrates that each of the observed C_T-depth profiles in the upper 150 m can be empirically described by a five parameter sigmoid function (Goyet and Brewer, in press). However the complex temporal variations in these functions cannot be easily modeled. Temporal variations of the biochemical processes within the water column are further confounded by changes in the physical circulation with season (Pedlosky, 1990). Unravelling these intricate biochemical, physical, and anthropogenic signals represents one of the most fascinating challenges for ocean scientists.

3 Potential changes in the biogeochemical cycles of the ocean driven by greenhouse gases.

3.1 Greenhouse gases increase

Signal recognition

If we are to quantify the anthropogenic CO_2 signal that has penetrated into the ocean, a careful protocol which accounts for the multifold variables contributing to C_T changes in the upper layer of the ocean needs to be established. The uptake of anthropogenic CO_2 by the ocean has long been recognized as an important process in the modulation of rising atmospheric CO_2 concentration (Brewer 1978; Broecker *et al.*, 1979). Yet, the quantification of the rate of this uptake and of the processes involved in the storage and transport of this carbon, are poorly documented. Simple calculations show that as the atmospheric CO_2 level has risen from its preindustrial value of 280 μatm to approximately 357 μatm today. Surface seawaters in striving to keep pace, must, through their carbonate equilibrium system, have risen in their dissolved total CO_2 content by some 40 μmol/kg. We do not have a 50 year measurement record of this oceanic signal, but virtually all ocean scientists would broadly agree with this calculation. If in pre-industrial time the surface ocean pCO_2, on the global scale was in equilibrium with the atmospheric pCO_2, then the ocean surface pCO_2 must now lag behind the atmosphere on the global scale by a mean value of about 8 μatm. Typically, oceanic surface waters contain about 2000 μmol/kg C_T today. As mentioned above, modern C_T measurement precision at sea by coulometric technique is close to 1.5 μmol/kg, thus a "contaminant" signal of 40 μmol/kg is theoretically measurable. In the natural world however, the amplitude of the seasonal variations of C_T in surface seawater vary significantly from place to place. It is close to 20 μmol/kg in the Sargasso Sea at Bermuda (BATS data report, 1991), or higher than 65 μmol/kg as observed at 47°N 20°W in the North Atlantic ocean (NABE data report, 1991). Furthermore, as the CO_2 content of the upper ocean rises the anthropogenic signal will increasingly intrude on observations of the natural carbon cycle. Recognition of the purely chemically forced signal will therefore be required if potential future changes in the biogeochemical cycles of the ocean driven by greenhouse gas induced temperature changes are to be observed. In order to begin to tackle this problem and provide a useful approximation to the fossil fuel signal, we apply a protocol first suggested by Brewer (1978) more than a decade ago to our earlier measurements and estimates of transport by ocean currents of A_T and C_T across the latitude 25°N in the North Atlantic ocean (Brewer *et al.*, 1989). This particular data set was chosen for two reasons:

1. The North Atlantic ocean is widely recognized as one of the major CO_2 sink areas for atmospheric CO_2, thus detection of the anthropogenic CO_2 signal is facilitated by its relatively large amplitude.

2. These data have recently been used to estimate the net southward CO_2 transport at 25°N (Brewer *et al.*, 1989), and this calculation can be enhanced by correcting for the transport of anthropogenic CO_2 by ocean currents at this latitude.

The protocol we use to estimate pre-industrial CO_2 concentrations assumes:

1. From a Lagrangian point of view, when the surface seawater sinks and is no longer in contact with the atmosphere, its CO_2 variations are produced only by mixing, and respiration; and variations of total alkalinity are the result of calcium carbonate dissolution and nitrate production,

$$\begin{aligned}(C_{106}H_{263}O_{116}N_{16}P) &+ 138O_2 === 106CO_2 + 16NO_3^- \\ &+ HPO_4^{2-} + 122H_2O + 18H^+ \end{aligned} \tag{15}$$

2. In surface seawater, total alkalinity normalized at salinity 35 remains constant with time (anthropogenic CO_2 does not affect total alkalinity), and seasonal variations are small. At depth, the variation of total alkalinity due to the formation /dissolution of calcium carbonate and respiration is then:

$$\Delta A_T = \Delta A_{TCaCO3} + (1 - 18)/138O_2 \tag{16}$$

Thus the total CO_2 concentration variations due to biological activity can be estimated from the relationship:

$$\Delta C_T = 106/138\Delta O_2 + 0.5\Delta A_{TCaCO3} \tag{17}$$

or

$$\Delta C_T = 0.8297\Delta O_2 + 0.5\Delta A_T \tag{18}$$

where ΔO_2 and ΔA_T represent the variations of O_2 and A_T respectively. The variations in oxygen concentration are given by the apparent oxygen utilization (AOU) (difference between the O_2 concentration of atmospherically equilibrated seawater (T, S) and the observed O_2 concentration), and the variations in total alkalinity are given by:

$$\Delta A_T = A_{T(o)} - A_{T(obs)} \tag{19}$$

where $A_{T(obs)}$ represents the observed A_T at depth D_i, and $A_{T(o)}$ represents A_T of this seawater (now at depth D_i) that it had when it was in contact with the atmosphere. For the North Atlantic surface waters, $A_{T(o)}$ can be computed from the following relationship (Brewer *et al.*, 1986) derived from the data set of the Transient Tracers in the Ocean (TTO) experiment in the North Atlantic Ocean (TTO, 1981):

$$A_{T(o)} = 50.56S + 547.0\mu\text{mol.kg}^{-1}. \tag{20}$$

The results of this calculation (equation 18) applied to our earlier measurements and estimates of A_T and C_T in several depth and isopycnal ranges can be used both to provide estimate of the preformed surface water pCO_2 value (pCO_2^{pre}, i.e., the value pCO_2 would have in the absence of

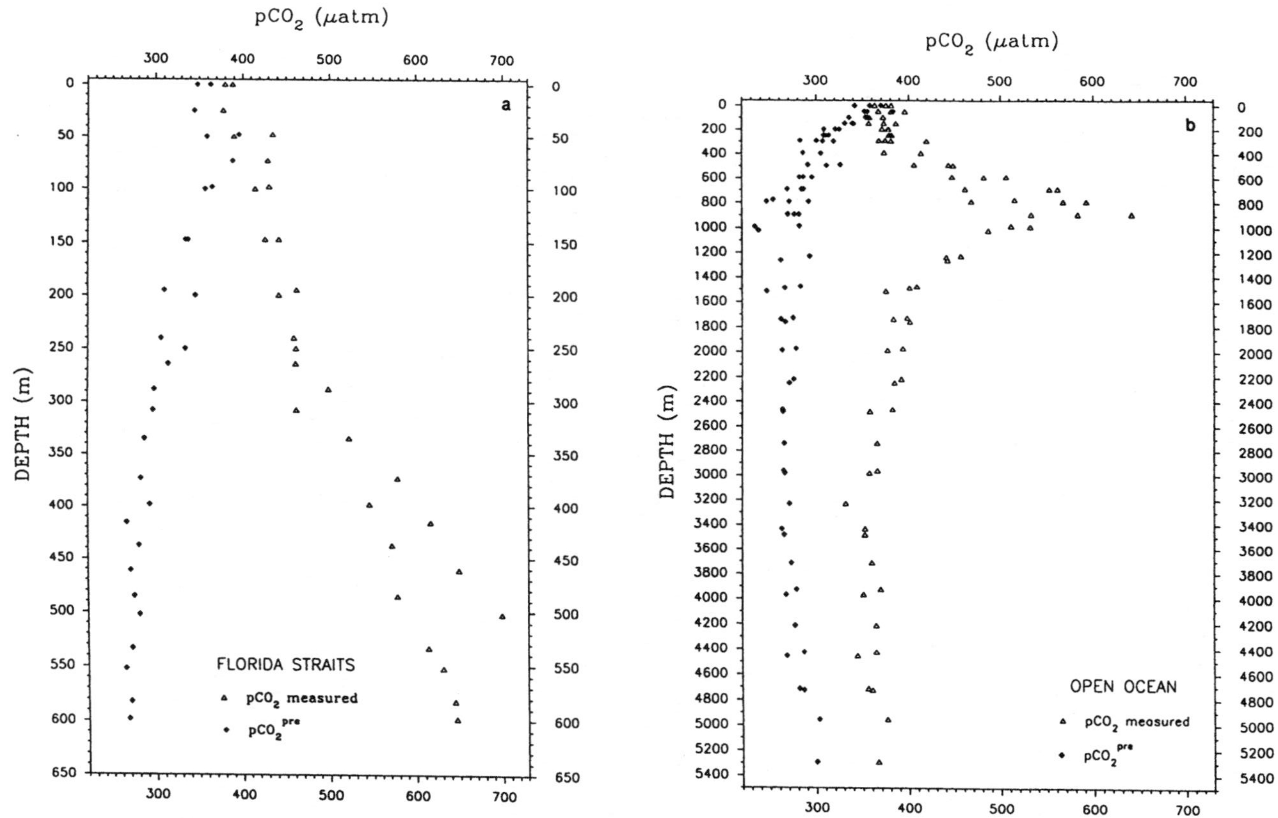

Figure 1: Present (Δ) and calculated (using equation 19,) pCO_2 depth profiles from two stations in the Florida Straits (a), and from two stations in the open ocean off the Bahamas (b).

anthropogenic CO_2 in the ocean and in the absence of exchanges between dissolved CO_2 and the marine biosphere) and to provide an estimate of the transport of anthropogenic CO_2 at latitude 25°N in the North Atlantic Ocean. Since by hypothesis, the preformed surface water pCO_2^{pre} is equal to the measured pCO_2 in the deep ocean corrected for changes due to biological activity, we computed pCO_2^{pre} from the computed difference (C_{Tobs} - ΔC_T), with the corresponding measured $A_{T(obs)}$ values using the most recent dissociation constants for carbonic acid (Goyet and Poisson, 1989). Figure 1 shows the results of these computations as well as the calculated present pCO_2 depth profiles, for two representative stations in the Florida Straits and two representative stations in the open ocean off the Bahamas. The present profiles of pCO_2 are typical of oceanic pCO_2 depth-profiles. In surface seawater pCO_2 values are close to the atmospheric equilibrium, and in subsurface, pCO_2 reaches a maximum which corresponds with the subsurface maximum of total inorganic carbon resulting from the so called "biological pump" (Volk and Hoffert, 1985). At depth, below the pCO_2 maximum, pCO_2 remains fairly constant. In the Florida Straits at depths below 350 m, and in the open ocean at depth below 1500 m, the mean of the pCO_2 values corrected for changes due to biological activity yield a pCO_2^{pre} of 272 ± 6 μatm and 271 ± 7 μatm, respectively. These values determined over a relatively large range of depths and in different areas suggest that either water mixing is efficient in homogenizing the water, or the ocean was initially homogenized and remained that way over a long time period. In the upper water column, such calculation cannot be used to determine pCO_2^{pre} since the water is in contact with the atmosphere and is therefore continuously exchanging CO_2 gas with the atmosphere. These pCO_2^{pre} values are remarkably close to the widely accepted estimate of 280 μatm for the pre-industrial atmosphere. If taken literally (and we do not claim accuracy at this level) they would indicate that the North Atlantic region must be a small pre-industrial CO_2 sink, and by implication some other region, such as the South Atlantic, may have been a small net CO_2 source.

In order to estimate the net pre-anthropogenic C_T transport by ocean currents at 25°N, we combined the C_T^{pre} values with estimates of oceanic mass transport (Bryden and Hall, 1980), used by Brewer *et al.*, (1989) to determine the present transport of carbon at this latitude. The assumption that water circulation did not change with time is made here in order to determine C_T fluxes as physicists would do for heat fluxes. The validity of this assumption, which may be questionable, is a problem faced by many physicists; we will not attempt to solve it here. The C_T^{pre} values (Table 1) were computed using the measured A_T values (hypothesis 2), and the estimated pCO_2^{pre} (272 μatm and 271 μatm). The

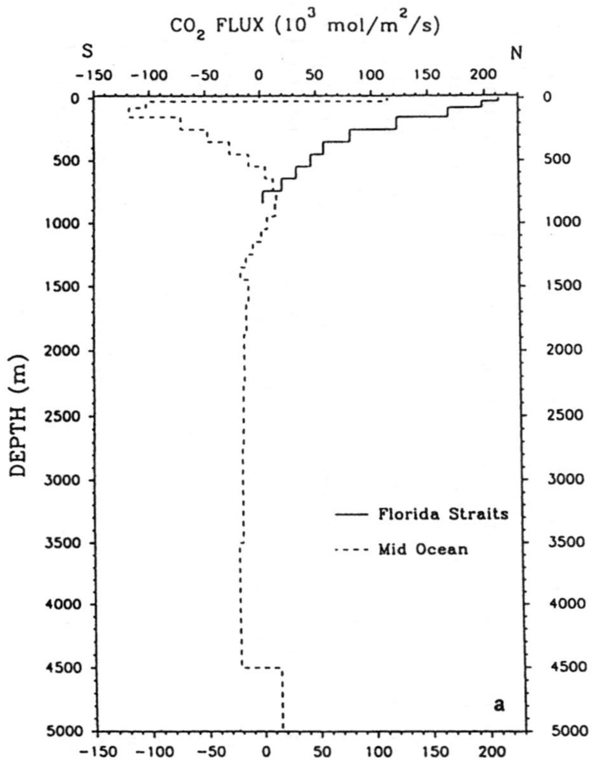

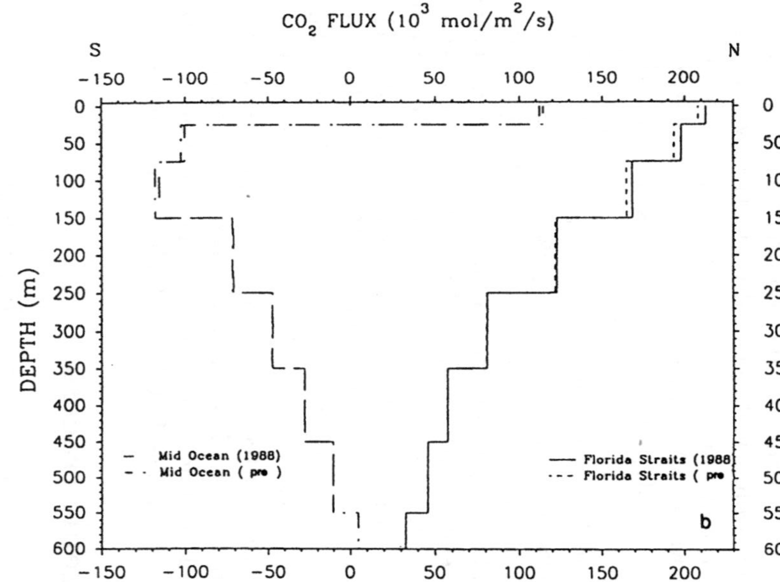

Figure 2: Present inorganic carbon transport by ocean currents in each depth interval, within the whole water column (a), Present and pre-anthropogenic carbon transport by ocean currents within the upper 600 m assuming the ocean circulation remained constant with time(b).

velocities calculated by Roemmich and Wunsch (1985) on a more rigorously sampled section would yield virtually identical values. In contrast to the present small southward CO_2 transport, the inferred total pre-anthropogenic southward CO_2 transport (Table 1) is relatively large: the northward transport of CO_2 by ocean currents across latitude 25°N is 63.12 10^6 mol/s, the southward transport is 64.26 10^6 mol/s, and the net southward flux is 1.14 10^6 mol/s or 0.42 gigatonnes of carbon per year. Figure 2 shows the respective components of the flux of carbon transported in each ocean depth interval for both the present CO_2 contaminated ocean and a CO_2 uncontaminated ocean. The difference in the CO_2 transport between the pre-industrial era (1.1 10^6 mol/s) and today (0.7 10^6 mol/s), from the variations in total dissolved inorganic carbon concentration alone, corresponds to an approximately 40% change. A formal assignment of errors here is difficult. However, these results are compatible with the recent result of Broecker and Peng (1992), which indicated that before the Industrial Revolution, 0.6 gigatonnes of carbon were annually transported by the North Atlantic deep water alone to the Southern hemisphere. The example used here illustrates the need for carefully sampled oceanic CO_2 sections which do not exist today, but which will be achieved by the JGOFS/WOCE combination.

4 Anthropogenic tracers

Under normal conditions chlorofluorocarbon compounds (CFCs) have low toxicity, are volatile and are chemically inert. They are therefore widely used as refrigerants, aerosol propellants, solvents and plastic foam blowing agents. Most of these purely anthropogenic gases are released into the atmosphere typically within a decade after their production. More than 90% of the production of two of these compounds, trichlorofluoromethane (F-11) and dichlorofluoromethane (F-12) occurs in the Northern hemisphere. However, since these gases are stable and mix rapidly within the lower atmosphere, their concentration in the troposphere is relatively uniform. The ratio of F-11/F-12 in the atmosphere is well known (Rasmussen and Khalil, 1986); it increased monotonically during the period 1930-1975 and has since remained about constant. As atmospheric CO_2 gas penetrates into the ocean, CFCs penetrate into the ocean by diffusion through the ocean-atmosphere interface. In the ocean, these compounds, which are chemically inert and stem from a purely anthropogenic origin, are ideal tracers. Their source function is well known, their input function is well constrained, and the pre-industrial background is zero. They can be used to better understand ocean mixing and circulation as well as to accurately determine the location and rate of penetration of an-

Depth (m)	Property C_T (μmol/ℓ)	Property C_T^{pre} (μmol/ℓ)	Transport (10^6 m^3/s)	Flux C_T (10^3/mol/s)	Flux C_T^{pre} (10^3/mol/s)
			Florida Straits		
0–25	2047	2002	2.6	5322	5205
25–75	2062	2017	4.8	9898	9682
75–150	2135	2090	5.91	12618	12352
150–250	2190	2168	5.59	12242	12119
250–350	2242	2231	3.60	8071	8032
350–450	2232	2232	2.56	5714	5714
450–550	2223	2223	2.05	4557	4557
550–650	2262	2262	1.44	3257	3257
650–750	2272	2272	0.87	1977	1977
750–850	2277	2277	0.10	228	228
			Total Flux	63882	63123
			Mid Ocean Section		
0–25	2104	2059	1.36	2861	2800
25–75	2112	2067	-2.43	-5132	-5023
75–150	2126	2081	-4.17	-8865	-8678
150–250	2148	2126	-3.34	-7174	-7101
250–350	2157	2146	-2.21	-4767	-4743
350–450	2172	2166	-1.29	-2802	-2802
450–550	2190	2187	-0.48	-1051	-1050
550–650	2205	2205	0.21	463	463
650–750	2230	2230	0.54	1204	1204
750–850	2243	2243	0.66	1480	1480
850–950	2251	2251	0.61	1373	1373
950–1050	2247	2247	0.28	629	629
1050–1150	2245	2245	0.04	90	90
1150–1250	2241	2241	-0.30	-672	-672
1350–1450	2236	2236	-0.82	-1834	-1834
1450–1625	2230	2230	-0.88	-1962	-1962
1625–1875	2232	2232	-1.49	-3326	-3326
1875–2250	2228	2228	-2.65	-5904	-5904
2250–2750	2236	2236	-3.75	-8385	-8385
2750–3500	2240	2240	-5.96	-13350	-13350
3500–4500	2245	2245	-9.55	-21440	-21440
4500–bottom	2254	2254	6.77	15260	15260
			Total Flux	-64601	-64260

Table 1: Estimates of present and pre-anthropogenic oceanic carbon transport in selected depth intervals for the Florida-Bahamas section and for the 25°N section. The layer transports are as calculated in Bryden and Hall (1980); the negative sign indicates southward flow.

thropogenic gases into the ocean. For instance, within the water column, pCO_2 can be written as the following sum:

$$pCO_2 = pCO_2^{pre} + pCO_2^{ant} + pCO_2^{bio} \tag{21}$$

where the superscripts pre, ant, and bio indicate preformed, anthropogenic and biological, respectively. Consequently at depths below the subsurface waters, correlations between CFC concentrations and the sum $pCO_2^{pre} + pCO_2^{ant}$ not only should exist but could also potentially be used to quantify and map the amount of anthropogenic CO_2 that currently penetrates into the ocean below the mixed layer depth, with the mixing and ventilation processes of intermediate and deep waters. For instance, recent CFCs measurements in the North Atlantic Ocean across a section from Iceland to 3°S along about 10°W, indicate that north of 35°N, CFCs (and consequently atmospheric CO_2 gas) penetrated to the deep ocean and that south of 20°N both F-11 and F-12 were below detection limits at depth below 500 m (Bullister, 1989), thus implying that in this ocean area anthropogenic CO_2 did not penetrate below the depth of 500m since 1930. Total CO_2 concentrations were not measured in this section along with the CFCs concentrations, but we can make use of the earlier TTO data set to determine a correlation between CFC concentrations and anthropogenic CO_2. Figure 3 and 4 display a section of pCO_2^{pre} + pCO_2^{ant} and F-11 (R. Gammon, personal communication), respectively, along about 45°W within latitudes 50°N to 30°N in the North Atlantic Ocean (from station 225 to station 235 of the TTO expedition, Figure 5). In Figure 6 we show the (pCO_2^{pre} + pCO_2^{ant}) - F-11 diagram for this section for points coincident with the isopycnal associated with Labrador Sea water at about the 1.5 km depth horizon. The correlation is linear and is described by:

$$pCO_2^{pre} + pCO_2^{ant} = 273 + 12.7x[F - 11] \tag{22}$$

with a correlation coefficient of 0.925. An estimate of the standard deviation indicates that on average the points may deviate from the regression line by 3.8 μatm. This (pCO_2^{pre} + pCO_2^{ant}) - F-11 relationship is apparently very tightly defined and implies that, in 1930, in surface waters of the North Atlantic Ocean, the sum of preformed and anthropogenic pCO_2 was 273 ± 3.8 μatm. The 1981 TTO North Atlantic CO_2 data set was obtained prior to the advent of coulometric titration procedures (see earlier section of this paper) and depended upon difficult potentiometric procedures (Bradshaw *et al.*, 1981). The absolute values of the background signal are therefore less well defined than the slope of the F-11/(pCO_2^{pre} + pCO_2^{ant}) line. Yet, this value is remarkably close to the known 280 μatm pre-industrial atmospheric value, and even closer to our estimates of 271 ± 7 μatm and 272 ± 6 μatm at 25°N, thus suggesting that most of the anthropogenic CO_2 that is currently in the ocean, penetrated the ocean since 1930. If we further assume that this relationship (Equation 22) is valid over the whole North Atlantic Ocean, then we can combine it with CFCs data sets such as the North Atlantic section shown by Bullister (1989) to estimate the amount of anthropogenic CO_2 that penetrated into the Atlantic Ocean along this section since the beginning of the CFCs emissions.

5 Temperature increase

"Naturally occuring greenhouse gases keep the Earth warm enough to be habitable. By increasing their concentrations, and by adding new greenhouse gases like chlorofluorocarbons (CFCs), humankind is capable of raising the global average annual-mean surface-air temperature (which, for simplicity, is referred to as the 'global temperature'), although we are uncertain about the rate at which this will occur. Strictly, this is an enhanced greenhouse effect - above that occuring due to natural greenhouse gas concentrations; the word 'enhanced' is usually omitted, but it should not be forgotten." IPCC (1990).

Current estimates from atmospheric models (Hansen et al., 1989), suggest that one of the climate impacts of the accumulation of the major greenhouse gases (CO_2, CFCs, N_2O, CH_4), will be to raise the air surface temperature by approximately 1.8°C over the next 50 years, which in turn will also warm the surface ocean and will have implications on the storage of CO_2 in the ocean. The storage of CO_2 in the ocean then depends on ocean currents and on the chemical properties of seawater. As Harvey (1926) and Redfield (1934) pointed out, although some local and seasonal variations exist, the ratio of concentrations of carbon, nitrate and phosphate (C/N/P) in the global ocean remains approximately constant. Hence, today, the most commonly used parameterization of the principal oceanic cycle representing the production and storage of CO_2 in the ocean (respiration/decomposition in the oceanic water column) is a simple exponential function of depth which is independent of both temperature and chemical composition of seawater. As a result, current model predictions of the impact of global warming on the ocean-atmosphere CO_2 system are likely to be inaccurate.

Virtually all chemical and biological reactions show an approximate rate change of a factor two for a 10°C change in temperature. For the oceanic decomposition reaction we lack accurate knowledge. The best estimates can be deducted from the work of sedimentary geochemists who have examined similar problems in anoxic systems. Their data clearly

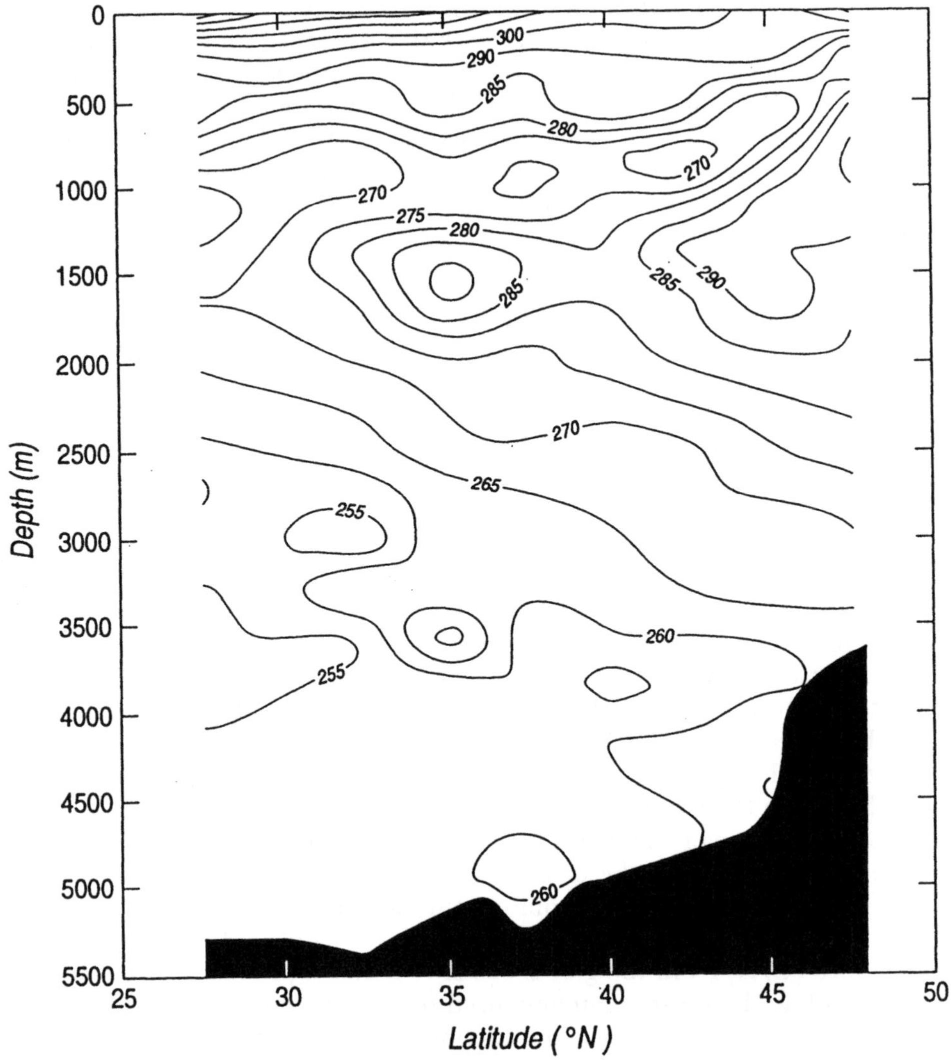

Figure 3: Latitudinal section of calculated $pCO_2^{pre} + pCO_2^{ant}$ from the TTO North Atlantic data set.

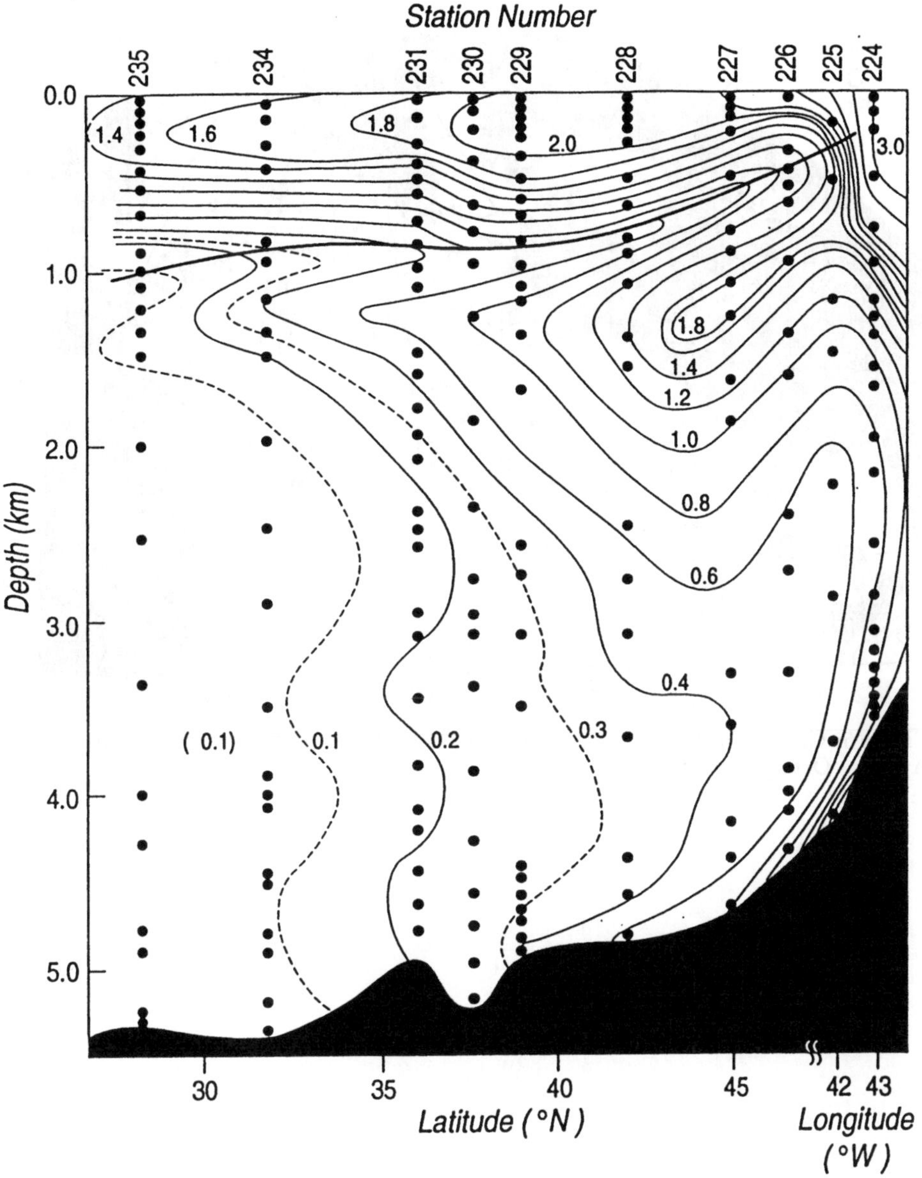

Figure 4: Latitudinal section of chlorofluorocarbon F-11 from the TTO North Atlantic data set.

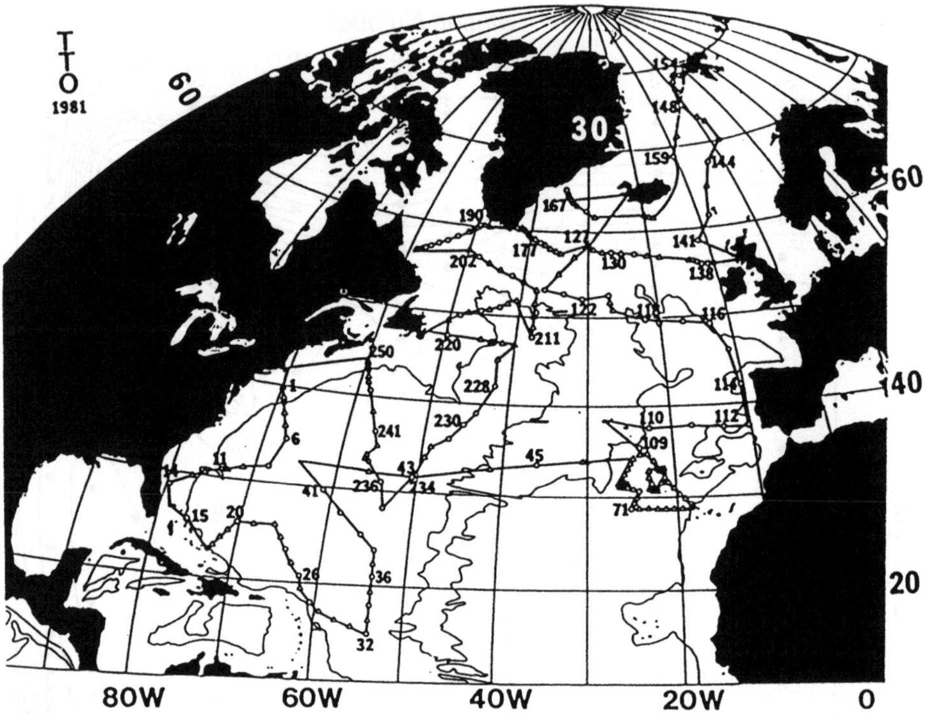

Figure 5: Geographic location of the stations 225 to 235 of the TTO North Atlantic cruise, April 1 - October 19, 1981.

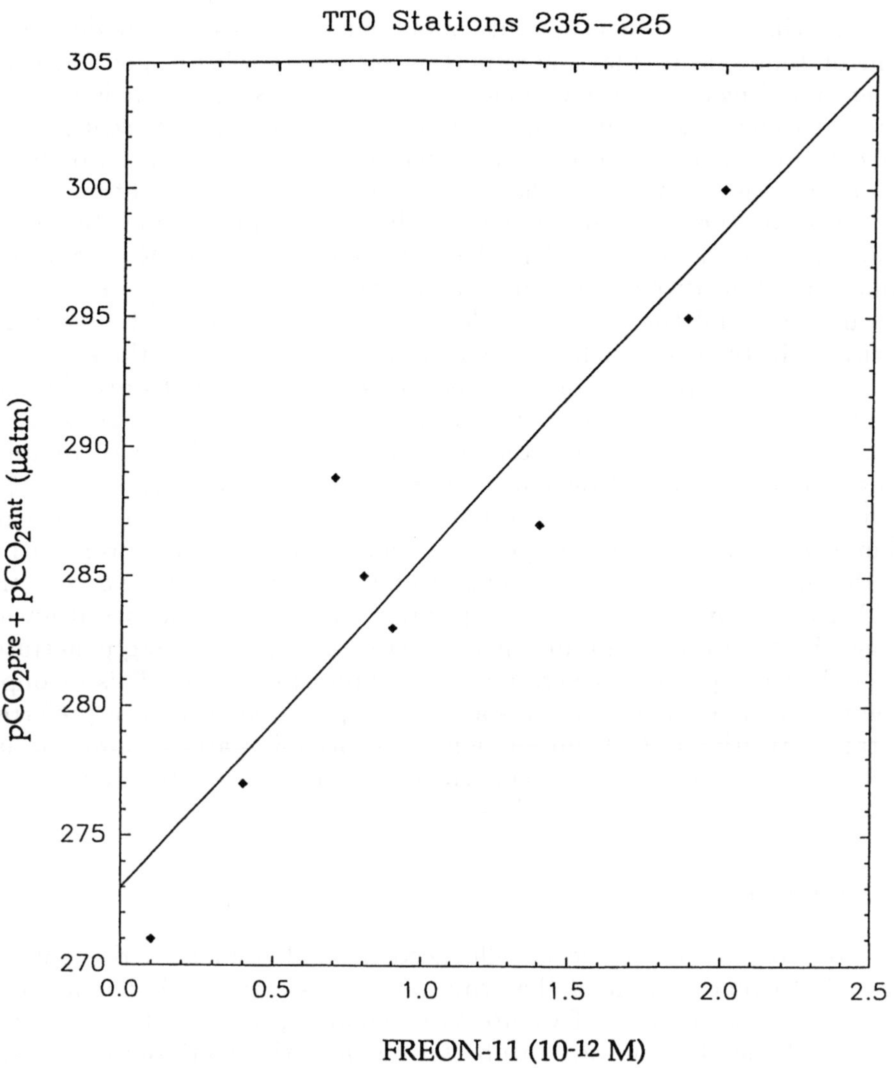

Figure 6: Correlation between pCO_2^{pre} + pCO_2^{ant} and concentration of chlorofluorocarbon F-11 for points coincident with the isopycnal associated with Labrador Sea water at about the 1.5 km depth horizon.

indicate that while the availability of organic matter exerts the dominant control, temperature plays an important secondary role (Jorgensen, 1977; Aller and Yingst, 1980; Westrich and Berner, 1988). Since it is unlikely that microbes in the sediments exhibit a fundamentally different physiological response than those in the water above, the effect of temperature would then be effectively to change the scale length of penetration of particle flux into the ocean, increasing this during a cooling period. Similarly, slower reaction kinetics for Dissolved Organic Carbon (DOC) would enhance the unoxidized reservoir of this material in the deep ocean. For a warming trend the reverse would be true. A substantial temperature variation in the ocean will have numerous consequences on the evolution of CO_2 fluxes across the ocean-atmosphere interface. Not only the balance of oceanic heat fluxes but also the rate of surface ocean respiration will be changed. If the reaction rate doubles for a 10 degree temperature change, then a 2 degree temperature change will produce a 15% variation in the organic carbon decomposition rate of the surface ocean. While no formal descriptions yet exist, the thermal anomaly propagating in the upper ocean may yield higher photosynthetic rates, whereas the cold deep ocean may hold decomposition rates at pre-industrial levels. A geochemical rate instability may persist until the warming penetrates to sufficient depth to accelerate the decomposition signal. This example suggests a highly non-linear response. We point out that the preferred parameterization of CO_2 production in the oceanic water column for numerical models will be as a concentration and temperature dependent reaction, not as an afunctional exponent of depth.

6 Conclusion

The oceanic carbon cycle is influenced by and influences the anthropogenic carbon content of the atmosphere. As a result, it has a direct impact on our climate. If we are to understand, document and predict climate change then we need to better quantify the spatiotemporal variations of carbon content in the ocean. As mentioned above, on a global scale, the pCO_2 balance between air and sea that must have held during the pre-anthropogenic era is now in disequilibrium by approximately 8 μatm. Is it experimentally possible to detect this small anthropogenic signal? What would be the necessary accuracy of the measurements? Since pCO_2^{sea} is strongly dependent upon surface seawater temperature (pCO_2 increases by approximately 13 μatm when the temperature rises by 1°C), the detection of 8 μatm in ΔpCO_2 would require accurate measurements of both ΔpCO_2 and surface seawater temperature. For temperature alone, an error of $\pm$ 0.5°C on the global average surface ocean temperature would mask most of the anthropogenic signal. Consequently, on a global scale, the temperature should be measured with an accuracy close to 0.1°C. For ΔpCO_2, the accuracy of the measurements should be better than 2 μatm. In practice, on a global average, pCO_2^{air} is well known within 1 μatm but pCO_2^{sea} is still relatively poorly known. Spatiotemporal sampling strategies would have to be improved to perform CO_2 measurements that could be correctly interpolated to yield a global average estimate of pCO_2^{sea} with an accuracy close to 1 μatm.

Acknowledgement:

Our work on oceanic carbon cycle is supported by the National Science Foundation under Grant OCE87-1461 and by the National Aeronautics and Space Administration. We thank R. Gammon for access to unpublished F-11 data. This is contribution No.8065 of the Woods Hole Oceanographic Institution.

References

ALMGREN T., D. DYRSSEN, M. STRANDBERG (1975). Determination of pH on the mole per kg sea water scale (Mw). *Deep-Sea Research,* **22,** 635-646

ALLER, R.C. AND J.Y YINGST (1980). Relationships between microbial distributions and the anaerobic decomposition of organic matter in surface sediments of Long Island Sound, N.Y, USA. *Marine Biology,* **56,** 29-42

BACASTOW R. AND E. MAIER-REIMER (1990). Ocean-circulation model of the carbon cycle. *Climate Dynamics Vol.4,* No.**2,** 95-125.

BATS DATA REPORT B-1A (1991). Bermuda Atlantic Time-Series study BATS 1-12, October 1988-September 1989. *Available from the U.S. JGOFS Planning and coordination office, WHOI, Woods Hole, MA 02543 USA.*

BEADSMORE, D.J. AND G.I PEARMAN (1987). Atmospheric carbon dioxide measurements in the Australian region: Data from surface observatories. *Tellus* **39B,** 42-66.

BRADSHAW A.L., P.G. BREWER, D.K. SHAFER AND R.T. WILLIAMS (1981). Measurements of total carbon dioxide and alkalinity by potentiometric titration in the GEOSECS program. *Earth and Planetary Science Letters,* **55,** 99-115

BREWER P.G., A.L. BRADSHAW AND R.T. WILLIAMS (1986). Measurements of total carbon dioxide and alkalinity in the North Atlantic Ocean in 1981. In: The changing carbon cycle. *A global analysis. Eds. J.R. Trabalka and D.E. Reichle, Springer-Verlag, New York, 348-370*

BREWER P.G., C. GOYET, AND D. DYRSSEN (1989). Carbon dioxide transport by ocean currents at 25°N latitude in the Atlantic Ocean. *Science* **246,** 477-479.

BREWER P.G. (1978). Direct observation of the oceanic CO_2 increase. *Geophysical Research Letters,* **Vol.5,** 997-1000

BROECKER H.C., J. PETERMANN AND W. SIEMS (1978). The influence of wind on CO_2 exchange in a wind-wave tunnel, including the effect of monolayers. *Journal of Marine Research,* **36,** 595-610.

BROECKER W.S., T. TAKAHASHI, H.J. SIMPSON, AND T.-H. PENG (1979). fate of fossil fuel carbon dioxide and the global carbon budget. *Science* **206,** 409-418

BROECKER W.S. (1974). "NO", A conservative water-mass tracer, *Earth Planet Science Letter,* **23,** 100-107.

BROECKER W.S. AND T-H PENG (1992). Interhemispheric transport of carbon dioxide by ocean circulation. *Nature,* **Vol 356,** 587-593

BRYDEN H.L. AND M.M. HALL, (1980). Heat transport by currents across 25°N latitude in the Atlantic Ocean. *Science* **207,** 884-886

BULLISTER J.L. (1989). Chorofluorocarbons as time-dependent tracers in the ocean. *Oceanography,* Vol.2, No.**2,** 12-17

BYRNE R.H. AND J.A. BRELAND (1989). High precision multiwavelength pH determinations in seawater using cresol red. *Deep-Sea Research,* Vol.36, No.**5,** 803-810

CONWAY T.J. AND P. TANS (1990). Atmospheric CO_2 concentrations-The NOAA/GMCC flask sampling network. *NDP-005/R1. Carbon dioxide Information Analysis Center, Oak Ridge National Laboratory, Oak Ridge, Tennessee.*

DEGRANDPRE M.D. (IN PRESS). Measurement of Seawater pCO_2 using a Renewable-Reagent Fiber Optic Sensor with Colorimeteric Detection. *Analytical Chemistry.*

DICKSON A.G. (1981). An exact definition of total alkalinity and a procedure for the estimation of alkalinity and total inorganic carbon from titration data. *Deep-Sea Research* **28A,** 609-623

DICKSON A.G. (1984). pH scales and proton-transfer reactions in saline media such as sea water. *Geochimica et Cosmochimica Acta,* **Vol.48,** 2299-2308.

DOE HANBOOK (1991). Handbook of methods for the analysis of the various parameters of the carbon dioxide system in sea water. *Eds. A.G. Dickson and C. Goyet*

DYRSSEN D. (1965). A gran titration of sea water on board SAGITTA. *Acta Chem. Scand.* **19,** 1265.

GOYET C. AND A. POISSON (1989). New determination of carbonic acid dissociation constants in sewater as a function of temperature and salinity. *Deep-Sea Research,* **Vol.36,** No.11, 1635-1654

GOYET C., C. BEAUVERGER, C. BRUNET AND A. POISSON (1991). Distribution of carbon dioxide partial pressure in surface waters of Southwest Indian Ocean. *Tellus,* **43B,** 1-11.

GOYET C., D.R. WALT, AND P.G. BREWER (1992). Development of a fiber optic sensor for measurement of pCO_2 in seawater: Design criteria and sea trials. *Deep-Sea Research,* **Vol.39,** No.6, 1015-1026

GOYET C., AND S.D. HACKER (1992). Procedure for calibration of a coulometric system used for total inorganic carbon measurements in seawater. *Marine Chemistry,* **38,** 37-51

GOYET C. AND P.G. BREWER (IN PRESS). Temporal variations of the properties of the carbonate system in the North Atlantic Ocean at 47°N 20°W: II. The daily variations of total disolved inorganic carbon in the upper 150 m. *Deep-Sea Research.*

HALL M.M. AND H.L. BRYDEN (1982). Direct estimates and mechanisms of ocean heat transport. *Deep-Sea Research* **29,** 339-359

HANSEN J., I. FUNG, A.LCIS, D. RIND, S. LBEDEFF, R. RUEDY AND G. RUSSELL (1988). Global climate changes as forecast by the Goddard Institute for Space Science three dimentional model. *J. Geophys. R.*, **93**, 9341-9364

HARVEY, H.W. (1926). Nitrates in the sea. *Marine Biol. Assn. U.K., Jour.*, **14**, 71-88, Plymouth.

INOUE H., Y. SUGIMURA, AND K. FUSHIMI (1987). pCO_2 and d13C in the air and surface sea water in the western North Pacific. *Tellus* **39B**, 228-242.

IPCC (1990). Climate Change: The IPCC Scientific Assessment. *J.T. Houghton, G.J. Jenkins, and J.J. Epharaums, eds. WMO/UNEP/Cambridge University Press, Cambridge, UK. 365pp.*

JOHNSON K.M., A.E. KING AND J.M. SIEBURTH (1985). Coulometric TCO_2 analyses for marine studies An introduction. *Marine Chemistry* **16**, 61-82

JOHNSON K.M., P.J. LEB. WILLIAMS, L. BRANDSTROM AND J.M. SIEBURTH (1987). Coulometric TCO_2 analysis for marine studies: automation and calibration. *Marine Chemistry* **21**, 117-133

JONES E.P. (1980). Gas exchange. Air-sea interaction instruments and methods. *Plenum Press.* 433-445

JONES E.P. AND S.D. SMITH (1977). Eddy correlation measurement of sea-air CO_2 flux. *J.G.R,* **82**, 5990-5992

JORGENSEN B.B. (1977). The sulfur cycle of a coastal marine sediment (Limfjorden, Denmark). *Limnol. Oceanogr.*, **22**, 814-832

KEELING C.D., R.B. BACASTOW, AND T.P. WHORF (1982). Measurements of the concentration of carbon dioxide at Mauna Loa Observatory, Hawaii. pp.377-385. In W.C. Clark (ed.), Carbon dioxide Review: 1982. *Oxford University Press, New York.*

KEELING CD, HEIMANN M, PIPER SC, 1989. A three dimensional model of atmospheric *CO_2 transport based on observed winds: 4. Analysis of the mean annual gradients of CO_2. In: Peterson DH (ed) Aspects of Climate Variability in the Pacific and the Western Americas. American Geophysical Union, Washington DC, 305–363*

KEELING C.D, AND T.P. WHORF (1991). Atmospheric CO_2 modern record: Mauna Loa. In: Trends 91. A compendium od data on global change. *Carbon dioxide information analysis center, Oak Ridge National Laboratory, 12-15.*

KOMHYR, W.D., R.H. GAMMON, T.B. HARRIS, L.S. WATERMAN, T.J. CONWAY, W.R. TAYLOR, AND K.W.THONING (1985). Global atmospheric CO_2 distribution and variations from 1968-1982 NOAA/GMCC CO_2 flask sample data. *Journal of Geophysical research*, Vol.90, **D3**, 5567-5596.

LISS P.S. (1983A). Gas transfer: Experiments and geochemical implications. In: Air-Sea exchange of gases and particles *(eds. P.S. Liss and W.G.N Slinn), Reidel, 241-298.*

LISS P.S. (1983B). The exchange of biogeochemically important gases across the air-sea interface. In: The major biogeochemical cycles and their interactions *(eds. B. Bolin and R.B. Cook), Wiley, 411-426.*

LISS P.S. AND L. MERLIVAT (1986). Air-Sea exchanges rates: introduction and synthesis. In: The role of air-sea exchange in geochemical cycling *(ed. P. Buat-Menard). NATO ASI series, D. Reidel, Dordrecht, 113-125.*

METZL N., C. BEAUVERGER, C. BRUNET, C. GOYET, AND A. POISSON (1991). Surface water pCO_2 in the western Indian sector of the southern ocean: A highly variable CO_2 source/sink region during the summer season. *Marine Chemistry,* **35,** 85-95.

MURPHY P.P., R.A. FEELY, R.H. GAMMON, D.E. HARRISON, K.C. KELLY, AND L.S. WATERMAN (1991). Assessment of the air-sea exchange of CO_2 in the South Pacific during Austral Autumn. *Journal of Geophysical Research,* **96, C11,** 20455-20465.

NABE DATA REPORT (1991). Data report ofthe North Atlantic Bloom Experiment April-July 1989. Process study data report P-1, February 1991. *Available from the NODC/U.S. JGOFS data management office, Woods Hole Oceanographic Institution, Woods Hole, MA 02543.*

PEARMAN G.I. AND D.J. BEARDSMORE (1984). Atmospheric carbon dioxide measurements in the Australian region: Ten years of aircraft data. *Tellus* **36b,** 1-24.

PEDLOSKY J. (1990). The dynamics of the oceanic subtropical gyres. *Science,* **248,** 316-322.

PENG, T.-H., T.TAKAHASHI, W.S. BROECKER, AND J. OLAFSSON (1987). Seasonal variability of carbon dioxide, nutrients and oxygen in the northern North Atlantic surface water: observations and a model, *Tellus,* **39,** 439-458.

PENG T.-H. AND T. TAKAHASHI (1990). In: Biogeochemistry of CO_2 and the greenhouse effect, *M.P. Farrel, Ed. (Am. Chem. Soc. Symp., CRC/LEWIS, Boca Raton, FL)*

RASMUSSEN, R.A. AND M.A.K. KHALIL, (1986). Atmospheric trace gases:trends and distributions over the last decade. *Science,* **232,** 1623-1624.

REDFIELD, A. C.(1934). On the proportions of organic derivatives in sea water and in their relation to the composition of plankton. *James Johnstone Memorial Volume, p. 176, Liverpool.*

ROEMMICH D. AND C. WUNSCH (1985). Two transatlantic sections: Meridional circulation and heat flux in the subtropical North Atlantic Ocean. *Deep-Sea Research* **32,** 619-664

SILLEN L.G. AND A.E. MARTEL (1964). Stability constants. *The Chemical Society, London 754pp. Supplement No. 1, 1971, 865pp.*

TANS P.P., I.Y. FUNG, AND T. TAKAHASHI (1990). Observational constraints on the global atmospheric CO_2 budget. *Science,* **Vol.247,** 1431-1438

TAKAHASHI T. (1961). Carbon dioxide in the atmosphere and in Atlantic ocean water. *Journal of Geophysical Research* **66,** 477-494

UNESCO (1987). Thermodynamics of carbon dioxide system in seawater. Report by the carbon dioxide subpanel of the JPOTS, UNESCO technical Papers in marine Science, **51,** 55.

UPSTILL-GODDARD, R.C., A.J.WATSON, P.S. LISS, AND M.I. LIDDICOAT (1990). Gas transfer velocities in lakes measured with SF6. *Tellus* **B42,** 364-377

VOLK T. AND M.I. HOFFERT (1985). Ocean Carbon Pumps: Analysis of relative strengths and efficiencies in ocean-driven atmospheric CO_2 changes. In: The Carbon Cycle and Atmospheric CO_2: Natural Variations Arhean to present. *E.T. Sundquist and W.S. Broecker, Editors. Geophysical Monograph 32.*

WATSON A.J., R.C. UPSTILL-GODDARD, AND P.S. LISS (1991). Air-sea gas exchange in rough and stormy seas measured by a dual-tracer technique. *Nature,* **Vol.349,** 145-147

WEISS R.F. (1974). Carbon dioxide in water and sewater: the solubility of a non ideal gas. *Marine Chemistry* **2**, 203-215.

WEISS R.F. (1981). Determinations of carbon dioxide and methane by dual catalyst flame ionization chromatography and nitrous oxide by electron capture chromatography. *J. Chromatogr. Sci.* **19**, 611-616

WESTRICH J.T. AND R.A. BERNER (1988). The effect of temperature on rates of sulfate reduction in marine sediments. *Geomicrobiol. J.* **6**, 99-117

THE GLOBAL CARBON CYCLE IN THE CLIMATE SYSTEM

MARTIN HEIMANN
Max-Planck-Institut für Meteorologie,
Hamburg, Federal Republic of Germany.

1 Introduction

The element carbon is one of the fundamental biogeochemical constituents of the "Earth system", comprising of life and its environment on our planet. Most life forms on Earth consist of organic carbon compounds, while large inorganic carbon reservoirs dominate the appearance of the environment. Carbon is continuously exchanged between these different pools, both by physico-chemical processes and through the activities of the biosphere.

The carbon cycle is also closely coupled to the physical climate system. Carbon in the atmosphere in the form of carbon dioxide (CO_2) is the second most dominant greenhouse gas (after water vapor) and thus affects by its presence the surface temperature of the Earth. Atmospheric carbon in the form of methane (CH_4) also represents an important greenhouse gas. Further, somewhat weaker couplings exist at the land surface and in the ocean, where the biosphere modifies physical properties, such as e.g. albedo, hydrological conductance, roughness and turbidity.

In the Earth's history, both climate and the carbon cycle have gone through big changes as witnessed by the paleographic records. Often, such as during the last glaciation, changes in both systems occurred almost simultaneously. The extent to which one or other system was forcing the companion system is one of the central, still unsolved, questions in climate research.

The activities of man have begun to seriously perturb the natural carbon cycle during the recent historical period. The use of fossil fuels for the generation of energy resulted in huge emissions of carbon dioxide into the atmosphere. The increasing population growth during the last two centuries induced large changes in land use as witnessed e.g. by the rapid disappearance of the tropical rainforests. Most of these land use changes are associated with net losses of stored carbon on land, and hence represent a second anthropogenic source of carbon dioxide. These CO_2 inputs have lead to a worldwide rise in the concentration of atmospheric carbon

NATO ASI Series, Vol. I 11
Modelling Oceanic Climate Interactions
Edited by J. Willebrand and D. L. T. Anderson

dioxide of more than 25% over pre-industrial levels, as documented by records extracted from air-bubbles in ice-cores and by direct atmospheric measurements since 1959. A rise in the atmospheric concentration of methane has also been recorded, which has almost doubled since the beginning of the last century.

If the recent rise in the concentraton of CO_2 and CH_4, were to continue into the future, a corresponding enhancement of the terrestrial greenhouse effect would result, which could lead to severe changes of the climate. In order to react to this potential threat by implementing of measures to curb the emissions, the dynamics of the carbon cycle must be understood.

A quick budget calculation shows that not all the CO_2 emitted by anthropogenic activities remains in the atmosphere. A substantial fraction is absorbed presumably by the carbon reservoirs of the ocean and the terrestrial biosphere. The exact partitioning of the uptake of excess carbon by these pools is rather uncertain and represents a key question of carbon cycle research. How the carbon cycle responds quantitatively to the prescribed present and future emissions constitutes the *direct perturbation problem.*

The second key question in current carbon cycle research is the *indirect perturbation problem*, i.e. the quantitative determination of the response of the carbon cycle to both natural or anthropogenically induced climate perturbations, as might be expected during the next centuries. It is this question that must be solved in order to realistically address the *coupled system problem*, i.e. the question of how the coupled climate-carbon system behaves in terms of its stability and sensitivity to external forcings. However, at present we can only really address the direct perturbation problem. Research has only recently started to focus on the indirect and the coupled problems.

The processes controlling the concentration of CH_4 are quite different from those of CO_2. The latter is chemically inert and is exchanged continuously with the much larger biospheric and oceanic carbon pools. On the other hand, CH_4 is controlled by atmospheric chemistry and by several, still poorly known, sources and sinks at the Earth's surface (Fung et al 1991). Consequently, the dynamic behavior and therefore the methods to describe and model the two gases are very different. Only the carbon cycle with respect to CO_2 is addressed here.

The present article has several objectives: It should provide a short overview of our current understanding of the global carbon cycle, of the basic concepts used to describe its dynamic behaviour by means of carbon models and of the tracer methods which are essential to calibrate and/or validate these models. Finally, two of the central controversies, the problems of the "missing sink" and the "northern hemisphere sink", are summarised.

This is a tutorial text and I have tried to adopt a non-specialist perspective. The article is intended to be self-contained, i.e. the main concepts should emerge without going back to the original literature. However, for sake of brevity this review is not exhaustive. The interested student is urged to consult the existing comprehensive and authoritative literature on the global carbon cycle [e.g. Bolin, 1979, 1981, 1986, Trabalka and Reichle, 1986, Siegenthaler, 1986, 1993, Keeling et al., 1989a, Heimann, 1993].

2 Overview of the carbon cycle

Figure 1 shows sizes and fluxes of carbon between the major carbon reservoirs that interact with the physical climate system on time scales of up to 10^5 years, *viz.* the atmosphere, the terrestrial biosphere and the oceans. Reservoir sizes are given in GtC (Gigatons of carbon, where 1 GtC $= 10^{12}$ kg of carbon) and fluxes in GtC yr^{-1}.

In the atmosphere, carbon exists primarily in the form of carbon dioxide, methane and carbon monoxide (CO). Table 1 lists the major carbon compounds, their current concentrations and estimated atmospheric carbon inventory. The concentrations are expressed as mixing ratios in parts per million by volume of dry air (ppmv). Also listed is the estimated atmospheric residence time of these gases, defined as the current atmospheric amount divided by the gross flux out of the atmosphere by all sink processes. The residence time, as defined here, reflects the average time a molecule remains in the atmosphere before being transferred to one of the other carbon reservoirs. It is important to note that the residence time is not necessarily equal to the decay time or lifetime of a perturbation input of these gases into the atmosphere. For CO_2, as shown in Section 3.3, the decay time of a perturbation is much larger ($\approx$ 50-200 years) because of exchanges with the oceanic and terrestrial biosphere reservoirs.

Both, CO_2 and CH_4 have long residence times compared to typical global atmospheric mixing rates which are on the order of one year. This implies that these two gases should exhibit fairly uniform atmospheric mixing ratio distributions. Consequently, as a driving agent for the physical climate system, only the globally averaged mixing ratio of these gases is relevant in calculating the exerted radiative forcing. However, as discussed in subsequent sections, the spatial and temporal patterns of their mixing ratio distributions is an important diagnostic tool since it reflects the large scale structure of the surface sources and sinks.

The ocean contains over 60 times more carbon than the atmosphere

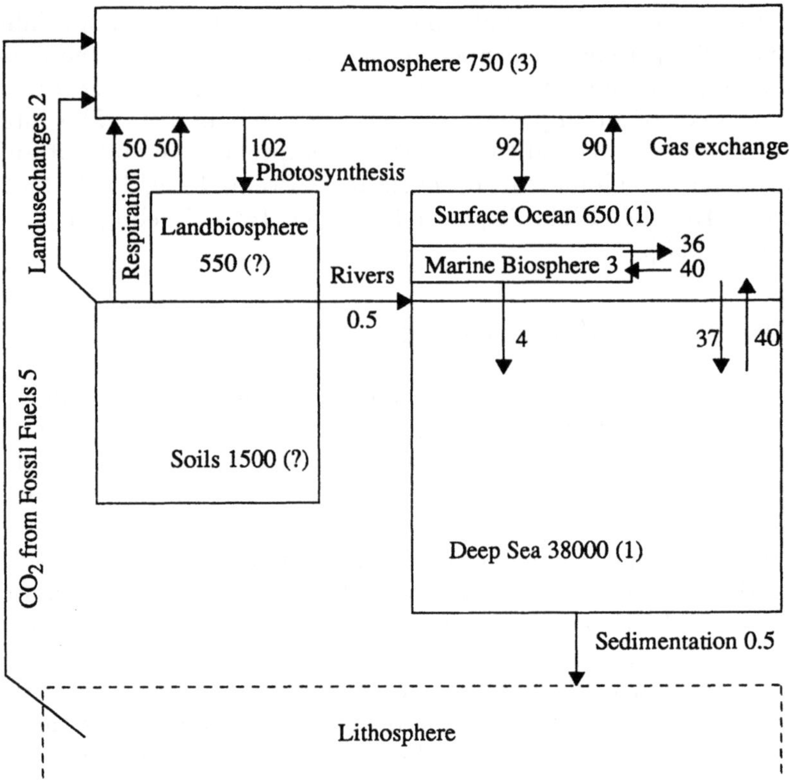

Figure 1: Global carbon pools, current (1980) inventory changes (in parentheses) and exchange fluxes. Poolsizes are given in GtC, inventory changes and exchange rates in GtC yr^{-1}.

Table 1: Major atmospheric gases containing carbon

Gas	Average Concentration (1990) ppmv	Inventory GtC	Residence time years
CO_2	353	749	3.7
CH_4	1.72	3.6	9.8
CO	≈ 0.1	≈ 0.2	0.17

Table 2: Carbon fluxes from the lithosphere to the atmosphere-ocean-land biosphere pools (in GtC yr^{-1}) [Sundquist, 1985, Arthur et al., 1985, Sarmiento and Sundquist, 1992].

Weathering	-0.1 – -0.2
Volcanos	< 0.1
Decarbonation	0.3 – 0.5
Sedimentation	-0.2 – -0.3

(Figure 1). The largest fraction is dissolved inorganic carbon (DIC), consisting of dissolved carbonic acid (H_2CO_3), bicarbonate (HCO_3^-) and carbonate ions ($CO_3^=$). Substantially smaller oceanic carbon pools are represented by the marine biosphere and dissolved organic carbon (DOC). The dynamics of these reservoirs and their role in the global carbon cycle is discussed in more detail by Goyet and Brewer [*this issue*] and by Maier-Reimer [*this issue*].

Approximately the same amount of carbon as in the atmosphere is contained in the living vegetation ($\approx$ 550 GtC). The dead biomass on land (i.e. litter and soils) comprises about twice as much carbon as the atmosphere ($\approx$ 1500 GtC).

If timescales beyond 10^5 years are considered then exchanges with the carbon pools in the lithosphere have to be taken into account. Estimated fluxes are given in Table 2 [Sundquist, 1985, Arthur et al., 1985, Sarmiento and Sundquist, 1992]. The magnitude of these fluxes do not exceed 0.5 GtC yr^{-1}.

Of particular importance today are the anthropogenic perturbations. During the decade 1980–89 the average emissions of CO_2 from the burning of fossil fuels and from gas flaring were approximately 5.4 GtC yr^{-1}. Estimates for the year 1990 give 6.0 GtC yr^{-1}. Table 3 lists the current global fuel mix [Marland and Boden, 1993] and some estimated reserves of fossil fuels [Sundquist, 1985]. There exist also minor industrial sources e.g. the production of cement releases CO_2 to the atmosphere. Current emissions are about 0.15 GtC yr^{-1}. The second anthropogenic source, the emissions from changes in land use and forest clearing is addressed below in Section 5.

Because of the anthropogenic CO_2 emissions the carbon cycle of today is not in a steady state. Figure 1 also lists (in parentheses) some estimates of the current inventory changes in the natural carbon reservoirs induced by the perturbation. One of the central tasks of carbon cycle research is to quantitatively determine these induced perturbations. Since they are very small compared to the carbon pool sizes, this task is not easy.

Table 3: Current carbon dioxide emissions from fossil fuel use [Marland and Boden, 1993] and estimated reserves, identified and ultimately recoverable [Sundquist, 1985]. The sum of the current emissions includes the contribution from cement manufacture and natural gas flaring.

	Emissions (1989) $GtC\ yr^{-1}$	Identified Reserves GtC	Ultimately Recoverable GtC
Coal	2.4	3200	6800
Natural Gas	1.0	40	140
Oil	2.4	90	200
Oil Shales/Sands	-	120	390
Sum	6.0	3400	7400

3 Dynamics

The dynamics of the carbon cycle are governed by the conservation law for the element carbon and by the processes controlling the exchange fluxes between the different carbon pools. Since the carbon cycle is imbedded in the physical climate system, any changes occurring in the latter are also likely to be reflected in fluctuations in the carbon cycle.

In the following I discuss from a modelling perspective the basics of the two most important exchange fluxes: air-sea gas exchange and surface exchange fluxes by photosynthesis and respiration.

3.1 Air-sea gas-exchange

The exchange flux of low and moderately soluble gases, such as CO_2, between the air and the sea is controlled by a thin (~100 μm) boundary layer on the water side of the interface, whithin which molecular diffusion exceeds transport by turbulent motions [Jähne et al., 1987]. The state of this boundary layer (its temperature and thickness as given by the local structure of micro turbulence) determines the gas transfer, given a driving concentration difference of dissolved CO_2 at the air-sea interface and below in the bulk mixed layer. In the presence of breaking waves entrained air bubbles contribute to the gas flux [Merlivat and Memery, 1983]. Formally, the net gas flux per unit area (F), i.e. the difference between the gross air-to-sea (F_{as}) and the gross sea-to-air gas transfer flux (F_{sa}), may be written as a bulk formula [e.g. Broecker and Peng, 1982]:

$$F = F_{as} - F_{sa} = w(\alpha P_a - C_m^*) \tag{1}$$

where P_a denotes the partial pressure of CO_2 in the atmosphere and the product αP_a the concentration of dissolved CO_2 at the air-sea interface. The parameter α is the solubility of CO_2 (the constant of Henry's law), and is a well determined function of temperature and salinity [Weiss, 1974]. The concentration of dissolved CO_2 in the mixed layer is C_m^*. The bulk coefficient w represents the "transfer velocity" or "piston velocity". Equation (1) is customarily written as

$$F = k_{ex}(P_a - P_m) \tag{2}$$

where the product $w\alpha = k_{ex}$ is termed the bulk coefficient for gas exchange, or"gas exchange coefficient", which is expressed in units of mol m^{-2} yr^{-1} μatm^{-1}. The CO_2 partial pressure of air in equilibrium with the sea water (often denoted as "P_{CO_2}") is given by $P_m = C_m^*/\alpha$.

Equation (2) forms the basis for both the empirical determination and the modelling of the gas-exchange flux. Both, P_a and P_m are readily measured or modelled, but the gas-exchange coefficient is poorly known.

Several microphysical models of the boundary layer have been constructed (e.g. Coantic [1986], Back and McCready [1988]), but, unfortunately, up to now these models do not permit a quantitative determination of the gas-exchange coefficient given a set of readily observable environmental parameters. Using an empirical approach, the gas-exchange coefficient may be specified as a simple function of windspeed and sea surface temperature which is fitted to measurements. However, direct observations of CO_2 gas-exchange fluxes under natural conditions suffer from several limitations [Broecker et al., 1986]. Likewise, the results of laboratory studies on wind/wave tunnels show considerable scatter and are not easily extrapolated to large-scale natural environments. The most promising approach so far has been to study other, more easily monitored, trace gases, such as ^{222}Rn or SF_6 and to translate the results to CO_2 by means of an assumed dependency on the molecular diffusivity.

Liss and Merlivat [1986] proposed an empirical, quantitative relationship between the gas-exchange coefficient and the windspeed at 10 m height, based partly on theoretical considerations and partly on the results of an experiment reported by Wanninkhof et al. [1985], who monitored the efflux of SF_6 from a lake after injection. Further lake experiments [Upstill-Goddard, 1990] and measurements by means of the "dual tracer technique" in the North Sea [Watson et al., 1991] partially confirmed the formula given by Liss and Merlivat [1986], however, another recent open ocean experiment on the Georges Bank revealed substantially higher values [Wanninkhof, 1993, unpublished manuscript].

There exists an independent constraint on the magnitude of the gas exchange coefficient, imposed from the observations of the global balance of natural and bomb produced radiocarbon [Broecker et al., 1986]. The global averaged value determined from this method, 0.061±0.010 mol m^{-2} yr^{-1} μatm^{-1}, is approximately 70% higher than the global average determined from the Liss and Merlivat [1986] formula and statistics of the distribution of windspeeds over the oceans from either satellite data [Etcheto and Merlivat, 1988] or meteorological analyses and ship reports [Heimann and Monfray, 1989]. Possible factors that might explain this discrepancy have been discussed by Watson [1993]. A satisfactory solution to the problem of the gas-exchange coefficient, however, still has to be found. A globally-averaged gas-exchange coefficient of 0.061 mol m^{-2} yr^{-1} μatm^{-1} corresponds to a global gross air-to-sea exchange flux of 90 GtC yr^{-1} based on an atmospheric CO_2 partial pressure of 353 μatm and an ocean surface area of 350 10^{12} m^2. The corresponding atmospheric air-sea exchange time (atmospheric CO_2 inventory divided by the gross air-sea flux) is 8.2 years.

In order to describe the air-sea gas-exchange flux in a model, the partial pressure of CO_2 in sea water, P_m, must be computed. By Henry's law, P_m is proportional to the concentration of dissolved carbonic acid, which, by chemical reactions, is exchanged with the other inorganic carbon components in surface waters, i.e. bicarbonate and carbonate. These chemical exchanges are very rapid compared to the gas exchange through the surface boundary layer. For modelling purposes one can therefore always assume a chemical equilibrium between the carbonate species and represent P_m as a nonlinear function of the surface concentration of total dissolved inorganic carbon, temperature, salinity and the concentration of other chemical constituents in sea water. This nonlinear function is computed by numerically solving the system of chemical reactions [Maier-Reimer, *this issue*].

For small perturbations P_m may be expanded around a background state (denoted with the subscript 0):

$$\begin{aligned} P_m(C_m, T, S, ...) &= P_{m0} + \left.\frac{\partial P_m}{\partial C_m}\right|_{\substack{C_m=C_{m0}\\ T=T_0}} (C_m - C_{m0}) + \left.\frac{\partial P_m}{\partial T}\right|_{\substack{C_m=C_{m0}\\ T=T_0}} (T - T_0) + ... \\ &\approx P_{m0}\left(1 + \xi\frac{C_m - C_{m0}}{C_{m0}} + \eta\frac{1}{P_{m0}}(T - T_0) + ...\right) \end{aligned} \quad (3)$$

where C_m denotes the concentration of DIC, T the sea surface temperature and S the salinity.

The quantity ξ as introduced in equation (3) denotes the buffer factor (or "Revelle factor") representing the fractional change of the equilibrium partial pressure of CO_2 for a fractional change in the concentration of DIC:

$$\xi = \frac{C_{m0}}{P_{m0}}\frac{\partial P_m}{\partial C_m} \quad (4)$$

For typical seawater conditions ξ has a value between 8 and 14, with a global average [Sundquist et al., 1979] of ~10. Thus, a 1% change of the concentration of DIC is associated with a change of ~10% in the corresponding equilibrium partial pressure of CO_2. A consequence of this is that effectively only one tenth of the oceanic DIC carbon pool is potentially available (in the final equilibrium) to absorb an additional atmospheric CO_2 input. Since the ocean contains about 60 times more carbon than the atmosphere the final repartition between air and sea will be approximately 1:6, i.e. about 16% of a CO_2 input will remain in the atmosphere, assuming no exchanges with other carbon reservoirs.

For large perturbations, higher order terms in the expansion of P_m must be taken into account. Alternatively, the representation of (4) may be retained (without higher order terms) but ξ redefined as a function ("variable buffer factor") of the inorganic carbon excess $C_m - C_{m0}$.

The second partial derivative in equation (3), η, represents the temperature sensitivity of P_m:

$$\eta = \frac{\partial P_m}{\partial T} \tag{5}$$

which has a value between 12 and 16 μatm K^{-1} in typical ocean waters.

As an example, if during the last ice age sea surface temperatures were lower by approximately 2 K the correspond equilibrium atmospheric CO_2 concentration would be ≈25 ppmv, assuming no changes in the chemical composition of the ocean. Evidently, the observed atmospheric decrease of more than 80 ppmv [Barnola et al.,1987] cannot be explained by temperature effects alone [Heinze et al., 1991, Broecker and Peng, 1993].

3.2 Exchange fluxes with the terrestrial biosphere

The terrestrial biosphere, comprised of living biomass and soil carbon, exchanges large quantities of carbon with the atmospheric reservoir (Figure 1). The carbon absorbed by photosynthesis of land plants ("Gross Primary Production") amounts to annually ≈ 100 GtC. About half of this flux is returned to the atmosphere by the autotrophic respiration of the plants. The remainder, termed "Net Primary Production" (NPP) of ≈ 60 GtC yr^{-1}, is transformed into organic carbon and built into plant tissue. Dead plant carbon is shed as litter, consumed by heterotrophic organisms and cycled back into the atmosphere as CO_2. Complex transformation processes convert some of the litter carbon to soil organic carbon which is oxidized only very slowly [Post, 1993]. Thus, for modelling purposes, the terrestrial biospheric carbon pool may be divided into three compartments: living biomass, litter and soil carbon. Furthermore, a distinction between herbaceous and woody plants because of different turnover times can be made [Esser, 1987]. A more realistic modelling of the complex controls of carbon transfers and storage in the terrestrial biosphere as functions of external parameters, such as climate variables, soil properties, ecosystem composition etc. represents a challenging task and is not reviewed here [Steffen et al., 1992].

In the global context one aspect is particularly important: How does the terrestrial biosphere carbon pool react to changes in the atmospheric CO_2 concentration? It is well known, that agricultural plants grow faster under enhanced CO_2 concentrations, due to the so-called "fertilization effect", at least as long as they are not limited by nutrients or water. It is still an open question, however, whether this effect also applies to natural ecosystems. It has been shown that plants under water stress respond favorably to enhanced CO_2 levels, because for the same amount of carbon exchanged the leaf stomata have to be open for a shorter time, thus resulting in reduced waterlosses by evapotranspiration. But it is not clear, if enhanced photosynthesis also leads to an increased storage of carbon in the terrestrial organic carbon pools or whether the fertilization effect merely stimulates a higher turnover rate [Strain and Thomas, 1992].

Hence, the issue of the fertilization effect is still controversial [Wisnievski and Lugo, 1992]. As discussed below in section 5 the assessment of the contemporary atmospheric CO_2 budget results in a substantial "missing sink" of the order of 1.5 GtC yr^{-1} during the decade of the 1980's. This imbalance is usually attributed to the CO_2 fertilization effect. However, other environmental factors might also induce changes in storage of carbon in natural terrestrial ecosystems, such as eutrophication [Peterson and Melillo, 1985] or climate changes [Esser, 1987].

For the purpose of modelling the terrestrial biosphere compartment in global carbon models, the fertilization effect usually is included as a single parameter, the so-called "beta factor" [Kohlmaier et al.,1989], which might also depend on other environmental variables [Esser, 1987]. It is defined as the coefficient of the linear term in the expansion of NPP as a function of the atmospheric CO_2 concentration:

$$F_{NPP} = F_{NPP,0}\left(1 + \beta\frac{P_a - P_{a0}}{P_{a0}} + \ldots\right) \qquad (6)$$

Empirical values for β exist only from a few selected observational studies [Strain and Thomas, 1992, Norby et al., 1992] which can not be extrapolated to the globe. In global carbon models the numerical value of the fertilization factor is often left as an adjustable parameter, which is determined by tuning the complete ocean-biosphere-atmosphere model to observations of the atmospheric CO_2 concentration increase during the last 30 years. This procedure is not very satisfactory, since any model defect will be projected into the magnitude of the fertilization factor and thus limit the predictive power of the model.

3.3 Models of the global carbon cycle

Simple but instructive models of the carbon cycle can be constructed by representing the different carbon pools with a few boxes and prescribing the exchange fluxes between them with simple (mostly first order) flux dynamics. The rate constants have to be specified either *a priori* or by tuning the distribution of tracers with dynamics similar to carbon, such as e.g. radiocarbon. Because such models are calibrated and/or validated by means of analogue tracer simulations, they may be used to address the direct perturbation problem (as defined in the introduction). However, they have a limited predictive capability because the feedbacks of the physical climate system on the carbon cycle are difficult to take into account. Furthermore, the opportunity to validate the models with direct observations is limited because the different carbon reservoirs are aggregated into large boxes.

A classic example of a model of this kind is given by the "box-diffusion" model of Oeschger et al. [1975]. For tutorial reasons this model is described here at some length, since it illustrates nicely the main features of the global carbon cycle dynamics, and allows analytical solutions for particular cases such as the steady state of natural radiocarbon, of the exponentially rising anthropogenic input of fossil fuel CO_2. A numerical version of the model, written in the Fortran programming language may be obtained from the author on request.

In the version of the box-diffusion model described here the global carbon cycle has a structure similar to that shown in Figure 1 but with a few simplifications. The ocean is represented by a single mixed layer of fixed height h_m overlying a one-dimensional vertical water column ("deep sea") within which tracers are mixed by eddy diffusion. The atmosphere and the terrestrial biosphere are described by single reservoirs that exchange carbon with first order dynamics. The model is a perturbation model, designed to study the response of the global carbon system to a bulk carbon or a carbon tracer perturbation. The state variables of the model are the perturbation carbon content in the atmosphere, $n_a(t)$, the terrestrial biosphere, $n_b(t)$, the ocean mixed layer, $n_m(t)$, and $c(z,t)$, the concentration of dissolved inorganic carbon in the deep sea, all expressed as deviations from their steady state background values. The equations of the box-diffusion model are given as follows (slightly modified from Oeschger et al. [1975]):

Atmosphere:

$$\dot{n}_a = \xi k_{ma} n_m - k_{am} n_a + k_{ba} n_b - \beta k_{ab} n_a + Q(t) \qquad (7)$$

Biosphere:

$$\dot{n}_b = \beta k_{ab} n_a - k_{ba} n_b \tag{8}$$

Ocean mixed layer:

$$\dot{n}_m = k_{am} n_a - \xi k_{ma} n_m + A_{oc} K \left. \frac{\partial c}{\partial z} \right|_{z=0} \tag{9}$$

Deep sea:

$$\frac{\partial c}{\partial t} = K \frac{\partial^2 c}{\partial z^2} \tag{10}$$

with the boundary conditions

$$c(z = 0, t) = c_m = \frac{n_m}{h_m A_{oc}} \tag{11}$$

$$\left. \frac{\partial c}{\partial z} \right|_{z=h_d} = 0 \tag{12}$$

The vertical coordinate in the deep sea compartment, z, runs from 0 at the base of the mixed layer to the globally averaged depth of the ocean, h_d. A_{oc} denotes the ocean surface area. $Q(t)$ represents the atmospheric CO_2 source e.g. from fossil fuel combustion. The buffer factor, ξ, and the fertilization factor, β, have been described in the previous section. Three further independent model parameters have to be specified: the vertical eddy diffusion coefficient, K, which is obtained by tuning the model to give a realistic penetration of bomb-radiocarbon in the ocean (see Section 4.1), the air-sea exchange time, $(k_{am})^{-1}$, (see Section 3.1) and the time constant for carbon exchange with the terrestrial biosphere, $(k_{ab})^{-1}$ (see Section 3.2). The other constants, k_{ma} and k_{ba}, are related to k_{am} and k_{ab}, respectively through the steady state sizes of the reservoirs of the atmosphere, N_a, the surface mixed layer, N_m, and the terrestrial biosphere, N_b:

$$k_{ma} = k_{am} \frac{N_a}{N_m} \tag{13}$$

and

$$k_{ba} = k_{ab} \frac{N_a}{N_b} \tag{14}$$

Standard numerical values of the model parameters as given by Siegenthaler [1983] are listed in Table 4.

The box-diffusion model has been extended to encompass polar outcrops [Siegenthaler, 1983] or more complicated configurations of the oceanic mixing patterns including a rudimentary thermohaline circulation [Siegenthaler and Joos, 1992, Shaffer and Sarmiento, 1992].

Table 4: Standard numerical values of the box-diffusion model parameters [Siegenthaler, 1983]

Quantity	Symbol	Value
Preindustrial size of atmosphere	N_a	594 GtC ($\doteq$ 280 ppmv)
Preindustrial size of biosphere	N_b	2.4 N_a
Preindustrial size of mixed layer	N_m	646 GtC
Mixed layer height	h_m	75 m
Depth of deep sea	h_d	3725 m
Buffer factor	ξ	10
Atmosphere-Biosphere exchange time	k_{ab}^{-1}	25 yr
Air-sea exchange time	k_{am}^{-1}	6.9 yr
Eddy diffusion constant	K	7685 m^2 yr^{-1}

During the last decade comprehensive high-resolution numerical carbon models have been developed. In advanced ocean models, the carbon pools are embedded in the three-dimensional flowfield from an oceanic circulation model [Maier-Reimer and Hasselmann, 1987, Maier-Reimer *this issue*, Sarmiento et al., 1992, Najjar et al 1992]. Advanced models of the terrestrial biosphere describe the cycling of carbon through the biospheric pools which are distributed on a two-dimensional global grid spanning the continents. The flux-rates are controlled by local properties, such as ecosystem structure, soil composition, climatic parameters (e.g. temperature and precipitation), but also statistics of land-use and agricultural practices [Esser, 1987].

For global considerations, the concept of impulse response or Greens function is useful [Oeschger and Heimann, 1983]. Assuming linear and time-invariant dynamics, we can define the impulse response function $h(\tau)$ as the fraction of carbon remaining in the atmosphere at time τ after a unit input at time zero. The atmospheric carbon content perturbation, $n_a(t)$ at time t resulting from any time dependent net source $Q(t)$ is then given by the convolution integral

$$n_a(t) = \int_0^\infty h(\tau)Q(t-\tau)d\tau \qquad (15)$$

If the carbon system is described by a series of boxes connected with first order exchange fluxes, the impulse response function can be shown to consist of a sum of decaying exponentials. The decay constants correspond to the eigenvalues of the linear differential equation system that describes the model. In the case of the box-diffusion model the impulse response function is more complicated but the Laplace transform of the impulse response is readily found from equations (7)-(10), resulting in

$$\tilde{h}(s) = \frac{1}{s}\left(1 + \beta\frac{k_{ab}}{s}\frac{1}{(1+k_{ba}/s)} + \frac{k_{am}}{s}\frac{1+\theta}{(1+\theta+\xi k_{ma}/s)}\right)^{-1} \tag{16}$$

where

$$\theta = \sqrt{\frac{K}{s}}h_m \tanh\left(\sqrt{\frac{s}{K}}h_d\right) \tag{17}$$

and s denotes the Laplace transform variable. No simple analytical form of the inverse Laplace transform of (16) can be found.

In the case of comprehensive carbon models, the impulse response function must be determined numerically. For quick calculations the function may be subsequently approximated by a finite sum of exponentials [Maier-Reimer and Hasselmann, 1987]:

$$h(\tau) = a_0 + \sum_{i=1}^{m} a_i e^{-\frac{\tau}{\tau_i}} \tag{18}$$

Because of the nonlinear ocean chemistry the impulse response determined numerically depends on the size of the initial perturbation. Table 5 lists the constant and the first four terms in the series (18) for the box-diffusion model, the Hamburg ocean carbon cycle model (OCCM) in its present version [Maier-Reimer, *this issue*] and the Princeton ocean model [Sarmiento et al., 1992], determined for a small initial input (25% of the preindustrial atmospheric CO_2 content). Note that these functions refer to models of the atmosphere-ocean system only that contain no terrestrial biosphere.

It is possible to qualitatively interpret the constants in Table 5 in terms of global oceanic carbon pools that respond to an atmospheric perturbation with the time-scale indicated by τ_i. Correspondingly, the shortest time constant might be interpreted as the response time of the atmosphere-mixed layer system, which is equilibrating within a little more than one year. The second, third and fourth time constant might reflect the adjustment with the upper and lower thermocline and the deep sea. The constant term a_0 describes the amount of CO_2 that remains in the model atmosphere. As explained in Section 3.1 based on the buffering capacity of the oceanic carbon system this should be about 16%. The Hamburg OCCM has a lower value, because it includes also exchanges with the sediment pools, which affect the long term partitioning between atmosphere and the ocean [Maier-Reimer, *this issue*]. In comparing the constants τ_i and a_i for different ocean models, one must realize that these have to be determined using a non-linear fitting procedure which is rather

Table 5: Numerical values of the constant and the first four terms in the exponential series approximation to the impulse response function of the box-diffusion model [Siegenthaler, 1983], the Hamburg OCCM [Maier-Reimer, *this issue*] and the Princeton model [Sarmiento et al., 1992] (see equation (18)).

	Box diffusion model		Hamburg OCCM		Princeton model	
i	a_i	$\tau_i(years)$	a_i	$\tau_i(years)$	a_i	$\tau_i(years)$
0	0.146		0.070		0.164	
1	0.148	1.6	0.084	1.6	0.059	1.0
2	0.193	10.1	0.079	17.6	0.229	10.3
3	0.251	46.3	0.101	71.9	0.302	60.8
4	0.262	241.	0.648	258.5	0.245	358.8

ill-conditioned in the present problem. Consequently, given a numerically determined response function, the constants of equation (18) depend very much on the number of terms allowed in the expansion.

One application of the impulse response function is the quick evaluation of the atmospheric CO_2 concentration rise resulting from a specified scenario of emissions using equation (15). If the input is given by an exponential function

$$Q(t) = Q_0 e^{\mu t} \tag{19}$$

representing, for example, the fossil fuel CO_2 source during the last 200 years, then equation (15) simplifies to

$$n_a = Q_0 e^{\mu t} \int_0^\infty h(\tau) e^{-\mu\tau} d\tau \tag{20}$$

In this case the atmospheric perturbation also increases exponentially with the same time constant. The atmospheric accumulation rate is given by $n_a \mu$ and hence the fraction of the emissions that remain in the atmosphere, i.e. the so-called "airborne fraction" is given by

$$\frac{\mu n_a(t)}{Q(t)} = \mu \int_0^\infty h(\tau) e^{-\mu\tau} d\tau = \mu \tilde{h}(\mu) \tag{21}$$

or, in other words, by the Laplace transform of the impulse response function evaluated at $s = \mu$ multiplied by the exponential rate constant μ. Assuming an exponential rate constant of $(35\ \mathrm{yr})^{-1}$ for the fossil fuel CO_2 input, the airborne fraction of the three models according to Table 5 results in 0.57 for the box-diffusion model, 0.74 for the Hamburg OCCM and 0.63 for the Princeton model. In all three cases no uptake by the terrestrial biosphere is assumed.

The impulse response function also permits a quick comparison between different models with respect to their global response characteristics to a perturbation. As an example Figure 2 displays the impulse

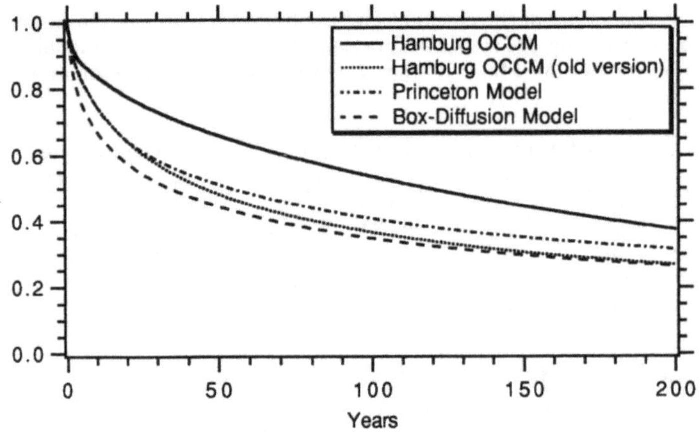

Figure 2: Impulse responses of different ocean carbon models.

response functions of the different models. For comparison, the response function of the first version of the Hamburg OCCM [Maier-Reimer and Hasselmann, 1987] is shown which exhibited a rather similar response characteristic to the box-diffusion model. The present version of the OCCM has a smaller ocean uptake.

The concept of the impulse response as outlined above has limitations. The assumption of linearity breaks down for large perturbations: for example, if the buffer factor increases significantly. The effect of the linearity assumption with respect to the ocean chemistry may be appreciated by comparing two scenario calculations with the box-diffusion model, one with a constant buffer factor and the other with a detailed carbonate chemistry calculation. In each case the model is forced with the past emissions of fossil fuel CO_2. The two model runs differ in 1990 by 0.3 ppmv, i.e. by less than 0.5% relative to the total perturbation of 70 ppmv since preindustrial times. If the calculation is extended to the end of the next century, prescribing an emissions scenario with a linearly increasing source reaching 22 GtC yr^{-1} in 2100, the difference between the two cases is almost 100 ppmv compared to a total atmospheric increase of 450 ppmv. However, up to excess CO_2 concentrations of 500 ppmv, which might be exceeded only in the second part of next century, the assumption of a constant buffer factor introduces errors of less than 10% on the projected atmospheric increases.

4 Tracers in the carbon cycle

Tracers play an important role in the study of the carbon cycle. Firstly, they provide an indispensable tool to assess certain constants in the carbon system (e.g. the globally averaged gas exchange rate). Secondly, their dispersal within the carbon system provides a means to validate carbon cycle models. The most important tracers are the stable carbon isotope ^{13}C and the radioactive ^{14}C.

Both carbon isotopes are governed by the same exchange processes as CO_2, but the isotopic *ratios*, $^{13}C/^{12}C$ and $^{14}C/C$, have different dynamical behaviour to bulk carbon for three main reasons:

1. An isotopic ratio perturbation is diluted in the carbon system without affecting the oceanic carbon chemistry. Hence the entire oceanic carbon pool is available as a reservoir within which the perturbation eventually will be diluted. Or, expressed differently, the oceanic reservoir is approximately 10 times larger for an isotopic ratio perturbation as compared to the case of a net CO_2 perturbation.

2. The isotopic signal is diluted into all carbon reservoirs with the gross exchange fluxes, independently of whether those reservoirs change in size with respect to bulk carbon or not. Exchanges and turnover times in the biosphere have to be taken into account for the modelling and the interpretation of the isotopic signal even if it is assumed that the biosphere does not respond to a rising CO_2 concentration (i.e. if the fertilization factor β is set to 0).

3. The differences in the atomic masses of the carbon isotopes imply slightly different physical properties (molecular diffusivity, chemical equilibria, solubility etc.) for ^{13}C and ^{14}C as compared to ^{12}C. This gives rise to fractionation effects which have to be taken into account.

As a consequence, the atmospheric impulse response function of an isotopic perturbation decays much more rapidly than the impulse response function for bulk carbon. Figure 3 shows the two response functions calculated with the box-diffusion model [Siegenthaler and Oeschger, 1987].

4.1 Radiocarbon

The radioactive carbon isotope ^{14}C has a lifetime of 8260 years. It is produced naturally by cosmic ray produced neutrons colliding with atmospheric nitrogen nuclei in the atmosphere. About two thirds of natural ^{14}C are produced in the stratosphere, the remaining third in the troposphere. The created ^{14}C nuclei quickly form carbon monoxide (^{14}CO)

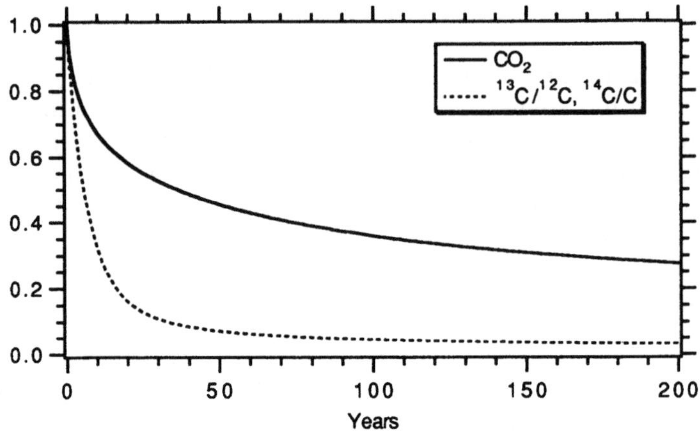

Figure 3: Impulse response for CO_2 (solid line) and for an isotopic impulse perturbation (dashed line), calculated with the box-diffusion model [Siegenthaler and Oeschger, 1987]

which subsequently is oxidized to $^{14}CO_2$. Thereafter the radiocarbon atoms are exchanged through the carbon reservoirs until the radioactive decay.

The background atmospheric natural $^{14}CO_2$ concentration is very small: the ratio of ^{14}C to C before the nuclear weapon tests was about 10^{-12} in the atmosphere. ^{14}C is measured conventionally with nuclear counters and recently by means of accelerator mass spectrometry (AMS).

The nuclear weapon tests in the 50's and early 60's injected a pulse of radiocarbon into the atmosphere. The ensuing atmospheric ^{14}C burden peaked around 1964 at a level approximately 80% above the natural pre-bomb level. Since then the atmospheric signal has been decreasing to a present level of $\approx$20% above pre-bomb values. This decrease is shown in Figure 4 [I. Levin, personal communication]. The initial seasonal cycles after the main input in 1962 reflect seasonal exchanges between the stratosphere and the troposphere. The seasonal cycles in the later part of the record arise from seasonal exchanges with the ocean and the terrestrial biosphere [Levin et al., 1992].

Figure 4 essentially shows the result of a truly global "experiment" to determine the impulse response of a carbon tracer (compare with Figure 3). However, a quantitative analysis is complicated by two problems: Firstly, the amount of radiocarbon delivered to the system by the nuclear weapon tests is not well known in magnitude. It was also not delivered in one single blast but by several bomb tests during several years. Secondly, while the oceanic uptake of radiocarbon has been traced in rather great

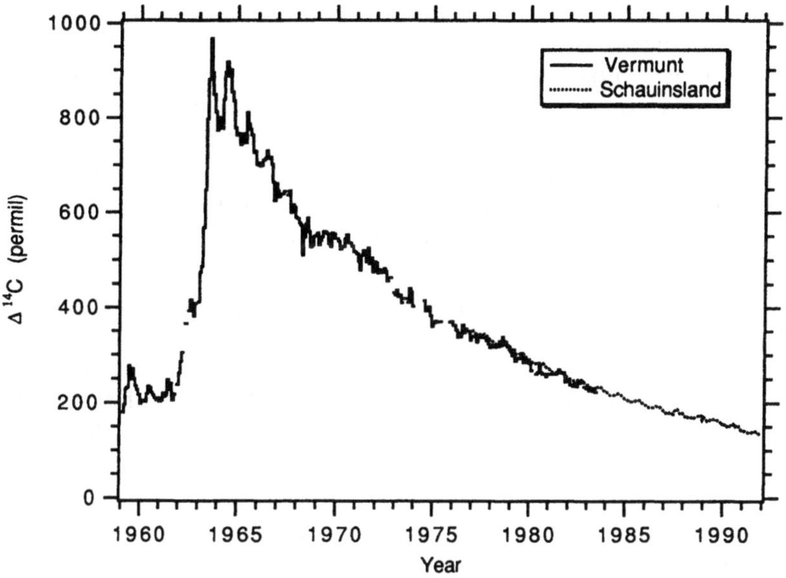

Figure 4: History of atmospheric radiocarbon since 1960 observed at the Vermunt (until 1983) and Schauinsland (since 1977) stations in Europe [Levin et al., 1992 and personal communication]. The isotopic ratios are expressed as an excess in ‰ relative to the pre-bomb natural atmospheric $^{14}C/C$ ratio.

detail [Broecker et al., 1985], the amount absorbed by the terrestrial biosphere is much more uncertain, as it depends on the gross exchange rates of carbon and on the turnover times of the various internal biospheric reservoirs.

The observed atmospheric record (Figure 4), however, can also be prescribed as the atmospheric boundary condition for an oceanic or a biospheric carbon model. The accurate prediction of the ^{14}C penetration into the ocean or the biospheric carbon reservoirs provides a severe test for the models. Alternatively, simple box models may be calibrated by tuning their coefficients to the observations of bomb radiocarbon in the ocean [Siegenthaler, 1983] or in soils. Potentially, the observed decrease in atmospheric radiocarbon in conjunction with an extended measurement program of radiocarbon in soils could lead to a constraint on soil turnover times and thus on the potential storage of excess carbon [Harrison and Broecker, 1993].

A further example for the use of radiocarbon to trace the dynamics of the global carbon cycle is given by the changes in the atmospheric $^{14}C/C$ ratio prior to the nuclear bomb tests. Time series of past variations of the $^{14}C/C$ ratio have been determined from measurements on tree-rings

[Stuiver and Quay, 1980, 1981]. The record for the last 120 years shows a pronounced decrease of ≈ -30‰ up to 1950 (the so-called "Suess effect"), which can be attributed to the addition of old, i.e. radiocarbon-free CO_2 from fossil fuels. In principle, this decrease could serve as an additional quantitative check on our understanding of how the global carbon cycle reacts to a perturbation. Indeed, carbon models predict a decrease of this size [Oeschger et al., 1975, Enting and Pearman, 1987]. However, changes in the cosmic ray production rate cannot be ruled out and might also have contributed to the signal.

On longer timescales the atmospheric $^{14}C/C$ ratio has fluctuated by several percent [Stuiver and Quay, 1980]. It is believed that these variations are caused by cosmic ray production rate variations caused by changes in the strength of the shielding magnetic field of the Earth, related partly to fluctuations in the solar wind and partly to changes in the dynamo currents of the Earths core. The variations in the production rate of ^{14}C can be reconstructed from measurements of the abundance of another cosmogenic isotope, ^{10}Be. This isotope is produced similarly to radiocarbon in the upper troposphere and stratosphere but is subsequently attached to aerosols and, once in the troposphere, quickly washed out to the surface. Thus, the ^{10}Be deposited in ice-cores indicates the production rate of ^{14}C which can be compared to the radiocarbon record obtained from the tree-rings. The relation between the two time-series reflects the dynamic exchanges of radiocarbon within the global carbon system [Beer et al., 1988].

The current seasonal and meridional distribution of the $^{14}C/C$ ratio in atmospheric CO_2 potentially might provide an additional constraint on the large-scale distribution of the carbon exchange fluxes at the Earths surface [Levin et al., 1992]. The radiocarbon-free CO_2 from fossil fuels is released primarily in the northern hemisphere (95%). We thus expect the northern hemisphere to be depleted relative to the southern hemisphere. Indeed, a decreasing gradient from tropical regions to the mid-latitudes of the northern hemisphere is observed. The higher latitudes in the southern hemisphere also show a relative ^{14}C depletion, although this might result from the CO_2 exchanges with the "older" surface ocean carbon in these latitudes.

4.2 The stable carbon isotope ^{13}C

Approximately 1% of all carbon atoms are of the ^{13}C stable isotope. The distribution of the $^{13}C/^{12}C$ isotope ratio in the carbon system is governed by fractionation effects. Since these effects are rather small, isotope ratios are commonly reported as a relative deviation, δ, from a standard ratio, and expressed in permil:

$$\delta = \frac{r}{r_s} - 1 \tag{22}$$

where r denotes the $^{13}C/^{12}C$ ratio of the sample and r_s that of a reference standard (which has a value of 0.0112372).

During photosynthesis the lighter carbon isotope, ^{12}C, is preferentially incorporated into plant tissue. Consequently, the $^{13}C/^{12}C$ ratio of organic carbon is smaller than in atmospheric CO_2. The relative depletion strongly depends on the type of plant. Two broad classes can be distinguished: Tissue of C4 plants (predominantly grasses) have a $^{13}C/^{12}C$ ratio of approximately 4‰ lower and C3 plants (most trees) of almost 18‰ lower than the atmosphere [Marino and McElroy, 1991]. In the global average C3 plants dominate greatly and the average depletion of organic material is about -18‰. No fractionation occurs when this organic carbon is oxidized back to inorganic forms. As a consequence CO_2 of biospheric origin is isotopically labeled with an isotopic ratio of approximately 18‰ lower than the atmosphere.

Fractionation effects also occur during air-sea gas-exchange. Oceanic carbon is enriched in ^{13}C relative to the atmosphere by approximately 10‰. However, kinetic fractionation effects occur when CO_2 is transferred between the air and the sea in both directions. As a consequence CO_2 of oceanic origin has a $^{13}C/^{12}C$ ratio close to the atmospheric value, depending to some extent on the local sea surface temperature [Mook, 1986].

Atmospheric CO_2 has a $^{13}C/^{12}C$ ratio of approximately -7.8‰ (1985) and is decreasing by about 0.02‰ yr^{-1} [Keeling et al., 1989a]. In preindustrial times the atmospheric δ was $\approx$ -6.5‰ and has been declining since then as recorded in ice-cores [Friedli et al., 1986]. This decline of $\approx$ -1.2‰ clearly reflects the input of CO_2 from fossil fuel sources which have an isotopic composition of approximately −27‰ [Tans, 1981].

5 The global budget and the "missing sink" problem

The atmospheric concentration of CO_2 has been rising during the last 200 years primarily because of the anthropogenic inputs from fossil fuel combustion and from changes in land use. Figure 5 shows the atmospheric record as obtained from ice-core measurements [Siegenthaler and Oeschger, 1987] and direct observations at the Mauna Loa station in Hawaii since 1959 [Keeling et al., 1989a]. How much of the CO_2 from the past emissions has remained in the atmosphere, and how much has been sequestered by the ocean and the terrestrial biospheric carbon pools? This question addresses the problem of the current atmospheric carbon budget, one of the key questions in carbon cycle research.

The atmospheric carbon budget may be expressed as

$$\frac{d}{dt}n_a(t) = Q_{foss}(t) + Q_{oc,net}(t) + Q_{b,net}(t) \tag{23}$$

where $\frac{d}{dt}n_a(t)$ denotes the rate of change of the atmospheric carbon inventory, Q_{foss} the anthropogenic CO_2 source from fossil fuel combustion and $Q_{oc,net}$ the net oceanic contribution (negative if the ocean represents a net sink). $Q_{b,net}$, the net biospheric source term includes the CO_2 source from biomass burning and changes in land use, but also any other biospheric source or sink, such as the sink due to the CO_2 fertilization effect.

The left hand side of equation (23) is readily determined from the atmospheric record shown in Figure 5 and $Q_{foss}(t)$ can be reconstructed from statistics of energy production. The net contributions from the ocean and the terrestrial biosphere, however, are much more difficult to assess. Basically, there exist four different approaches:

1. indirect assessment of the oceanic and biospheric budget by means of carbon model simulations;
2. direct measurements of air-surface carbon fluxes and extrapolation to the entire surface of the Earth;
3. direct measurements of the rate of change of the carbon content in the oceanic and biospheric carbon pools;
4. observations of tracers or other substances that are coupled to the carbon cycle (^{13}C and atmospheric oxygen).

5.1 Model derived budget estimates

The indirect approach by means of model simulations sofar has provided the most satisfactory estimates of the current atmospheric budget and is discussed first. In this approach an ocean carbon model is run through the industrial period by prescribing the atmospheric CO_2 concentration from observations (Figure 5) as a boundary condition. The ocean model thus calculates the oceanic component, $Q_{oc,net}(t)$ as a function of time. Equation (23) is then solved for the net biospheric component, $Q_{b,net}(t)$.

Figure 6 shows the result from such an exercise. Plotted are the oceanic components as calculated with the Hamburg OCCM, the fossil fuel source, and the net biospheric component obtained from the budget equation (23). If the ocean model is realistic and if the statistics of fossil

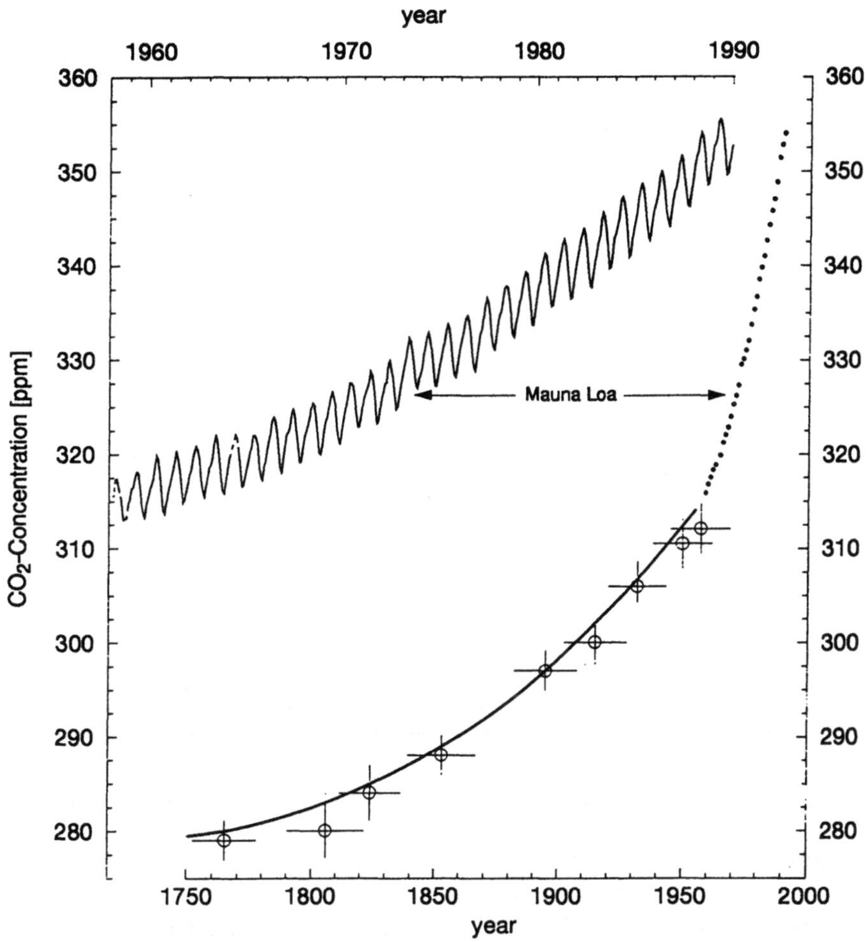

Figure 5: History of the atmospheric CO_2 concentration. Upper graph (left and upper axes): Direct observations from Mauna Loa, Hawaii [Keeling et al., 1989a], lower graph (bottom and right axes): Measurements from the Siple ice-core (circles with errorbars) [Friedli et al., 1986] joined by a smooth curve and extended after 1959 by the annual means of the Mauna Loa record (dots)

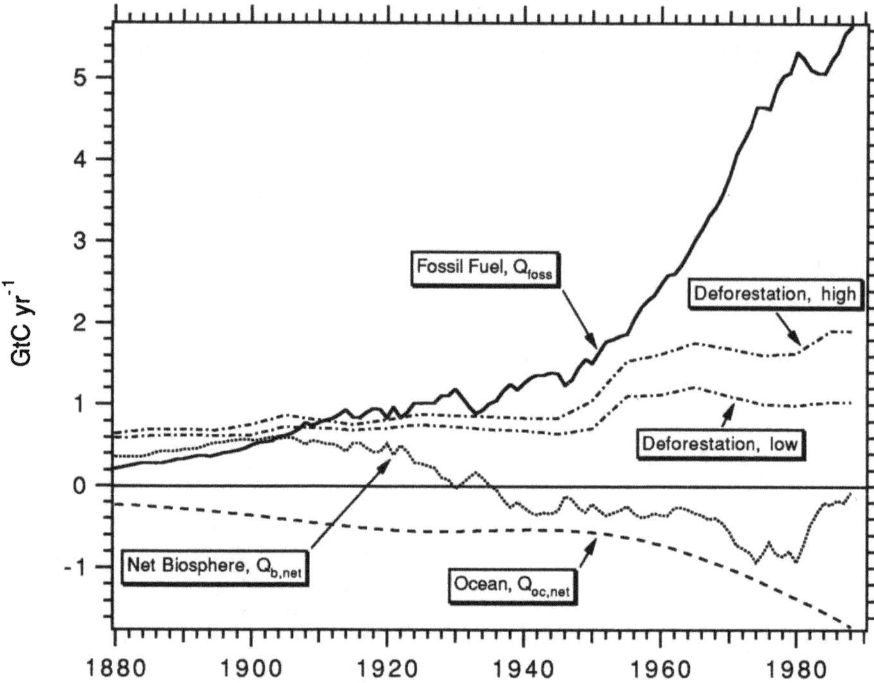

Figure 6: Source terms of the atmospheric CO_2 budget equation (23) and estimated CO_2 emissions from deforestation and changes in land use [Houghton 1993]

fuel CO_2 emissions are correctly estimated, then the terrestrial biosphere is inferred to have been a net source to the atmosphere until about the 1930's and a sink for CO_2 since then. During the decade of the 1980's the Hamburg OCCM predicts an ocean uptake of $\approx$ 1.5 GtC yr^{-1} requiring a net biospheric sink of $\approx$ 0.5 GtC yr^{-1} in order to close the atmospheric budget.

Obviously, the size of this biospheric sink determined by this method depends on the ocean model. As indicated in Section 3.3 above, the Hamburg OCCM has a relatively small uptake capacity as compared to other state-of-the-art models. Taking uncertainties in the ocean models and in the emission statistics into account, the atmospheric budget for the 1980's is given in the top four rows of Table 6 [Houghton et al., 1990]. Accordingly, $Q_{b,net}$ is inferred to have been almost 0 with rather large ($\pm$ 1 GtC yr^{-1}) error margins.

In principle the same procedure could also be performed with a model of the terrestrial biosphere. However, the state of affairs in terrestrial biosphere modelling is much less advanced compared to ocean models. Furthermore, contrary to the oceans there is no extended biospheric bomb

Table 6: Global atmospheric CO_2 budget for the period 1980-89 in GtC yr^{-1} [Houghton et al., 1990]

Fossil fuel emissions	Q_{foss}	5.4 ± 0.5
Atmospheric increase	$\frac{d}{dt}n_a$	-3.4 ± 0.2
Ocean uptake	Q_{oc}	-2.0 ± 0.8
Net biosphere	$Q_{b,net}$	0.0 ± 1.0
Land use and deforestation		1.6 ± 1.0
"Missing sink"		-1.6 ± 1.4

radiocarbon dataset that would allow tests (or even calibration) of terrestrial biosphere models.

Fossil fuel CO_2 is not the only anthropogenic source. A significant CO_2 input also arises from changes in land use and deforestation, predominantly in the Tropics. This source can be compiled from statistics of land use, taking into account different agricultural practices and also regrowth of secondary forests on cleared land areas [Houghton et al., 1987]. Two scenarios of the CO_2 emissions from this source are also shown in Figure 6, bracketing to some extent the considerable uncertainties in these estimates [Houghton, 1993]. Quantitatively, the emissions from changes in land use amount to 1.6 ± 1 GtC yr^{-1} during the 1980's (Table 6). If we accept this estimate we are left with an imbalance of -1.6 ± 1.4 GtC yr^{-1} in the global budget (Table 6). This difference constitutes the so-called "missing sink".

For a long time the missing sink was sought in the ocean and research efforts were directed at identifying additional oceanic processes which could sequester excess carbon (e.g. organic carbon burial on coastal shelfs [Walsh et al.,1981]). Recently, mostly in connection with the northern hemisphere sink problem (see Section 6), the emphasis has shifted to the biosphere and in particular the concept of CO_2 fertilization is taken more seriously [Wisnievski and Lugo, 1992].

5.2 Observations of surface carbon fluxes

Direct measurements of carbon fluxes suffer from several complications. The main problem is that, in order to determine the global budget, *net* fluxes have to be determined, which in most locations are very small differences of large gross exchange fluxes. Air-sea fluxes over the ocean can be deduced from measurements of the partial pressure difference between air and surface waters, using equation (2). This requires the knowledge of the local gas-exchange coefficient which, as discussed in section (3.1), is only poorly known. Furthermore, local fluxes vary substantially in space and time, thus necessitating high spatial and temporal sampling [Garçon et al., 1992]. Therefore, estimates of the net global oceanic carbon sink from direct observations have large uncertainties [Takahashi et al., 1991, Tans et al.,1990].

Direct observations of CO_2 fluxes between the atmosphere and the terrestrial biosphere can be made with several methods, e.g. by the eddy correlation technique. The enormous heterogeneity of the terrestrial biosphere in space and time, however, makes a reliable extrapolation to the continental and global scale very uncertain.

5.3 Observations of carbon inventory changes

Direct observation of the rate of change in carbon content in the oceanic and terrestrial biospheric carbon pools also suffer from the problem of determining a very small signal against large spatial and temporal background variability. Typical oceanic DIC concentrations are $\approx$2000 mmolC m^{-3}. Model estimates of the rate of change induced by the anthropogenic perturbation in surface waters are on the order of 1 mmolC m^{-3} yr^{-1}. Therefore, on a 10 year timescale, variations in DIC would have to be measured with an accuracy of better than 1% in order to determine the carbon balance of a particular oceanic region. Such an accuracy can be obtained, but with a substantial effort in sampling [Keeling, 1993b]. The determination of the global oceanic budget by this approach, however, is not feasible.

Changes in the land biosphere carbon pool are also very small. If 1 GtC yr^{-1} were absorbed in the living vegetation ($\approx$550 GtC) this corresponds to an annual change of less than 0.2%. Even by extended tree-ring studies such a small signal would be very hard to detect and translate into a global budget estimate without very large uncertainty ranges. If 1 GtC yr^{-1} were stored in soils which contain $\approx$1500 GtC, the corresponding relative changes would be even smaller.

5.4 The budgets of ^{13}C and oxygen

The most promising model-independent approach to assess the current global carbon budget is by observations of other trace substances that are coupled to the carbon cycle. Recently, Quay et al. [1992] have proposed a method to determine the global budget from repeated measurements of vertical profiles of the $^{13}C/^{12}C$ ratio in DIC. Such observations would permit the determination of the rate of change of the oceanic $^{13}C/^{12}C$ ratio which, together with the observed changes in the atmospheric $^{13}C/^{12}C$ ratio, allow the corresponding global carbon budget to be estimated. From a preliminary analysis, based on a sparse dataset containing only 7 stations in the Pacific ocean sampled in the early 1970's and again in 1990, Quay et al. [1992] determined an oceanic sink of 2.1±0.8 GtC yr^{-1}.

The essential difference compared to the direct determination of changes in the oceanic DIC concentration lies in the sensitivity of the proposed method: As discussed in Section 4 the isotopic ratio changes are not influenced by the chemical buffering reactions of the oceanic carbon system. Therefore the isotopic signal of the anthropogenic perturbation in the ocean is larger than the corresponding relative concentration change in DIC. Of course, the required analytical quality of the measurement program is still rather high. But one can estimate, that the uncertainty range of the current oceanic sink may be reduced within 10 years to $\approx$0.5 GtC yr^{-1} by this method, if an extended measurement program is vigorously pursued [Heimann and Maier-Reimer, 1993].

Yet another very promising approach has been demonstrated recently by Keeling and Shertz [1992], which is based on measurements of the trend in atmospheric molecular oxygen. Oxygen in many respects is complementary to carbon. It is consumed during combustion and respiration and is released during carbon uptake by photosynthesis of the biota. The crucial difference compared to carbon is that the oceanic oxygen pool is completely negligible and can be ignored in the global oxygen balance. Measurements of the temporal trend in atmospheric oxygen together with the known O:C atomic relations during combustion, respiration and photosynthesis permit the establishment of global oxygen and carbon balances. The precision requirements for the measurements are very high, however: The atmospheric oxygen concentration is expected to decrease by 1–2 ppmv per year. Compared to an atmospheric background concentration of 20% this corresponds to a relative decrease by 10^{-5} per year. The currently available measurements covering little more than one year limit the accuracy of the method. It is expected, however, that within 10 years the most accurate estimate of the global carbon balance will be achieved by this technique.

6 The problem of the "northern hemisphere sink"

During the last three decades a world wide network of stations has been established which can monitor the atmospheric CO_2 concentration with rather high accuracy. All the stations record basically the same increasing trend as seen at the Mauna Loa station (Figure 5). Superimposed on the trend are seasonal variations, which are most pronounced in the northern hemisphere and generated by the seasonality of the terrestrial biosphere in midlatitudes [Keeling et al., 1989a]. If the trend and the sea-

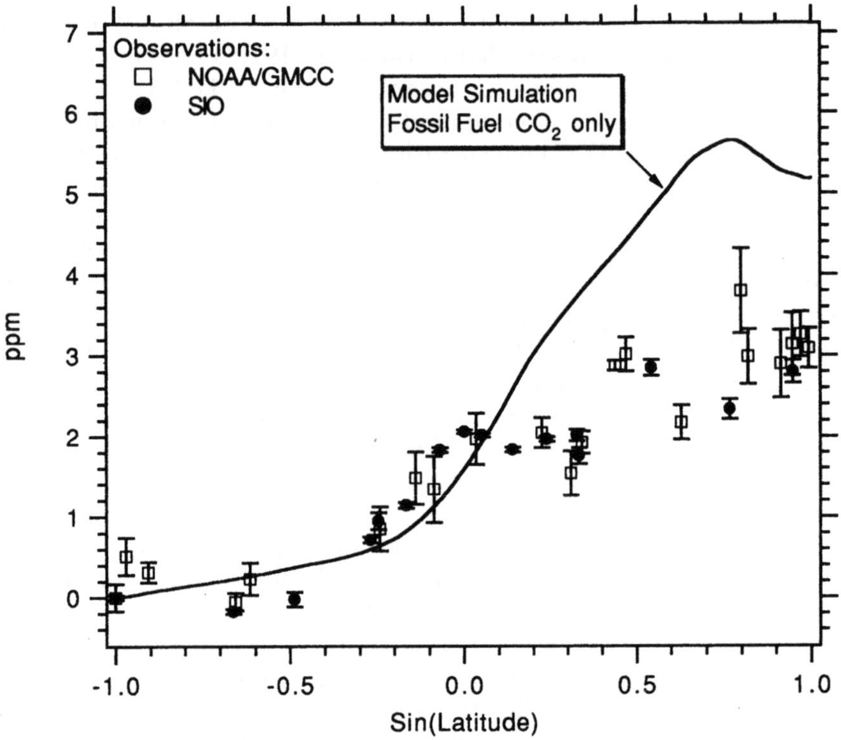

Figure 7: Observations of the annual mean CO_2 concentration as a function of latitude (1980-1987), plotted as deviations from the South Pole station [Convay et al., 1988, Keeling et al., 1989a]. The solid line shows the zonally averaged surface concentration resulting from a three-dimensional model simulation with fossil fuel as only source (see text).

sonal variations are removed and the mean annual CO_2 concentrations are plotted versus the latitude of the observing stations, a meridional concentration gradient emerges. Stations in the Arctic exhibit approximately 3 ppmv higher concentrations than stations in the midlatitudes of the southern hemisphere. Figure 7 shows this meridional concentration gradient for the years 1980–87, where the observations from two observation networks (run by the National Oceanographic and Atmospheric Administration [Convay et al., 1988] and the Scripps Institution of Oceanography [Keeling et al., 1989a]) are plotted as differences with respect to the concentration measurements at the South Pole. The errorbars shown in Figure 7 reflect the interannual variability 1980–89 of the annual means. The monitoring stations are located mostly in oceanic regions, far from local sources and sinks.

Over 95% of the fossil fuel CO_2 emissions take place in the midlatitudes of the northern hemisphere. As a consequence, one would expect a meridional concentration gradient to develop. The observed gradient, however, also shows the atmospheric signature of other localized sources and sinks. For example, there is an indication that the equatorial region exhibits higher CO_2 concentrations as compared to the adjacent regions to the north and south. This is most apparent in the data from the SIO stations (shown as solid circles in Figure 7) which are located in the central Pacific ocean. Possibly this concentration "bump" reflects the equatorial region where waters with high partial pressures in CO_2 upwell to the surface and constitute a natural source of outgassing CO_2.

Obviously, the data displayed in Figure 7 provide some important information on the large scale distribution of sources and sinks of atmospheric carbon dioxide. The quantitative assessment of this information, however, requires that the atmospheric transport is taken into account. Two studies using three-dimensional atmospheric transport models recently have been performed in order to analyze this meridional gradient [Keeling et al., 1989b, Tans et al., 1990]. The essential result of both studies is that a substantial sink (of the order of 1–2 GtC yr^{-1}) must exist in the midlatitudes of the northern hemisphere. This is illustrated by the curve shown in Figure 7 labeled "Fossil fuel only". This curve shows the predicted meridional gradient using a three-dimensional atmospheric transport model in which only the fossil fuel source is operating and the CO_2 is allowed to accumulate everywhere in the model atmosphere. The predicted gradient between the high latitudes of both hemispheres results in approximately 5 ppmv — significantly larger than the observed $\approx$3 ppmv. The discrepancy with the observations can only be reconciled if a strong sink is operating in the northern hemisphere which offsets 40% of the effect of the fossil fuel source. Of course, this inference crucially depends on the assumption that the atmospheric models accurately describe the large scale transport. Fortunately, there exist other atmospheric tracers (in particular ^{85}Kr and CFC-11) with simple well-known sources and sinks that can be used to independently validate the models atmospheric transport [Prather et al., 1987, Jacob et al., 1987, Heimann and Keeling, 1989].

6.1 Two interpretations

The current network of CO_2 monitoring stations unfortunately is not dense enough to resolve more precisely the location of the northern hemisphere sink. Consequently, in modelling the atmospheric CO_2 distribution, additional information on the spatial distribution of sources and sinks must be used. It is here, where the studies by Keeling et al. [1989b] and Tans et al. [1990] differ markedly.

Tans et al. [1990] included observations of the partial pressure difference between air and sea from the north Atlantic and Pacific oceans in their analysis, to estimate the magnitude of the northern hemisphere oceanic sinks. The existing data (together with estimates of the gas exchange coefficient, see Section 3.1) yield only a marginal northern hemispheric ocean uptake (0.3–0.7 GtC yr^{-1}). Consequently, the northern hemisphere sink is assumed to be located in the temperate latitudes over the continents, and attributed to the terrestrial biosphere. If such a large terrestrial sink (2-3 GtC yr^{-1}) is operating then a much smaller global oceanic sink is needed to close the global carbon budget. Indeed, the preferred scenario of Tans et al. [1990] results in a global oceanic sink of less than 0.8 GtC yr^{-1}. The meridional distribution of the assumed sources and sinks, fossil, oceanic and biospheric is depicted in five broad latitude bands in Figure 8.

It must be recognized, that the meridional gradient not only reflects the net perturbation sources and sinks induced by the anthropogenic excess CO_2, but also any natural steady-state source-sink distributions which, when integrated over the entire globe, add up to zero. In the study of Keeling et al. [1989b] it is assumed that the northern hemisphere sink is caused to a large extent by the global asymmetric thermohaline circulation of the oceans. In this scenario approximately 1 GtC yr^{-1} of CO_2 are assumed to be absorbed in the North Atlantic ocean, transferred to the south at depth and released back to the atmosphere south of 40S. Indeed, such a meridional oceanic transfer of carbon can be traced in the interior of the Atlantic [Broecker and Peng, 1992]. However, the magnitude of such a postulated sink in the North Atlantic is not compatible with the observations of the air-sea partial pressure differences used by Tans et al. [1990]. The meridional distribution of this scenario is also shown in Figure 8.

Two further observational circumstances support the oceanic interpretation. Firstly, the analysis of the meridional gradient can also be performed with data from the early and late 1960's. It is found that the magnitude of the northern hemisphere sink did not appreciably change during the last 30 years (see Figure 9) as would be expected if the sink reflected a feature of the natural background carbon cycle. On the other hand, if the northern hemisphere sink were a result of the fertilization effect, its size should have augmented substantially because of the atmospheric CO_2 increase from the early 1960's to the 1980's.

Secondly, observations of the meridional gradient of the $^{13}C/^{12}C$ ratio of atmospheric CO_2 are not compatible with a large northern hemisphere

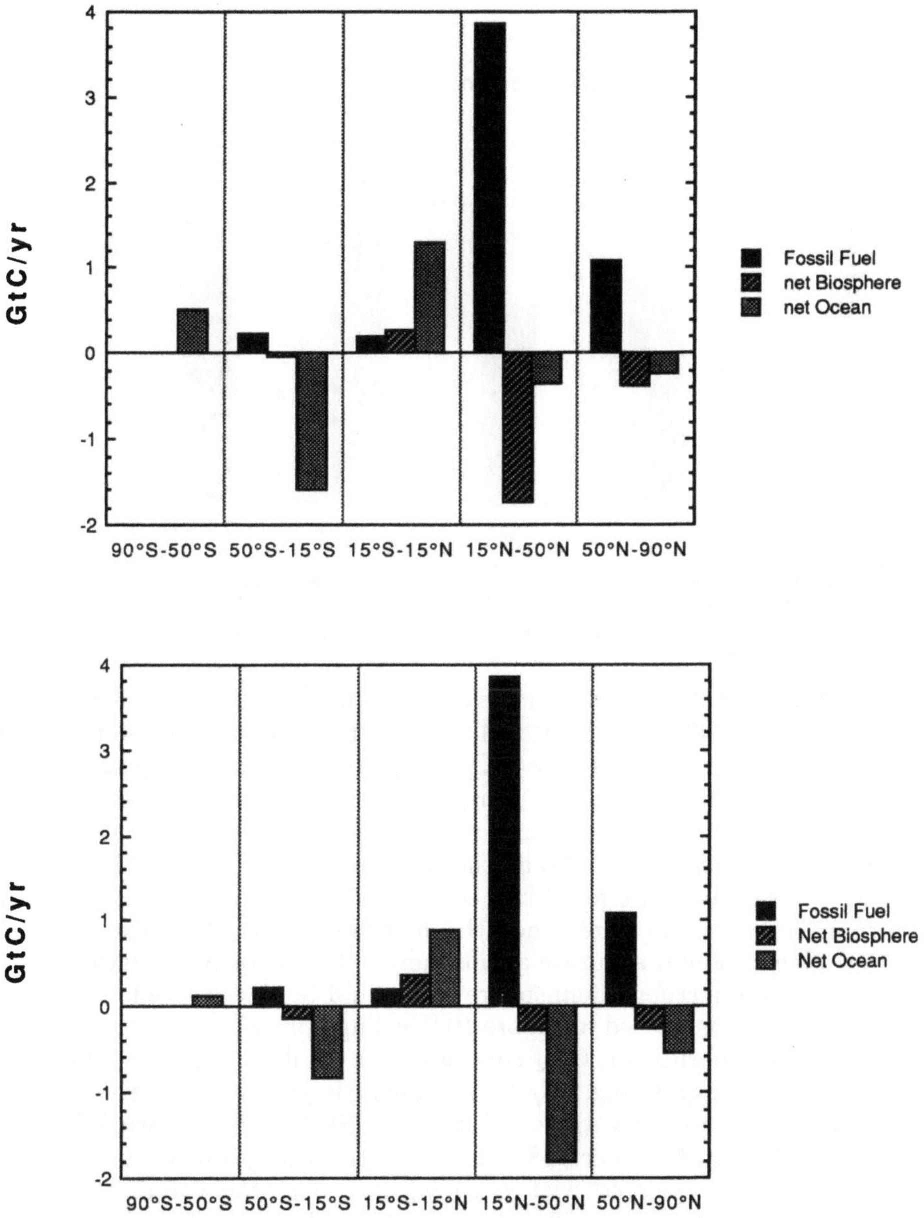

Figure 8: Summary of the meridional distribution of the source/sink scenarios of Tans et al. [1990] (upper panel) and of Keeling et al. [1989b] (lower panel).

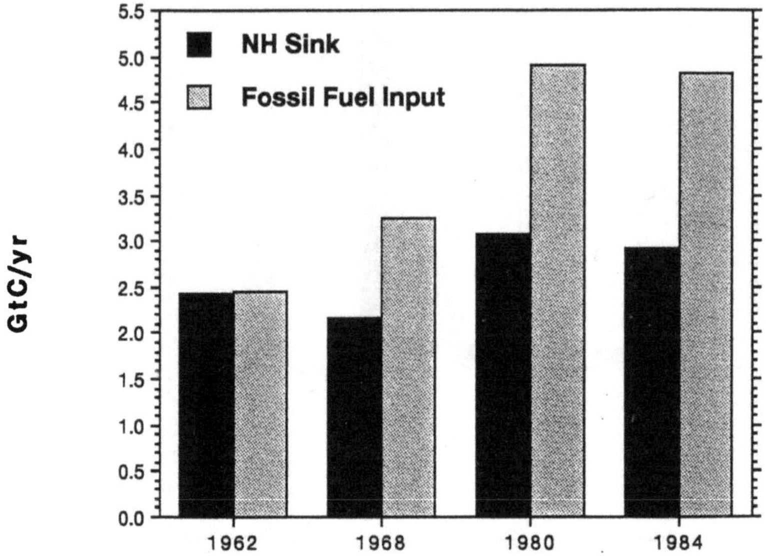

Figure 9: History of the northern hemisphere sink as determined from the atmospheric meridional CO_2 concentration gradient [Keeling et al., 1989b]. For comparison the magnitude of the fossil fuel CO_2 source in the northern hemisphere is also shown.

biospheric sink. Since fossil fuel is depleted in ^{13}C relative to atmospheric CO_2 by about 20‰ (see Section 4.2) the anthropogenic emissions impose a meridional gradient in the $^{13}C/^{12}C$ ratio with more negative values in the northern hemisphere. Biospheric carbon is also depleted in ^{13}C, implying that a large biospheric sink in the northern hemisphere would offset to a considerable extent the meridional gradient in the $^{13}C/^{12}C$ ratio generated by the fossil fuel source. On the other hand, an oceanic carbon sink does not influence the meridional $^{13}C/^{12}C$ ratio gradient. Thus, given enough accurate atmospheric $^{13}C/^{12}C$ ratio observations the nature of the northern hemisphere sink could be determined.

This is demonstrated in Figure 10 [Keeling, 1993a]. The left-hand panels show the meridional CO_2 concentration gradient and the right-hand panels the corresponding $^{13}C/^{12}C$ isotope ratio gradient in the central Pacific ocean. The observations are from the SIO station network [Keeling et al., 1989a] for the year 1984. The isotope ratio gradients in the right-hand panels are plotted with reversed axes, i.e. the values increase in the downward direction. The two model simulations refer to an oceanic scenario, i.e. with a large North Atlantic sink (upper panels) and to a scenario with a large biospheric sink in the temperate latitudes (lower panels), a case similar to the preferred scenarios of Tans et al. [1990]. Clearly, the oceanic sink scenario fits the isotopic data better.

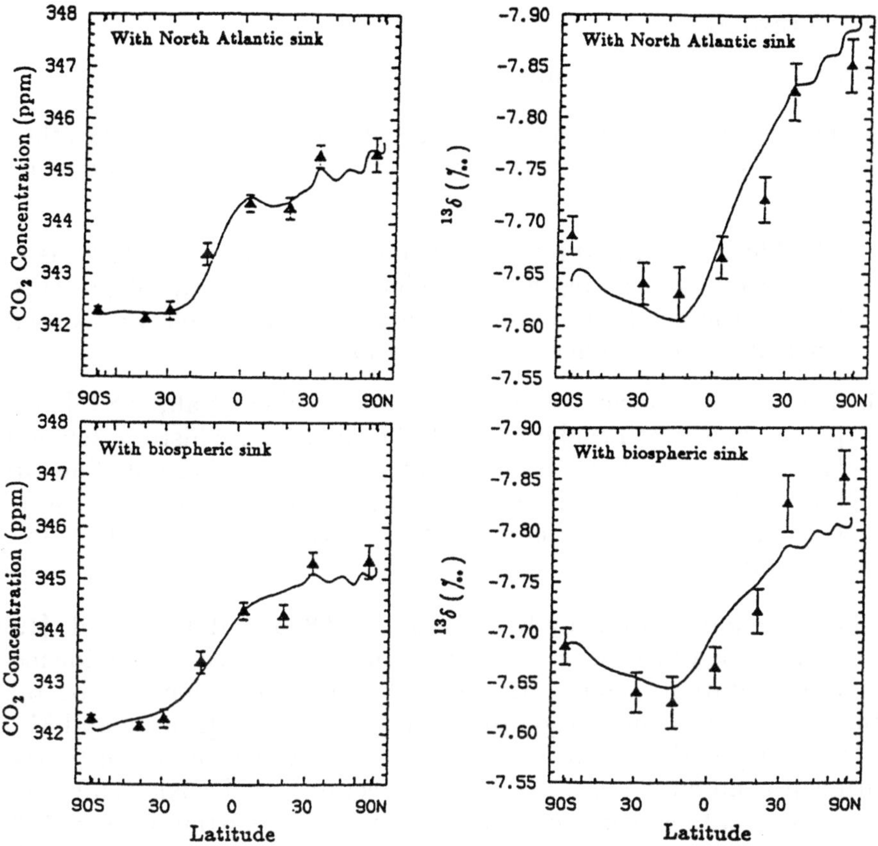

Figure 10: Meridional profiles in 1984, near 160°W. Left panels: concentration, right panels: $^{13}C/^{12}C$ ratio expressed in ‰. The data are from the SIO observation network [Keeling et al., 1989a], errorbars indicate ± 2 standard errors. The solid lines show the simulation result from a three-dimensional atmospheric transport model with a scenario including a large oceanic sink in the North Atlantic (upper panels) and a large terrestrial biospheric sink (lower panels) [Keeling 1993a, Keeling et al., 1989b].

6.2 Summary

The controversy may be summarized as follows:

1. The observed meridional CO_2 concentration gradient implies the existence of a substantial (1–2 GtC yr^{-1}) sink in the northern hemisphere.
2. The biospheric interpretation of this sink [Tans et al., 1990] is supported by the existing air-sea differences of the partial pressure of CO_2 in the northern oceans but it is not compatible with observations of the atmospheric $^{13}C/^{12}C$ ratio.
3. The oceanic interpretation of this sink [Keeling et al., 1989b] is supported by the atmospheric $^{13}C/^{12}C$ ratio data, its history during the last 30 years and the transport by the oceanic thermohaline circulation. The oceanic interpretation, however, is not compatible with the existing air-sea differences of the partial pressure of CO_2 in the northern oceans.

Since the publications by Keeling et al. [1989b] and Tans et al. [1990] several investigations have scrutinized individual pieces of evidence in the argumentation chains. Several additional effects have to be taken into account: the contribution from carbon monoxide [Enting and Mansbridge, 1991], carbon transfer from the continent to the ocean by rivers [Sarmiento and Sundquist, 1992] and corrections for the skin temperature effect in estimating air-sea carbon fluxes from CO_2 partial pressure differences [Robertson and Watson, 1992]. So far none of these modifications has eliminated the controversy on the nature of the northern hemisphere sink.

References

ARTHUR MA, DEAN WE, SCHLANGER SO, 1985: Variations in the global carbon cycle during the Cretaceous related to climate, volcanism and changes in atmospheric CO_2. In: ET Sundquist, WS Broecker, (eds.) *The carbon cycle and atmospheric CO_2: Natural variations archean to present*, American Geophysical Union, Washington DC, 504–530.

BARNOLA JM, RAYNAUD D, KOROTKEVITCH YS, LORIUS C, 1987: Vostok ice core: A 160,000 year record of atmospheric CO_2. *Nature* **329** 408–414.

BEER J, SIEGENTHALER U, BONANI G, FINKEL RC, OESCHGER H, SUTTER M, WÖLFLI W, 1988: ^{10}Be in the Camp Century ice core: information on past solar activity and geomagnetism. *Nature***331** 675–679

BLACK DD, MCCREADY MJ, 1988: Effect of small-wavelength waves on gas transfer across the ocean surface. *J. Geophys. Res.* **93** 5143–5152

BOLIN B (ED), 1981: SCOPE 16: Carbon cycle modelling. *John Wiley and Sons, New York*

BOLIN B, 1986: How much CO_2 will remain in the atmosphere? *In: Bolin B, Döös BR, Jäger J, Warrick RA (eds), SCOPE 29: The greenhouse effect, climatic change, and ecosystems. John Wiley and Sons, New York*

BOLIN B, DEGENS ET, KEMPE S, KETNER P (EDS), 1979: SCOPE 13: The global carbon cycle. *John Wiley and Sons, New York*

BROECKER WS, PENG T-H, 1982: Tracers in the sea. *Eldigio Press, Lamont Doherty Geological Observatory, Palisades, New York*

BROECKER WS, PENG T-H, ÖSTLUND G, STUIVER M, 1985: The distribution of bomb-radiocarbon in the ocean. *J. Geophys. Res.* **90** 6953–6970.

BROECKER WS, LEDWELL JR, TAKAHASHI T, WEISS R, MERLIVAT L, MEMERY L, PENG T-H, JÄHNE B, MÜNNICH KO , 1986:; Isotopic versus micrometeorologic ocean CO_2 fluxes: A serious conflict.; J. Geophys. Res.; 91; 10517–10527.

BROECKER WS, PENG T-H, 1992: Interhemispheric transport of carbon dioxide by ocean circulation. *Nature* **356** 587–589

BROECKER WS, PENG T-H, 1993: What caused the glacial to interglacial CO_2 change? *Heimann M (ed) The global carbon cycle. Springer-Verlag, Heidelberg, in press*

COANTIC M, 1986: A model of gas transfer across air-water interfaces with capillary waves. *J. Geophys. Res.* **91** 3925–3943

CONVAY TJ, TANS PP, WATERMAN LS, THONING KW, MASARIE KA, GAMMON RH, 1988: Atmospheric carbon dioxide measurements in the remote global troposphere, 1981–1984.; Tellus; 40B; 81-115

ENTING IG, PEARMAN GI, 1987: Description of a one-dimensional carbon cycle model calibrated using techniques of constrained inversion. *Tellus* **39B** 459–476

ENTING IG, MANSBRIDGE JV, 1991: Latitudinal distribution of sources and sinks of CO_2: results of an inversion study. *Tellus* **B43** 156-170

ESSER G, 1987: Sensitivity of global carbon pools and fluxes to human and potential climatic impacts. *Tellus* **39B** 245–260

ETCHETO J, MERLIVAT L, 1988: Satellite determination of the carbon dioxide exchange coefficient at the ocean-atmosphere interface: a first step. *J. Geophys. Res.* **93** 15669–15678

FRIEDLI H, LÖTSCHER H, OESCHGER H, SIEGENTHALER U, STAUFFER B, 1986: Ice record of the $^{13}C/^{12}C$ ratio of atmospheric CO_2 in the past two centuries. *Nature* **324** 237–238

FUNG IY, JOHN J, LERNER J, MATHEWS E, PRATHER M, STEELE LP, FRASER PJ 1991: The three-dimensional model synthesis of the global methane cycle. *J. Geophys Res.* **96,** 13033-13065.

GARÇON V, THOMAS F, WONG CS, MINSTER JF, 1992: Gaining insight into the seasonal variability of CO_2 at O.W.S.P. using an upper ocean model. *Deep Sea Res.* **30** 921–938

HARRISON K, BROECKER WS, 1993: A strategy for estimating the impact of CO_2 fertilization on soil carbon storage. *Global Biogeochemical Cycles, in press*

HEIMANN M (ED) 1993: The global carbon cycle. *Springer-Verlag, Heidelberg, in press*

HEIMANN M, KEELING CD, 1989: A three dimensional model of atmospheric CO_2 transport based on observed winds: 2. Model description and simulated tracer experiments. *In: Peterson DH (ed) Aspects of Climate Variability in the Pacific and the Western Americas. American Geophysical Union, Washington DC, 277–303*

HEIMANN M, MONFRAY P, 1989: Spatial and temporal variation of the gas exchange coefficient for CO_2: 1. Data analysis and global validation. *Report No. 31, Max-Planck-Institut für Meteorologie, Hamburg*

HEIMANN M, MAIER-REIMER E, 1993: On the relations between the uptake of CO_2 and its carbon isotopes by the ocean. *Glob. Biogeochem. Cycl., submitted*

HEINZE C, MAIER-REIMER E, WINN K, 1991: Glacial pCO_2 reduction by the world ocean: experiments with the Hamburg carbon cycle model. *Paleooceanography* **6** 395–430

HOUGHTON JT, JENKINS GT, EPHRAUMS JJ, 1990: Climate change, the IPCC scientific assessment. *Cambridge U. Press, Cambridge*

HOUGHTON JT, 1993: Changes in terrestrial carbon over the last 135 years. *In: Heimann M (ed) The global carbon cycle. Springer-Verlag, Heidelberg, in press*

HOUGHTON RA, BOONE RD, FRUCI JR, HOBBIE JE, MELILLO JM, PALM CA, PETERSON BJ, SHAVER GR, WOODWELL GM, MOORE B, SKOLE DL, MYERS N, 1987: The flux of carbon from terrestrial ecosystems to the atmosphere in 1980 due to changes in land use: Geographic distribution of the global flux. Tellus 39B 122–139

JACOB, DJ, PRATHER MJ, WOFSY SC, MCELROY MB, 1987: Atmospheric distribution of ^{85}Kr simulated with a general circulation model. *J. Geophys. Res.* **92** 6614–6626

JÄHNE B, MÜNNICH KO, BÖSINGER R, DUTZI A, HUBER W, LIBNER P, 1987: On the parameters influencing air-water gas exchange. *J. Geophys. Res.* **92** 1937–1949

KEELING CD, 1993A: Global observations of atmospheric CO_2. *In: Heimann M (ed) The global carbon cycle. Springer-Verlag, Heidelberg, in press*

KEELING CD, 1993B: Surface ocean CO_2. *In: Heimann M (ed) The global carbon cycle. Springer-Verlag, Heidelberg, in press*

KEELING CD, BACASTOW RB, CARTER AF, PIPER SC, WHORF TP, HEIMANN M, MOOK WG, ROELOFFZEN H, 1989a: A three dimensional model of atmospheric CO_2 transport based on observed winds: 1. Analysis of observational data. In: Peterson DH (ed) Aspects of Climate Variability in the Pacific and the Western Americas. American Geophysical Union, Washington DC, 165–236

KEELING CD, HEIMANN M, PIPER SC, 1989B: A three dimensional model of atmospheric CO_2 transport based on observed winds: 4. Analysis of the mean annual gradients of CO_2. *In: Peterson DH (ed) Aspects of Climate Variability in the Pacific and the Western Americas. American Geophysical Union, Washington DC, 305–363*

KEELING RF, SHERTZ SR, 1992: Seasonal and interannual variations in atmospheric oxygen and implications for the global carbon cycle. *Nature* **358** 723–727

KOHLMAIER GH, SIRÉ E, JANECEK A, KEELING CD, PIPER SC, REVELLE R, 1989: Modelling of the seasonal contribution of a CO_2 fertilization effect of the terrestial vegetation to the amplitude increase in atmospheric CO_2 at Mauna Loa Observatory. *Tellus* **41B** 487–510

LEVIN I, BÖSINGER R, BONANI G, FRANCEY R, KROMER B, MÜNNICH KO, SUTTER M, TRIVETT NBA, WÖLFLI W, 1992: Radiocarbon in atmospheric carbon dioxide and methane: global distribution and trends. In: Taylor (ed) Radiocarbon after four decades: an interdisciplinary perspective. Springer-Verlag, Heidelberg

LISS PS, MERLIVAT L, 1986: Air-sea gas exchange rates: Introduction and synthesis. *In: Buat-Ménard P (ed) The role of air-sea exchange in geochemical cycling. D. Reidel Publishing Company, London*

MAIER-REIMER E, HASSELMANN K, 1987: Transport and storage of CO_2 in the ocean — an inorganic ocean-circulation carbon cycle model. *Clim. Dyn.* **2** 63–90

MARINO BD, MCELROY MB, 1991: Isotopic composition of atmospheric CO_2 inferred from carbon in C4 plant cellulose. *Nature* **349** 127–131

MARLAND G, BODEN T, 1993: The magnitude and distribution of fossil-fuel-related carbon releases. *In: Heimann M (ed) The global carbon cycle. Springer-Verlag, Heidelberg, in press*

MERLIVAT L, MEMERY L, 1983: Gasexchange rates across an air-water interface: experimental results and modelling of bubble contribution to transfer. *J. Geophys. Res.* **88** 707-724

MOOK WG, 1986: ^{13}C in atmospheric CO_2. *Netherland J. of Sea Res.* **20** 211–223

NAJJAR RG, SARMIENTO JL, TOGGWEILER JR 1992.: Downward transport and fate of organic matter in the ocean: simulations with a general circulation model. *Global Biogeochemical Cycles,* **6** 45-67.

NORBY RJ, GUNDERSON CA, WULLSCHLEGER SD, O'NEILL EK, MCCRACKEN MK, 1992: Productivity and compensatory response of yellow-poplar trees in elevated CO_2. *Nature* **357** 322–324

OESCHGER H, SIEGENTHALER U, SCHOTTERER U, GUGELMANN A, 1975: A box diffusion model to study the carbon dioxide exchange in Nature *Tellus* **27** 168–192

OESCHGER H, HEIMANN M, 1983: Uncertainties of predictions of future atmospheric CO_2 concentrations. *J. Geophys. Res.* **88C** 1258–1262

PETERSON BJ, MELILLO JM, 1985: The potential storage of carbon caused by eutrophication of the biosphere. *Tellus* **37B** 117–127

POST WM, 1993: Organic carbon in soil and the global carbon cycle. In: Heimann M (ed) *The global carbon cycle. Springer-Verlag, Heidelberg, in press*

PRATHER MJ, MCELROY MB, WOFSY SC, RUSSELL G, RIND D, 1987: Chemistry of the global troposphere: fluorocarbons as tracers of air motion. *J. Geophys. Res.* **91** 6671–6681

QUAY PD, TILBROOK B, WONG CS, 1992: Oceanic uptake of fossil fuel CO_2: Carbon-13 evidence. *Science* **256** 74–79

ROBERTSON JE, WATSON AJ, 1992: Thermal skin effect of the surface ocean and its implications for CO_2 uptake. *Nature* **358** 738–739

SARMIENTO JL, SUNDQUIST ET, 1992: Revised budget for the oceanic uptake of anthropogenic carbon dioxide. *Nature* **356** 589–593

SARMIENTO JL, ORR C, SIEGENTHALER U, 1992: A perturbation simulation of CO_2 uptake in an ocean general circulation model. *J. Geophys. Res.* **97** 3621–3645

SHAFFER G, SARMIENTO JL, 1992: Biogeochemical cycling in the global ocean 1: A new, analytical model with continuous vertical reolution and high latitude dynamics. *J. Geophys. Res., in press*

SIEGENTHALER U, 1983: Uptake of excess CO_2 by an outcrop-diffusion model of the ocean. *J. Geophys. Res.* **88** 3599–3608

SIEGENTHALER U, 1986: Carbon dioxide: It's natural cycle and anthropogenic perturbations. *In: Buat-Ménard P (ed) The role of air-sea exchange in geochemical cycling. D. Reidel Publishing Company, London*

SIEGENTHALER U, 1993: Modelling the present-day oceanic carbon cycle. *In: Heimann M (ed) The global carbon cycle. Springer-Verlag, Heidelberg, in press*

SIEGENTHALER U, OESCHGER H, 1987: Biospheric CO_2 sources emissions during the past 200 years reconstructed by deconvolution of ice core data. *Tellus* **39B** 140–154

SIEGENTHALER U, JOOS F, 1992: Use of a simple model for studying oceanic tracer distributions and the global carbon cycle. *Tellus* **44B** 186–207

STEFFEN WL, WALKER BH, INGRAM JI, KOCH GW (EDS) 1992: Global change and terrestrial ecosystems: The operational plan. *IGBP Report No. 21, IGBP, Royal Swedish Academy of Sciences, Stockholm*

STRAIN BR, THOMAS RB, 1992: Field measurements of CO_2 enhancement and climate change in natural vegetation. *Water, Air, and Soil Pollution* **64** 45–60

STUIVER M, QUAY PD, 1980: Changes in atmospheric carbon-14 attributed to a variable sun. *Science* **207** 11-12

STUIVER M, QUAY PD, 1981: Atmospheric ^{14}C changes resulting from fossil fuel CO_2 release and cosmic ray flux variability. *Earth and Planetary Science Lett.* **53** 349–362

SUNDQUIST ET, 1985: Geological perspectives on carbon dioxide and the carbon cycle. *In: Sundquist ET, Broecker WS (eds) The carbon cycle and atmospheric* CO_2*: Natural variations archean to present. American Geophysical Union, Washington DC*

SUNDQUIST ET, PLUMMER LN, WIGLEY TML, 1979: Carbon dioxide in the ocean surface: The homogeneous buffer factor. *Science* **204** 1203–1205

TAKAHASHI T, GODDARD J, CHIPMAN DW, SUTHERLAND SC, MATHIEU G, 1991: Assessment of carbon dioxide sink/source in the North Pacific Ocean: seasonal and geographic variability, 1984–1989. *Final Technical Report for Cont. 19X-SC428C, Department of Energy, Lamong-Doherty Geol. Observatory, Palisades, NY*

TANS PP, 1981: $^{13}C/^{12}C$ of industrial CO_2. *In: Bolin B (ed) SCOPE 16: Carbon cycle modelling. John Wiley and Sons, New York, 127–129*

TANS PP, FUNG IY, TAKAHASHI T, 1990: Observational constraints on the global atmospheric CO_2 budget. *Science* **247** 1431–1438

TRABALKA JR, REICHLE DE (EDS) 1986: The changing carbon cycle: A global analysis. *Springer-Verlag, New York*

UPSTILL-GODDARD RC, WATSON AJ, LISS PS, LIDDICOAT MI, 1990: Gas transfer velocities in lakes measured with SF_6. *Tellus* **42B** 364–377

WALSH JJ, ROWE GT, IVERSON RL, MCROY CP, 1981: Biological export of shelf carbon is a sink of the global CO_2 cycle. *Nature* **291** 196–201

WATSON AJ, 1993: Air-sea gas exchange and carbon dioxide. *In: Heimann M (ed) The global carbon cycle. Springer-Verlag, Heidelberg, in press*

WATSON AJ, UPSTILL-GODDARD RC, LISS PS, 1991: Air-sea gas exchange in rough and stormy seas measured by a dual-tracer technique. *Nature* **349** 145–147

WEISS RF, 1974: Carbon dioxide in water and seawater – the solubility of a non-ideal gas. *Mar. Chem.* **2** 203–215

WISNIEVSKI J, LUGO AE (EDS) 1992 Natural sinks of CO_2. *Kluwer Academic Publishers, Dordrecht, The Netherlands*

OCEAN BIOLOGY, TRACE METALS AND CLIMATE

MICHAEL WHITFIELD
Plymouth Marine Laboratory
Plymouth, United Kingdom

1 Introduction

Our planet is more appropriately described as 'ocean' rather than 'earth', and the world's climate is dependent to a large degree on the exchange of heat and volatiles across the sea surface. Current concerns over man's impact on climate are largely focussed on the inexorable build-up of carbon dioxide and other greenhouse gases in the atmosphere as a result of man's activities. The oceans presently act as a significant sink for both heat and carbon dioxide, whose transfer from the atmosphere to the oceans is modulated by the presence of life. In addition, the biological cycles influence the transport of particulate carbon from the ocean surface to the deep ocean where it can be sequestered out of contact with the atmosphere. As a consequence, the response of life in the oceans to climate change is recorded in the ocean sediments by the accumulation of biogenic debris.

The comparison of this record with the climatic information obtained from ice cores (Barnola *et al.* 1987) suggests that oceanic productivity, atmospheric carbon dioxide content, and temperature have changed dramatically and in concert during episodes of glacial and interglacial climate over the past 150 thousand years. The purpose of this chapter is to show in outline how life in the oceans interacts with the climate and how a study of trace metal distributions can help to quantify these processes in the contemporary oceans and thereby aid our interpretation of the historical record.

2 Components of the Biological System

2.1 The Mobile Pasture

The biology of the oceans is driven by a *mobile pasture* of photosynthetic bacteria and drifting microscopic algae (the phytoplankton) which grow in the sunlit surface layers of the ocean (the euphotic zone). This zone forms a skin only a 100 metres or so in thickness over an ocean whose average depth is 4.5 kilometres. The productivity of this pasture and

NATO ASI Series, Vol. I 11
Modelling Oceanic Climate Interactions
Edited by J. Willebrand and D. L. T. Anderson

its disposition in space and time are dictated by the supply of light and nutrients (Parsons *et al.* 1984, Herring *et al.* 1990). These factors depend on the vertical stability of the water column which is in turn dependent on the balance struck between the input of heat to the surface and the stirring of these surface layers by the wind (Archer 1990). The intensity of phytoplankton growth is therefore intimately related to climatic variables.

In temperate and high latitudes there is a strong seasonal signal with a bloom of algal growth occurring in spring as soon as the surface layers are stabilised by the sun's warmth and the phytoplankton are able to utilise the nutrients brought to the surface by winter storms. The presence of the phytoplankton, absorbing and scattering incoming radiation, effectively reduces the depth to which heat can penetrate the ocean surface layers. This helps to accelerate the stabilisation of the surface waters and, in turn, reduces the rate of transfer of heat from the atmosphere to the oceans (Archer 1990). As the season progresses, the surface layers become progressively more stable and the phytoplankton in these latitudes exhaust the available nutrients that they contain. As this happens production moves down to take advantage of the nutrients diffusing through to the seasonal thermocline or being stirred up from deeper water by storm events. During photosynthesis, plants remove carbon dioxide from the water and release oxygen. Since the transfer of carbon dioxide from the atmosphere to the ocean is directly dependent on the partial pressure differences between the atmosphere and the ocean surface (Liss and Merlivat 1986), the mobile pasture modulates the ocean-atmosphere exchange of carbon dioxide (Watson *et al.* 1991).

During the fixation of carbon, the phytoplankton take up essential nutrients such as nitrogen and phosphorus and also a range of trace elements in defined ratios to the carbon concentration. These "Redfield" ratios provide a stoichoimetry that imprints a signature on the composition of the deep ocean waters (Broecker and Peng 1982, Whitfield and Turner 1987). The efficiency with which biological processes can remove carbon from the surface layers is critically dependent on the Redfield ratios of carbon to the limiting nutrients nitrogen and phosphorous in the growing cells. These ratios are quite variable between different species (see for example Moal *et al.* 1987) and any shifts in climate that influence the population structure can also feed back on the uptake of carbon dioxide in this way.

The impact of phytoplankton on the partial pressure of carbon dioxide is also dependent upon the way in which these organisms fix carbon. If the carbon is used to build up organic material, this results directly in a reduction in the partial pressure of carbon dioxide. The diatoms which usually constitute the early spring bloom are prime examples of this kind of uptake (Watson *et al.* 1991). However coccolithophorids, which usually appear later in the season, also use carbon to produce calcite shells. This process actually reduces the alkalinity of the surface water and increases carbon dioxide partial pressure.

$$Ca^{2+} + 2HCO_3^- \rightleftharpoons CaCO_3 + H_2O + CO_2$$

Recent detailed studies of coccolithophorid blooms in the North Atlantic suggest that this effect can substantially reduce the lowering of carbon dioxide partial pressure in the surface layers by photosynthesis. Tiny platelets of calcite (liths), which are released as these algae grow, can turn the water a milky white. This increases the reflectivity of the surface waters to such a degree that the blooms are visible from space. These algae also release dimethyl sulphide (DMS) when they die or are grazed. On release into the atmosphere, the DMS is oxidised to yield sulphuric acid and methyl sulphonic acid (MSA) which act as very effective cloud condensation nucleii, producing highly reflecting clouds in the troposphere and thereby exerting a further influence on the radiation budget (Charlson *et al.* 1987, Savoie and Prospero 1989). MSA has been detected in ice cores and has been shown to be inversely correlated with atmospheric carbon dioxide levels but directly correlated to palaeo-indicators of ocean productivity.

The particles produced by the mobile pasture tend to be small (usually less than 5 microns in diameter) and they settle at a rate of less than 1 metre a day. Such particles make up more than 98% of the mass of particulate matter in the water and constitute more than 99% of the surface area available for the uptake of trace metals. However, they contribute less than 2% to the vertical flux of particulate material from the surface waters to the deep ocean (Clegg and Whitfield 1990). The major flux of carbon from the surface to the deep water is driven by the larger particles (greater than 50 microns in diameter) which result from the grazing of phytoplankton by small animals (the herbivorous zooplankton) and sink at 100 to 1000 metres per day.

2.2 The Particle Pump

While the onset of a phytoplankton bloom is dependent on the availability of light and nutrients, the demise of the bloom is usually controlled by grazing pressure. A bewildering array of zooplankton eat the phytoplankton and repackage their remains as fecal pellets and aggregates which fall rapidly from the euphotic zone into the deep ocean. The ratio between the flux of carbon that leaves the euphotic zone and the gross primary production is known as the *f-ratio* and this effectively controls the export of carbon to the deep ocean. The relationship between the f-ratio and the gross primary production is a complex one (Berger *et al.* 1989) since it depends upon the extent of coupling between the phytoplankton and zooplankton populations and on the size structure of these populations. The picture is further complicated by the fact that a significant fraction of the biologically fixed organic carbon is released into the water as dissolved organic matter (DOM). The concentration of DOM in sea water is still a matter of some controversy (Sugimura and Suzuki 1988) but, if the more recent higher estimates are to be believed, the transport of carbon out of the euphotic zone in this form may be nearly as great as the transport by particulate matter. A resolution of this matter awaits a concensus on the analytical methods for determining dissolved organic carbon and nitrogen in sea water.

In a simplistic fashion one might assume that smaller phytoplankton would be eaten by smaller zooplankton which would in turn produce smaller particles which would sink more slowly and would therefore be more prone to bacterial decay before they left the euphotic zone, yielding low f-ratios. The currently most popular hypothesis suggests that there are two extreme conditions. In well mixed systems such as those occurring in temperate and high latitude zones in spring and early summer, or in tropical upwelling zones, the rapid supply of nutrients results in bloom conditions. In such a situation the plant and animal communities tend to have low diversity and high growth rates which yield an efficient export of carbon with f-ratios approaching 0.5. At the other extreme, persistent hydrodynamic conditions can result in well-stratified waters, as frequently observed in tropical and sub-tropical regions. Under such circumstances, the biological populations achieve a steady state where a wide diversity of species that specialise in the exploitation of low nutrient levels are able to institute a parsimonious economy in which there is extensive recycling of the organic material produced. There is only a small leakage of organic matter from the euphotic zone (f-ratios less than 0.1) in these stable systems. Clearly then, the export of carbon from the surface layers to the deep water is highly dependent on climatic conditions. Changes that would result in more persistent stratification and decreased upwelling, such as those that might accompany global warming, could reduce the efficiency of the ocean as a sink for atmospheric carbon dioxide thereby introducing positive feedback. Another factor that is important here is the mass ratio of organic carbon to carbonate carbon in sinking particles (the *rain ratio*). The higher this ratio, the more efficiently is carbon dioxide partial pressure reduced in the surface layers by biological activity. In broad terms the rain ratio ranges from 10 in well mixed waters to 0.1 in highly stratified waters.

From the moment that the particles are formed they are being degraded by biological action, releasing carbon dioxide and nutrients and consuming oxygen. Once the particles leave the euphotic zone this process becomes dominant and the particle flux acts as a pump transferring the carbon dioxide from the surface layer to the deep ocean. The particulate breakdown results in high carbon dioxide and nutrient contents in the deep water. When this water upwells to the surface along the eastern ocean boundaries or in equatorial regions, it is warmed and will therefore release carbon dioxide on contact with the atmosphere. However, the higher level of nutrients that these waters contain will trigger vigorous phytoplankton growth which in turn reduces the return flux of carbon dioxide into the atmosphere.

Although the biological productivity of large areas of the ocean appears to be currently controlled by the availability of the major nutrients, nitrogen and phosphorus, there are regions at high latitude (typically greater than 60°N or 60°S) where the phytoplankton growth does not seasonally exhaust the nutrient supply in the surface layers. General circulation models suggest that these regions are likely to be most susceptible to climate change and, since other factors clearly control productivity, they are also the regions where the biological system itself is likely to be more sensitive to the corresponding modifications of the physical forcing functions. At least three hypotheses have been put forward to explain the lack of nutrient limitation at these latitudes.

1. *Light limitation.* The turbulent mixing that occurs at these latitudes throughout the year and the short season during which light levels are adequate for photosynthesis effectively restrict the growth of phytoplankton below the optimum level.

2. *Grazing limitation.* The zooplankton populations at these latitudes have evolved strategies of overwintering and rapid growth that enable them to graze down the phytoplankton as soon as they are produced, thereby preventing them from using up all the available nutrients.

3. *Other limiting nutrients.* Here the suggestion is that nitrogen and phosphorus are not necessarily the only limiting nutrients and that trace metals might be essential for photosynthesis. At high latitudes, elements such as iron and manganese may be in short supply in regions that are not adequately fed by the input of trace metals from air-borne particles or from river run-off.

It is important that the relative contributions of these various controls should be understood since these regions of the ocean constitute a major oceanic sink for atmospheric carbon dioxide. The solubility of carbon dioxide in sea water increases rapidly with decreasing temperature and the cold, dense water produced at the surface in these regions has the capability of carrying with it a considerable burden of atmospheric CO_2 as it sinks to the ocean depths. Since these regions are physically active, the contact time between the atmosphere and the seawater is relatively short and the efficiency of the physical dissolution of carbon dioxide is relatively low (less than 50%, Volk and Hoffert 1985, Volk and Liu 1988). Any biological process that reduces the partial pressure of carbon dioxide and thereby enhances the efficiency of this solubility process will therefore have a significant effect on the uptake of carbon dioxide by the oceans.

The stable atmospheric partial pressure of carbon dioxide during the 1,000 years prior to the industrial revolution, as indicated by ice core measurements, suggests that the oceanic biological system has remained close to steady state. There is as yet no significant evidence to suggest that this system has been perturbed by man. However, it is clear that if we are to assess the current ability of the oceans to take up carbon dioxide we must take into account the biological processes that currently modulate the surface partial pressure of carbon dioxide. Only by designing our sampling to take such factors into account can we get the true average partial pressure with which to calculate oceanic uptake by physico-chemical processes.

It is also clear that the biological system itself is intimately interrelated with the climatic system so that it is necessary to quantify the behaviour of the biological system if we are to be able to predict the behaviour of the ocean-atmosphere system once climate change gets underway. The interrelationship between biological processes and climatic factors provides significant potential for feedback in terms of the controls on both the mobile pasture and on the removal of fixed carbon from the surface layers into the deep ocean (Figure 1, Sarmiento 1990).

Such feedbacks are of potential importance since the overall biological cycle holds within the ocean a quantity of carbon which is approximately twice that currently present in the atmosphere.

Finally we need to be able to interpret and understand the contemporary biological mechanism if we are to unravel the record of climate change that is stored in the ocean sediments for comparison with the ice core data. These data sets can provide us with an understanding of the long term behaviour of the ocean-climate system.

3 Trace Elements and the Biological Particle Cycle

The biological processes which fix carbon and take up nutrients such as nitrogen and phosphorus also incorporate a wide range of trace elements

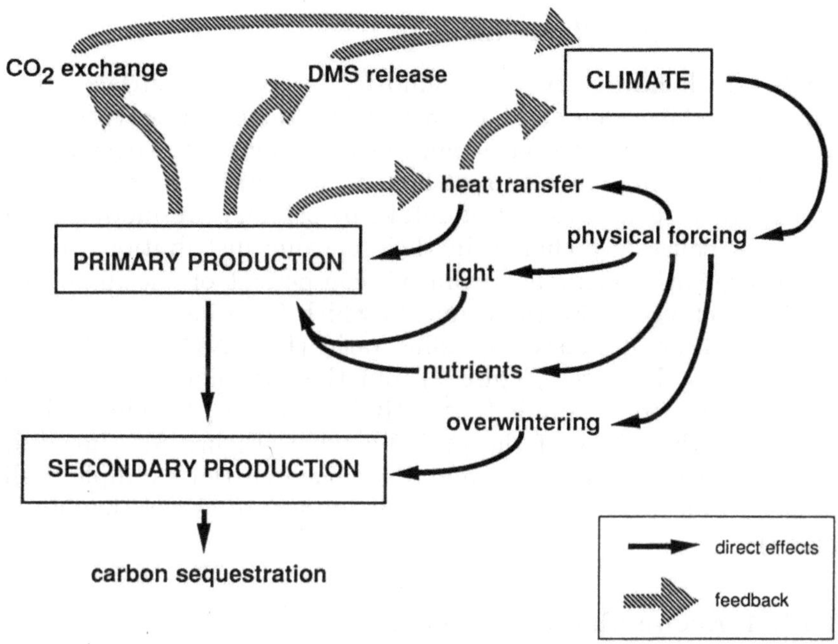

Figure 1: Potential feedbacks involving biological processes in the ocean-climate system.

into the particulate organic material. These elements are consequently removed from the surface layers and released in deeper water from the organic or mineral phases in the plant and animal remains. The particle pump thereby influences the distribution within the oceans of most elements in the periodic table.

Studies of these trace element cycles can tell us much about the dynamics of the biological particle cycle and can also provide signatures that help us unravel the historical record that is accumulated in the ocean sediments. Fortunately, there is a rational pattern behind the incorporation of trace elements into biological material in the oceans and this provides a framework against which we can interpret the behaviour of the various elements (Whitfield and Turner 1987). Assuming for the time being that transport of particulate matter into the deep ocean is a one dimensional process we can divide the elements into three broad categories according to the shape of the concentration versus depth profiles that they exhibit in the deep ocean.

3.1 Accumulated Elements

Elements which have very weak affinities for particle surfaces tend to accumulate in the oceans and have very long residence times (greater than 10^5 years). These elements have high concentrations in the oceans relative to their crustal abundance and, although many of them are required by biological processes, the extent of uptake by primary production is barely sufficient to cause any change in their oceanic distribution (Table 1). They consequently form the "salinity" component of the dissolved sea salt. In this category are the elements which form inert gas cations, or anions with a small charge to radius ratio (Li^+, Na^+, K^+, Rb^+, Cs^+, Mg^{2+}, F^-, Cl^-, Br^-, see Figure 2) and the oxyanions of small highly charged cations (for example SO_4^{2-}, HCO_3^-, $B(OH)_4^-$). These elements tend to form the wings of the periodic table (Figure 3). In addition, some of the heavier elements mimic this behaviour (for example thallium (I) acting as an inert gas cation and molybdenum (VI) behaving as an oxyanion).

TABLE 1 Accumulated Elements

CHARACTERISTICS	TYPICAL DEPTH PROFILE
- Weak particle interactions - High concentrations relative to crustal abundance - Long oceanic residence times (millions of years) - Uniform distribution; 'Salinity component' of dissolved salts - Geochemical control	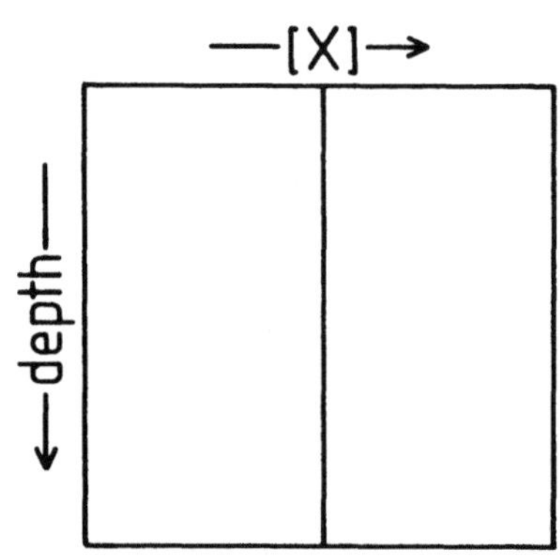

An unusual and interesting element in this category is uranium which forms the uranyl oxycation in seawater (UO_2^{2+}). This cation forms a very strong complex with the carbonate ions which holds it in solution. Uranium therefore exhibits a constant concentration ratio to the major

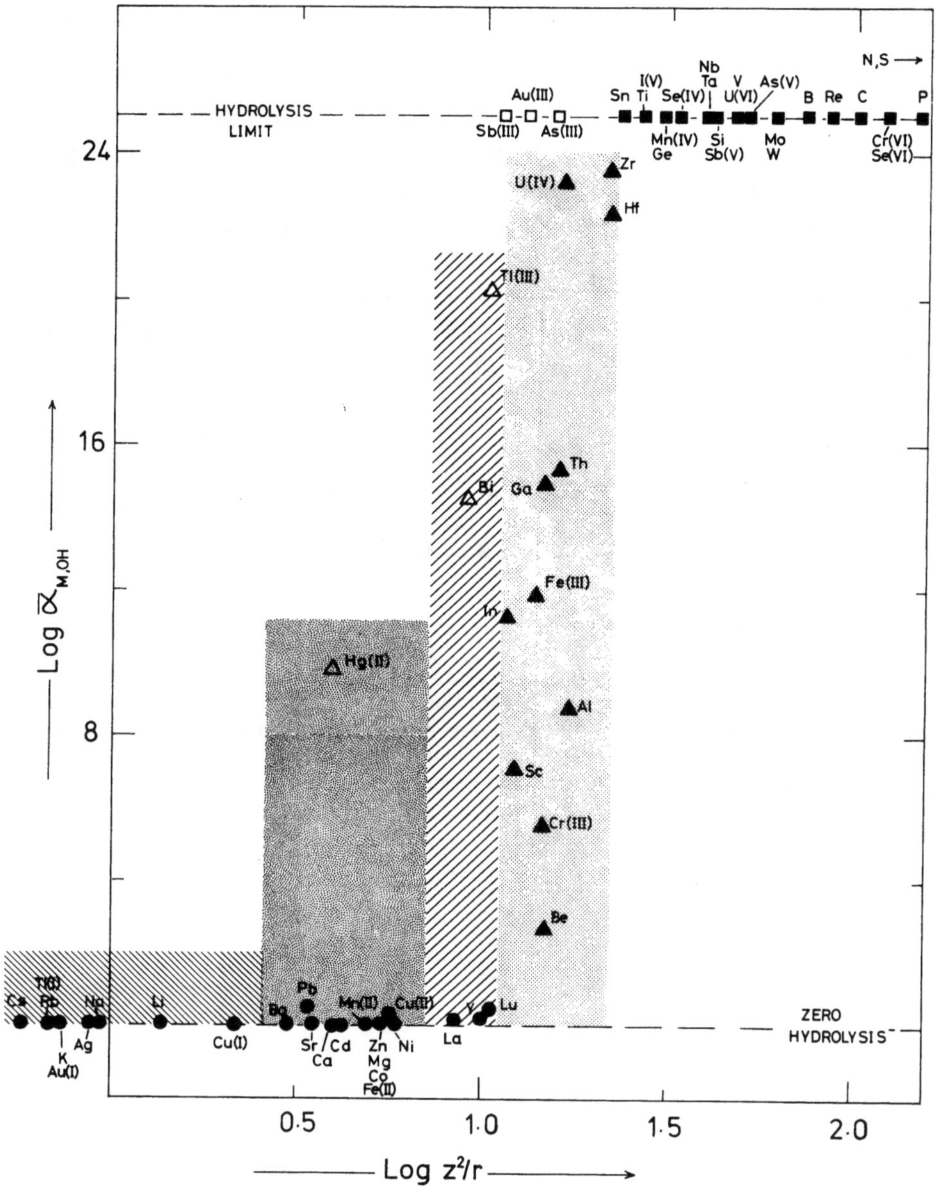

Figure 2: 'Born Diagram'. The ordinate ($log\overline{\alpha}m_{OH}$) is a measure of the intensity of hydrolysis of the elemental cation. The abscissa ($log z^2/r$) is a measure of the intensity of electrostatic interaction between the elemental cation and water molecules. The shaded bars represent four broad categories of elements which (reading from left to right) are characterised by: monovalent cations ($z^2/r < 2.5$), divalent cations ($2.5 < z^2/r < 7$), trivalent cations ($7 < z^2/r < 11$) and strongly hydrolysed cations ($11 < z^2/r < 23$). The abrupt shift in $log\overline{\alpha}m_{OH}$ at $z^2/r = 23$ indicates a rapid increase in the strength of interaction of the cations with water molecules. Figure adapted from Turner *et al.* (1981).

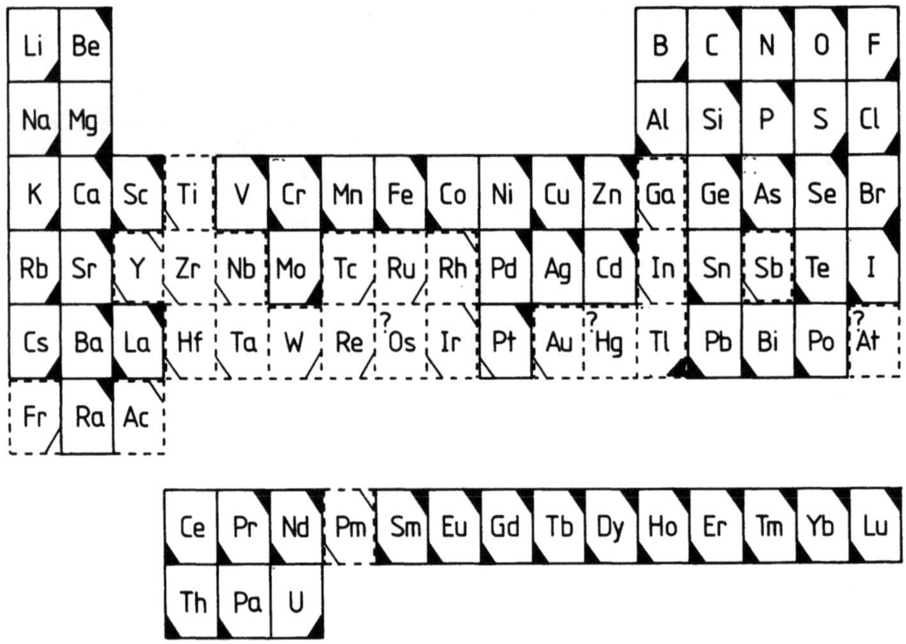

Figure 3: Periodic table showing the elements classified according to the nature of the vertical concentration profiles they exhibit in the oceans (Tables 1 to 3). The categories are indicated by corner shadings as follows: lower right - accumulated elements, upper right - recycled elements, lower left - scavenged elements. Elements whose oceanic behaviour has not been investigated are enclosed in dashed lines and their predicted behaviour is shown by open corner markings. Figure adapted from Whitfield and Turner (1987).

cations which form the "salinity component" of the oceanic salt content. It is consequently possible to calculate uranium concentrations with a high accuracy from salinity measurements. This is of considerable significance since some of the daughter elements that are released from uranium by radioactive decay are useful markers of the particle cycle. We will consider thorium in this context in a later section. In general the accumulated elements are removed from seawater slowly through trapping in interstial waters or by interactions at hydrothermal sites.

3.2 Recycled Elements

Elements which are actively taken up during formation of the biological particles in the surface layers exhibit concentration versus depth profiles which are characterised by low concentrations in the surface waters with a progressive increase with depth as the particles are broken down on their passage to the ocean sediment and their components are released into solution (Table 2). These elements have intermediate concentrations and intermediate residence times. In these elements, too, there are systematic groupings in the periodic table (Figure 3).

The most obvious elements in this category are those which perform essential biological functions (C, N, P, Se, I, Cr, Fe, Ni, Cu, Zn, V). A number of other elements are closely related to these biologically important elements and are taken up by the growing phytoplankton in the absence of any efficient discriminatory mechanism. Examples are provided by Ge (*cf.* Si), As (P), and possibly Ag (Cu), Cd (Zn), Pd and Pt (Ni).

TABLE 2 Recycled Elements

CHARACTERISTICS	TYPICAL DEPTH PROFILE
- Actively taken up by organisms - Intermediate concentrations - Intermediate oceanic residence times (tens of thousands of years) - Systematic variations in distribution in space and time - Biological control	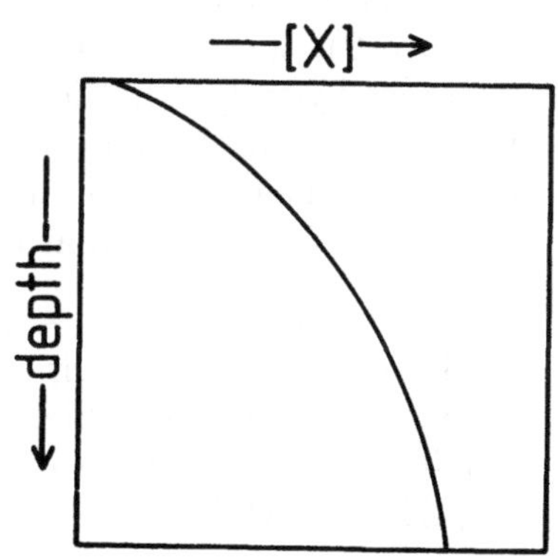

Other elements are attracted to the particle surfaces produced by the growing phytoplankton and are scavenged in this way (Be, Sc, the lanthanides), or else they are incorporated into the mineral phases which are produced by organisms living in the surface layers (Ca, Sr, Ba, Si, Fe, Mn). Apart from these major mineral producing elements, most of the elements in this category are taken up by the organic matter produced on photosynthesis and show strong Redfield ratios to the organic carbon content (Collier and Edmond 1984, Morel and Hudson 1985, Jickells and Burton 1988, Morel *et al.* 1991).

3.3 Scavenged Elements

Within this category fall the elements that are most effectively removed from seawater during formation of particulate matter. Such elements have very low concentrations in the seawater relative to their crustal abundance and have correspondingly short residence times, of the order of decades to centuries. Because of this, their distribution within the oceans is heterogeneous, but it is usually characterised by a surface maximum concentration which decreases rapidly with depth (Table 3). At least three categories of particulate removal can be discerned within this group of elements.

1. *Electrostatic Scavenging.* The interaction of cations in seawater with the water molecules surrounding them and with particulate matter can be characterised by the intensity of the electrostatic field at the cation surface. This can be related in a semi-quantitative fashion (Turner *et al.* 1981) to the ratio of the square of the surface charge (z^2) to the cation radius (r). Elements which have relatively high z^2/r ratios (Pb, Cu, Be, Sc and the lanthanides,Figure 2) are, therefore, efficiently removed by adsorption onto particle surfaces.

2. *Hydrolytic Scavenging.* When the z^2/r ratio exceeds a critical value (Figure 2), the cation interacts intensely with water molecules and becomes hydrolysed. Elements which are strongly hydrolysed in this way (for example Th, Al, Bi, Sn, Cr(III), Te) also appear to be strongly adsorbed into particle surfaces and are very effectively removed by particle formation and transport.

3. *Oxidative Scavenging.* A range of transition metals have variable oxidation states and their chemistry in seawater is such that the lower oxidation state is soluble, whereas the higher oxidation state forms insoluble oxides or hydroxides. Typical examples are provided by Fe (II/III), Mn(II/IV), Co(II/III), Ce(III/IV), Ir(II/IV) and Pt(II/IV),

although Pb, Sn, Bi and Pa might also be removed in this fashion (Whitfield and Turner 1987).

TABLE 3 Scavenged Elements

CHARACTERISTICS	TYPICAL DEPTH PROFILE
- Very strong particle interactions - Low concentrations relative to crustal abundance - Short oceanic residence times (decades to centuries) - Heterogeneous distribution - Biologically mediated control	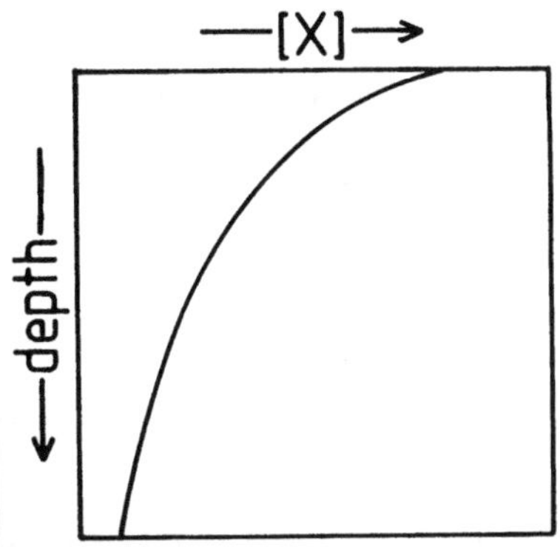

The scavenged elements themselves form coherent groups on the periodic table (Figure 3) which point to the underlying chemical rationale behind their behaviour. In this context it is also interesting to note that the classification of the elements into accumulated, recycled and scavenged categories can also be characterised according to the chemical bonding properties of the element concerned (Whitfield and Turner 1987) in a way which also provides a simple classification of the chemical speciation of the elements in seawater (Turner *et al.* 1981).

3.4 The Oceanic Conveyor Belt

Before we look in detail at the distribution of selected elements in the ocean, it is important to move away from a one dimensional view of the particle cycle and consider the impact of ocean circulation on the varying patterns of distribution of the elements between the different oceans. Taking a much simplified view (Broecker and Peng 1982), the bulk of the deep ocean circulation is driven by the formation of cold saline water in the North Atlantic. This dense water forms a deep current which runs

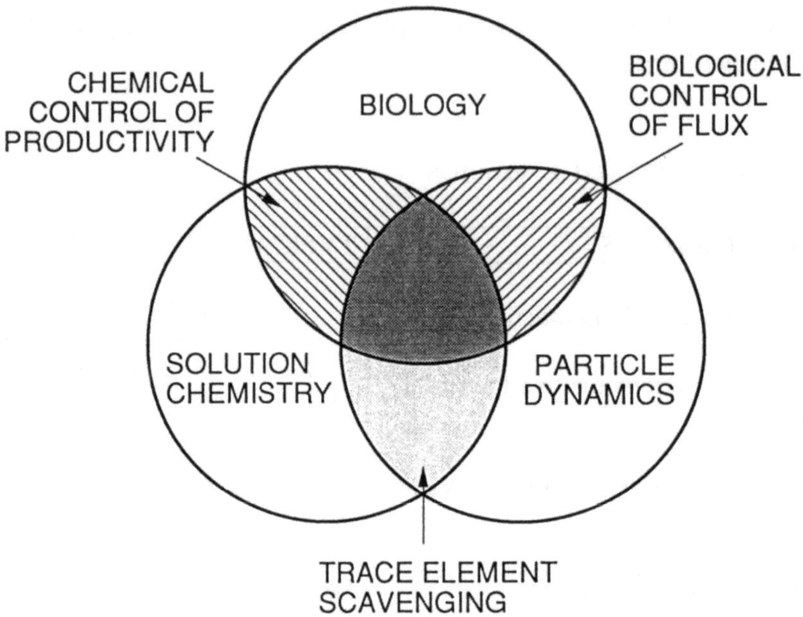

Figure 4: The interrelationships between solution chemistry, primary production and particle production. The areas of intersection of the circles of influence represent control processes.

along the ocean floor from north to south, displacing deep water *en route* and giving rise to tropical upwelling zones as the thermohaline circulation interacts with that driven by the wind.

The North Atlantic deep water is augmented by additional dense water during its west to east passage round the Antarctic circumpolar current. This provides an additional source of deep water which flows northwards, both in the Indian and the Pacific Ocean, and on upwelling drives the southwards and westwards passage of a warm surface current. This current flowing round the tip of South Africa augments the northward flow of warm surface water in the Atlantic which eventually contributes to the Gulf Stream. This then completes the cycle by its north eastward transport across the Atlantic.

During its passage, the water acts as an *oceanic conveyor belt* picking up and accentuating the signals provided by the vertical transport of particulate material through the particle pump. As a consequence, the concentration of recycled elements in deeper waters is enriched *en route* from the Atlantic through to the Pacific as it is augmented by the breakdown products of successive particle fluxes. In contrast, the concentration of scavenged elements is depleted in the deeper waters as they are successively stripped out by the rain of particulate matter *en route* from the Atlantic to the Pacific.

The picture is complicated to some degree by the much larger ratio of riverine and air-borne input to the ocean surface area in the Atlantic compared to the Pacific and by the large anthropogenic inputs into the northern basin of the Atlantic. Nonetheless the composition of the deep water and consequently the composition of the particulate matter reflects to a large degree the ocean conveyor belt circulation pattern.

3.5 Feedback and Control

So far we have seen that the biological processes in the ocean drive an active particle cycle that links together the chemical cycling of most of the elements in the periodic table. This particle cycle also exhibits intimate connections with climatically driven physical forcing functions and there is ample scope for feedback in this system (Figure 4).

Trace elements may be incorporated into the particle cycle either directly through the biological synthesis of organic matter, or indirectly by physico-chemical adsorption on to the particle surfaces once they are formed. Furthermore, to complete the cycle of interactions, the trace metals may themselves exert some control over biological productivity if they act as essential elements in the fixation of organic carbon or if they are close to their toxic limits.

In the real system, of course, all of these processes interact to form an intricate network (represented by the dark area, Figure 4) which can introduce a second level of feedback into biological cycles (Bruland *et al.* 1991) and thereby exert a degree of control on ocean-atmosphere interactions. In the next section we will explore the possibility of unravelling some of these interactions by following the behaviour of selected trace metals through the biogeochemical cycle.

4 Trace Metal Case Studies

4.1 Scavenged Elements

The scavenged elements (Table 3, Figure 3) are particularly useful for understanding the dynamics of the particle cycle since their distribution is largely controlled by physico-chemical interactions with the particle surfaces and they are not directly involved in biological processes. This enables quite simple models to be established. Radioactive isotopes are the most useful members of this group since the decay of these elements

can be used to determine the timescales of the processes controlling the particle cycle.
(i) *Thorium* has isotopes whose half lives range from 24 days (^{234}Th) to 10^{10} years (^{232}Th) and has proved especially useful in this regard. The isotopes ^{234}Th and ^{230}Th result from the decay of uranium isotopes by the following sequence:

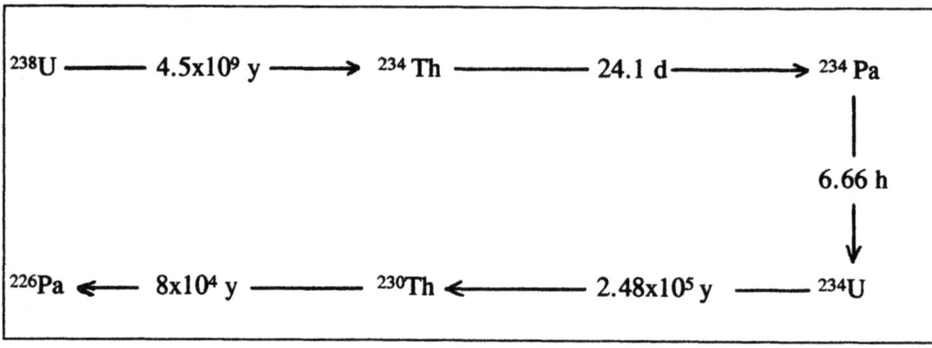

and ^{228}Th arises from the ^{232}Th decay via the following sequence:

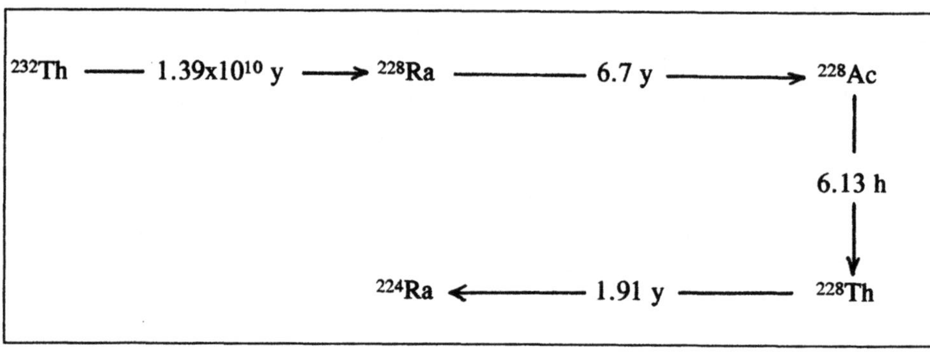

Since uranium is an accumulated element, the source function of the ^{234}Th and ^{230}Th isotopes can be accurately determined from the salinity. The source of ^{232}Th is detrital mineral fragments being transported from the continents. ^{228}Th results from the breakdown of ^{228}Ra which is produced in the surface layers and which leaches from the sediments. The first step in using thorium as a tracer is to define a simple particle model (Figure 5).

We can divide the particle population crudely into small non-sinking particles (P_s, g m^{-3}, diameter<50 microns) that constitute most of the surface area and particle mass and large particles (P_l, g m^{-3}, diameter>50

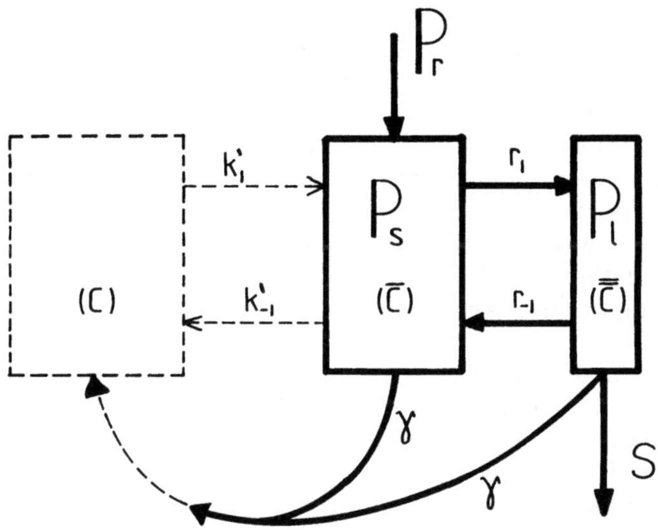

Figure 5: Elementary particle model. The concentrations of thorium in the dissolved, small particle and large particle reservoirs are represented by C (dpm m^{-3}), $\bar{C}$, and $\bar{\bar{C}}$ respectively (dpm g^{-1}). The other terms are defined in the text. Adapted from Clegg and Whitfield (1990).

microns) that have significant sinking rates and provide most of the vertical flux. The model (eq (1)-(3), Table 4) will have to consider the production of small particles by biological processes (P_r, g m^{-2} y^{-1}) and the processes of aggregation and disaggregation (r_1, r_{-1}, y^{-1}) that define the formation and breakdown of large particles by physical and biological processes. Both the large and the small particles can be re-dissolved by physical and biological processes (γ) and the large particle population at any given depth is depleted as the particles sink into deeper water (S, m y^{-1}). Thorium will equilibrate with the small particles since they also constitute most of the surface area in the particle component. The model (Figure 5) will have to consider the adsorption/desorption rate constants (k_1, k_{-1}, y^{-1}) that define the equilibration of the dissolved thorium with the small particles.

The small particle population can be effectively sampled using *in situ* pumping systems that pull large volumes of water through a series of filters. The smallest filter used is normally 0.4 microns in pore size and this effectively provides an operational definition for "dissolved" thorium. There is increasing evidence that a substantial proportion of the thorium that passes through these filters is not truly dissolved, but is adsorbed on to bacteria or colloidal material (Honeyman and Santschi 1982). Nonetheless, we will pursue the simple model. Large particle concentrations are normally determined by intercepting the vertical flux of material with sediment traps. There has been widespread criticism of the efficiency of such traps (see for example, Buesseler 1991) but they remain the most direct method we have of determining the large particle population. *In situ* photographic and holographic techniques are being actively pursued and these might eventually provide a more accurate method of determining particle concentrations. However, the material sampled by the sediment traps will still be required for determining the total particle mass and the particle composition. The problems of sampling and characterising marine particles have recently been reviewed (Spencer and Hurd 1991).

With information on the distribution of the various thorium isotopes between the dissolved, the large particle and the small particle phases it is possible to establish a model of particle dynamics. This approach was taken by Bacon and Anderson (1982) and by Nozaki *et al.* (1987) to determine adsorption and desorption rate constants and aggregation and disaggregation rate constants. Detailed studies of the mobile pasture and the biological particle pump are now being undertaken under the auspices of the JGOFS (Joint Global Ocean Flux Study) Programme and these enable independent estimates to be made of some of the biologically controlled parameters in the particle model. Since the wide range of parameters required to define the steady state model (Table 4) has rarely been collected at a single site, the model is a mosaic of information, largely gathered from areas of moderately high primary production (mesotrophic areas) in the Eastern Pacific that are biologically close to steady state for significant periods of the year (Martin *et al.* 1987). From such sites it is possible to estimate primary production, export flux, mineralisation rates, and particle composition as a function of depth. This information can be combined with data on particle sinking rates and on the mass ratio of large to small particles to provide estimates of the particle aggregation and disaggregation rate constants using the equations in Table 4. The isotope data then enable the adsorption and desorption rate constants to be determined.

TABLE 4. Steady State Thorium Model
(Clegg and Whitfield 1990,1991)

1. PARTICLE CYCLING

For small particles

$$dP_s/dt = 0 = P_r - P_s(\gamma + r_1) + r_{-1}P_l + K_v(d^2P_s/dz^2)$$

For large particles

$$dP_l/dt = 0 = r_1P_s - P_l(\gamma + r_{-1}) - dF/dz$$

For the particle flux

$$F = P_l.S$$

2. TRACE METAL CYCLING

For dissolved fraction

$$dC/dt = 0 = P - C(\lambda + k_1') + \bar{C}(k_{-1}' + \gamma) + \bar{\bar{C}}\gamma + K_v(d^2C/dz^2)$$

For small particle fraction

$$d\bar{C}/dt = 0 = Ck_1' + \bar{\bar{C}}r_{-1} - \bar{C}(\lambda + k_{-1}' + \gamma + r_1) + K_v(d^2\bar{C}/dz^2)$$

For large particle fraction

$$d\bar{\bar{C}}/dt = 0 = \bar{C}r_1 - \bar{\bar{C}}(r_{-1} + \lambda + \gamma) - S(d\bar{\bar{C}}/dz)$$

λ = Radioactive decay constant
K_v = Vertical eddy diffusion coefficient ($m^2\ y^{-1}$)

Model calculations based on such parameterisations (Clegg and Whitfield 1990) show that the use of ^{228}Th or ^{230}Th data in comparison to ^{234}Th distributions provides the most sensitive means of determining adsorption and desorption rate constants. The use of ^{232}Th and ^{228}Th data provides a sensitive means of estimating remineralisation rates since both of these isotopes have surface sources. The aggregation and disaggregation rate constants for the particle cycle can be most accurately determined in the deep water by a study of the relationship between the ^{230}Th and the ^{232}Th distributions. By combining information on the biological particle cycle with the radioisotope distribution it is therefore possible to focus on aspects of the particle cycle that might otherwise be difficult to study by direct measurements. This approach has proved successful in modelling the distribution of a range of thorium isotopes at various sites in the Pacific (Clegg and Whitfield 1991; Clegg, Bacon and Whitfield 1991). In situations where the primary production is intense in the euphotic zone it is necessary to increase the adsorption rate constant to obtain a consistent interpretation of the vertical profiles of particulate and dissolved thorium (Figure 6). This suggests either that the particles being produced during primary production are more active in adsorbing thorium, or that there is some active uptake of thorium into the growing cells in a manner analogous to that of the uptake of the recycled elements.

The application of such models remains difficult because there are very few studies where the particle cycle itself is adequately characterised by direct experimental measurements. For example, the determination of remineralisation rates by dark incubation studies, or aggregation rates by grazing studies would broaden the scope for testing the consistency between the biological measurements and the chemical measurements of isotope distributions. Since the thorium appears to be adsorbed onto particle surfaces irrespective of their composition or origin, problems also arise in the conversion of the concentration of particulate organic carbon, that is usually determined, to the total mass of the particles being formed (Bishop 1991). This value will clearly depend upon the rain ratio and is likely to be variable from place to place and from time to time. The models in their simplest form also assume a steady state and this condition is rarely fulfilled, particularly in areas such as the North Atlantic which show a strong seasonal cycle in primary production. In an analysis of ^{234}Th data from the JGOFS North Atlantic Bloom Experiment (NABE) Buesseler *et al.* (1992) have developed an elegant but simple non-steady state model to relate the ^{234}Th distributions to changes in primary production during the spring bloom. Their model considers a thorium uptake on to a single particle population ($C_p = \bar{C} + \bar{\bar{C}}$, Table 4) which is depleted by vertical transport (Table 5).

The model for the removal of particulate matter was tested by estimating the Redfield ratios of particulate organic carbon to thorium and to nitrogen from the material obtained from the sediment traps and from *in situ* pump samples and thereby estimating the flux of nitrogen in the upper 35 m of the water column. Their model calculations (Table 5) were

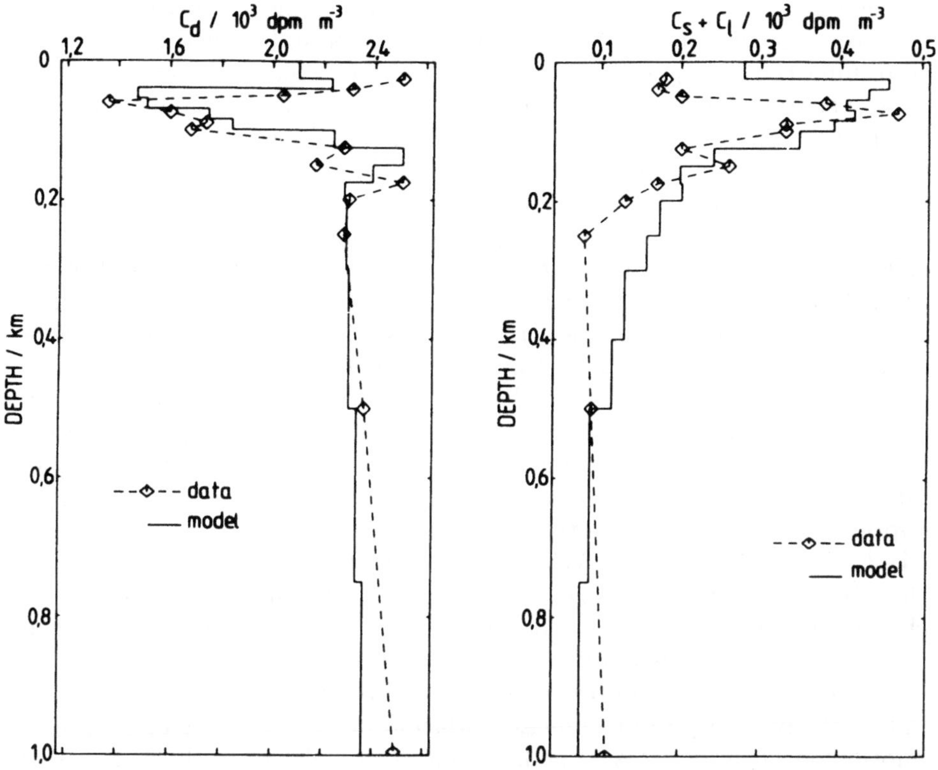

Figure 6: Observed data on the vertical distribution of dissolved (left-hand plot) and particulate (right-hand plot) ^{234}Th from the East Pacific (VERTEX Station 4) compared to the model calculations. The model assumes that the adsorption rate constant (k_1) is twice as great during primary production in the top 75 metres at depth. Details of the calculations will be found in Clegg and Whitfield (1991) from which this figure has been adapted.

consistent with those determined from an independent nitrogen budget and were consistently higher than those that would be estimated by determining the nitrogen flux from the total particulate nitrogen captured by the sediment traps. This supports Buesseler's earlier contention that sediment traps tend to underestimate particle fluxes and that the most reliable values can be obtained from such calculations of the radiotracer budgets (Buesseler 1991).

The combined particle and radiotracer models of Clegg and Whitfield (1990) can also be applied to this data set and indicate that the thorium budget calculated by Buesseler *et al.* (1992) is consistent with the direct measurements of the biological particle cycle at this site. The experience so far suggests that the characteristic parameters controlling the thorium cycle (notably the adsorption, desorption rate constants and the aggregation, disaggregation rate constants) do not vary greatly from site to site and that consistent models of the deep water transport of particulate carbon below 500 m can be effectively established once the particle export flux through this horizon can be defined.

(ii) *Iron and Manganese.* Whereas thorium tends to be scavenged from the water column via adsorption onto particles, iron and manganese are chemically precipitated as solid phases once they are oxidised to their higher oxidation states. Both elements are produced in the divalent, reduced form by biological processes either directly in anoxic sediments, or indirectly by the photochemical oxidation of organic matter in the euphotic zone (Sunda *et al.* 1983, Kuma *et al.* 1992). Both elements are also introduced into the surface layer by the transport of airborne dust. Consequently, a number of different processes have to be introduced to the particle model to explain the distribution of the dissolved element. A further factor that is of considerable importance is that these elements are essential for photosynthesis. Oxidative scavenging reduces their concentrations to very low levels in the open ocean and it has been suggested that both metals might limit primary production in certain areas of the ocean (Bruland *et al.* 1991, Morel and Hudson 1985, Morel *et al.* 1991).

TABLE 5. Non-Steady State Thorium Model
Buesseler *et al.* (1992)

A. Model Description

For dissolved ^{234}Th

$$dC/dt = \text{[rate of production] - [rate of decay] - [net uptake rate]}$$
$$= P - C\lambda - Ck'_1 + C_p k'_{-1}$$

For particulate ^{234}Th

$$dC_p/dt = \text{[net rate of uptake] - [rate of decay] - [particle loss]}$$
$$= Ck'_1 - C_p k'_{-1} - C_p\lambda - S(dC_p/dz)$$

B. Nitrogen Fluxes (mmol m^{-2} d^{-1})

Dates of observations (1989)	Observed Flux	Calculated Flux	
		low [a]	high [b]
24 April to 5 May	1.4	0.9	2.8
5 May to 19 May	7.4	2.6	6.5
19 May to 30 May	0.9	1.4	3.5

a 'Redfield' ratios from 150m traps
b 'Redfield' ratios from pump or trap data at 35m

The measurement of *manganese* in seawater is relatively straightforward provided that precautions are taken to avoid contamination. Techniques have even been developed to measure manganese in situ from submersibles (Chin *et al.* 1992). As a consequence, a large body of data is now being accumulated on the distribution of manganese in the oceans. A detailed section of manganese concentrations was measured in the North East Pacific off the coast of central California during the VERTEX programme (Martin *et al.* 1985). This data set shows clearly the heterogeneous distribution of manganese and provides a good test of our understanding of the processes controlling the partitioning of trace metals between the dissolved and solid phases. Johnson (1991) has modelled the distribution of dissolved manganese in this section using a simplified particle model containing only one class of particles. He estimated the vertical flux of particulate matter from the data of Martin *et al.* (1987) using an approach closely similar to that already described for thorium.

A characteristic feature of vertical profiles of manganese in this region is an enhanced level of manganese in the oxygen depleted water immediately below the thermocline, at a depth of 600 metres or so. Johnson explored two hypotheses for the generation of this maximum. The first assumed that the maximum was caused by the horizontal advection of manganese from biologically active sediment at this level along the shelf edge. The second assumed that the scavenging of manganese on to particles falling through the oxygen minimum zone was less effective at low oxygen levels. Both models were able to reproduce the gross features of the manganese distribution within the section although neither model could account adequately for the relatively high levels of manganese in this oxygen minimum zone at distances up to 2,000 kilometres from the shore. However, this exercise shows that the simple particle-water interaction model could be used for studying the distribution of elements other than thorium and that the main features of the trace metal distribution could be reproduced. Again, the application of a more detailed model is limited by the range of information available.

Iron follows a similar set of transformations as manganese, but it is much more readily oxidised and precipitated than manganese. The differences in the kinetics of the iron and manganese cycles have been discussed by Landing and Bruland (1987). The analysis of iron is more difficult than for manganese because of the much greater problems of contamination. However, some detailed studies have been carried out of the distribution of iron. Furthermore, there is evidence to show that, at high latitudes where primary production appears to cease before all of the nitrogen and phosphorus are utilised, primary production can be enhanced by the addition of small quantities of iron. The first observations of this kind were reported from samples obtained from Station P in the Northeast Pacific off the Alaskan coast (Martin and Fitzwater 1988, Martin and Gordon 1988, Martin *et al.* 1989). Further experimental studies have been made of samples taken from Antarctic waters by Martin *et al.* (1990A) and by de Baar *et al.* (1989) although not always with such clear-cut results Measurements in the Antarctic also suggest that productivity might be related to the levels of iron in the water. For example, the highly productive Gerlache Strait (3 gms of carbon per meter square per day) has iron concentrations in excess of 7 nanomolar which are introduced by run off from the land, whereas the Drake Passage only supports a productivity of 0.1 gms of carbon per metre square per day with iron concentrations less that 0.2 nanomolar (Martin *et al.* 1990A).

Davies (1990) has estimated that 1 megatonne of iron dispersed into 3.2 cubic kilometres of sunlit but iron depleted ocean could remove 6.4 gigatonnes of carbon if all the nitrate was utilised. Such considerations have given rise to the suggestion that the natural biogeochemical cycle may be manipulated to enable the oceanic system to absorb more of the anthropogenic carbon dioxide currently accumulating in the atmosphere. It is however a long extrapolation from experiments carried out in containers on board ship to the dynamics of natural ecosystems in the open ocean (see the discussion by Banse 1990, Martin et al. 1990B, Bruland *et al.* 1991).

A joint US JGOFS and EPA study has suggested that an experiment could be carried out on a mesoscale over an area of 100 square kilometres or so in a site just south of the Galapagos Islands. A vessel would distribute 500 kg per day of soluble iron for 5 to 7 day period. The iron solution would be mixed with a tracer such as sulphur hexafluoride which would enable the fertilized water body to be identified. The behaviour of the ecosystem would then be monitored in the marked and unmarked water bodies at adjacent sites. Even if such an experiment were successful and iron fertilisation on a large scale was demonstrated unequivocally, the wisdom of using such a technological fix to remove carbon dioxide from the atmosphere must be questioned given our ignorance of the overall dynamics of the biological and geochemical cycles and of the longer term implications of such manipulations.

4.2 Recycled Elements

The main process removing recycled elements from the surface layers of the ocean is active uptake by phytoplankton, either as part of the photosynthetic process or because they closely resemble elements that are directly required. Such elements are therefore interesting either because of the control they exert on productivity (Davies 1983, Morel *et al.* 1991, Bruland *et al.*1991), or because their presence in the sediment can tell us something about the productivity in ancient oceans.

(i) *Copper* is a particularly interesting element since it is close to its toxic limit in the contemporary oceans. Relatively small enhancements in the total copper concentration have been shown to reduce the efficiency of carbon uptake by open ocean phytoplankton (Knauer and Martin 1983) and to inhibit the growth of bacteria (Sunda and Guillard 1976). Experi-

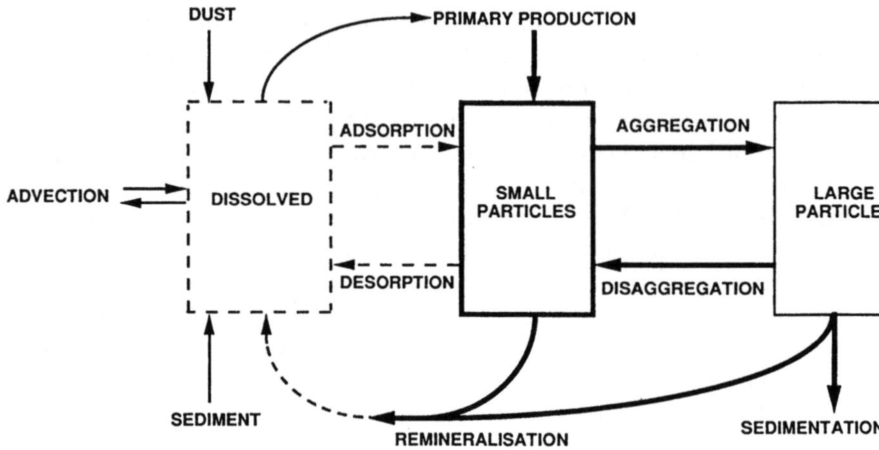

Figure 7: An elaboration of the simple particle model to incorporate elements actively involved in the biological particle cycle. Of especial importance is the feedback introduced into the particle cycle by the influence of trace metal availability on primary production (see also Figure 4).

ments with cultures indicate that it is not the total copper concentration that is critical here, but it is the concentration of free ionic copper which controls toxicity (Morel and Morel-Laurens 1983). In the oceans the concentration of free-copper appears to be largely controlled by the presence of organic compounds produced by phytoplankton. These compounds complex the free copper and make it unavailable for biological uptake, thereby providing a measure of control on its toxicity and a source of feedback in the particle cycle (Fisher and Fabris 1982, Coale and Bruland 1988, Bruland *et al.* 1991). The sources of copper seem to be similar to those of manganese in that a strong input from the sediment has been observed and there are situations, particularly in the Indian Ocean, where an airborne source is apparent (Saager *et al.* 1992). The concentration of copper in the deep ocean is closely correlated with that of silicon. The shift in ratios of copper to silicon on moving from the Altantic through to the Pacific Ocean suggest that copper is scavenged on to the particulate phase in the deep ocean (Nolting *et al* 1991). This is consistent with its relatively high charge to ionic radius ratio (Figure 2). A model for copper distributions similar to that outlined for manganese (Figure 7) would be extremely valuable in helping to unravel the relationships between trace metal chemistry and productivity in the oceans. The formulation of such a model awaits the accumulation of a larger database for copper distribution in the ocean.

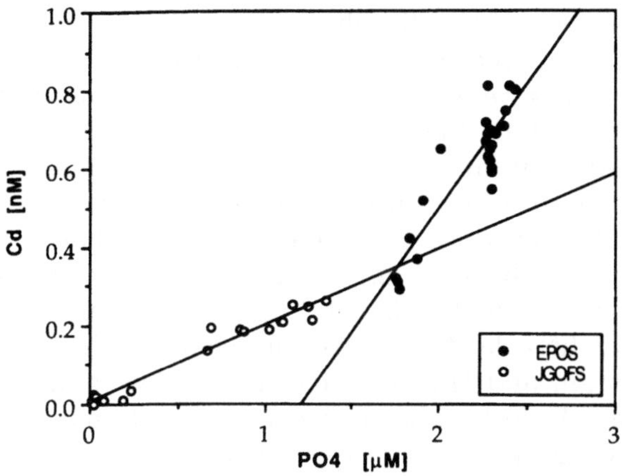

Figure 8: The influence of ambient phosphate concentration on the Cadmium-Phosphate relationship. Adapted from Nolting *et al.* (1991). EPOS data were collected from 57°S, 49°W and JGOFS data from 33°N, 20°W.

(ii) *Cadmium* is an example of a trace metal that is closely involved with the biological cycle although it is not considered to perform any essential function and it is present in seawater at concentrations far below the toxic level. Nonetheless, the concentration of cadmium is closely correlated with that of phosphate in seawater throughout the world's oceans (Boyle *et al.* 1976, Knauer and Martin 1981, Hunter and Ho 1991, Saager *et al.* 1992). Cadmium is also able to substitute for calcium in calcium carbonate shells and these shells can be effectively preserved in the sediment over wide areas of the ocean. This has led to the suggestion that cadmium assimilated into the biogenic calcium carbonate could be used as a proxy for the estimation of phosphate levels in ancient oceans (Boyle 1988). Foraminifera are microscopic organisms which build calcium carbonate shells. There are species of foraminifera that drift with the oceanic plankton and other species which live on the sediment surface. If these different species are picked out from the sedimentary record and carefully cleaned, the cadmium concentrations can be determined and the data used to estimate the phosphate concentrations at the time of their formation. The $^{13}C/^{12}C$ ratios of the foraminiferan shells can also be used to estimate the fractionation of carbon in the ancient oceans and therefore provide a measure of the effectiveness of the biological pump (Boyle 1986). The involvement of cadmium in the biological cycle therefore provides an elegant tool for relating changes in the properties of material accumulating in the ocean sediment with the changes in atmospheric composition that can be observed in ice cores.

Closer investigation of the distribution of cadmium and phosphorus in the world oceans has revealed systematic variations in the parameters for the straight line plots between the two concentrations (Hunter and Ho 1991, Nolting *et al.* 1991, Figure 8, Table 6). Different slopes and intercepts are observed in situations where both cadmium and phosphorus are effectively stripped out of the surface layers than at high latitudes where the primary production does not deplete the phosphorus concentration to such an extent. Nolting *et al.* (1991) suggest that for surface phosphorus concentrations greater than 1.3 micromoles per kilogram, such as those observed at high latitudes, the intercept should be close to -0.25 nanomoles cadmium and the slope 0.4 nanomoles cadmium per micromole phosphorus. At lower phosphorus concentrations the intercept approaches zero and the slope 0.2. It appears that in phosphorus-rich waters a different cadmium assimilation mechanism is switched on, possibly as the phytoplankton produce cellular polyphosphates. These studies indicate the care that must be taken in using trace metal data to unravel the palaeoceanographic record.

TABLE 6. Cadmium-Phosphate Relationships

$$Cd\ (nM) = a\ PO_4\ (uM) + b$$

Location	a	b	Mean [PO_4]	Reference
Tasman Sea	0.38	-0.01	1.2	Hunter and Ho (1991)
Antarctica	0.63	-0.78	2.0	Nolting *et al* (1991)
N Atlantic	0.19	0.035	<1	Nolting *et al* (1991)
N Atlantic 34.06N 66.07W	0.25	-0.02	<1	Bruland and Franks (1983)
Indian Ocean 14.30N 67.00E	0.50	-0.46	2.4 (100 - 4000m)	Saager *et al* (1992)
14.30N 67.00E	0.15	-0.05	0.4 (<100m)	
22.30N 60.40E	0.87	-1.26	2.6 (50 - 3000m)	
22.30N 60.40E	0.16	-0.01	0.6 (<50m)	

4.3 Accumulated Elements

Although the group 2A cations (calcium, strontium and barium, Figure 3) are accumulated in the oceans, they are also involved in biological cycles as components of biogenic minerals. As a consequence there is a slight surface depletion in the concentration profiles of these elements. *Barium* is of particular interest in this group because it can form barium sulphate which is close to its solubility limits in seawater. The phytoplankton assimilate sulphur as an essential component of a number of amino acids. This sulphur exhibits a characteristic 'Redfield' concentration ratio to carbon and nitrogen. As the phytoplankton fall out of the surface layer, either as dead cells or in fecal detritus, the organic sulphur compounds are oxidised and sulphate can be released. As the sulphate concentration rises within the decaying particulate material, barium sulphate can be precipitated. Since barium is an accumulated element its concentration relative to salinity is approximately constant throughout the world's oceans, and the extent to which barium sulphate is precipitated can be considered as an indicator of the production of excess sulphate. Given the stoichiometric relationship between sulphur, nitrogen and carbon in phytoplankton cells, a relationship is established between the flux of particulate barium and the flux of particulate organic nitrogen and carbon from the euphotic zone. This signal is apparently retained during the deposition process. Sediment trap fluxes of barium sulphate (barites) therefore appear to be directly related to the utilisation of nitrate in the euphotic zone and to the particulate organic carbon flux. Barium in the settled sediments has consequently been proposed as an indicator of palaeo productivity (Dehairs *et al.* 1991, Stroobants *et al.* 1991). This relationship is not universally observed but it seems to be most robust where diatoms form a significant component of the phytoplankton population. It is possible that silica frustules provide a template for the deposition of barium sulphate, or that their presence in the sinking particulate material ensures rapid removal from the surface and accumulation in the sediment. It is clear that much further work will need to be done to enable a quantitative indicator for palaeo productivity to be derived from barium distributions in sediment.

5 Conclusion

The distribution and productivity of the ocean's mobile pasture are controlled by physical forcing functions in the surface layers of the ocean that are strongly influenced by climate change. The mobile pasture is in turn able to influence climate (a) by modulating the air-sea exchange of carbon dioxide (b) by influencing the uptake of heat by the oceans and (c) by influencing the nucleation of clouds over open ocean areas by the release of dimethyl sulphide. These processes provide the first level of feedback in the biological system (Figure 1). The grazing of this pasture by zooplankton provides a flux of particulate material into the deep ocean

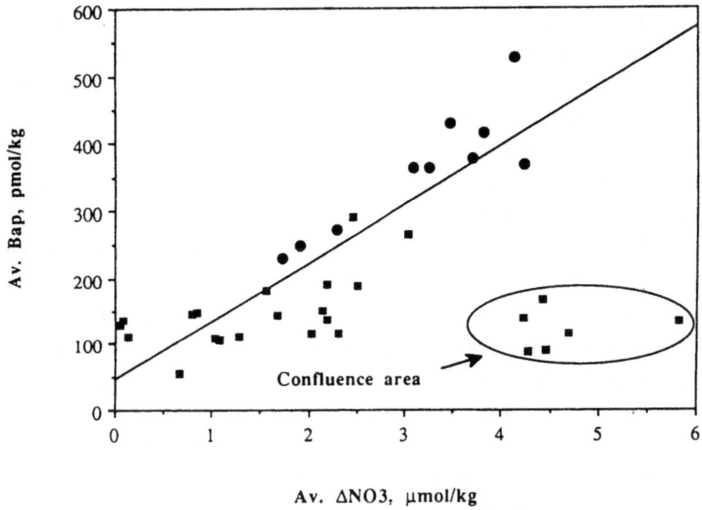

Figure 9: The relationship observed between particulate barium (Bap) and the utilisation of nitrate in the euphotic zone (Δ NO_3) in samples taken from 61°S, 32°E (circles) and 60°S, 90°W (squares). The confluence area is the zone between the Scotia Front and the pack ice. Adapted from Dehairs *et al.* 1991.

which controls the long term sequestration of carbon and which carries to the sediment a signal of the processes occurring in the surface layers. The involvement of trace metals in this important cycle of events can be described in a rational fashion by considering the inorganic properties of the elements as summarised by the Periodic Table (Figure 3).

Scavenged elements which adsorb very effectively on to the surface of particles can be used to investigate the dynamics of the particle cycle (Figure 5), especially where the elements themselves are subject to radioactive decay. As well as marking the progress of the particle cycle, some of the scavenged trace elements (notably manganese and iron, and possibly cobalt) are also essential nutrients. Under these circumstances the trace metals themselves play an active role in controlling the dynamics of the particle cycle thereby introducing a second level of feedback (Figures 4 and 7). An understanding of the behaviour of these elements is essential if we are to define the long term controls on the intensity of the biological pump.

Most of the elements that are taken up by phytoplankton during photosynthesis act as recycled elements and show characteristic surface depletion and deep water accumulation. Some elements are essential nutrients and others (such as copper and possibly zinc) might be close to their toxic limits. The interaction of these elements with the biological cycle must be investigated if we are to understand the interaction between the oceans and climate over long timescales. To complicate matters, there is considerable evidence to suggest that the impacts of the various trace metals on primary production are not mutually independent. For example, the presence of a second trace metal may enhance or suppress the effect of a toxic metal on growth (Bruland *et al.* 1991). The implications of these effects for the dynamics of the particle cycle have not yet been investigated.

Some elements, such as cadmium, show strong correlations with the essential nutrients. If they are also assimilated into biogenic minerals in sedimenting material, they can be used, with care, as proxy indicators of primary production in ancient oceans. The accumulated element barium can also be used in this fashion, but for very different reasons.

Work on thorium is beginning to show that a coherent model can be constructed which incorporates both the biological properties of the particle cycle and the distribution of trace elements. As such models are refined, the interactions between the trace metal cycles and the biological cycles will be more clearly delineated. This will enable us to improve our quantitative understanding of the interaction of biological processes with climate change, both in the contemporary system and in ancient oceans.

References

ARCHER D (1990). Modelling pCO_2 in the upper ocean: a review of relevant physical, chemical and biological processes. *US Department of Energy Publication No DOE/RL-01830T-H5.*

BACON MP and ANDERSON RF (1982). Distribution of thorium isotopes between dissolved and particulate forms in the deep sea. *Journal of Geophysical Research* **87**, 2045-2056.

BANSE K (1990). Does iron really limit phytoplankton production in the offshore sub-Arctic Pacific? *Limnology and Oceanography* **35**, 772-775.

BARNOLA JM RAYNAUD D KOROTKEVICH YS and LORIUS C (1987). Vostok ice core provides 160,000 year record of atmospheric CO2. *Nature* **329**, 408-414.

BERGER WH, SMETACEK VS and WEFER G (1989). Ocean productivity and paleo productivity - an overview. In: Productivity of the Ocean: Present and Past. *Editors WH Berger, BS Smetacek and G Wefer. Wiley and Sons, pages 1-34.*

BISHOP JKB (1991). Getting good weight in marine particles: analysis and characterisation. Geophysical Monograph No 63 (DW Spencer and D Heard editors) pages 229-234 American Geophysical Union Washington DC.

BOYLE EA, SCLATER FR and EDMOND JM (1976). On the marine geochemistry of cadmium. *Nature* **263**, 42-44.

BOYLE EA (1986). Paired carbon isotope and cadmium data from benthic foraminifera: implications for changes in oceanic phosphorus, oceanic circulation and oceanic carbon dioxide. *Geochimica et Cosmochimica Act* **50**, 265-276.

BOYLE EA (1988). Cadmium: chemical tracer of deep water paleo-oceanography. *Paleo-oceanography* **3**, 471-489.

BROECKER WS and PENG TH (1982). Tracers in the sea. *Eldigio Press Palisades New York 690 pages.*

BRULAND KW, DONAT JR and HUTCHINS DA (1991). Interactive influences of bioactive trace metals on biological production in oceanic waters. *Limnology and Oceanography* **36**, 1555-1557.

BRULAND KW and FRANKS RP (1983). Mn, Ni, Cu, Zn and Cd in the western north Atlantic In: Trace Metals in Seawater (CS Wong, EA Boyle, KW Bruland, JD Burton and ED Goldberg editors). *Plenum Press New York, pages 395-414.*

BUESSELER KO, BACON MP, COCHRAN JK and LIVINGSTON HD (1992). Carbon and nitrogen export during the JGOFS North Atlantic Bloom Experiment estimated from ^{234}Th and ^{238}U disequilibria. *Deep Sea Research (In press).*

BUESSELER KO (1991). Do upper ocean sediment traps provide an accurate record of particle flux? *Nature* **353**, 420-423.

CHARLSON RJ, LOVELOCK JE, ANDREAE MO and WARREN SG (1987). Oceanic plankton, atmospheric sulphur, cloud albedo and climate. *Nature* **326**, 655-661.

CHIN CS, JOHNSTON KS and COALE KH (1992). Spectrophotometric determination of dissolved manganese in natural water with 1-(2-pyridylazo)-2-naphthol: application to analysis in situ in hydrothermal phemes. *Marine Chemistry* **37**, 65-82.

CLEGG SL and WHITFIELD M (1990). A generalised particle model for the scavenging of trace metals in the open ocean. I Particle cycling. *Deep Sea Research* **37**, 809-832.

CLEGG SL and WHITFIELD M (1991). A generalised model for the scavenging of trace metals in the open ocean. II Thorium scavenging. *Deep Sea Rearch* **38**, 91-120.

CLEGG S, BACON M and WHITFIELD M (1991). Application of a generalised scavenging model to thorium isotope and particle data at equatorial and high latitude sites in the Pacific Ocean. *Journal of Geophysical Research* **99**, 20655-20670.

COALE KH and BRULAND KW (1988). Copper complexation in the north western Pacific. *Limnology and Oceanography* **33**, 1084-1101.

COLLIER RW and EDMOND JM (1984). The trace element geochemistry of marine biogenic particulate matter. *Progress in Oceanography* **30**, 113-199.

DAVIES AG (1983). The effects of heavy metals upon natural marine phytoplankton populations. *Progress in Phycology Research* **2**, 113-141.

DAVIES AG (1990). Taking a cool look at iron. *Nature* **345**, 114-115.

DE BAAR HJW, BUMA AGJ, JACQUES J, NOLTING RF and TREGUER PJ (1989). Iron and manganese effects on phytoplankton growth. *Berichte zur Polarforschung* **65**, 34-43.

DEHAIRS F, STROOBANTS N and GOEYENS L (1991). Suspended barite as a tracer of biological activity in the southern ocean. *Marine Chemistry* **35**, 399-410.

FISHER MS and FABRIS JG (1982). Complexation of copper, zinc and cadmium by metabolites excreted from marine diatoms. *Marine Chemistry* **11**, 245-255.

HERRING PJ, CAMPBELL AK, WHITFIELD M and MADDOCK L (1990). Light and Life in the Sea, *Cambridge University Press 357 pages.*

HONEYMAN BD and SANTSCHI PH (1989). A Brownian pumping model for oceanic trace metal scavenging: evidence from thorium isotopes. *Journal of Marine Research* **47,** 951-992.

HUNTER KA and HO FWT (1991). Phosphorus-cadmium cycling in the North East Tasman Sea 35-40°S. *Marine Chemistry* **33,** 279-298.

JICKELLS TD and BURTON JD (1988). Cobalt, copper, manganese and nickel in the Sargaso Sea. *Marine Chemistry* **23,** 131-144.

JOHNSON KS (1991). *Personal communication.*

KNAUER and MARTIN (1983). In: Trace Metals in Seawater (CS Wong, EA Boyle, KW Bruland, JD Burton and ED Goldberg editors). *Plenum Press New York pages 825-840.*

KNAUER GA and MARTIN JH (1981). Phosphorus-cadmium cycling in northeast Pacific waters. *Journal of Marine Research* **39,** 65-76.

KUMA K, NAKABAYASHI S, SUZUKI Y, KUDO I and MATSUNAGA K (1992). Photoreduction of Fe (III) by dissolved organic substances and existence of Fe (II) in seawater during spring blooms. *Marine Chemistry* **37,** 15-28.

LANDING WM and BRULAND KW (1987). The contrasting biogeochemistry of iron and manganese in the Pacific Ocean. *Geochimica et Cosmochimica Acta* **51,** 29-43.

LISS PS and MERLIVAT L (1986). Air-sea gas exchange rates: introduction and synthesis. In: "The Role of Air-Sea Exchange in Geochemical Cycling" (P Buat-Menard editor). *D Reidel.*

MARTIN JH, GORDON RM, FITZWATER S and BROENKOW WW (1989). VERTEX: Phytoplantkon/iron studies in the Gulf of Alaska. *Deep Sea Research* **36,** 649-680.

MARTIN JH, KNAUER GA, KARL DN and BROENKOW WW (1987). VERTEX: Carbon cycling in the North East Pacific. *Deep Sea Research* **34,** 267-286.

MARTIN JH, KNAUER GA and BROENKOW WW (1985). VERTEX: The lateral transport of manganese in the North East Pacific. *Deep Sea Research* **32,** 1405-1427.

MARTIN JH, FITZWATER FE and GORDON RM (1990A). Iron deficiency limits phytoplantkon growth in Antarctic waters. *Global Biogeochemical Cycles* **4,** 5-12.

MARTIN JH, BROENKOW WW, FITZWATER SE and GORDON RN (1990B). Yes it does: A reply to the comment by Banse. *Limnology and Oceanography* **35,** 775-778.

MARTIN JH and FITZWATER SE (1988). Iron deficiency limits phytoplankton growth in the North East Pacific Sub Arctic. *Nature* **331,** 341-343.

MARTIN JH and GORDON RM (1988). North East Pacific iron distributions in relation to phytoplankton productivity. *Deep Sea Research* **35,** 177-196.

MOAL J, MARTIN-JEZEQUEL V, HARRIS RP, SUAMAIN J-F and POULET SA (1987). Interspecific and intraspecific variability of the chemical composition of marine phytoplantkon. *Oceanologica Acta* **10,** 339-346.

MOREL FMM and HUDSON RJM (1985). A geobiological cycle of trace elements in aquatic systems: Redfield revisited. In: "Chemical Processes in Lakes" (W Stumm editor). *Wiley New York, pages 251-281.*

MOREL FMM and MOREL-LAURENS NML (1983). Trace metals and plankton in the oceans: facts and speculations. In: "Trace Metals in the Seawater" (CS Wong, EA Boyle, KW Bruland, JD Burton and ED Goldberg editors). *Plenum Press New York pages 841-869.*

MOREL FMM, HUDSON RJM and PRICE NM (1991). Limitation of productivity by trace metals in the sea. *Limnology and Oceanography* **36**, 1742-1755.

NOLTING RF, DE BAAR HJW, VAN BENNEKON AJ and MASSON A (1991). Cadmium, copper and iron in Scotia Sea, Weddell Sea and Weddell/Scotia Confluence (Antarctica). *Marine Chemistry* **35**, 219-243.

NOZAKI Y, YANG H-S and YAMADA M (1987). Scavenging of thorium in the oceans. *Journal of Geophysical Research* **87**, 2045-2056.

PARSONS TR, TAKAHASHI N and HARGRAVE B (1984). Biological oceanographic processes, third edition. *Pergamon Press Oxford 330 pp.*

SAAGER PA, DEBAAR HJW and HOWLAND RJ (1992). Cadmium, zinc, nickel and copper in the Indian Ocean. *Deep Sea Research* **39**, 9-35.

SARMIENTO JL (1990). Role of ocean carbon cycle in determining atmospheric pCO_2. In: Global Climate Feedbacks. *Pages 83-93. US Department of Energy Publication No CONF-9006134.*

SAVOIE DL and PROSPERO JN (1989). Comparisons of oceanic and continental sources of non sea salt sulphate over the Pacific Ocean. *Nature* **339**, 685-687.

SPENCER DW and HURD D (EDITORS) (1991). Marine particles: analysis and characterisation. Geophysical Monogram No 63. *American Geophysical Union Washington DC.*

STROOBANTS N, DEHAIRS F, GOEYENS L, VANDERHUIJDEN N and VAN GRIEKEN R (1991). Barite formation in the southern ocean water column. *Marine Chemistry* **35**, 411-421.

SUGIMURA Y and SUZUKI Y (1988). A high temperature catalytic oxidation method of non-volatile dissolved organic carbon in the seawater by direct injection of liquid sample. *Marine Chemistry* **24**, 105-131.

SUNDA WG, HUNTSMAN SA and HARVEY GR (1983). Photoreduction of manganese oxides in the seawater and geochemical and biochemical implications. *Nature* **301**, 234-239.

SUNDA WG and GUILLARD RRL (1976). The relationship between cupric iron activity and the toxicity of copper to phytoplankton. *Journal of Marine Research* **34**, 511-529.

TURNER DR, WHITFIELD M and DICKSON AG (1981). The equilibrium speciation of dissolved components in freshwater and seawater at 25°C and 1 atmosphere pressure. *Geochimica et Cosmochimica Acta* **45**, 855-881.

VOLK T and HOFFERT MI (1985). Ocean carbon pumps: analysis of relative strengths and efficiencies in ocean atmosphere driven atmospheric pCO_2 changes. *In: The Carbon Cycle and Atmospheric CO_2, natural variations Archean to present.*

SUNDQUIVIST ET and BROECKER WS (EDS). Pages 93-110. *AGU Washington.*

VOLK T and LIU Z (1988). Controls of CO2 sources and sinks in the Earth's scale surface ocean: temperature and nutrients. *Global Biogeochemical Cycles* **2**, 73-89.

WATSON AJ , ROBINSON C , ROBERTSON JE , WILLIAMS PLEB , and FASHAM MJ (1991). Spatial variability in the sink for atmospheric carbon dioxide in the N. Atlantic. *Nature* **350**, 50-53.

WHITFIELD M and TURNER DR (1987). The role of particles in regulating the composition of seawater. In: "Aquatic Surface Chemistry: Chemical Processes at the Particle-Water Interface" *(W Stumm editor). John Wiley and Sons. Pages 457-493.*

MODELLING CLIMATIC INTERACTIONS OF THE MARINE BIOTA

A.H. TAYLOR
Plymouth Marine Laboratory
Plymouth PL1 3DH, U.K.

Abstract

The marine ecosystem is strongly driven by physical processes. In turn it influences the physical environment by hindering the penetration of radiation down the water column, and altering the chemical balances within the ocean and atmosphere; examples of these processes are described. This paper discusses the problem of incorporating these couplings in ocean models. Different methods of representing physical structure and physical forcing are reviewed and the difficulties of deciding on ecosystem compartments discussed. Although there are no universal equations governing population growth, Wiegert's equations provide a convenient summary of most of the processes that have to be considered in formulating any model. The terms making up the equations are described, e.g. how limiting resources are treated, and the problems of estimating parameter values are discussed. Finally, a modelling investigation of data from the JGOFS North Atlantic Bloom Experiment in 1989 is presented as an illustrative case study. It shows how, in spite of the uncertainties associated with defining an appropriate model structure and assigning parameter values, an ecosystem model can reproduce and explain many of the features that are observed. There are also cautionary conclusions that arise from the study. The model illustrates how the impact of a climatic perturbation may be inaccurately predicted by a simpler model. Further, examination of observations from the Continuous Plankton Recorder Survey suggests that this model, which was developed for the NE Atlantic, may not be applied even to the NW Atlantic without some changes to take account of species differences.

1 Introduction

The close coupling between the marine ecosystem and its physical forcing has been demonstrated in many modelling investigations starting with that of Riley *et al.* (1949). Phytoplankton growth in the surface layers of the oceans is critically dependent on the physical mixing processes which determine the light and nutrient environment (Tett and Edwards,

NATO ASI Series, Vol. I 11
Modelling Oceanic Climate Interactions
Edited by J. Willebrand and D. L. T. Anderson

1984). At temperate and polar latitudes, the combination of a deep surface mixed layer (due to surface cooling and wind stirring) and low solar radiation during the winter months results in the inhibition of phytoplankton growth because of light limitation. When the seasonal pycnocline becomes established in the spring or early summer, plant cells are held in the upper 20-60 m of the water column and the vernal bloom occurs until terminated by depletion of inorganic nutrients, particularly silicate and nitrate. During the summer months, the standing stocks of phytoplankton are determined by the extent to which physical mixing processes maintain an upward flux of nutrients into the euphotic zone (Eppley and Peterson, 1979). At low latitudes, light often penetrates to sufficient depth for a deep chlorophyll maximum to form in the pycnocline immediately above the nutrient pool. This dependence of phytoplankton production on the physical background has been explored quantitatively in a variety of one-dimensional (depth/time) models e.g.: Radach and Maier-Reimer (1974), Jamart *et al.* (1977, 1979), Radach (1980), Kiefer and Kremer (1981), Tett (1986), Woods and Onken (1982), Fasham *et al.* (1983), Klein and Coste (1984), Evans and Parslow (1985), Garside (1985), Parker (1986), Taylor *et al.* (1986), Musgrave *et al.* (1988), Taylor (1988), Fasham *et al.* (1990), Taylor and Joint (1990), Taylor *et al.* (1991) and Taylor *et al.* (1992).

Changes in climatic forcing are the source of many long-term trends in the plankton. Thus, zooplankton and phytoplankton abundances in the NE Atlantic and North Sea closely follow latitudinal shifts in the north wall of the Gulf Stream in the vicinity of the U.S. coast, a relationship that probably arises from variations in the tracks of weather systems (Taylor *et al.*, 1992). While the marine ecosystem is driven by physical processes, the oceanic biota in turn influence the physical environment. The two major ways the biological system affects its environment are: (A) hindering the penetration of radiation down the water column and hence modifying the heat budget, and (B) altering the chemical balances within the ocean and atmosphere.

1.1 Attenuation of Irradiance

Dense phytoplankton populations such as those occurring during spring blooms can be associated with a shallowing of the mixed layer. The Kraus-Turner model of the mixed layer can be used to demonstrate that this effect can sometimes be significant. In the Kraus-Turner model, the depth h of the mixed layer during the period it is getting shallower is given by (Kraus and Turner 1967; Kraus 1972):

$$h = \frac{2(W - D + \lambda J_o)}{(J_o - Y)}. \tag{1}$$

J_o is the solar irradiance at the sea surface and Y is the net surface heat loss by evaporation, sensible heat flux and infrared radiation. In J_o and Y these energy flows have each been divided by both the specific heat (c_p) and the density (ρ_s) of seawater. W is the work of the wind on the sea surface and D is the dissipation of turbulent energy within the mixed layer; each of these has been divided by ($g\alpha\rho_s$), g being the acceleration due to gravity and α the expansion coefficient for sea water. λ^{-1} is the extinction coefficient of the water. The temperature T_s of the mixed layer is given by:

$$\frac{dT_s}{dt} = \frac{(J_o - Y)}{h}. \tag{2}$$

Following Kraus (1972) we assume:

$$2(W - D) = \frac{C\rho_s u^3}{(\rho_s c_p)}, \tag{3}$$

and use the value for C (2.5) determined by Kato and Phillips (1969). If the friction velocity u is 0.012 m^{-1} (a wind speed of about 8 m s^{-1}), λ is 10 m, $(\rho_s c_p)Y$ is 80 W m^{-2} and $(\rho_s c_p)J_o$ is 200 W m^{-2}, the mixed layer depth h, calculated from (1), is 34 m. A phytoplankton bloom with a chlorophyll concentration of 6 mg m^{-3} can reduce λ to 5 m, and this reduces h to 18 m whereupon equation (2) indicates that the rate of warming will be doubled. During May and June 1991 a bloom of the very reflective coccolithophorid *Emiliania huxleyi* was observed in the North Atlantic in which λ was reduced to 3 m or less. Assuming a value of 3 m and allowing for 20% of the incident radiation being reflected from the surface, h is predicted to fall to 15 m. Shallowing of the mixed layer of about this magnitude was observed in the region of the bloom.

Simonot *et al.* (1988) have investigated the impact of shading by phytoplankton on thermal structure by coupling the mixed layer model of Gaspar (1988) to a primary production model adapted from Agoumi (1985). In a simulation of two seasonal cycles at Ocean Weather Station Romeo, they observed that the typical spring and autumn errors of predicted sea surface temperatures (up to 1°C) from the mixed layer model sometimes reversed sign in the thermodynamic-biological-optical version of the model, indicating that the phytoplankton seasonal cycle had a significant impact on the sea surface temperature prediction. Sathyendranath *et al.* (1991) have estimated that biological heating can contribute up to 4°C per month in the Arabian Sea during the period from

August to September, an area with a maximum observed cooling rate due to upwelling of $\sim$ 2.5°C per month. Biological processes may therefore influence weather patterns in this region, for Kershaw (1985) has shown that a knowledge of anomalies of SST in the Arabian Sea allows the more accurate prediction of the evolution of atmospheric depressions associated with the monsoon.

1.2 Chemical Balance of Ocean and Atmosphere

Carbon dioxide consumption and production are the major chemical means through which the marine ecosystem may affect the climate, phytoplankton growth having a substantial impact on CO_2 concentrations in the surface waters of the ocean and hence on those in the atmosphere. Other gases which are recognised as influencing climate are nitrous oxide (N_2O), methane (CH_4), and dimethylsulphide (DMS) which has been implicated in cloud formation (Charlson *et al.*, 1987). The relationship between dissolved CO_2 and the growth of plant cells is illustrated by observations during the Joint Global Ocean Flux Study (JGOFS) North Atlantic Bloom Experiment in 1989 (Fig. 1, Watson *et al.*, 1991); the variations in chlorophyll-a and pCO_2 observed during a meridional section through the bloom being in agreement with the predictions of a simple phytoplankton-nutrient model which allows for the latitudinal variation of the spring bloom (Taylor *et al.*, 1992). Observed day-night cycles of CO_2 uptake and oxygen release were also in agreement with model predictions (Taylor *et al.*, 1991; Taylor and Stephens, 1993).

One group of phytoplankton, the coccolithophorids, are calcifying organisms, precipitating CO_2 as calcium carbonate. This process:

$$Ca^{2+} + 2HCO_3^- -------- > CaCO_3 + 2H^+ + CO_2,$$

although reducing the total inorganic carbon content of the water, reduces the alkalinity and increases the partial pressure of carbon dioxide (pCO_2). This simplistic representation of the calcification process implies that for each mole of $CaCO_3$ deposited a mole of CO_2 is liberated. While this is nearly true in freshwater, the process is reduced in buffered seawater and only about 0.6 moles of CO_2 are liberated per mole of $CaCO_3$ deposited. As the transfer of CO_2 across the sea surface is determined by the pCO_2 in the water, calcification leads to either reduced take-up of CO_2 by the ocean or outgassing of CO_2 (Taylor *et al.*, 1991). During June 1991 a coccolithophore bloom along 20°W between 60 and 63°N (Holligan *et al.*, submitted) was associated with TCO_2 and alkalinity reductions of $\sim$ 20 μ mol / kg. Although further south away from the bloom the pCO_2 had been reduced to 300 μ atm., in the region of the bloom it was raised close to the atmospheric value, 350 μ atm.

These are the main mechanisms connecting the oceanic ecosystem to the climate which should be included in models and the following sections will discuss how these processes may be represented. Section 2 discusses different treatments of vertical structure, turbulent mixing processes and the way radiation varies with depth. Methods for incorporating the necessary biological processes are described in section 3. Finally, section 4 presents a case study from the North Atlantic which illustrates many of these techniques, shows some of the complexity of the biological system, indicates what can be achieved by ecosystem modelling, and provides examples of the problems and limitations of this kind of modelling.

2 Physical Forcing

The last section has shown how phytoplankton growth may influence the climate, and that the development of these populations is to a large extent determined by the vertical structure of the oceanic water column and the penetration of light. How should these processes be represented in models?

2.1 Vertical Mixing and Advection

The depth of mixing is not constant during the day but fluctuates as the buoyant convection varies (Woods, 1980, 1985). While wind mixing stirs the upper 10 - 20m at all times of the day, the convection due to surface cooling is a source of turbulent kinetic energy whose penetration is deeper and much more variable. At night, convection reaches throughout the mixed layer and contributes significantly to the deepening of the layer by turbulent entrainment; but during most of the day the heat input by the sun exceeds the heat loss by evaporation, sensible heat flux and long-wave radiation, so that convection only extends to shallow depths. When there is a net warming an estimate of these depths is provided by the thermal compensation depth, the depth over which the irradiance absorbed equals the rate of surface heat loss; Woods (1980) has calculated the proportion of the day during which this depth is more than 10m for different sites and seasons. These processes, which are particularly significant at the onset of the spring bloom (Taylor and Stephens, 1993), can be represented in mixed layer models of the Kraus-Turner (1967) type. Woods and Onken (1982) and Wolf and Woods (1988) have modelled phytoplankton growth in the presence of daily and seasonal changes in vertical mixing by combining a one-dimensional model of the surface mixed layer with a Lagrangian ensemble model tracking the vertical random walks of individual phytoplankters. They argue that light-dependent growth of phytoplankton will depend on the historical trajectory of the cells within the turbulent mixed layer and so this Lagrangian approach is necessary. The vertical distribution of the phytoplankton is obtained by ensemble averaging at each depth zone. While this approach may sometimes have advantages, it may not be needed in general. Lande and Lewis (1989) have analysed bulk property (Eulerian) and single cell (Lagrangian) models of photoadaptation and photosynthesis by phytoplankton undergoing vertical turbulent mixing in the upper ocean. Evaluation of the models with experimental data on a small oceanic diatom, *Thalassiosira pseudonana*, shows that photosynthetic rates at particular depths predicted by the two models differ by less than about 1%.

The most frequently used method for representing mixing and transport in the vertical is by means of advection-diffusion equations (e.g. Riley *et al.*, 1949; Radach and Maier-Reimer, 1974; Jamart *et al.*, 1977, 1979; Radach, 1980; Tett and Edwards, 1984; Tett, 1986; Fasham *et al.*, 1983; Garside, 1985; Parker, 1986 and Taylor *et al.*, 1986), these equations being solved by finite-difference methods, and many aspects of the vertical distribution, e.g. deep chlorophyll maxima, can be reproduced and investigated in this way. Turbulent transport appears in these models as eddy diffusion coefficients which can be estimated from the meteorological conditions and the density structure. An alternative approach for representing changing vertical structure is a mixed layer model of the Kraus-Turner type. The Kraus-Turner (1967) model does not include eddy-diffusion processes which may make only a relatively small contribution to the heat budget. However, these terms can be much more important to the biological system, providing the main nutrient supply into the nutrient depleted mixed layer during summer (Taylor *et al.*, 1986; Taylor, 1988).

Because nutrient and particle concentrations are commonly uniform across the mixed layer, the layer is often treated as a single slab whose thickness and content may vary with time, the layer either being closed or resting on an abiotic source of nutrients (Evans and Parslow, 1985; Taylor *et al.* 1986; Fasham *et al.*, 1990; Taylor and Joint, 1990; Taylor *et al.*, 1992). This approximation can be improved by adding a second layer in the thermocline region to the system (Taylor *et al.*, 1991), and one or two layer approximations can be shown to replicate the major features of continuous 1D models (Taylor *et al.*, 1986; Taylor *et al.*, 1988). Turbulent mixing between layers is parameterised simply as the product of a mixing coefficient and the difference in concentration between the layers. However, the choice of a value for the mixing coefficient is not straightforward. Evans and Parslow (1985) used a value of 3 m d^{-1} and Evans (1988) explored values all the way down to zero. Fasham *et al.* (1990) used 0.1 m d^{-1} in the vicinity of Bermuda, while further north in the Atlantic Taylor *et al.* (1991, 1992) have used a mixing coefficient which increased with the depth of the mixed layer from a summer minimum value of 0.4 to 0.6 m d^{-1}.

2.2 Irradiance

Most of the solar radiation crossing the sea surface never penetrates to any depth, about 50% (infrared) being lost in the first few centimetres of water. Of the remainder, the photosynthetic active radiation (PAR), red light penetrates much less deeply than green light. If good optical data are available the light can be decomposed into a large number of wavebands with different attenuation coefficients (Morel, 1988; Sathyendranath and Platt, 1988). Light intensity down the water column is often approximated as a simple exponential and this has the advantage that the average primary production within a layer can be represented by an analytical expression (Vollenweider, 1965). A useful compromise between these extremes is to divide the light into two bands (green and red) which decay at different rates, an approximation which has been shown to give good fits to observed irradiance-depth plots (Fasham *et al.*, 1983; Carr, 1986). When estimating attenuation coefficients it is also necessary to take into account the light absorbed by the phytoplankton itself. This is parameterised by a self-shading coefficient which will depend on the wavelength of the light. For example, if this wavelength dependence is neglected, P mg m^{-3} of chlorophyll increase the attenuation by about 0.016 P m^{-1} (Carr, 1986).

3 Model Structure

3.1 Compartments

Ecosystem models consist of compartments which may be biotic (e.g. phytoplankton, zooplankton) or abiotic (e.g. nitrate, detritus). To build a model it is first necessary to define the compartments and then specify the equations governing the transfer of material between compartments. The choice of compartments is guided by the scientific questions that the model is created to address, and the ability of the model to predict observations and to increase understanding. It has to be borne in mind that an increase in complexity may not lead to closer agreement with observations; the predictions of the more complex version may be limited by the availability of the data needed to drive it and by its more complicated dynamics. The best approach may therefore be to begin with

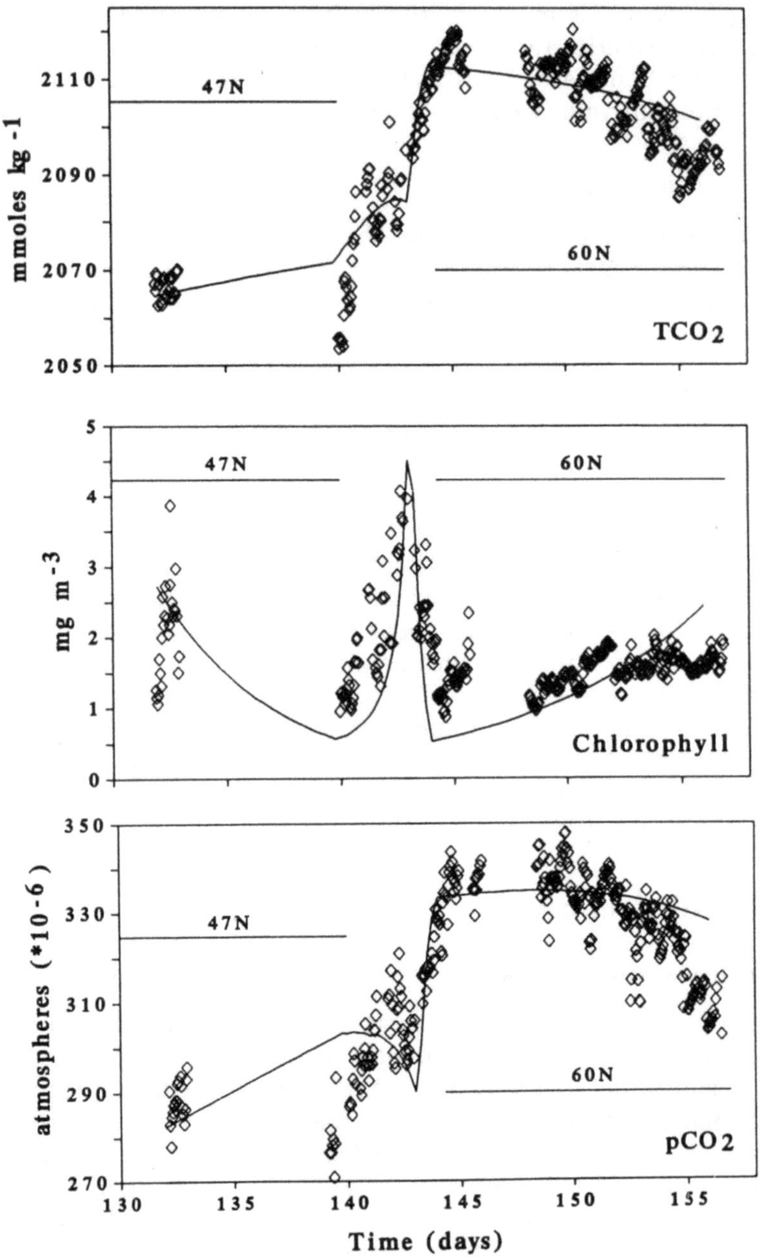

Figure 1: Changes in pCO_2, chlorophyll a and TCO_2 along 20°W during the U.K. BOFS cruise 12 May to 5 June 1989 (Watson *et al.*, 1991; Taylor *et al.*, 1992; the chlorophyll a data have been recalibrated). The continuous lines are the model predictions.

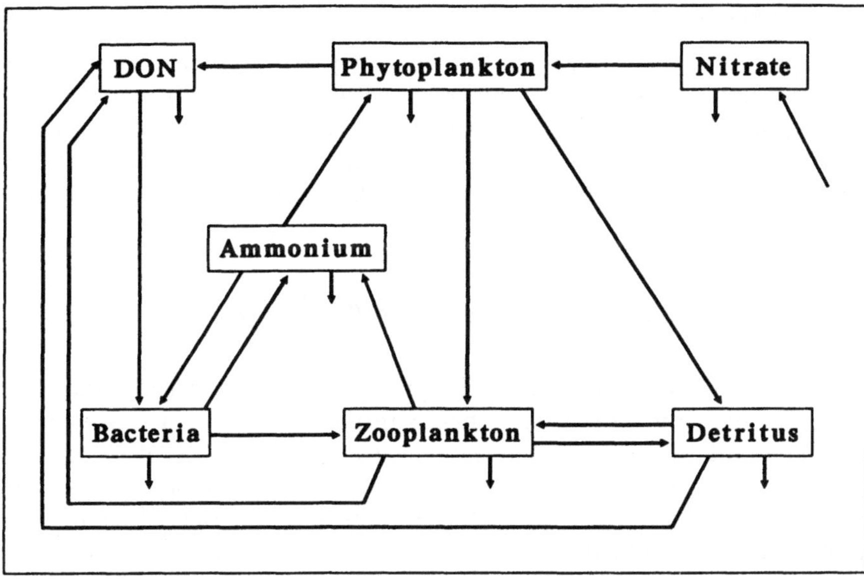

Figure 2: A schematic diagram showing the compartments and inter-compartmental flows of the Fasham *et al.* (1991) model of the mixed layer ecosystem. The short arrows just below each box represents inputs and outputs between the mixed layer and ocean layers below.

a model containing the minimum number of compartments that might tackle the problem in hand and only add further compartments as necessity dictates. Thus, the annual changes of carbon dioxide and oxygen concentrations in the mixed layer of the NE Atlantic have been reproduced by means of a model with only one phytoplankton population and one nutrient (Fig. 1, Taylor *et al.*, 1991, 1992); Wroblewski *et al.* (1988) have used a model with one phytoplankton, one zooplankton and one nutrient to simulate features of the distribution of phytoplankton across the North Atlantic basin; and the slightly more complex system of Fasham *et al.* (1990), shown in Fig. 2, is being used to simulate the carbon and nitrogen budgets of the North Atlantic (Sarmiento *et al.*, 1989). When combining groups of organisms into model compartments it is best not to combine organisms with very different turnover times (turnover time is defined as the organism pool size divided by the flux of biomass through that pool, Fasham, in press). Turnover time is closely coupled to growth rate, which in turn is often found to be dependent on organism size and so size-related compartments have been a feature of several models (Steele and Frost, 1977; Moloney *et al.*, 1991).

3.2 Model Dynamics

Although there are no explicit universal equations that can be used to determine how material is transferred between compartments a general set of equations governing population growth, which can accommodate most of the important processes, have been formalised by Wiegert (1979, see also Fasham, in press). The rate of change of any species or abiotic source X_j is given by:

$$\frac{dX_j}{dt} = \sum_{i=1}^{m}(e_{ij}\tau_j p_{ij} f_{ij} X_j) - (\mu_j + \phi_j + \rho_j)X_j - \sum_{k=1}^{m}(\tau_k p_{jk} f_{jk} X_k). \quad (4)$$

These expressions summarise the processes that have to be considered in formulating any model. The first summation in eq. 4 represents the assimilated ingestion or uptake by species j from all the other modelled species or abiotic sources. The middle term represents losses due to physiological causes, death or external factors (e.g. grazing) that are not explicitly included within the model. The last summation represents the predation on species j by other species. The coefficients in the equations are defined as follows:

e_{ij} is the assimilation efficiency of species j using resource i,
τ_j is the maximum specific ingestion / uptake rate of species j ,
p_{ij} is the preference of species j for resource i ,
f_{ij} is the limitation of ingestion / uptake of species i by resource j ,
μ_j is the specific loss rate due to natural or externally imposed mortality,
ϕ_j is the specific loss rate due to excretion,
ρ_j is the specific loss rate due to respiration.

The loss terms in Wiegert's equations are linear in biomass. The mortality of the highest predator in a model can be thought of as a closure term, and Steele and Henderson (1992) have recently proposed that this closure term might take on other forms, such as a quadratic or hyperbolic. The form of this term may be significant in the system dynamics.

Although the formal equations (4) provide a general summary of the processes to be considered, the coefficients in eq. (4) may depend on a variety of physiological and behavioural interactions making them non-linear functions of the species or abiotic sources. The equations therefore are not as well-defined as the physical equations of motion. It is common to simplify most of these coefficients to either constant values, functions of time or space, or functions of the physical forcing. The choice of simplifications varies between models. One group of coefficients in which there is more agreement about functional dependence is those relating to limiting factors.

3.3 Limiting Factors (f_{ij})

Light is one of the most important factors which limits the growth of phytoplankton. Most models calculate a light-dependent photosynthesis rate by using one of the many empirical expressions for the photosynthesis-light (P-I) curve. This curve is frequently parameterised by its initial slope, α , and the maximum photosynthetic rate P_{max} , together with another parameter if photoinhibition is taken into account (Platt *et al.*, 1980). While there are a number of equations that may be used (Platt *et al.*, 1977; Ledermann and Tett, 1981) when modelling annual cycles the choice of α and P_{max} is probably more critical than small differences in the shape of the P-I curve (Fasham, in press). Phytoplankton are capable of partially adapting to varying light levels implying adjustments to α and P_{max} in response to changes in ambient light (Geider and Platt, 1986); this may need to be considered for cells below the mixed layer.

Having calculated the light-dependent growth most models proceed by multiplying this function by a second function representing nutrient limitation. However, if it is considered that photosynthesis and nutrient uptake are separate processes that provide the substrates carbon and nitrogen for growth, an alternative approach is to use Liebig's law of the minimum by setting the extent of growth limitation at each depth to whichever of the light and nutrient limitations is the more severe (Parsons and Kessler, 1987). It is, as yet, unclear which of these approaches is more accurate, perhaps because models should treat the physiological processes more realistically (e.g. Fasham and Platt, 1985; Lancelot and Billen, 1986; Sakshaug *et al.*, 1989).

Liebig's law is also used when there is more than one nutrient that may be limiting (e.g. silicate, nitrate or phosphate), the appropriate limiting function is calculated for each nutrient and the growth rate is multiplied by the lowest value obtained. The limitation function most commonly employed is the Michaelis-Menten relationship:

$$\frac{n}{(n+n_h)}, \tag{5}$$

in which n is the concentration of the limiting nutrient and n_h is the half-saturation constant. A complication arises if the nutrient can occur in different forms for which there are different preferences. Perhaps the most important case is that of nitrogen which is one of the most common limiting nutrients in the ocean. Nitrogen may be present as nitrate and ammonium (and also as organic nitrogen), and it is energetically more favourable for an algal cell to take up ammonium than nitrate. Further, it appears that nitrate uptake in many algae may be inhibited in the presence of fairly low concentrations of ammonium (Dortch, 1990, who also points out that the interaction between ammonium and nitrate uptake may be too uncertain to model reliably). Accurate representation of the limitation by low concentrations of nitrate and ammonia in models is needed because algal primary production divides into two forms (Dugdale and Goering, 1967; Eppley and Peterson, 1979): "new" production fuelled by nitrate-nitrogen that is brought into the mixed layer by physical mixing processes, and "regenerated" production, fuelled mainly by ammonium-nitrogen that is recycled within the mixed layer by zooplankton and bacteria. This partitioning can be measured directly by ^{15}N uptake techniques and so used to constrain models. At equilibrium new production will balance physical transfers of carbon out of the mixed layer. A second reason for accurate representation of the limitation process is the possibility that ammonium inhibition of nitrate uptake may be involved in the lack of utilisation of nitrate, and possible reduced carbon dioxide consumption, at high latitudes.

The simplest expression for nitrogen limitation is obtained by adding together the Michaelis-Menten ratios for nitrogen and ammonium using a suitable weighting to represent the ammonium inhibition (Wroblewski, 1977; Fasham *et al.*, 1990):

$$\frac{Ne^{-\Psi A}}{(N+N_h)} + \frac{A}{(A+A_h)}. \tag{6}$$

N and N_h are the nitrate concentration and half-saturation constant, respectively, and A and A_h are the corresponding values for ammonium. Ψ is an empirical constant (1.5 (m mol NH_4 m^{-3})$^{-1}$) representing inhibition of nitrate uptake by ammonium. In the absence of this inhibition ($\Psi = 0$), expression (6) has the unrealistic property that when nitrate is absent the growth will be no more than half its maximum whatever ammonium is present. Alternative methods for treating multiple resources have been discussed by Harris (1974), O'Neill *et al.* (1989) and Legovic (1989). A formulation which allows either excess nitrogen or excess ammonium to produce a maximum growth rate when there is no inhibition is (Legovic, 1989; Taylor and Joint, 1990):

$$\frac{\left(\frac{Ne^{-\Psi A}}{N_h} + \frac{A}{A_h}\right)}{D}, \tag{7}$$

in which D can be either:

$$D = 1 + \frac{N}{N_h} + \frac{A}{A_h}, \tag{8}$$

implying that ammonium inhibition reduces the growth rate below its maximum even in the presence of excess nitrate, or:

$$D = 1 + \frac{Ne^{-\Psi A}}{N_h} + \frac{A}{A_h}. \tag{9}$$

for which ammonium inhibition is less pronounced when nitrate is present in abundance. Whichever formulation (6) or (7) is used it is likely that A_h should be considerably less than N_h (Dortch, 1990; Taylor and Joint, 1990); setting these equal as in Fasham *et al.* (1990) can lead to higher ammonium concentrations and an overestimate of ammonium inhibition. An alternative treatment of limiting nutrients allows the cells to consume and store excess nutrients when they are available. This is observed experimentally and has been incorporated into models (Tett and Edwards, 1984; Tett, 1986).

Food limitation of grazers is also commonly described by the Michaelis-Menten formula (5) and if a grazer has more than one food source available its preferences for different food types must be considered. Different ways of taking account of feeding preferences have been discussed by Harris (1974), O'Neill *et al.* (1989), Legovic (1989) Fasham *et al.* (1990) and Fasham (in press). An additional aspect which has to be considered when modelling grazers and which is not contained explicitly in Wiegert's equations is swimming behaviour. Many species of zooplankton migrate vertically, often diurnally (Longhurst and Williams, 1979), and it has been shown that migrating zooplankton may export primary production from the mixed layer by grazing at night at shallow depths and then excreting at deeper depths during the day (Longhurst *et al.*, 1990).

3.4 Parameter Values

The estimation of parameter values for a biological model presents many problems. A model will often require estimates for a large number of parameters (even a relatively simple model such as Fig. 2 may have 20 or so), and many of these will be either difficult to measure or inherently variable. Even after an extensive review of the literature there will almost certainly be some parameters for which good estimates cannot be obtained. A study of the model's sensitivity to individual parameter values can be useful if it shows that the values assigned to some of the parameters are not too critical. Further, if the model's results are excessively sensitive to any particular parameter it may indicate that the process concerned has been too crudely represented. The usefulness of a sensitivity study may be limited because the sensitivity of any parameter can depend on the values assigned to other parameters.

Often many features of the model's behaviour are common to quite a wide range of parameter values, and so the model can be useful in spite of the uncertainties in the parameters. When a model having many poorly known parameters is compared with observations there is a temptation to assume that it might be tuned to fit almost any data. Experience with attempting to match models to data shows that this is not necessarily true, however, for most parameters are subject to constraints: e.g. they have limited observational ranges, they cannot be negative or they are fractions and so must lie between 0 and 1. Therefore it is always of value to compare model output with observations as thoroughly as possible in order to test whether the model behaves as nature appears to.

An important problem associated with the assignment of parameter values is that of survival through the winter. This may be critical in models used for climatic prediction as population levels during the summer can be very dependent on overwintering processes (Colebrook, 1985). Showing that the predicted annual cycle is stable when the model simulates several successive years demonstrates that the model is mathematically internally consistent. However, ensuring that populations survive the winter in such calculations, especially at high latitudes, frequently necessitates either reducing the mortality during the winter or constraining the abundance by a lower threshold. Either of these alternatives is merely a means of parameterising some unknown overwintering strategy (e.g. resting states or vertical migration), for it may well be that processes occurring through the winter are not described by the same equations as those during the rest of the year. Predicted population fluctuations from year to year will be sensitive to the form of the parameterisation in the winter.

4 A Case Study from the NE Atlantic

4.1 The Model

During the U.K. Biogeochemical Ocean Flux Study, as part of the JGOFS North Atlantic Bloom Experiment in 1989, surveys were carried out from 47 to 60°N along 20°W between May and August. These observations reveal the changes of phytoplankton and nutrients through that spring and summer and show how the timing of these changes varied along the section. The latitudinal variation of these seasonal changes has been replicated in a simple model which is driven by physical forcing (Taylor *et al.*, in press). The model is one-dimensional with a single layer, the surface mixed layer, whose dynamics and light regime are represented as in Taylor *et al.* (1991). Within this layer is a time-dependent ecosystem (Fig. 3) composed of four classes of phytoplankton: diatoms (P_d), dinoflagellates (P_n), phytoflagellates (P_f), picophytoplankton (P_c); three classes of heterotrophs: free bacteria (B), heteroflagellates (H_f), microzooplankton ($> 15\mu$ m, C_T); four nutrients: silicate (S), nitrate (N), ammonium (A), metabolisable dissolved organic carbon (DOC, D); and detritus (D_T). The equations describing these variables are an extension of those of Taylor and Joint (1990):

$$\frac{dP_d}{dt} = (a_d\phi - m_d - v_d - k)P_d, \tag{10}$$

$$\frac{dP_n}{dt} = (a_{nNA}\phi_{NAn} - m_n - v_n - k)P_n, \tag{11}$$

$$\frac{dP_f}{dt} = (a_{fNA}\phi_{NAf} - m_f - a_{fH}H_f - v_f - k)P_f, \tag{12}$$

$$\frac{dP_c}{dt} = (a_{cNA}\phi_{NAc} - m_c - a_{cC}C_T - v_c - k)P_c, \tag{13}$$

$$\frac{dB}{dt} = (a_{BD}\phi_D - m_B - a_{BH}H_f - k)B, \tag{14}$$

$$\frac{dH_f}{dt} = (a_{Hc}P_c + a_{HB}B + a_{HT}D_T - a_{HH}H_f - v_H - k)H_f, \tag{15}$$

$$\frac{dC_T}{dt} = (a_{Cf}P_f + a_{CH}H_f + a_{CT}D_T - a_{CC}C_T - v_C - k)C_T, \tag{16}$$

$$\begin{aligned} \frac{dD}{dt} &= a_{Dd}\phi P_d + a_{Dn}\phi_{NAn}P_n + a_{Df}\phi_{NAf}P_f + a_{Dc}\phi_{NAc}P_c \\ &- a_{DB}\phi_D B + m_{Dd}P_d + m_{Dn}P_n + m_{Df}P_f \\ &+ m_{Dc}P_c + m_{DH}H_f + m_{DC}C_T + m_{DT}D_T \\ &+ (a_{DHc}P_c + a_{DHB}B + a_{DHT}D_T + a_{DHH}H_f)H_f \\ &+ (a_{DCf}P_f + a_{DCH}H_f + a_{DCT}D_T + a_{DCC}C_T)C_T \\ &+ k(D_o - D), \end{aligned} \tag{17}$$

$$\frac{dS}{dt} = -a_{Sd}\phi P_d + k(S_o - S), \tag{18}$$

$$\begin{aligned} \frac{dN}{dt} &= -a_{Nd}\phi\phi_{Nr}P_d - a_{NAn}\phi_{Nn}P_n - a_{NAf}\phi_{Nf}P_f - a_{NAc}\phi_{Nc}P_c \\ &- a_{NDB}\phi_D B + k(N_o - N), \end{aligned} \tag{19}$$

$$\begin{aligned} \frac{dA}{dt} &= -a_{Ad}\phi\phi_{Ar}P_d - a_{NAn}\phi_{An}P_n - a_{NAf}\phi_{Af}P_f - a_{NAc}\phi_{Ac}P_c \\ &- a_{ADB}\phi_D B + m_{Ad}P_d + m_{An}P_n + m_{Af}P_f + m_{Ac}P_c \\ &+ m_{AH}H_f + m_{AC}C_T + m_{AT}D_T + k(A_o - A), \end{aligned} \tag{20}$$

$$\begin{aligned} \frac{dD_T}{dt} &= m_{Td}P_d + m_{Tn}P_n + m_{Tf}P_f + m_{Tc}P_c \\ &+ m_{TH}H_f + m_{TC}C_T + m_{TT}D_T \\ &+ \{a_{THc}P_c + a_{THB}B + (a_{TH} - a_{THT})D_T + a_{THH}H_f\}H_f \end{aligned}$$

$$+ \{a_{TCf}P_f + a_{TCH}H_f + (a_{TH} - a_{TCT})D_T + a_{TCC}C_T\}C_T$$
$$+ v_T D_T + k(D_{To} - D_T). \quad (21)$$

Each coefficient a_x describes a rate of transfer between state variables. Thus, a_{DHB} is the constant representing the rate of DOC (D) production resulting from heteroflagellates (H) grazing bacteria (B). The subscripts associated with the other state variables are diatoms (d), dinoflagellates (n), phytoflagellates (f), picophytoplankton (c). In the same way each coefficient m_x is a transfer or loss associated with the externally imposed mortality which represents natural mortality and grazing by mesograzers. Each parameter v_x is the loss rate from the mixed layer due to sinking, and is therefore the appropriate sinking speed divided by the depth of the mixed layer. The constant k is the rate of turbulent exchange through the bottom of the surface layer and has dimensions of $[\text{TIME}]^{-1}$. The equations assume that the living components are all absent below the mixed layer while silicate, nitrate, ammonia, DOC and detritus have the constant concentrations S_o, N_o, A_o, D_o and D_{To}, respectively.

The main processes encompassed within these equations are summarized below; some of the parameter values appear in Tables 1 and 2.

The phytoplankton equations (10-13). These consider phytoplankton growth, and losses due to grazing, sinking and vertical mixing. Light limitation in the lower part of the mixed layer is incorporated into a_{dNA}, a_{nNA}, a_{fNA} and a_{cNA} using a two waveband approximation as in Taylor *et al.* (1991). Each of the phytoplankton can utilise both nitrate and ammonia with nitrogen limitation expressed in each case (x) by:

$$\phi_{xNA} = \phi_{Nx} + \phi_{Ax},$$

where $\phi_{Nx} = (N/N_{hx})/(1 + N/N_{hx} + A/A_{hx})$ and $\phi_{Ax} = (A/A_{hx})/(1 + N/N_{hx} + A/A_{hx})$. N_{hx} and A_{hx} are the appropriate half-saturation constants for nitrate and ammonia. In the calculations reported here, each of these is the same for all phytoplankton groups. Diatoms may also be limited by silicate according to the relation: $\phi_S = S/(S + S_h)$. The nutrient limitation is then given by ϕ which is the smaller of ϕ_S and ϕ_{nNA}. Nitrogen uptake by diatoms is then calculated by $\phi_{Nr} = \phi_{Nn}/\phi_{nNA}$ and $\phi_{Ar} = \phi_{An}/\phi_{nNA}$. When silicate becomes depleted the diatom population declines abruptly, perhaps partly due to the formation of resting spores and increased sinking rates. This has been represented in the model by increasing the sinking velocity of diatoms by a factor when the silicate concentration falls below twice S_h (a factor of 8 was used but could have been larger).

The bacterial equation (14). This describes bacterial growth, and losses due to grazing, sinking and turbulent mixing out of the surface

TABLE 1 Parameters used in the model

Maximum growth rate of diatoms	d^{-1}	0.93	
Maximum growth rate of dinoflagellates	d^{-1}	0.33	
Maximum growth rate of phytoflagellates	d^{-1}	1.5	
Maximum growth rate of picophytoplankton	d^{-1}	1.5	
Grazing rate of heteroflagellates	$(mg\ C)^{-1}\ d^{-1}$	0.2	
Grazing rate of microzooplankton	$(mg\ C)^{-1}\ d^{-1}$	0.06	
Silicate half saturation constant	$mmol\ m^{-3}$	0.3	
Nitrate half saturation constant	$mmol\ m^{-3}$	0.3	
Ammonia half saturation constant	$mmol\ m^{-3}$	0.005	
Sinking speed of diatoms	$m\ d^{-1}$	1.5	
Sinking speed of dinoflagellates	$m\ d^{-1}$	0	
Sinking speed of phytoflagellates	$m\ d^{-1}$	0	
Sinking speed of picophytoplankton	$m\ d^{-1}$	0	
	Units	**47°N**	**60°N**
Maximum temperature	°C	16.5	9
Minimum temperature	°C	9.5	6.5
Maximum mixed layer depth	m	350	500
Minimum mixed layer depth	m	28	45
Vertical mixing coefficient	$m\ d^{-1}$	0.1	0.2
Extinction coefficient (green)	m^{-1}	0.053	0.062
Deep nitrate concentration	$mmol\ m^{-3}$	7	10
Deep silicate concentration	$mmol\ m^{-3}$	4	6

Parameter values for intermediate latitudes are obtained from these by linear interpolation. All other parameters of the model are those of Taylor and Joint (1990), where sources are provided. Parameters involving diatoms or dinoflagellates correspond to those for large phytoplankton and those involving phytoflagellates, or microzooplankton correspond to those for picophytoplankton or heteroflagellates. The sinking speed of diatoms increases to 12 m/d when the silicate is depleted.

TABLE 2 Externally imposed mesozooplankton grazing rates

At 47°N

	Jan	Feb	Mar	Apr	May	Jun	Jul	Aug	Sep	Oct	Nov	Dec
Grazing requirements (mg C m^{-3} d^{-1})												
	.01	.01	.01	.06	.30	.36	.12	.12	.12	.02	.01	.01
Percentage of grazing from diatoms, dinoflagellates, phytoflagellates, picophytoplankton, bacteria, heteroflagellates and microzooplankton												
Dia	14	14	14	14	17	17	17	17	17	14	14	14
Din	14	14	14	14	17	17	17	17	17	14	14	14
Pfl	14	14	14	14	20	20	20	20	20	14	14	14
Pic	14	14	14	14	5	5	5	5	5	14	14	14
Bac	14	14	14	14	5	5	5	5	5	14	14	14
Hfl	14	14	14	14	17	17	17	17	17	14	14	14
Mzo	14	14	14	14	17	17	17	17	17	14	14	14

At 60°N

	Jan	Feb	Mar	Apr	May	Jun	Jul	Aug	Sep	Oct	Nov	Dec
Grazing requirements (mg C m^{-3} d^{-1})												
	.01	.01	.01	.06	.60	.72	.24	.24	.24	.02	.01	.01
Percentage of grazing from diatoms, dinoflagellates, phytoflagellates, picophytoplankton, bacteria, heteroflagellates and microzooplankton												
Dia	14	14	14	14	20	20	20	20	20	14	14	14
Din	14	14	14	14	20	20	20	20	20	14	14	14
Pfl	14	14	14	14	10	10	10	10	10	14	14	14
Pic	14	14	14	14	5	5	5	5	5	14	14	14
Bac	14	14	14	14	5	5	5	5	5	14	14	14
Hfl	14	14	14	14	20	20	20	20	20	14	14	14
Mzo	14	14	14	14	20	20	20	20	20	14	14	14

layer. The bacteria are assumed to be always limited by the availability of DOC, this limitation being represented by: $\phi_D = D/(D + D_h)$, where D_h is a half-saturation constant.

The heteroflagellate and microzooplankton equations (15 and 16). The growth rates of heteroflagellates and microzooplankton are each assumed to be proportional to the concentrations of the 3 food sources: picophytoplankton, bacteria and detritus in the case of heteroflagellates, and phytoflagellates, heteroflagellates and detritus in the case of microzooplankton. Losses occur due to grazing, sinking, turbulent mixing and cannibalism. Grazing of the groups on their own kind is formulated in the same way as grazing on other components. In each self-grazing term a_{xx} is the difference between the growth rate arising from grazing of x and the rate of consumption of x.

The DOC equation (17). The first 4 terms describe the leakage of DOC by bacterial growth. The remaining terms treat production of DOC during grazing ('sloppy feeding'), DOC production by breakdown of detritus, and the turbulent transport of DOC.

The silicate equation (18). Silicate uptake by the diatoms and the physical transport of silicate from depth are considered.

The nitrate equation (19). Nitrate uptake by the 4 groups of phytoplankton and by bacteria, and the physical transport of nitrate from depth are considered.

The ammonia equation (20). Uptake of ammonia by the 4 groups of phytoplankton and by the bacteria, and mixing through the bottom of the mixed layer are treated as in the nitrate equation. The remaining terms treat the excretion of ammonia associated with respiration.

The detritus equation (21). Detritus (faecal pellets) is produced in the course of all grazing (i.e. grazing by heteroflagellates, microzooplankton and external mesograzers). It is lost by means of sinking, mixing, grazing by heteroflagellates, microzooplankton and mesograzers, and breakdown by attached bacteria. The constant m_{TT} encapsulates the effects of mesograzers and of attached bacteria.

The calculation of many of the transfer coefficients in eqs. (10) - (21) is described in Taylor and Joint (1990); interactions involving the additional components of the model were formulated in the same way. Physiological differences between phytoplankton populations were minimised, the phytoplankton groups differing only in their growth and sinking rates, their trophic relationships, and in the requirement of diatoms for silicate. Respiration rates were temperature dependent with a maximum value of 0.1 day^{-1}. Mesozooplankton grazing was represented as a seasonally varying mortality, estimated from observations during 1989 (Table 2, Morales *et al.*, 1991), of between 0.01 and 0.06 day^{-1}.

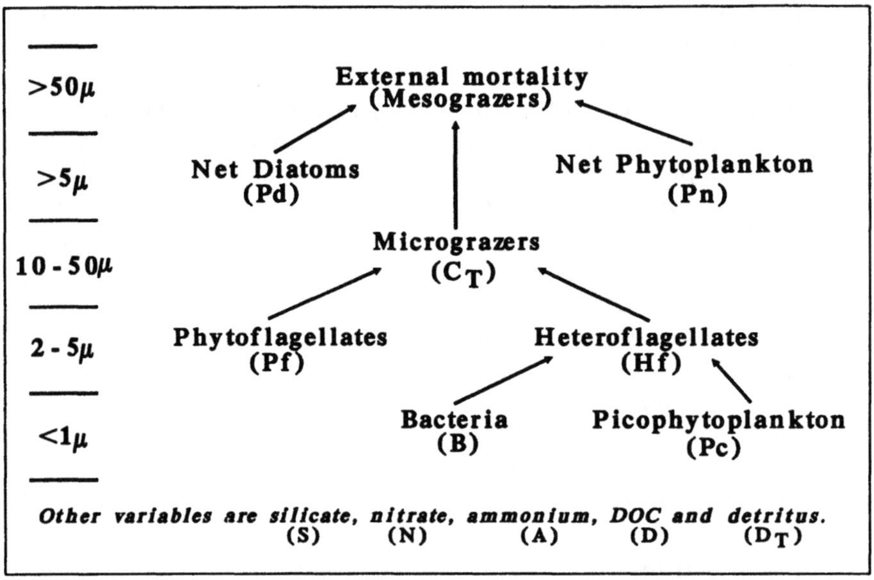

Figure 3: The ecosystem structure used in the NE Atlantic model of Taylor *et al.* (in press). The arrows indicate transfer of carbon and nutrients by grazing. Approximate sizes of the components of the ecosystem are indicated.

Obtaining a complete set of parameter values for the model is not straightforward. Transfers between all the compartments in Fig. 3 require about 60 to 70 parameters, depending on how many are set equal (even the relatively simple system in Fig. 2 needs 20 to 30 parameters). Sources and values for most of the physiological parameters have been given by Taylor and Joint (1990), and these are the values employed here (see Appendix) together with the additional data in Tables 1 and 2. The transfer coefficients in eqs. 10 to 21 were calculated from these parameters as in Taylor and Joint (1990).

Model calculations were carried out initially at each of the four latitudes studied during the BOFS field program along 20°W, namely: 47, 52, 56 and 60°N; each of these calculations beginning on 1 January with 0.002 mg C m^{-3} of diatoms, 1 mg C m^{-3} of each of the other phytoplankton, 1.6 mg C m^{-3} of each of the heterotrophs, and with nutrient concentrations at the values imposed below the mixed layer. Actual population sizes during the winter are not known; these three values gave abundances which agreed with those observed early in the season. These winter values are applied as lower thresholds throughout the year and, as the abundances fall to these levels in the winter, the annual cycle repeats in subsequent years. Only the physical environment was varied between latitudes; physiological and trophic parameters were assumed to be unchanged. The model needs an unbroken series of irradiance and mixed layer depth data from January to August at each latitude simulated, and historical data was used to provide this coverage. Irradiance at the sea surface was estimated using cloud cover spatially interpolated from long-term mean data at Ocean Weather Stations I, J and K (Bunker, 1976). Estimating mixed layer depths is complicated by mesoscale variability and by the difficulty in defining the depth of mixing during the important early part of the season. The model uses the simple procedure (Taylor *et al.*, 1991) of assuming the mixed layer depths are proportional to those at 35°N given by Levitus (1982), the proportionality being chosen to ensure that the minimum and maximum depths are those appropriate for the latitude concerned (Table 1). The development of the seasonal thermocline at 47 and 60°N is illustrated in Fig. 5C which also indicates the mixed layer depths used in the model. Horizontal transports were not included in the calculations; earlier studies (Taylor *et al.*, 1991, 1992) have indicated that the basic geographical features of the seasonal cycle can be reproduced even when advection is neglected.

4.2 Results

For comparison with the observations each simulation was sampled at the same times and places as the field data (Fig. 4, 5A and 6B,C). The BOFS program in 1989 consisted of three legs: a northward survey from 47 to 60°N between 11 May and 5 June, a southward survey from 60 to 47°N between 14 June and 8 July and a northward survey from 47 to 60°N between 18 July and 5 August. The times spent at the end latitudes of the legs are indicated on Fig. 4. Figure 6B shows southward and northward surveys carried out during the U.S. JGOFS program.

The model results were compared with: microscope counts of four groups of organisms (Fig. 4A), chlorophyll-a measurements and primary production rates from size-fractionated samples (Fig. 4B and 4C), and silicate and nitrate concentrations (Fig. 6B,C, Slagle and Heimerdinger, 1991). The comparisons show how the measurements during the three cruises, spread across a range of latitudes and times, sampled different phases of a seasonal succession (Fig. 5B) whose latitudinal variation is driven by the physical and chemical environment. When reduced vertical mixing and increased irradiance allowed phytoplankton populations to grow at the end of the winter the first groups to do so were those with the fastest intrinsic growth rates, the phytoflagellates and the picophytoplankton. The increases of these groups (about day 147 at 47°N, Figs 4B and C) were soon halted by the rapid growth of herbivorous microzooplankton and zooflagellates whereupon they were succeeded by diatoms, the next fastest growers (about day 138 at 52 and 56°N, Fig. 4A, and day 167 at 60°N, Figs. 4B and 4C). Diatoms depleted the silicate and consumed some of the nitrate (Fig. 6B). With silicate depletion the diatom bloom ended (about day 128 at 47°N, the decaying bloom continuing to utilise silicate and nitrate until about day 138). The diatom bloom was briefly replaced by phytoflagellates and picophytoplankton before these in turn were replaced by the slower growing dinoflagellates (all four latitudes after day 175, Fig 4A). Between day 138 and day 152 the silicate concentration at 47°N was slightly replenished by physical mixing while nitrate consumption continued.

The plankton samples shown in Fig. 4 were mixed layer averages collected at only four latitudes. Figure 5A illustrates the spatial variability of the region from a series of surface samples passing through the spring bloom during May, collected with a spacing of 1/4° of latitude. The model reproduces the basic pattern in this series namely a diatom bloom at its peak near 54°N which is increasingly replaced by phytoflagellates and photosynthetic dinoflagellates further south. It also indicates high flagellate numbers at the northern end of the series which the observations

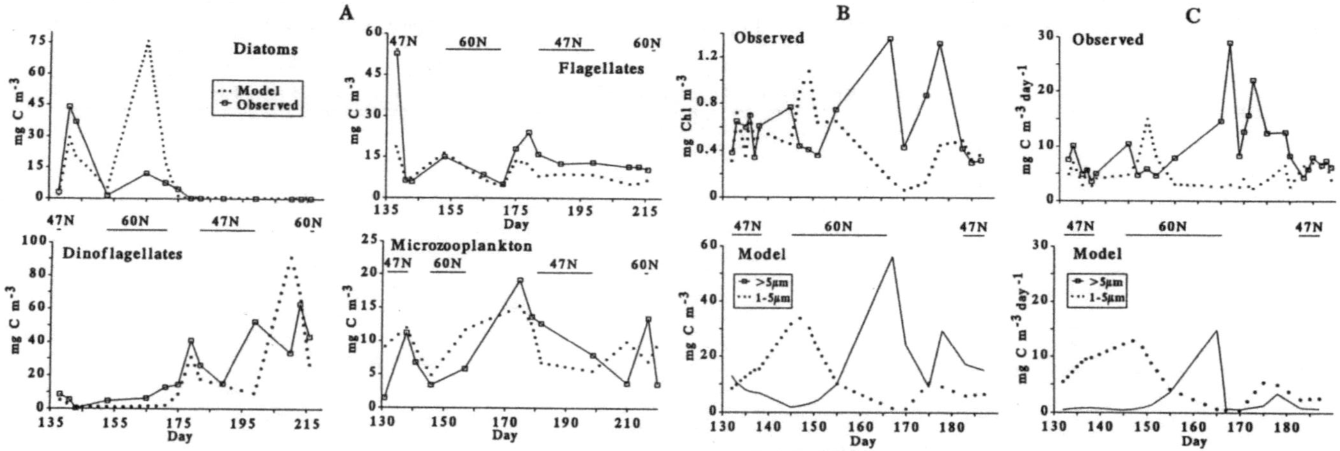

Figure 4: **A** Observed (solid) and predicted (dotted) abundances of diatoms, dinoflagellates, flagellates and microzooplankton during JGOFS cruises in 1989. **B** Observed (upper) and predicted (lower) phytoplankton abundance, the fraction $> 5\mu m$ is shown by a solid line and the 1-5μm fraction by a dotted line. The upper and lower scales used for the graphs have been chosen on the basis of a carbon-chlorophyll ratio of 40. **C** Observed and predicted phytoplankton production rates shown in the same way as in B. Production rates of the fraction $< 1\mu$ m are predicted to be 2-4 mg C m^{-3} d^{-1} which is the range of the observations. In each part of the figure the days spent at the end stations, 47 and 60°N, are indicated to illustrate the spatial coverage. All the observations could be in error by up to ± 25%.

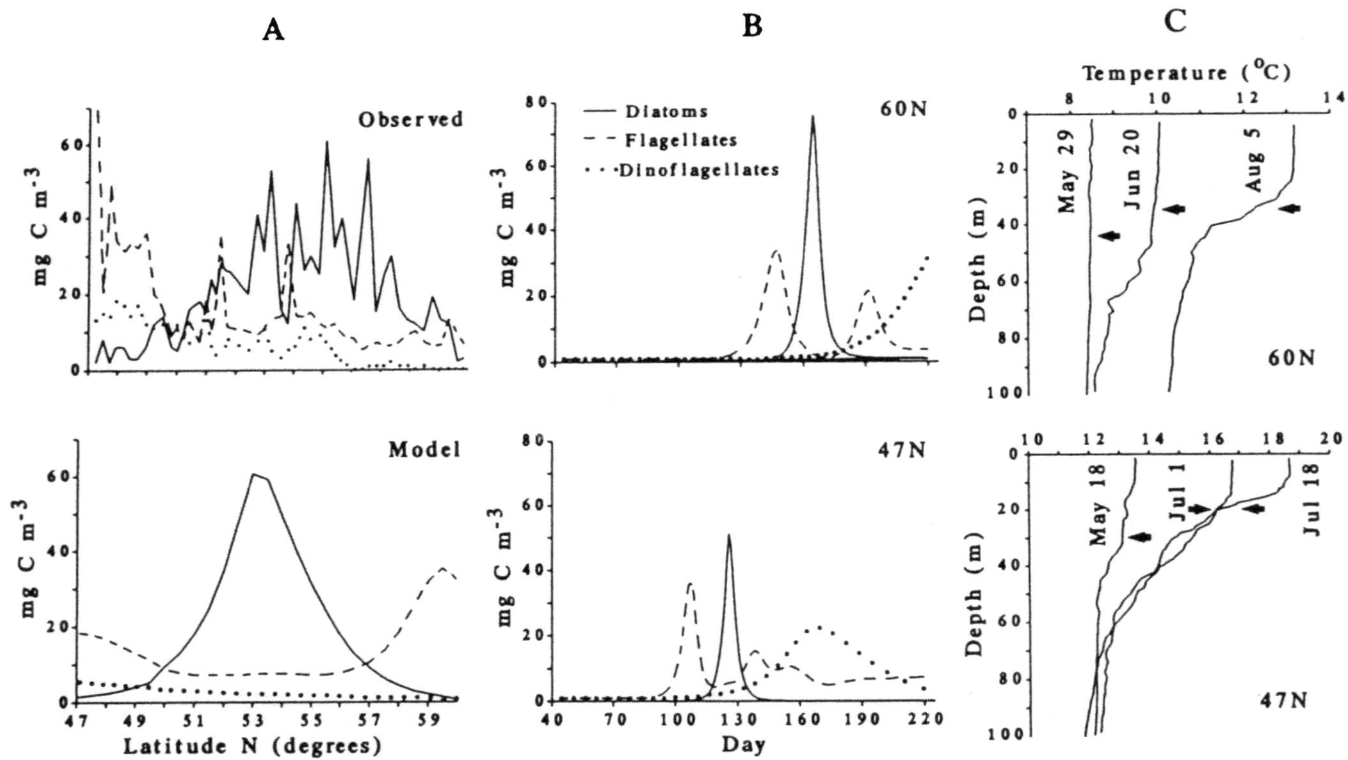

Figure 5: **A** Observed (upper) and predicted (lower) abundances of phytoplankton groups between 47 and 60°N during 19 to 24 May, 1989, and **B** the predicted abundances of the same groups during the spring and summer at 47 and 60°N. Diatoms are indicated by a solid line, dinoflagellates by a dotted line and flagellates by a broken line (only phytoflagellates are included in B). **C** Temperature profiles at 47 and 60°N observed during the JGOFS cruises in 1989, the arrows indicate the corresponding mixed layer depths in the model.

do not show. This occurs because the phytoflagellate peak preceding the diatom bloom lasts too long at 60°N in the simulation, a result that can be seen in Figs. 4B and C at about day 145.

Figures 4, 5A and 6B,C demonstrate that the model proposed here replicates the way in which the different groups were observed to appear and disappear during 1989, the predicted values being close to those observed, and also reproduces the observed nutrient concentrations. The seasonal changes at individual latitudes suggest the nature of nutrient utilisation in the region (Fig. 6A). While nitrate and silicate are both reduced to low values at 47°N, only silicate is depleted at 60°N. This result is robust to changes in the values in Table 1 which alter the timings and amplitudes of the populations but do not disturb the seasonal succession in Fig. 5B; it is robust to errors in estimating mixed layer depths (Fig. 5C). Takahashi *et al.* (1985) have observed silicate depletion without nitrate consumption at 64°N in the North Atlantic during 1983 and 1984. In our calculations, nitrate is not depleted at northern latitudes due to the greater depth of the mixed layer, more intense vertical mixing and the shorter season, so that the phytoplankton have more nutrients to utilise and a shorter time in which to do it. Grazing has also contributed to limiting phytoplankton growth but the rates that have been quoted are lower at high latitudes and so cannot account for the latitudinal gradient in nitrate consumption.

The possibility that in parts of the oceans phytoplankton may be prevented from using all the nitrate and phosphate that is available because it is limited by a trace nutrient, iron, was raised by Martin *et al.* (1988, 1989) initially for the sub-arctic and central equatorial Pacific and the Southern Ocean but also for the northeast Atlantic (Martin *et al.*, 1990). Banse (1990, 1991) has criticised the interpretation of the iron limitation experiments (as have Dugdale and Wilkerson, 1990) suggesting that light limitation, due to the depth of mixing, and microzooplankton grazing are the most acceptable explanations for the low phytoplankton standing stocks. Price *et al.* (1991) have shown that in equatorial Pacific waters iron is in short supply and limits the rate of nitrate and nitrite utilisation, but does not severely limit the growth rate of the indigenous phytoplankton community. The model calculations suggest that there is no need to invoke limitation by a trace nutrient such as iron to reproduce the events in the NE Atlantic during 1989.

Figures 7 and 8 try to address the question of how important community changes are to climate modelling by using this model to predict the CO_2 changes in the BOFS data set of Fig. 1. The model was run at each quarter degree of latitude using the same initial conditions and parameter values as previously, and the output sampled at the times and places of

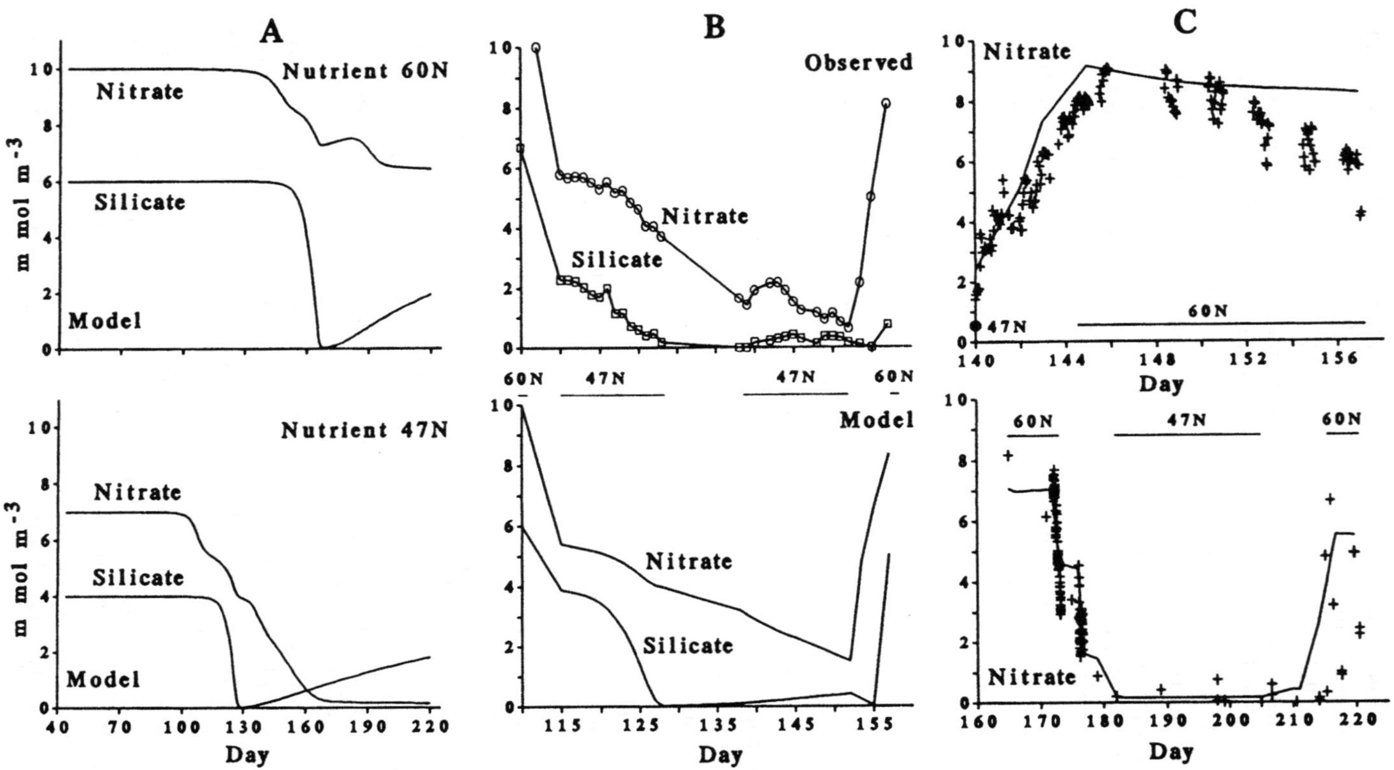

Figure 6: **A** Predicted concentrations of silicate and nitrate during the spring and summer at 47 and 60°N. **B** Observed (upper) and predicted (lower) surface concentrations of silicate and nitrate during U.S. JGOFS cruises in 1989 (Slagle and Heimerdinger, 1991). **C** Predicted (solid line) and observed (crosses) surface concentrations of nitrate during the BOFS cruises in 1989. The days spent at the end stations, 47 and 60°N, are indicated on B and C to illustrate the spatial coverage.

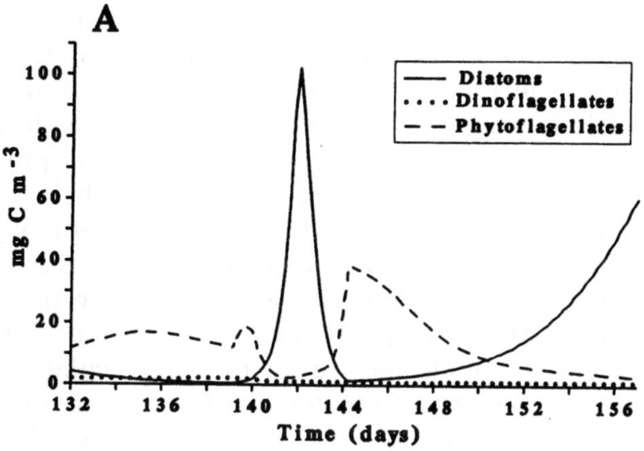

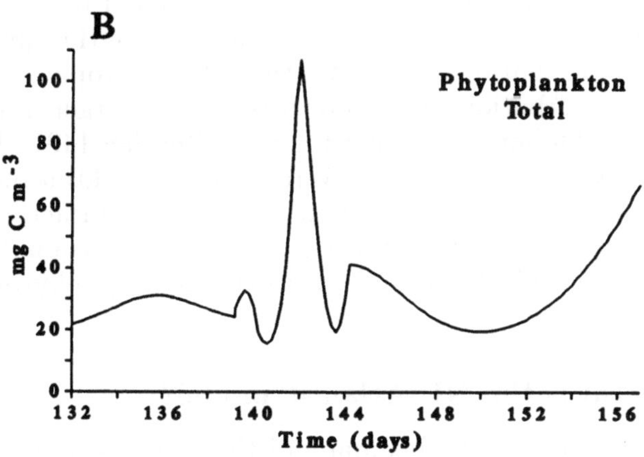

Figure 7: **A** Predicted abundances of diatoms, dinoflagellates and phytoflagellates along 20°W during the U.K. BOFS cruise 12 May to 5 June 1989. **B** Predicted total phytoplankton abundance along 20°W (c.f. Fig. 1).

Fig. 1, with carbon dioxide uptake and release calculated as in Taylor *et al.* (1991). The main variations in phytoplankton abundance along the section are shown in Fig. 7A, the total abundance of all the phytoplankton (Fig. 7B) showing a variation along the section that is similar to that of the measured chlorophyll concentrations. The predicted distribution of pCO_2 along the section is similar to that obtained from the model grouping all the phytoplankton together as chlorophyll (Fig. 1). The effects of such an aggregation are illustrated by Fig. 8B which shows the distribution of pCO_2 calculated using only a single phytoplankton group, represented by the total phytoplankton abundance (the sum of the four phytoplankton populations given by the calculation shown in Fig. 7B), which is assumed to be growing at the average growth rate of the different populations. This growth rate was obtained by averaging the instantaneous specific growth rates of the individual phytoplankton groups. The distribution of pCO_2 is similar in shape to, but 5 - 10 μ atm. lower than, that in Fig. 8A. When stratification begins one month earlier each method of calculating pCO_2 predicts lower values along the section (Fig. 8 A and B), the pCO_2 falling from 20 - 70 μ atm. below atmospheric to 40 - 80 μ atm. below. The spring bloom at 60°N now peaks earlier at day 149. Although the change in the distribution predicted is similar in the two calculations (Fig. 8C) the total phytoplankton calculation predicts a perturbation that is up to 10 μ atm. larger immediately following the spring bloom (about day 142 and after day 149). This error when calculating pCO_2 using only an average phytoplankton population, which would overestimate of CO_2 fluxes across the sea surface by at least 20%, is attributable to the spring diatom bloom being terminated by silica depletion more sharply than happens in the 'average phytoplankton' calculation.

4.3 How general is the NE Atlantic model?

The model just presented explains many of the features observed in the NE Atlantic during 1989 and provides some indication of the importance of seasonal succession to CO_2 variations in the region. But how representative of other times and other places are the events observed in 1989? Partial answers to these questions are provided by data from the Continuous Plankton Recorder Survey, a routine, monthly survey of the plankton of the North Atlantic ocean and the North Sea that has been carried out since 1948 using Continuous Plankton Recorders towed by merchant ships and Ocean Weather Ships on regular routes (Glover, 1967; Oceanographic Laboratory, Edinburgh, 1973). Samples are collected at the standard depth of 10m and, as far as possible, the organisms are iden-

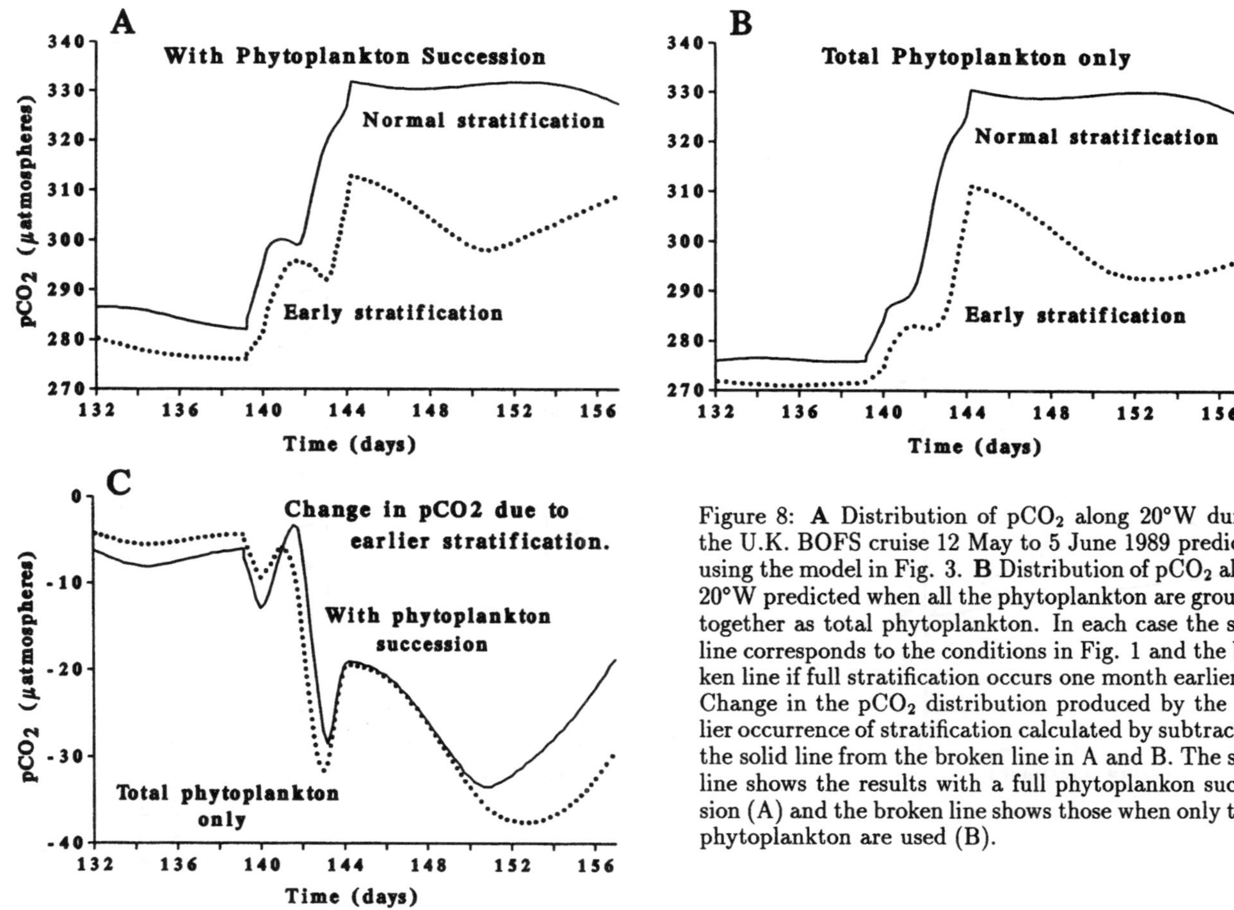

Figure 8: **A** Distribution of pCO_2 along 20°W during the U.K. BOFS cruise 12 May to 5 June 1989 predicted using the model in Fig. 3. **B** Distribution of pCO_2 along 20°W predicted when all the phytoplankton are grouped together as total phytoplankton. In each case the solid line corresponds to the conditions in Fig. 1 and the broken line if full stratification occurs one month earlier. **C** Change in the pCO_2 distribution produced by the earlier occurrence of stratification calculated by subtracting the solid line from the broken line in A and B. The solid line shows the results with a full phytoplankon succession (A) and the broken line shows those when only total phytoplankton are used (B).

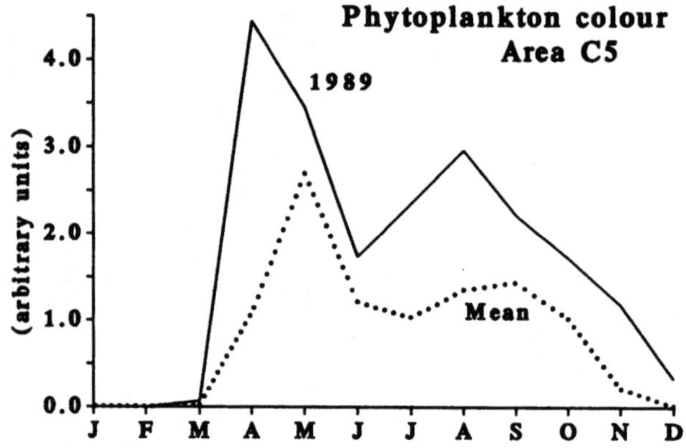

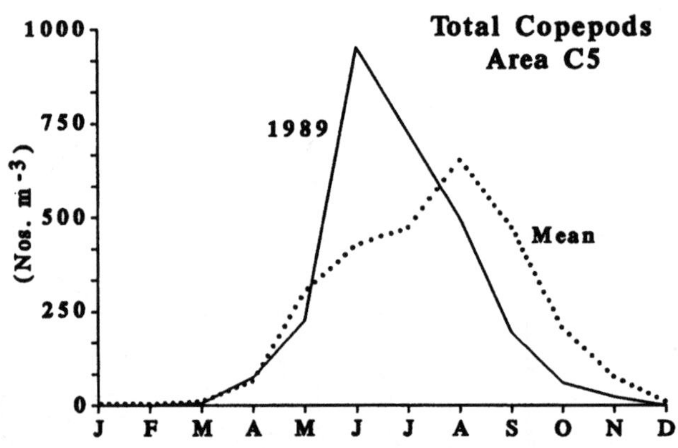

Figure 9: Seasonal changes of phytoplankton colour and the numbers of copepods observed by the Continuous Plankton Recorder Survey in two areas of the North Atlantic (Fig. 10). The solid line shows the values in 1989 and the broken line the average values for the period 1948 to 1991.

tified to species. In Fig. 9 data from the Survey collected during 1989 are compared with data averaged over the period 1948 to 1991 for an area close to 20°W (Fig. 10). The two variables shown, phytoplankton colour which is an estimate based on a visual assessment of the green coloration of the sample silks and the total number of copepods, each show higher than usual abundances during the time of the BOFS surveys, May to August 1989. Comparison with other years shows that the differences in Fig. 9 are not exceptional so that the events in 1989 may be typical of many years. However, they may not be representative of other areas in the North Atlantic.

The diatom bloom that occurred along 20°W during 1989 was composed of *Hyalochaete-Chaetoceros* species, a group which is a common over this region but almost absent over the western side of the ocean. Figure 10 shows that the *Chaetoceros* species are almost absent where the copepod *Calanus finmarchicus* is one of its main grazers and abundant where *Calanus helgolandicus* occurs instead. Although the two grazers are morphologically virtually identical, in the open waters of the North Atlantic they show quite distinct distributions. In spite of their morphological similarity it is known that they have very different life histories (Williams and Conway, 1980, 1984). *Calanus finmarchicus* overwinters by moving down into deep cold water and living, very slowly, on food reserves accumulated during the late autumn. *Calanus helgolandicus*, on the other hand, can only overwinter in warmer or shallower water where it can breed, however slowly, through the winter. The geographical distributions of the two species are determined to a large extent by these different overwintering strategies. Figure 11 compares the seasonal changes in an area of the NE Atlantic where the spring bloom occurs as observed in 1989 with those in an area of the NW Atlantic where *Calanus finmarchicus* is abundant. In this area the phytoplankton is much less abundant and there is no spring bloom so that the model just described would not be appropriate. Further, it seems likely that any model attempting to simulate the whole North Atlantic area may need to take some account of the behaviour of these two species of *Calanus* for a shift in their distribution patterns in response to a climatic change might significantly perturb the carbon budget of the region.

4.4 Conclusions from the case study

Incorporating simple ecosystem models into general circulation models (e.g. Sarmiento *et al.* 1989), or even using such models over entire ocean basins (Wroblewski *et al.* 1988), may indicate some of the processes responsible for geographical and seasonal differences, for simple models

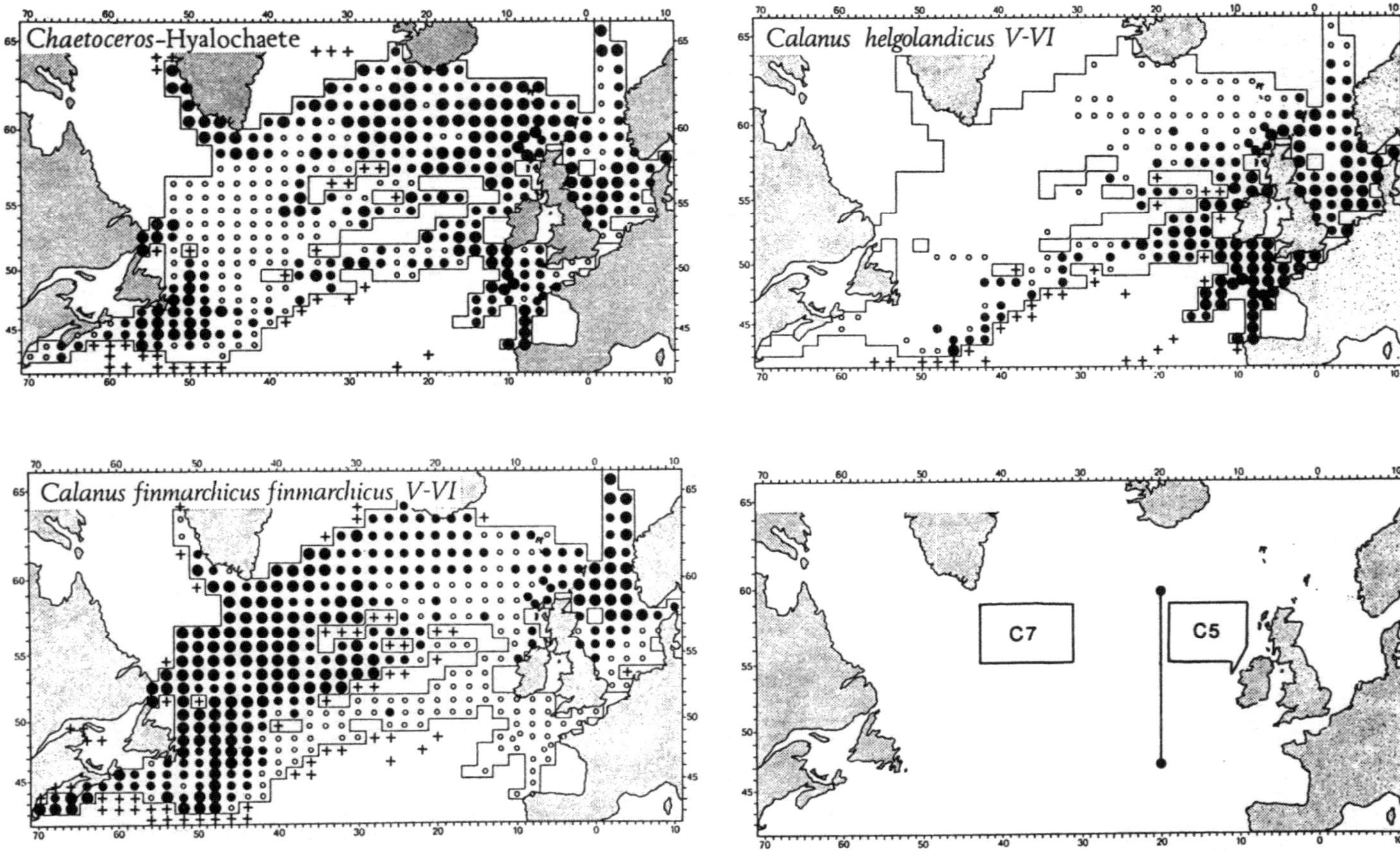

Figure 10: Distributions from the Continuous Plankton Recorder Survey (Oceanographic Laboratory, Edingurgh, 1973) of *Chaetoceros* (Hyalochaete) Ehrenberg (Gran), *Calanus finmarchicus finmarchicus* (Gunnerus) and *Calanus Helgolandicus* Claus. A chart showing the areas C5 and C7 and the BOFS transect from 47 to 60°N is included.

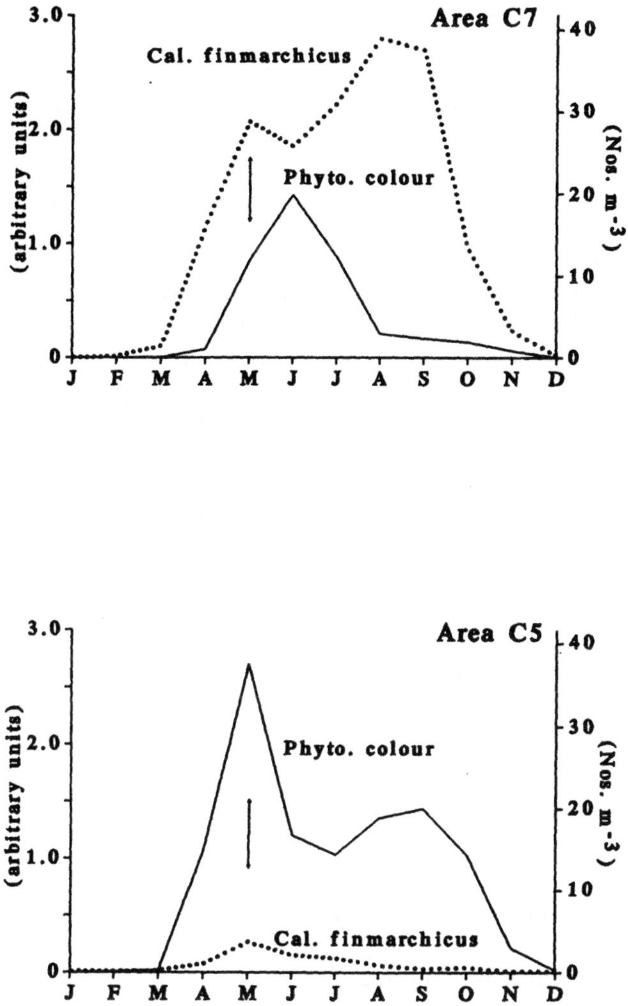

Figure 11: Seasonal changes of phytoplankton colour (solid line) and the abundance of *Calanus finmarchicus* (broken line) observed by the Continuous Plankton Recorder Survey in two areas of the North Atlantic (Fig. 10). To aid comparison, the time of the spring bloom in area C5 is indicated on each graph by a vertical bar.

can reproduce and explain many features that are observed (e.g. Fig. 1). However, community structure and behaviour much more complex than such simple systems consider (c.f. Figs. 3 and 5) may be important when investigating the impact of any climatic perturbation (Fig. 8). This complexity may also limit the geographical range over which any individual model may be applied (Figs. 10 and 11 suggest that the model in Fig, 3 may require modification even for use in the NW Atlantic).

ACKNOWLEDGEMENTS

This work forms part of the Biogeochemical Ocean Flux Study which is the U.K. contribution to the Joint Global Ocean Flux Study. The phytoplankton data in Figs. 4 and 5 were provided by Derek Harbour (Plymouth Marine Laboratory) and microzooplankton data by Peter Burkill (Plymouth Marine Laboratory) and Elaine Edwards (Southampton University). The chlorophyll and production measurements were provided by Ian Joint and Alan Pomroy (Plymouth Marine Laboratory). This is BOFS contribution no. 115.

APPENDIX: PARAMETER VALUES FOR NE ATLANTIC MODEL

Ammonium concentration below the mixed layer:	0
DOC concentration below the mixed layer:	0
Detritus concentration below the mixed layer: (The pool of refractory DOC is not considered)	0
Fraction of diatom, dinoflagellate, phytoflagellate and picophytoplankton carbon leaked as DOC:	0.1
C:N ratio of diatoms, dinoflagellates, phytoflagellates and picophytoplankton:(wt:wt)	5.6
C:N ratio of bacteria:	4.5
Fraction of all mortality of diatoms, dinoflagellates, phytoflagellates and picophytoplankton which becomes DOC:	0.1

Fraction of dead diatom, dinoflagellate, phytoflagellate and picophytoplankton nitrogen that becomes ammonium:	0.5
Fraction of all mortality of diatoms, dinoflagellates, phytoflagellates and picophytoplankton which becomes detritus:	0.25
Maximum specific growth rate of heterotrophic bacteria:	4.2 d^{-1}
DOC consumed per unit of bacterial carbon produced: (mg bacterial C)$^{-1}$ d^{-1}	2.5 mg C
Fraction of bacterial nitrogen obtained from inorganic nitrogen:	0.5
Fraction of bacterial inorganic nitrogen demand which is nitrate: (assuming that dissimilatory utilisation of nitrate does not occur in pelagic systems and that ammonium, rather than nitrate, is utilised by heterotrophic growth)	0
Fraction of DON consumed by bacteria that is excreted as ammonium:	0.4
Fraction of all bacterial mortality that becomes DOC:	0.1
Fraction of dead bacterial nitrogen becoming ammonium:	0.5
Fraction of all mortality of bacteria which becomes detritus:	0.25

Fraction of picophytoplankton, bacterial and detritus biomass grazed per unit time by one unit of zooflagellates:	0.2 d^{-1}
Fraction of phytoflagellate, zooflagellate and detritus biomass grazed per unit time by one unit of microzooplankton:	0.06 d^{-1}
Efficiency of conversion of picophytoplankton and bacterial carbon to zooflagellate carbon:	0.3
Efficiency of conversion of phytoflagellate and zooflagellate carbon to microzooplankton carbon:	0.3
Efficiency of conversion of detrital carbon to zooflagellate or microzooplankton carbon:	0.1
Fraction of detritus losses breaking down to become DOC:	0.1
Fraction of consumed detritus nitrogen that becomes ammonium:	0.5
Rate of detritus loss due to attached bacteria and macrograzers:	0.1 d^{-1}
Half saturation constant for DOC limited growth by heterotrophic bacteria:	12 mg C m^{-3}
Fraction of zooflagellates photosynthesising:	0

REFERENCES

AGOUMI A (1985) Modelisation de l'ecosysteme pelagique en Manche,Etude de l'influence des phenomenes physiques sur le systeme planctonique, *These de Doctorat d'Etat, 321pp., Univ. Pierre et Marie Curie, Paris.*

BANSE K (1990) Does iron really limit phytoplankton production in the offshore subarctic Pacific? *Limnol Oceanogr* **35,** 772-775

BANSE K (1991) Iron availability, nitrate uptake, and export new production in the subarctic Pacific. *J Geophys Res* **96,** 741-748

BUNKER AF (1976) Computation of surface energy flux and annual air-sea interaction cycles of the North Atlantic Ocean. *Mon Wea Rev* **104,** 1122-1140

CARR MR (1986) Modelling the attenuation of broad band light down a water column. *The Statistician* **35**, 325-333

CHARLSON RJ, LOVELOCK JE, ANDREAE MO, WARREN SG (1987) Oceanic phytoplankton, atmospheric sulfur, cloud albedo and climate. *Nature* **326**, 655-661

COLEBROOK JM (1985) Continuous Plankton Records: overwintering and annual fluctuations in the abundance of zooplankton. *Mar Biol* **84**, 261-265

DORTCH Q (1990) The interaction between ammonium and nitrate uptake in phytoplankton. *Mar Ecol Prog Ser* **61**, 183-201

DUGDALE RC, GOERING JJ (1967) Uptake of new and regenerated forms of nitrogen in primary production. *Limnol Oceanogr* **12**, 196-206

DUGDALE RC, WILKERSON FP (1990) Iron addition experiments in the Antarctic: a reanalysis. *Global Biogeochem Cycles* **4**, 13-19

EPPLEY RW, PETERSON BJ (1979) Particulate organic matter flux and planktonic new production in the deep ocean. *Nature* **182**, 677-680

EVANS GT (1988) A framework for discussing seasonal succession and coexistence of phytoplankton species. *Limnol Oceanogr* **33**, 1027-1036

EVANS GT, PARSLOW JS (1985) A model of annual plankton cycles. *Biol Oceanogr* **3**, 327-347

FASHAM MJR (IN PRESS) Modelling the marine biota. *In: Heimann M (ed) The Global Carbon Cycle, Springer-Verlag, Berlin.*

FASHAM MJR, PLATT T (1985) Photosynthetic response of phytoplankton to light: a physiological model. *Proc Roy Soc Lond B* **219**, 355-370

FASHAM MJR, HOLLIGAN PM, PUGH PR (1983) The spatial and temporal development of the spring bloom in the Celtic Sea, April 1979. *Prog Oceanogr* **12**, 87-145

FASHAM MJR, DUCKLOW HW, MCKELVIE SM (1990) A nitrogen-based model of plankton dynamics in the ocean mixed layer. *J Mar Res* **48**, 591-639

GARSIDE C (1985) The vertical distribution of nitrate in open ocean surface water. *Deep Sea Res* **32**, 723-732

GASPAR P (1988) Modelling the seasonal cycle of the upper ocean. *J Phys Oceanogr* **18**, 161-180

GEIDER RJ, PLATT T (1986) A mechanistic model of photoadaptation in microalgae. *Mar Ecol Prog Ser* **30**, 85-92

GLOVER RS (1967) The continuous plankton survey of the north Atlantic. *Symp Zool Soc Lond* **19**, 189-210

HARRIS JRW (1974) The kinetics of polyphagy. *In Usher MB, Williamson MH (eds) Ecological Stability, Chapman and Hall Ltd, London, p 123-139*

HOLLIGAN PM, FERNANDEZ E, AIKEN J, BALCH WM, BOYD P, BURKILL PH, FINCH M, GROOM SB, MALIN G, MULLER K, PURDIE DA, ROBINSON C, TREES C, TURNER SM, VAN DER WAL P A biogeochemical study of the coccolithophore, Emiliania huxleyi, in the North Atlantic. *Submitted to Deep-Sea Res*

JAMART BM, WINTER DF, BANSE K, ANDERSON GC, LAM RK (1977) A theoretical study of phytoplankton growth and nutrient distribution in the Pacific Ocean off the northwestern U.S. coast. *Deep-Sea Res.* **24**, 753-773

JAMART BM, WINTER DF, BANSE K (1979) A sensitivity analysis of a mathematical model of phytoplankton growth and nutrient distribution in the Pacific Ocean off the northwestern U.S. coast. *J Plankt Res* **1**, 267-290

KATO H, PHILLIPS OM (1969) On the penetration of a turbulent layer into stratified fluid. *J Fluid Mech* **37**, 643-655

KERSHAW R (1985) Onset of the south west monsoon and sea surface temperature anomalies in the Arabian Sea. *Nature* **315**, 561-563

KIEFER DA, KREMER JN (1981) Origins of vertical patterns of phytoplankton and nutrients in the temperate, open ocean: a stratigraphic hypothesis. *Deep-Sea Res* **28**, 1087-1105

KLEIN P, COSTE B (1984) Effects of wind-stress variability on nutrient transport into the mixed layer. *Deep-Sea Res* **31**, 21-38

KRAUS EB (1972) Atmosphere-ocean interaction, *Clarendon Press, Oxford*

KRAUS EB, TURNER JS (1967) A one-dimensional model of the seasonal thermocline. II The general theory and its consequences. *Tellus* **13**, 98-105

LANCELOT C, BILLEN G (1986) Carbon-nitrogen relationships in nutrient metabolism of coastal marine ecosystems. *Adv Aquatic Microbiol* **3**, 263-321

LANDE R, LEWIS MR (1989) Models of photoadaptation and photosynthesis by algal cells in a turbulent mixed layer. *Deep-Sea Res* **36**, 1161-1175

LEDERMAN TC, TETT P (1981) Problems in modelling the photosynthesis light relationship. *Bot Mar* **24**, 125-134

LEGOVIC T (1989) Predation in food webs. *Ecol Modell* **48**, 267-276

LEVITUS S (1982) Climatological atlas of the world ocean, *NOAA Prof Pap 13, US Govt Print Off, Washington, DC*

LONGHURST AR, BEDO AW, HARRISON WG, HEAD EJH, SAMEOTO DD (1990) Vertical flux of respiratory carbon by oceanic diel migrant biota. *Deep-Sea Res* **37**, 685-694

LONGHURST AR, WILLIAMS R (1979) Material for plankton modelling: vertical distribution of Atlantic zooplankton in summer. *J Plankt Res* **1**, 1-28

MARTIN JH, FITZWATER SE (1988) Iron deficiency limits phytoplankton growth in the north-east Pacific subarctic. *Nature* **331**, 341-343

MARTIN JH, FITZWATER SE, GORDON RM (1990) Iron deficiency limits growth in Antarctic waters. *Global Biogeochem Cycles* **4**, 5-12

MARTIN JH, GORDON RM, FITZWATER SE, BROENKOW WW (1989) VERTEX phytoplankton / iron studies in the Gulf of Alaska. *Deep-Sea Res* **36**, 649-680

MOLONEY CL, FIELD JG (1991) The size-based dynamics of plankton food webs 1. a simulation model of carbon and nitrogen flows. *J Plankt Res* **13(5)**, 1003-1038

MORALES CE, BEDO A, HARRIS RP, TRANTER PRG (1991) Grazing of copepod assemblages in the north-east Atlantic: the importance of the small size fraction. *J Plankt Res* **13**, 455-472

MOREL A (1988) Optical modeling of the upper ocean in relation to its biogenuous matter content (case I waters). *J Geophys Res* **93 (C9)**, 10749-10768

MUSGRAVE DL, CHOU J, JENKINS, WJ (1988) Application of a model of upper ocean physics for studying seasonal cycles of oxygen. *J Geophys Res* **93**, 15679-15700

OCEANOGRAPHIC LABORATORY, EDINBURGH (1973) Continuous plankton records: a plankton atlas of the north Atlantic and the North Sea. *Bull Mar Ecol* **6**, 251-273

O'NEILL RV, ANGELIS DL, PASTOR JJ, JACKSON BJ, POST WM (1989) Multiple nutrient limitations in ecological models. *Ecol Modell* **46**, 147-163

PARKER RA (1986) Simulating the development of chlorophyll maxima in the Celtic Sea. *Ecol Modell* **33**, 1-11

PARSONS TP, KESSLER TA (1987) An ecosystem model for the assessment of plankton production in relation to the survival of young fish. *J Plankt Res* **9**, 125-137

PLATT T, DENMAN KL, JASSBY AD (1977) Modeling the productivity of phytoplankton. *In: Goldberg ED, McCave IN, O'Brien JJ, Steele JH (eds) The Sea, vol 6, Wiley-Interscience, New York, p 807-856*

PLATT T, GALLEGOS CL, HARRISON WG (1980) Photoinhibition of photosynthesis in natural assemblages of marine phytoplankton. *J Mar Res* **38**, 687-701

PRICE NM, ANDERSON LF, MOREL FMM (1991) Iron and nitrogen nutrition of equatorial Pacific plankton. *Deep-Sea Res* **38**, 1361-1378

RADACH G, MAIER-REIMER E (1974) The vertical structure of phytoplankton growth dynamics, a mathematical model. *Mem Soc R Sci Liege Ser 6,* **7**, 113-146

RADACH G (1980) Preliminary simulations of the phytoplankton and phosphate dynamics during FLEX '76 with a simple two-component model. *Meteor Forschungsergeb (Reihe A)* **22**, 151-164

RILEY GA, STOMMEL H, BUMPUS DF (1949) Quantitative ecology of the plankton of the western North Atlantic. *Bull Bingham Oceanogr Collect* **12**, 1-169

SAKSHAUG E, KIEFER DA, ANDRESEN K (1989) A steady state description of growth and light absorption in the marine planktonic diatom Skeletonema costatum. *Limnol Oceanogr* **34**, 198-205

SARMIENTO JL, FASHAM MJR, SIEGENTHALER U, NAJJAR R, TOGGWEILER JR (1989) Models of chemical cycling in the ocean: *Progress report II, Tech Rep 6, Ocean Tracers Lab, Princeton Univ, Princeton, N J, 46pp*

SATHYENDRANATH S, GOUVEIA AD, SHETYE SR, RAVINDRAN P, PLATT T (1991) Biological control of surface temperature in the Arabian Sea. *Nature* **349**, 54-56

SATHYENDRANATH S, PLATT T (1988) The spectral irradiance field at the surface and in the interior of the ocean: a model for applications in oceanography and remote sensing. *J Geophys Res* **93**, 9270-9280

SIMONOT J-Y, DOLLINGER E, LE TREUT H. (1988) Thermodynamic-biological-optical coupling in the oceanic mixed layer. *J Geophys Res* **93(C7)**, 8193-8202

SLAGLE R, HEIMERDINGER G (EDS) (1991) U S Joint Global Ocean Flux Study North Atlantic Bloom Experiment Process Study Data Report P-1, April - July, 1989. *U S JGOFS Data Management Office, National Oceanographic Data Center*

STEELE JH, FROST BW (1977) The structure of plankton communities. *Phil Trans R Soc B* **280**, 485-534

STEELE JH, HENDERSON EW (1992) The role of predation in plankton models. *J Plankt Res* **14(1)**, 157-172

TAKAHASHI T, OLAFSSON J, BROECKER WS, GODDARD J, CHIPMAN DW, WHITE J (1985) Seasonal variability of the carbon-nutrient chemistry in the ocean areas west and north of Iceland. *Rit Fisk* **9**, 20-36

TAYLOR AH (1988) Characteristic properties of models for the vertical distribution of phytoplankton under stratification. *Ecol Modell* **40**, 175-199

TAYLOR AH, JOINT IR (1990) A steady-state analysis of the 'microbial loop' in stratified systems. *Mar Ecol Prog Ser* **59**, 1-17

TAYLOR AH, STEPHENS JA (1993) Diurnal variations of convective mixing and the spring bloom of phytoplankton. *Deep-Sea Res* **40** 389-408

TAYLOR AH, COLEBROOK JM, STEPHENS JA, BAKER NG (1992) Latitudinal displacements of the Gulf Stream and the abundance of plankton in the NE Atlantic. *J Mar Biol Ass* **72**, 919-921

TAYLOR AH, HARRIS JRW, AIKEN J (1986) The interaction of physical and biological processes in a model of the vertical distribution of phytoplankton under stratification. *In: Nihoul JCJ (ed) Marine Interfaces Ecohydrodynamics, Elsevier Oceanogr Ser vol 42, Elsevier, New York, p 313-330*

TAYLOR AH, WATSON AJ, ROBERTSON JE (1992) The influence of the spring phytoplankton bloom on carbon dioxide and oxygen concentrations in the surface waters of the northeast Atlantic during 1989. *Deep-Sea Res* **39**, 137-152

TAYLOR AH, HARBOUR DS, HARRIS RP, BURKILL PH, EDWARDS ES Seasonal succession in the pelagic ecosystem of the North Atlantic and the utilisation of nitrogen. *In press.*

TAYLOR AH, WATSON AJ, AINSWORTH M, ROBERTSON JE, TURNER DR (1991) A modelling investigation of the role of phytoplankton in the balance of carbon at the surface of the North Atlantic. *Global Biogeochem Cycles* **5**, 151-171

TETT P, EDWARDS A (1984) Mixing and plankton: an interdisciplinary theme in oceanography. *Oceanogr Mar Biol* **22**, 99-123

TETT P, EDWARDS A, JONES K (1986) A model for the growth of shelf-sea phytoplankton in summer. *Est Coast Shelf Sci* **23**, 641-672

VOLLENWEIDER RA (1965) Calculation models of photosynthesis-depth curves and some implications regarding day rate estimates in primary production measurement. *In Goldman CR (ed) Primary Production in Aquatic Environments, University of California Press, Berkeley, p 425-457*

WATSON AJ, ROBINSON C, ROBERTSON JE, WILLIAMS PJLEB, FASHAM MJR (1991) Spatial variability in the sink for atmospheric carbon dioxide in the North Atlantic. *Nature* **350**, 50-53

WIEGERT RG (1979) Population models: experimental tools for the analysis of ecosystems. *In: Horn DJ, Mitchell R, Stairs GR (eds) Proceeding of colloquium on analysis of ecosystems, Ohio State University Press, p 239-275*

WILLIAMS R, CONWAY DVP (1980) Vertical distribution of Calanus finmarchicus and C. helgolandicus. *Mar Biol* **60**, 57-61

WILLIAMS R, CONWAY DVP (1984) Vertical distribution and seasonal and diurnal migration of Calanus helgolandicus in the Celtic Sea. *Mar Biol* **79**, 63-73

WOLF KU, WOODS JD (1988) Lagrangian simulation of primary production in the physical environment - the deep chlorophyll maximum and nutricline. *In: Rothschild BJ (ed) Towards a theory of biological-physical interactions in the World Ocean, D Reidel, Dordrecht, p 51-70*

WOODS JD (1980) Diurnal and seasonal variation of convection in the wind-mixed layer of the ocean. *Quart J R Meteor Soc* **106**, 379-394

WOODS JD (1985) The physics of thermocline ventilation. *In: Nihoul JCJ (ed) Coupled ocean-atmosphere models, Elsevier Oceanogr Ser vol 40, Elsevier, Amsterdam, p 543-590*

WOODS JD, ONKEN R (1982) Diurnal variation and primary production in the ocean - - preliminary results of a Langrangian ensemble model. *J Plankt Res* **4**, 735-756

WROBLEWSKI JS (1977) A model of phytoplankton plume formation during variable Oregon upwelling. *J Mar Res* **35**, 357-394

WROBLEWSKI JS, SARMIENTO JL, FLIERL GR (1988) An ocean basin scale model of plankton dynamics in the North Atlantic, I, Solutions for the climatological oceanographic conditions in May. *Global Biogeochem Cycles* **2**, 199-218

DESIGN OF A 3D BIOGEOCHEMICAL TRACER MODEL FOR THE OCEAN

ERNST MAIER-REIMER
Max-Planck Inst. f. Meteorologie
Hamburg, FRG

Abstract

A state-of the art report is given of the Hamburg model of the oceanic carbon cycle. The model treats geochemical tracers important to the carbon cycle advected by the currents of a general circulation model. The model is an extension of the Bacastow and Maier-Reimer (1990) model. It is based on a more realistic current field and includes a mechanism of lysocline - sediment interaction with alkalinity, dissolved inorganic carbon, phosphate, oxygen and silicium as prognostic variables. The carbon variables are defined for ^{12}C, $\delta^{13}C$, *and* $\Delta^{14}C$ *separately.*

1 Introduction

The increase of carbon dioxide in the atmosphere due to the combustion of fossil fuels poses a major concern for the discussion in the scientific community as well as in public newspapers and in governmental policy. At present, it is estimated that approximately one third of the CO_2 emissions enter the ocean. Simplified box-models for the sequestering of CO_2 by the ocean have been developed almost 20 years ago (Oeschger *et al.*,1975,). Traditionally, the fluxes of water between the boxes are calibrated by fitting the results to the distribution of known tracers, primarily to radiocarbon (i.e. the deviation $\Delta^{14}C$ of the mixing ratio of ^{14}C with respect to the stable ^{12}C from a standard sample). Radiocarbon is produced naturally in the upper atmosphere by cosmic rays from nitrogen atoms. It has a mean life-time of 8267 years. Another source was provided by the big nuclear weapon tests prior to the test ban treaty in 1963. The decrease in the atmosphere (Nydal and Lövseth, 1983) and spreading in the ocean give sharp constraints on circulation models. Unfortunately, in the fitting procedure, the calibration to the natural radiocarbon gives quite different answers from the fit of the distribution of bomb radiocarbon (Siegenthaler, 1983).

With the development of three dimensional models of the physical circulation of the ocean (Bryan, 1969), the base was given for a treatment of carbon in the sea as a three dimensional continuum. Maier-Reimer and

NATO ASI Series, Vol. I 11
Modelling Oceanic Climate Interactions
Edited by J. Willebrand and D. L. T. Anderson

Hasselmann (1987) studied the uptake of fossil fuel CO_2 by the ocean in a purely inorganic model with spatially homogeneous alkalinity (see appendix for a discussion of the definition and importance of alkalinity). The main shortcoming of this model (apart from an insufficient formation of North Atlantic deep water) was the complete neglect of all biological processes which are responsible for appr. 80% of the gradients of dissolved CO_2 in the ocean. Consequently, the basic state on which the anthropogenic perturbation was imposed, could not be compared with the real distribution of CO_2 in the ocean.

The role of biological processes in the uptake of fossil fuel CO_2 by the ocean is often misinterpreted. The naive sketch that the excess CO_2 in the atmosphere is brought into the depth of the ocean by the biological pump is in striking contradiction to the fact that the CO_2-content of the atmosphere has remained almost perfectly constant during the ten thousand years preceding the onset of anthropogenic emissions. The natural biological pumps are in equilibrium with the upwelling of remineralized nutrients. In regions of high productivity, the CO_2-content of the surface water is depleted by only 15% of the deep sea value. It is hard to imagine that an increase of 10% of dissolved CO_2 (the increase that would be in equilibrium with a doubling in the atmosphere (cf. Revelle and Suess, 1957)) could substantially improve the utilization of nutrients. The strength of the gradients of dissolved CO_2 in the ocean, however, is due primarily to biological processes; thermodynamical effects alone could explain only one fifth of the observed gradients. Without the biological pumps, the ocean would be chemically quite different from today's real ocean even in the abiotic components. A direct effect of the biological pump on the uptake of CO_2 can be expected only from changes in the pumping mechanisms. Such changes could result in a modified thermodynamic field through changes in the plankton communities or through a changed residence time of particulate matter in the deep ocean (cf. Heinze *et al.*, 1991).

An indirect effect of biology on the uptake of fossil fuel CO_2 is provided by the existence of sediment layers of which the upper ten or twenty centimeters are bioturbated and, thus, still in close contact with the overlying water. Incidentally, this pool is estimated to contain the same amount of carbon as the present estimates of coal resources (Broecker and Takahashi, 1977). During the uptake of excess CO_2 the surface water becomes more acid. When it reaches the sediment layer, it tends to dissolve the sediment and, consequently, to neutralize the acidification as well as the excess pCO_2. In the modern ocean, carbonate sediments are preserved - in the global average - at depths shallower than 4 km. In equilibrium with a doubled CO_2 content in the atmosphere, the concentration of CO_3

ions in the seawater will be halved (cf appendix) and the preservation level ("lysocline") will rise by several kilometers and thus expose the corresponding sediment layers to dissolution.

Another approach of 3-D modelling has been presented by Sarmiento *et al.*, (1992). They treated the anthropogenic perturbation of CO_2 in the ocean as a perturbation of a basic state, regardless of how it may look like. In this approach, chemistry appears only in terms of buffer factors which give the transformation of changes in pCO_2 to changes in the concentration of dissolved CO_2. The validation of their model is done by similar perturbation experiments for the radiocarbon (Toggweiler *et al.*, 1989) and for a "perturbation phosphate" (Najjar *et al.*, 1992). Although these models gave reliable results for specific questions, it is difficult to derive a global geochemical flux sketch from these sensitivity experiments.

The need for realistic three-dimensional models of the oceanic carbon cycle is, thus, likely to be with us for a long time, yet. Entering the decade of the Joint Global Ocean Flux Studies (JGOFS), such models appear to be even more urgently required for the interpretation - and dynamical interpolation - of the data. In return, JGOFS will provide a unique dataset for the calibration and validation of such models.

Geochemical modelling depends crucially on the assumptions about physical circulation; on the other hand, physical modellers can gain much insight into their efforts from an inspection of the simulation of geochemical tracers (Maier-Reimer *et al.*, 1992). It may appear as a waste of computer time to take an elaborate model of the carbon cycle for the study of processes where the biological processes, most probably, play no role at all. The distributions of the isotopes of carbon (Kroopnick, 1985, Östlund and Stuiver, 1980, and Stuiver and Östlund, 1980), however, provide stronger constraints on modelling efforts than the traditional temperature and salinity (Levitus, 1982) distributions (cf. Fiadeiro, 1982 and Moore and Bjorkström, 1986). For such reasons, I strongly advocate the withdrawal of the walls between physical and geochemical modellers.

In a previous paper (Bacastow and Maier-Reimer,1990) we presented results based on our state of art of 1988. The most obvious weakness of that model was the almost complete absence of the Atlantic conveyor belt structure which was seen most pronouncedly in a much too high ^{14}C - age in the Atlantic. In a recent paper (Maier-Reimer *et al.* 1992) it was pointed out that this weak Atlantic overturning was an effect of inappropriately imposed thermal boundary conditions rather than an intrinsic property of the model. In this paper I describe the results of our carbon cycle model embedded in a physical circulation field which fits the ideas about the conveyor belt much better than our previous circulation field.

Some of the results have been published already in a proceedings volume of a NATO ARW (Maier-Reimer and Bacastow, 1990); in that contribution, however, some errors have occurred in the graphic representation of the results; moreover, I was much too enthusiastic about the ability of our model to produce horizontal gradients. Consequently, the gradients were overestimated, as seen primarily in the distribution of surface phosphate.

2 The Model

The formulation of the biological processes is not so straightforward as in models of physical circulation. The biological activity in the real ocean is extremely heterogeneous in space and time; there are, up to now, no clear rules concerning its dependence on environmental parameters, like temperature, light, turbulence, nutrient supply, etc. The only strict rule is the conservation of inventories of the different chemical components. Most of the processes that constitute the model are derived qualitatively from a close inspection of the distributions of tracers; the strength follows from a tuning process by comparison of computed distributions with the observed fields. Once the basic structure of the distributions is accepted, the details of any biological process formulation can be discussed within the sensitivity of the model distribution with respect to that parameter.

Figs. 1 a,b display the distribution of phosphate measured during GEOSECS (Bainbridge, 1980 and Craig *et al.*, 1981) along sections in the western Atlantic and Pacific. The asymmetry between Atlantic and Pacific appears in almost all other distributions, too, but with different relative amplitudes. In the Atlantic, the biological products are quickly exported towards the circumpolar current. In the Pacific, the deep circulation feeds into the northern basin; the biological products are more or less trapped. For biogenic material, the northern Atlantic acts as a repellor, the northern Pacific as an attractor. These effects are seen in all figures displaying the model results. The North Atlantic component is characterized by low phosphate, low silicate, high $\delta^{13}C$, high $\Delta^{14}C$. The opposing extremes are in the mid-depth waters of the Northern Pacific.

Fig. 1c displays p.144 of "Tracers in the Sea" (Broecker and Peng, 1982). It shows scatter diagrams of several variables on the isopycnal $\sigma_4 = 45.8$. In the individual basins, the variables are almost perfectly correlated. The differences in the slopes of the regressions is explained partly by mixing of the different compositions of deep water sources.

The model is based on a 15 layer version of the ATOS1 experiment of Maier-Reimer *et al.*, (1992). The main difference to the circulation field of Bacastow and Maier-Reimer lies in the incorporation of a simple sea-ice model which is driven by the COADS (Woodruff *et al.*, 1987) data set of

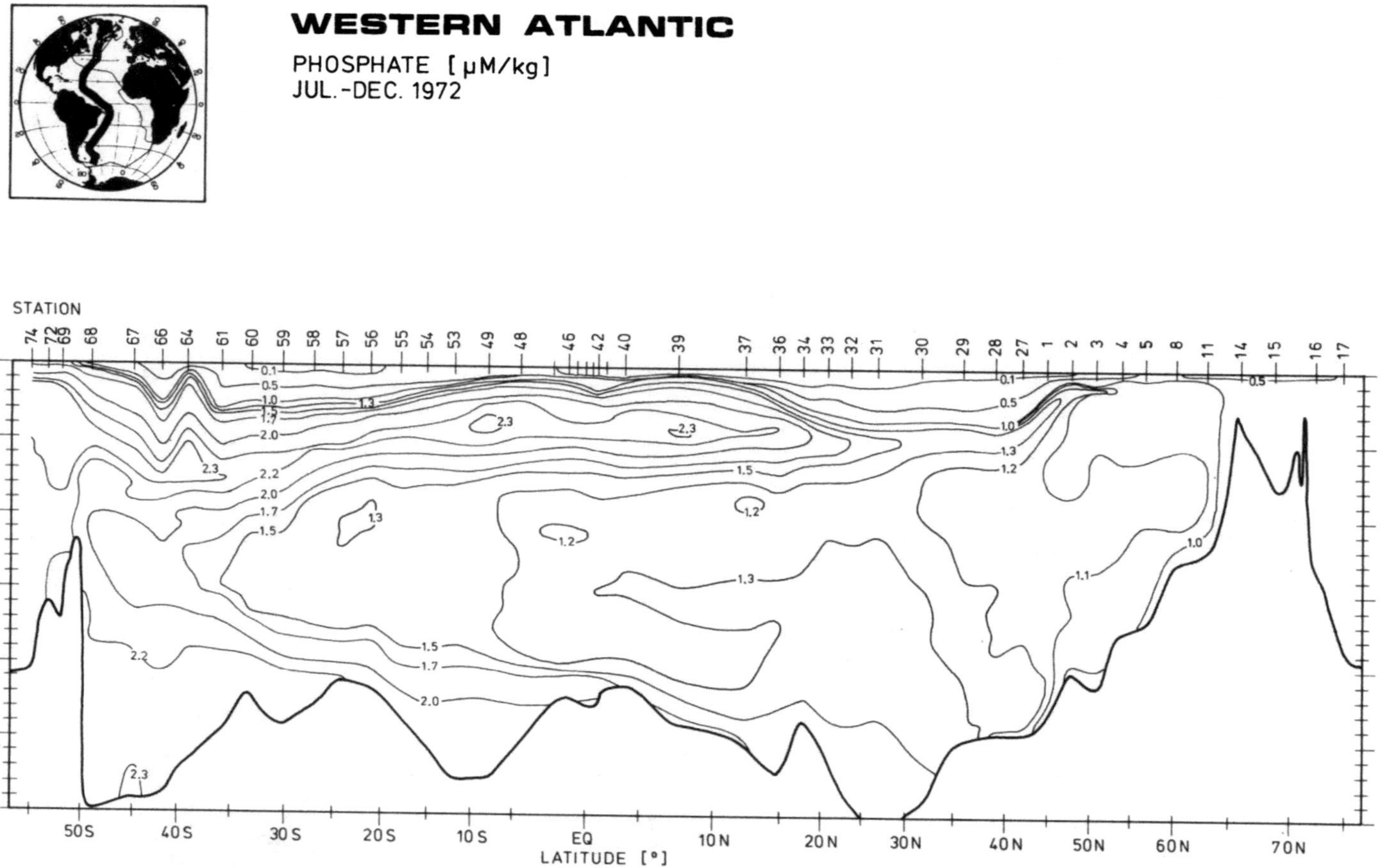

Figure 1a: GEOSECS Atlantic phosphate

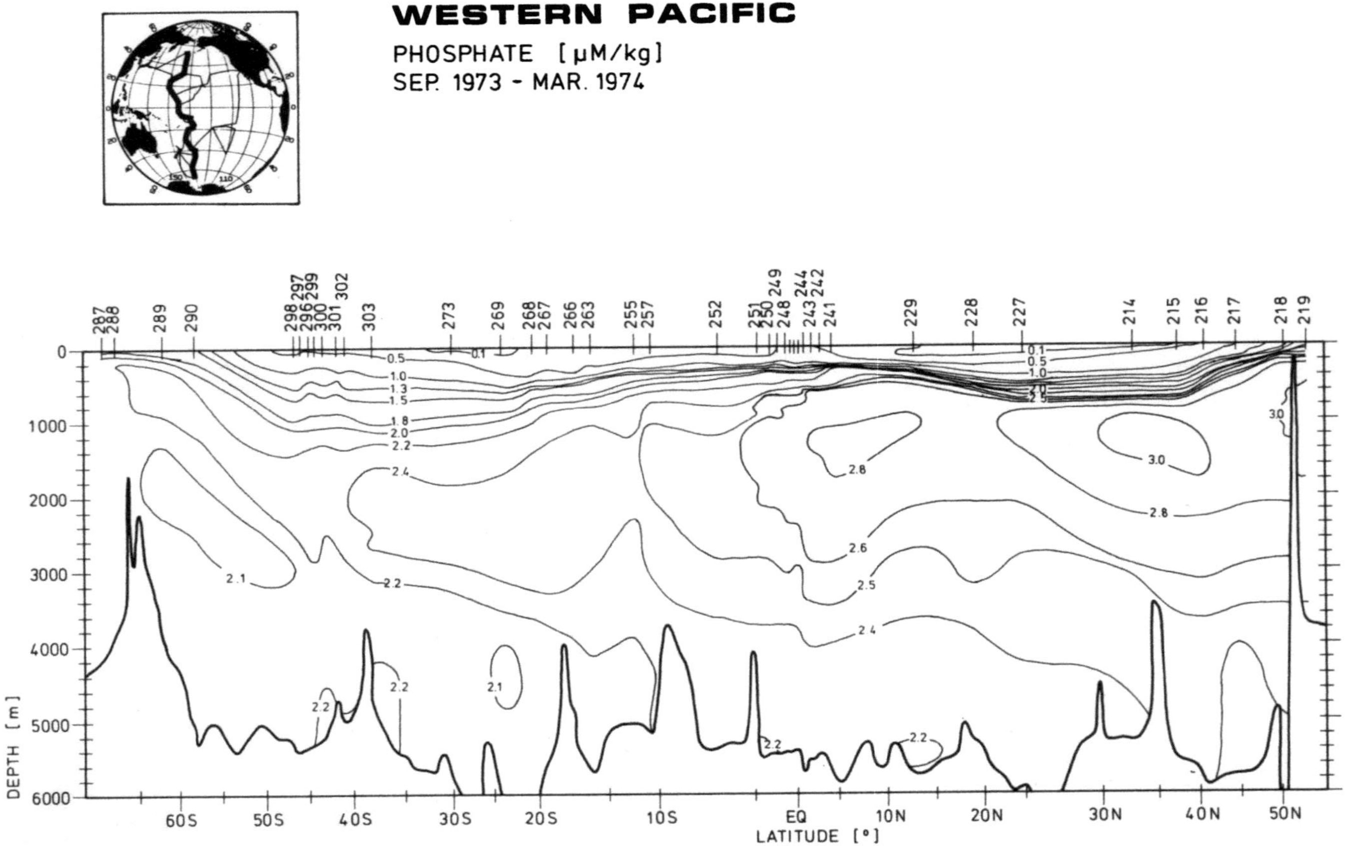

Figure 1b: GEOSECS Pacific phosphate

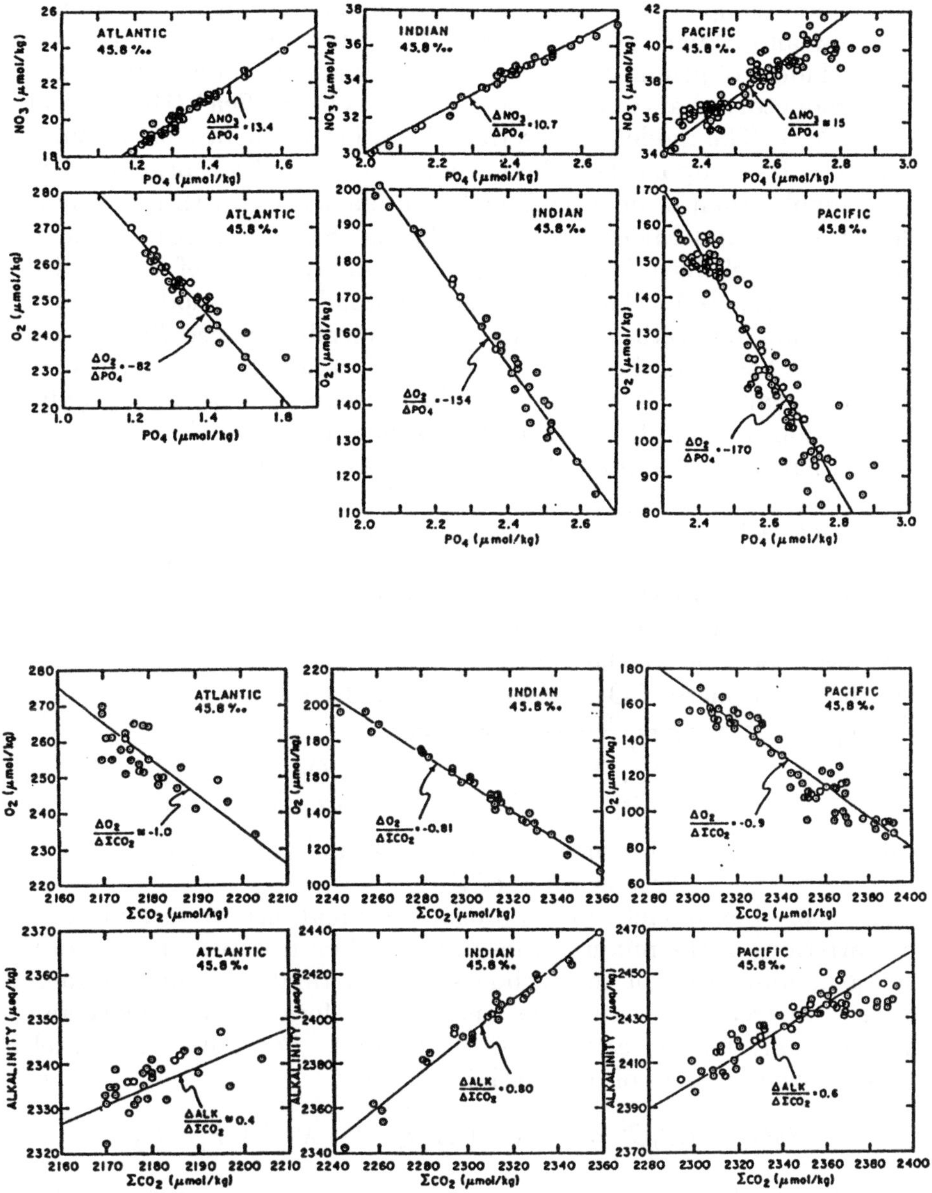

Figure 1c: Intercorrelation of several geochemical tracers (from Broecker and Peng, 1983)

atmospheric temperature. The carbon cycle model is an extension of the Bacastow and Maier-Reimer model (1990). It follows closely generally accepted ideas of primary productivity and fluxes as summarized, e.g., in the introductory chapter in the book "Tracers in the Sea" (Broecker and Peng, 1982). The model, as described here, has deliberate shortcomings in most of its components. The formulation of plankton productivity is much simpler than in in special models of plankton growth (cf. Dugdale 1967, Radach and Maier-Reimer, 1975, and Fasham 1985). Similarly, elaborate models have been developed for the dynamics of sediment deposits (Archer, 1991). The model is conceptually not a model of biology in the ocean, but rather a model of biogenically induced fluxes in the ocean. It is based on the assumption of perfect stoichiometric constancy of organic material ("Redfield Ocean"). Consequently, we do not address the question which nutrient really limits the production. We assume one limiting nutrient with units of carbon equivalents. Just before discussion of the results, this tracer is divided by 122 which is the global average relation of carbon to phosphate atoms in biomass (the "Redfield" ratio, redetermined by Takahashi *et al.*, 1985) and interpreted as phosphate. By this terminology, the complicated interactions between different components of the nitrogen cycle are disregarded.
Prognostic variables of the model are
- in the atmosphere: CO_2, $^{13}CO_2$, $^{14}CO_2$, and O_2
- in the ocean: $\Sigma^{12}CO_2$, $\Sigma^{13}CO_2$, $\Sigma^{14}CO_2$,
  $PO^{12}C$, $PO^{13}C$, $PO^{14}C$, $Ca^{12}CO_3$, $Ca^{13}CO_3$, $Ca^{14}CO_3$,
  alkalinity, phosphate, O_2, borate, and silicate.
- at the sea floor: sediment layers of $CaCO_3$ and organic material for all three isotopes of C.

(In many box models, direct prognostic equations are formulated for the variations of the mixing ratios of the carbon isotopes; in view of the strong nonlinearities of the carbon system, I think it is more straightforward to treat concentrations and to transform on the geochemical standard only for the analysis of the results). Formally, the model is characterized by a set of differential equations:

$$\partial_t C^i + u_j \partial_j C^i = A^h \Delta C^i + A^v \partial_{zz} C^i - \lambda^i C^i + F^i_j C^i C_j$$

where C^i stands for any of the tracers, u is the advection velocity, A denotes the diffusion operator, λ denotes tracer specific sink/source terms like decay of ^{14}C, and F denotes the nonlinear interaction functionals between the tracers. As boundary condition we impose zero fluxes across all boundaries for all tracers but oxygen and the carbon isotopes which are subject to exchange with the atmospheric variables. For all bio related tracers we do not assume any restoring to observed values. Consequently, the comparison of the tracer fields with observations gives a mixed information for evaluation of the physical model as well as of the geochemical model. Total borate is assumed to be constant; it does not appear as an individual tracer. The partitioning between the different states of dissociation, and, thus, its contribution to the total alkalinity, follows from the law of mass action for the boron atoms and for the carbon atoms. The dissociation constants are taken from Culberson and Pytkowicz (1968), while the solubility product for calcium carbonate is from Edmond and Gieskes (1970). We emphasize that the exact values of these constants are still in debate. Any systematic error in the coefficients would reflect in errors in the pCO_2 - ΣCO_2 - alkalinity system. An inspection of the results of the model, however, indicates that the uncertainties of the coefficients are not crucial for the model. The formulation of the biogeochemical processes is not straight-forward; the available data do not allow, up to now, a unique fit of functional relationships. In most processes we assume a parameter of maximum interaction which is reduced by limiting factors of the environment.

2.1 Gas Exchange

The Liss and Merlivat (1987) parametrization, which is primarily dependent on the wind speed, is inconsistent in the global average with the mean gas exchange coefficient derived from the inventories of bomb produced radiocarbon in the ocean (Broecker *et al.*, 1985). For the experiment described in this paper, a constant value $\lambda = 0.06 mol/(m^2 a \mu atm)$ was chosen, matching Broecker's estimate. With this λ, the air-sea difference of the CO_2 partial pressure ΔpCO_2 changes according

$$\frac{\partial \Delta pCO_2}{\partial t} = -\lambda \Delta pCO_2$$

The fractionation of the carbon isotopes stems from a lower fugacity F, i.e. the tendency to leave the water, of the heavier isotopes:

$$F^{13}/F^{12} = (1.02389 - 9.483/T)$$

and $F^{14}/F^{12} = (F^{13}/F^{12})^2$ (Mook,1986), where T is the SST in K. For oxygen, the gas exchange operates an order of magnitude faster and so we assume the surface oxygen to be at the temperature dependent saturation level.

2.2 New Production

Depending on the phosphate concentration, the new (or "export") production E of soft tissue material is written:

$$E = rP(P/(P + P0)),$$

where P is the concentration of phosphate, assuming a lower productivity rate in nutrient-poor water, and r depends on the temperature T (in Centigrades), the solar angle Δ, and the actual depth of the mixed layer as represented in the model by the depth Cd of the convective column according to

$$r = r_0 cos\Delta(50m/Cd)(T + 2)/(T + 10).$$

We did not assume any kind of iron limitation as proposed by several authors (e.g. Martin *et al.*, 1990). E effects a transformation from P and ΣCO_2 to the POC pools. The alkalinity is increased by 16 E due to the incorporation of nitrate into soft tissue. An inspection of phosphate vs. $\delta^{13}C$ scatter diagrams (Broecker and Peng, 1982)in the deep ocean indicates that the implementation of heavy carbon isotopes into POC should be proportional to 0.98 times the local isotopic ratios of ^{13}C and, consequently, 0.9604 for ^{14}C, resp., in ΣCO_2. Newer results, (Rau *et al.*,1991) however, indicate that this fractionation effect depends on the local concentration of dissolved gaseous CO_2, i.e. strongly on temperature. In the model described here, their functional relationship is tested:

$$F_P^{13} = 1 + 0.001(((3.4 - 17log(C_{aq}))/1000 + 1)(996.1 - 0.3C_{aq}) - 999)$$

(It turns out that a constant fractionation yields a better agreement with data). The newly formed POC quantities are released in the depth z according a penetration flux $E(100m/z)^{0.8}$ for $z > 100m$ (Berger *et al.*, 1989, based on Suess, 1980). The flux reaching the bottom is added to an organic sediment layer from where remineralization occurs with a time constant of 0.1/month. The organic sediment in the real ocean stems primarily from continental runoffs. In the experiment described here, the organic sediment pool is less than 15 Gigatons; for the global carbon budget, this plays no role. In the real ocean, a substantial part of the dead matter may be transformed into slowly decaying compounds of dissolved organic carbon (DOC). The nature of this DOC, however, is highly uncertain. Different methods of measurements give quite different results for its concentration (Williams and Druffel, 1987, and Sugimura and Suzuki,1988). For the modeller it could serve to remove some flaws in the phosphate distribution (cf. Bacastow and Maier-Reimer, 1991) but a direct comparison with data is impossible. I take it therefore as rather premature to incorporate it routinely into the model.

2.3 Remineralization

Provided there is enough oxygen we describe remineralization by a time constant of 50 months. During this process oxygen and alkalinity are reduced according to the relative Redfield (122 C:177 O_2) ratios. The remineralized POC is added to the local concentrations of phosphate and ΣCO_2 in all three isotopes.

We do not consider processes like denitrification or sulfate transformation to H_2S. The resulting distribution of POC just gives an indication where these processes should take place. We do not consider, too, the crystallization of apatite which could take place in concentrations of more than $3\mu mol/kg$ of phosphate (Atlas and Pytkowicz, 1977).

2.4 Shell Material

We assume silicate to be the preferred concrete to build shells. The redissolution of silicate is weaker than of carbonate. Thus, the export of silicate shells is connected with the gross primary productivity rather than with the new production, the main process of our model. With a (rather arbitrarily chosen) constant relation of 1:10 between new and gross production we define a potential for silicate export production $P_{Si} = Si/(Si+Si_0)$, where Si_0 denotes the (poorly known) solubility product of opal. The productivity of carbonate shells is reduced in low temperature regions. With α = 3/K and Tc = 5°C, we define a potential for carbonate formation

$$P_{CO_3} = exp(\alpha(T - Tc))/(1 + exp(\alpha(T - Tc)))$$

which yields a strong suppression of calcite formation below 2^oC. The Fractionation between carbonate and silicate producers is then:

$$F_{Si} = P_{Si}/(P_{Si} + P_{CO_3})$$

and F_{CO_3} = 1 - F_{Si}. Silicate and carbonate are released to the interior ocean with an exponential penetration depth of 2 km for carbonate and 4 km for silicate. During the formation of carbonate shells there is no fractionation between carbon isotopes.

2.5 Calcite Sediment-Carbonate System

With our model we cannot solve the mystery of why there is no precipitation of calcite from regions of high supersaturation. We take the solubility product to be dependent on salinity, temperature, and pressure according to Broecker and Takahashi (1978) and compute locally the degree of saturation. We assume that equilibrium is established as a result of continuously-acting crystallization and dissolution processes where the recombination is somehow blocked. In supersaturated water with positive δCO_3, the excess carbonate ion concentration, the monthly dissolution rate is written:

$$1 - \delta CO_3/(|\delta CO_3| + \delta CO_{30}),$$

with $\delta CO_{30} = 10\mu$mol/kg. In undersaturated water we assume complete dissolution within on month. Spontaneous crystallization is zero in the interior ocean and 0.2/month in the bottom layers of the model. Here, the newly formed calcite is transformed into the sediment layer. The sediment layer represents only the bioturbated upper 10cm of the overall sediment. Sediment thickness is limited at 2 kmol/m^2 of carbon (i.e. 200 kg/m^2 of $CaCO_3$); higher values are taken out. The resulting loss of carbon and alkalinity out of the system is in equilibrium with a compensating prescribed amount of river inflow along the coasts in the surface layer of 0.1 Gt/y carbon in the form of calcite with a zero content of $\delta^{13}C$.

3 The Stationary (preindustrial) State

3.1 The Nutrient - Oxygen Cycle

The surface distribution of the variables is mainly determined by the structure of the surface temperature, which results from the prescribed atmospheric temperature, and by the vertical component of water flow, which reflects primarily the input of vorticity by the prescribed windfield. Fig.2a shows the annual mean vertical velocity at the base of the uppermost model layer (at 50m). The dominant features are the strong equatorial upwelling in the Atlantic and Pacific, moderate upwelling along the western coasts and in the ACC, and downwelling in the subtropical gyres. Many regional structures such as the West African coastal upwelling cannot be resolved in such a coarse model. Similar structures are seen also in the distributions of surface phosphate (Fig.3a) and surface silicate (Fig.3b). Distinct differences in the phosphate and silicate patterns are in arctic regions. Here, obviously, the productivity is not limited by nutrients; the reduction of the productivity, as derived from observations, too, (Koblents-Mishke,1977), may be caused by low light intensity as well as by high turbulence (cf. Radach and Maier-Reimer,1975). The structure depends sensitively on the choice of the time constant for maximum productivity. The agreement of fig.2b (a factor of ten between maximum and minimum) with the observed structure (cf. Reid, pers. comm.) was obtained with $r_0 = 1/(4 months)$. Figs. 4a,b and 5a,b show sections

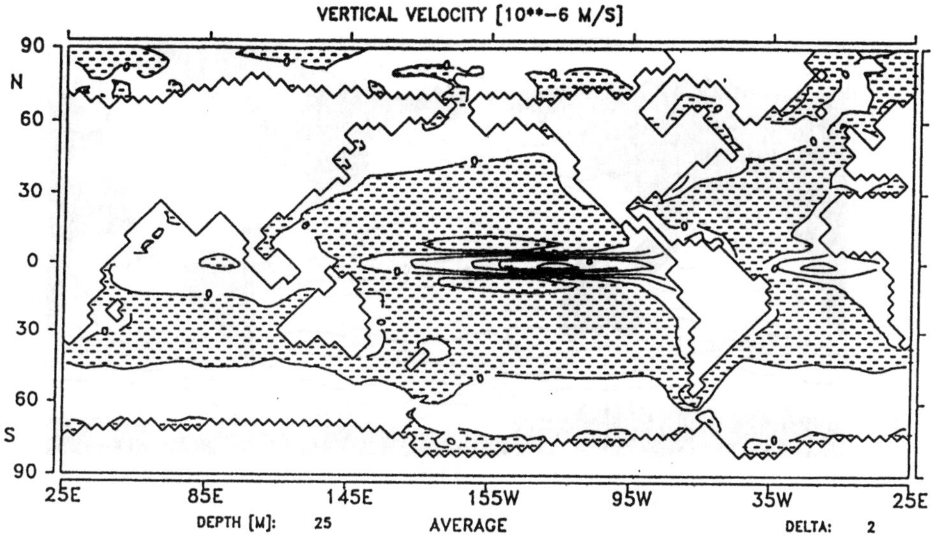

Figure 2a: Annual mean vertical velocity at the base of the uppermost layer, primarily the imposed Ekman velocity, smoothed by friction

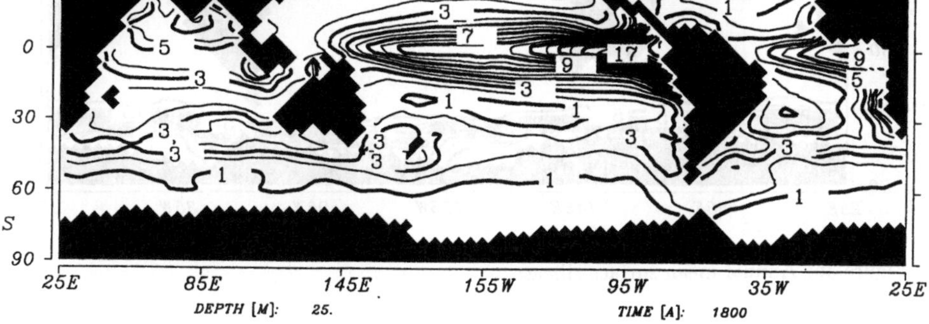

Figure 2b: Annual mean export production of organic carbon. The structures reflect those of Figure 2a, but smoothed. In high latitudes, the productivity is limited by light and temperature

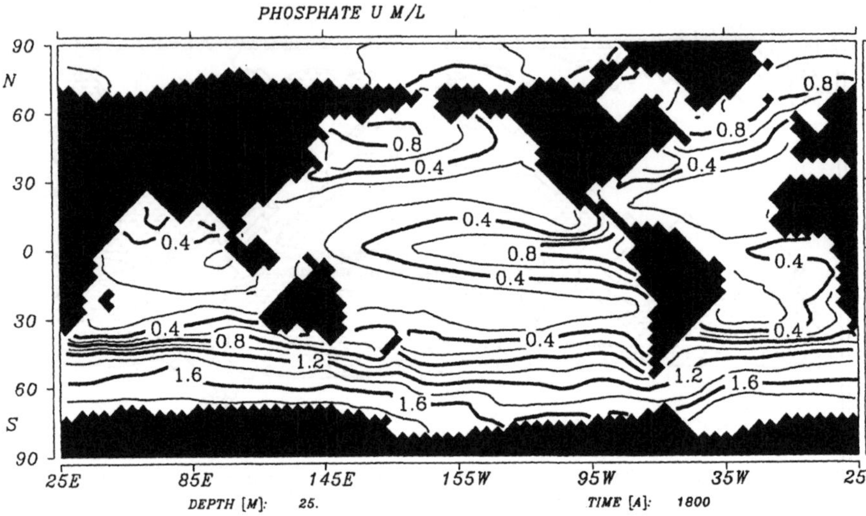

Figure 3a: Annual mean surface phosphate.

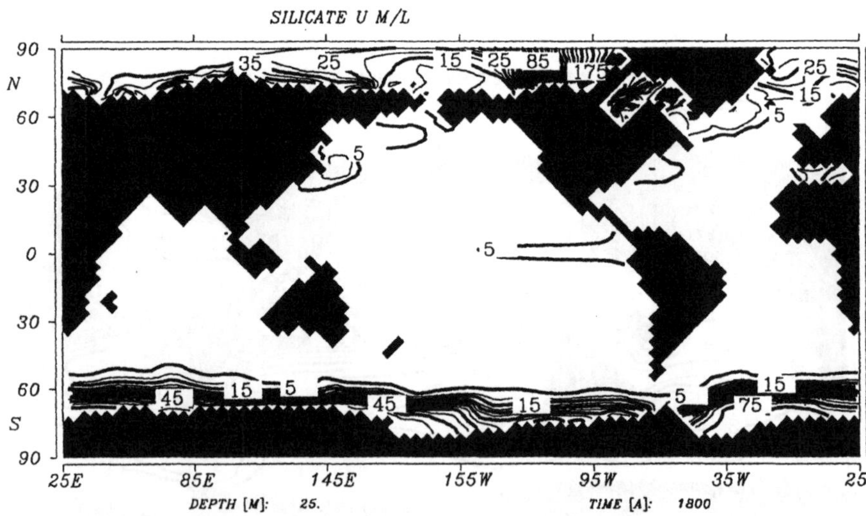

Figure 3b: Annual mean surface silicate

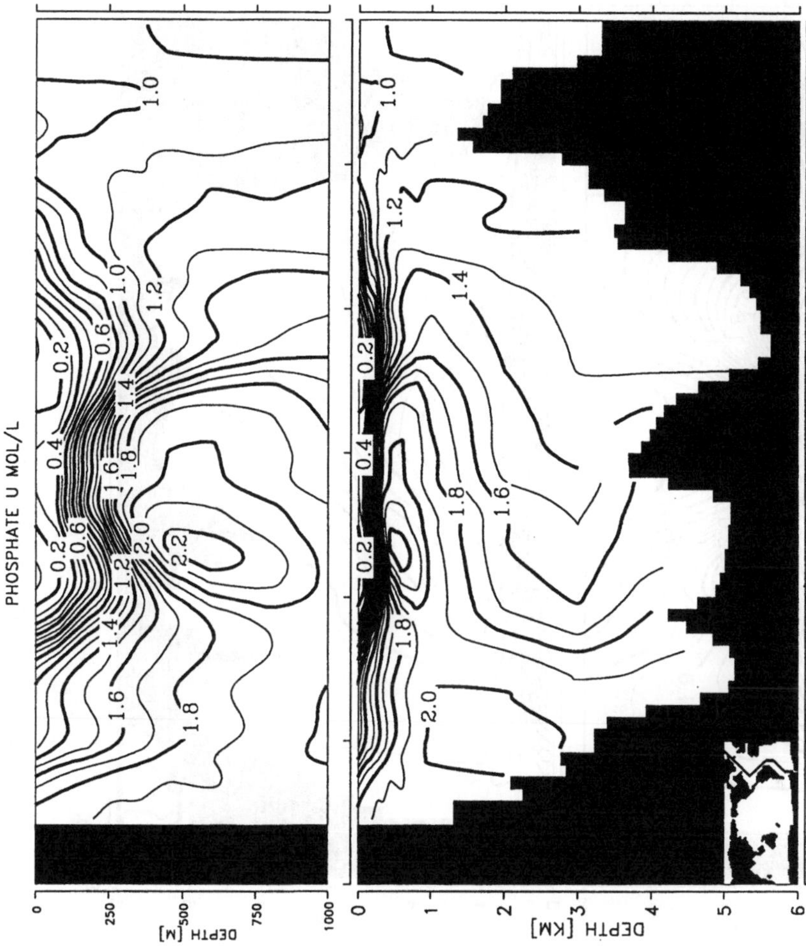

Figure 4a: Phosphate in the western Atlantic

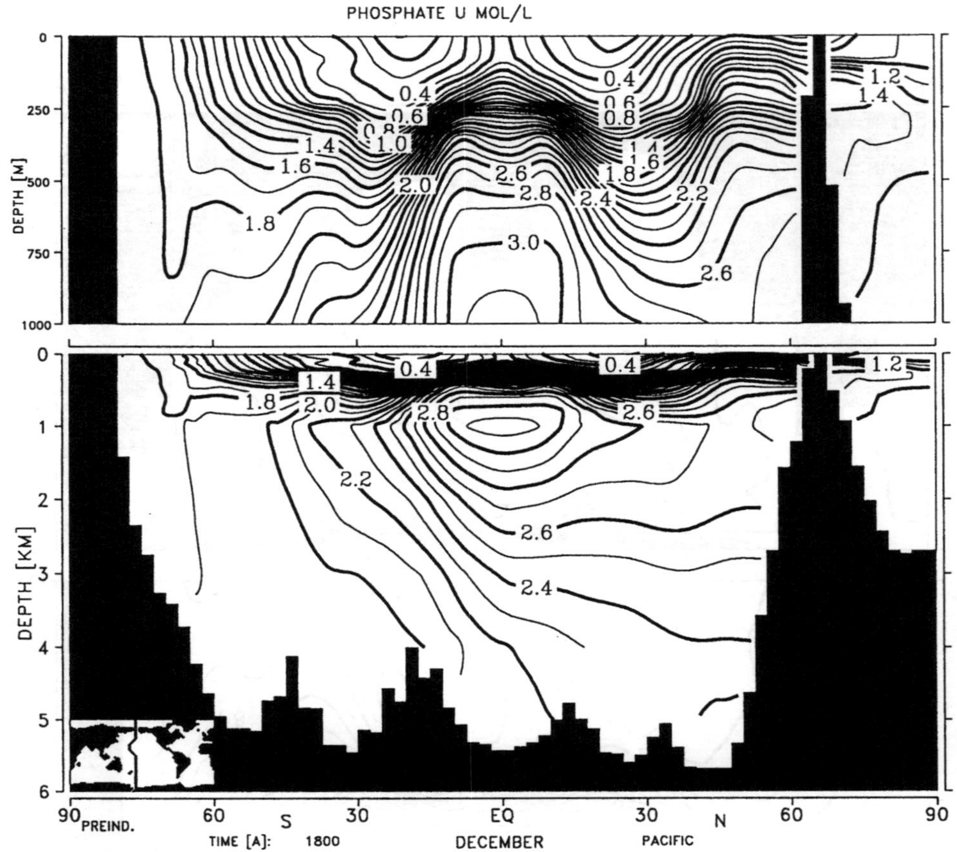

Figure 4b: Phosphate in the western Pacific

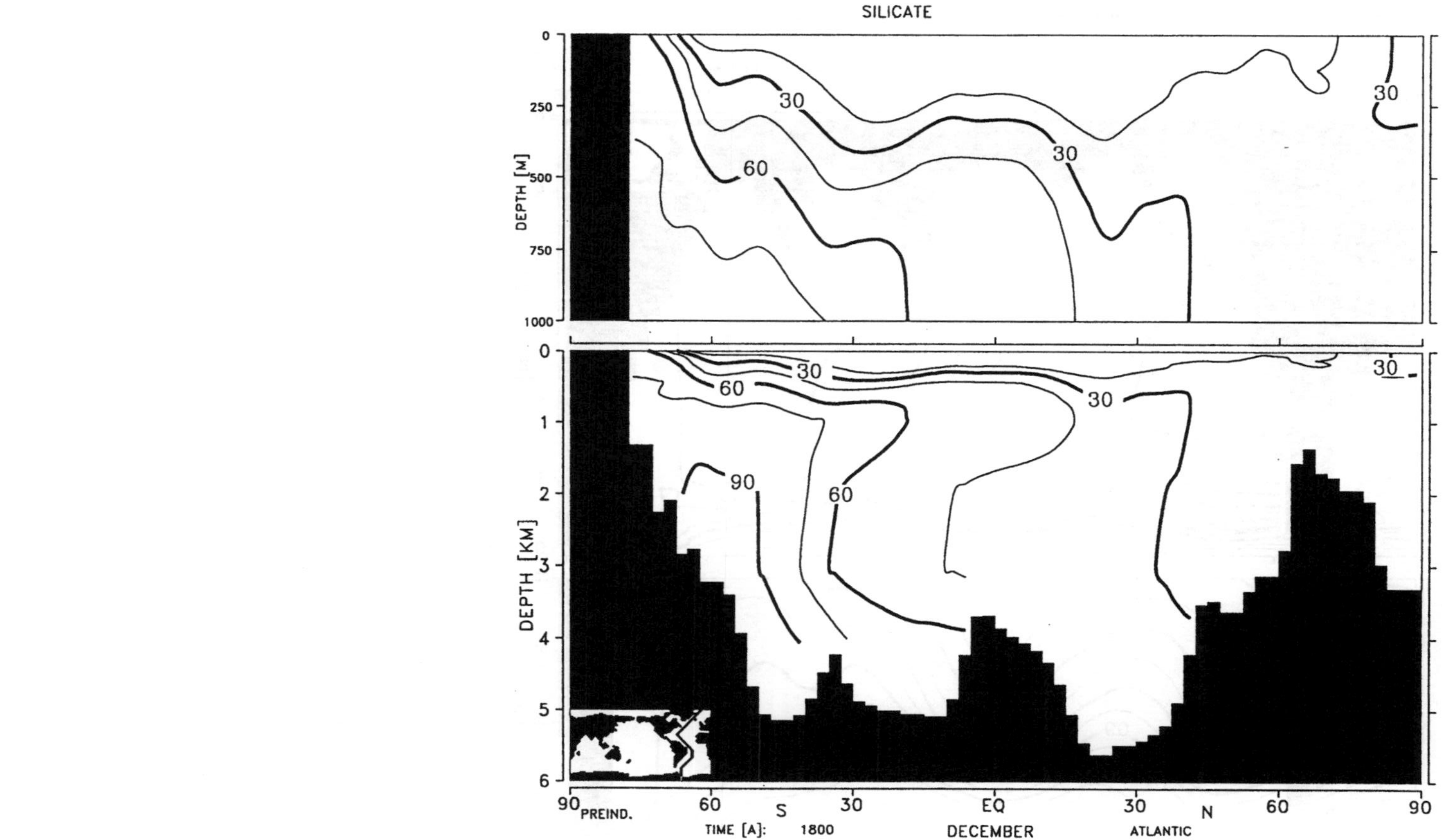

Figure 5a: Silicate in the western Atlantic

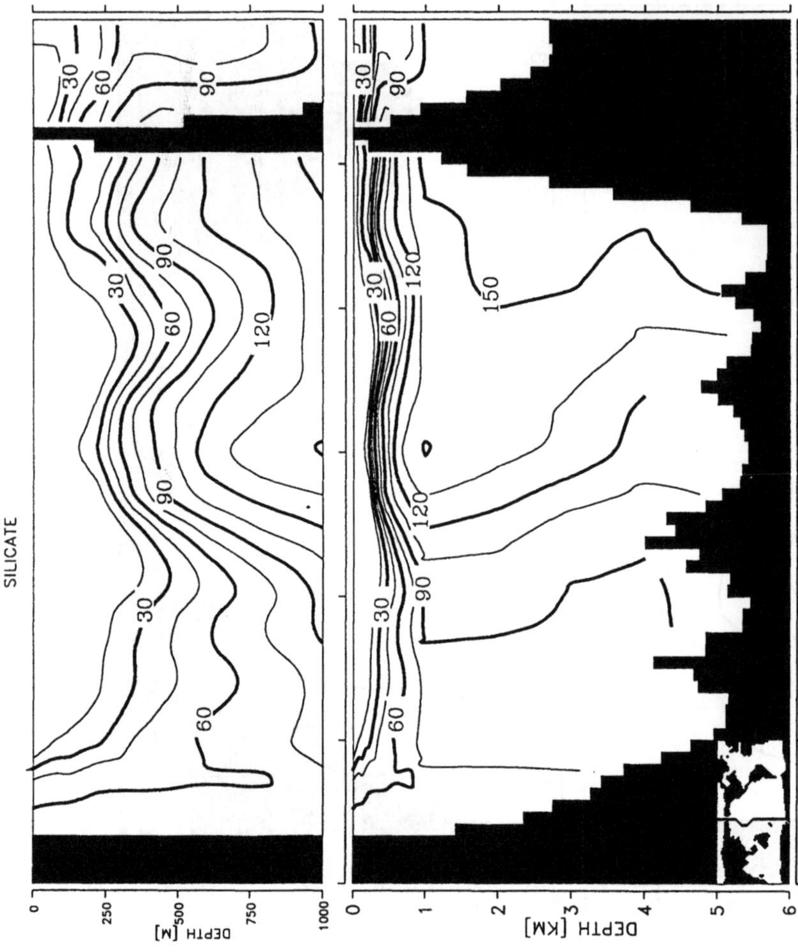

Figure 5b: Silicate in the western Pacific

of phosphate and silicate along the western Atlantic and western Pacific (The sections were chosen to coincide roughly with the GEOSECS sections). The "conveyor belt" structure of the circulation model is clearly reflected in the pronounced difference between Atlantic and Pacific. In contrast to the gradients in the surface distribution, the deep gradients are rather insensitive to the details of the productivity mechanism. With $r_0 = 1/\text{month}$ (the value of Maier-Reimer and Bacastow, 1990), the relation between maximum and minimum phosphate concentration is in the order of 10^4 , whereas the corresponding relation in the deep oceans increase only from 2.5 to 3. For silicate, a much stronger surface depletion than for phosphate was assumed. A comparison with data, however, indicates that the factor of 10 between the removal rates for silicate and phosphate together with the exponential sinking depth of 2km is still too weak: The resulting factor between North Pacific and North Atlantic is only 8 instead of 10 as in the GEOSECS sections; the surface concentration around Antarctica goes up to 75 mmol/l instead of 50. Figs 6a-d show the horizontal distribution of phosphate and silicate at the depth of 1km (the level of absolute oxygen minimum) and at 3 km (At this depth, the spreading of NADW into the global ocean is most pronounced). An unrealistic maximum is seen in the deep eastern Pacific, where the concentrations go up to 300μmol/l. This may be due to the strong equatorial upwelling which yields an overestimate of the equatorial productivity; another explanation could be the assumption of a closed silicate cycle. In the real ocean, a constant balance is maintained between loss of silicate in the sediments and river input from weathered rocks. The distribution of dissolved oxygen (Fig.7 a-d) has partly similar structure to the phosphate distributions, especially in the deep Pacific, where the phosphate maximum is found in almost anoxic water. The Atlantic is generally higher in oxygen, corresponding to the lower phosphate content which overrules the lower solubility due to the higher temperature. The correlation between oxygen and phosphate is shown in fig.8a: In the Indian and in the Pacific ocean most of the values are on the line defined by the assumed constant Redfield ratio. In the Atlantic, however, the scatter diagram is dominated by the mixing between the two end members of NADW and WSBW. Fig.8b shows the distribution of

$$PO_4^* = PO_4 + O_2/177 - 1.95$$

a quantity - in which the effects of the details of the remineralisation are cancelled out - that can be regarded as a physical tracer (Broecker *et al.*, 1991). The value 1.95 is arbitrarily subtracted in order to obtain similar numbers to those for phosphate. The distribution is dominated by mixing effects between the endmembers NADW (North Atlantic Deep

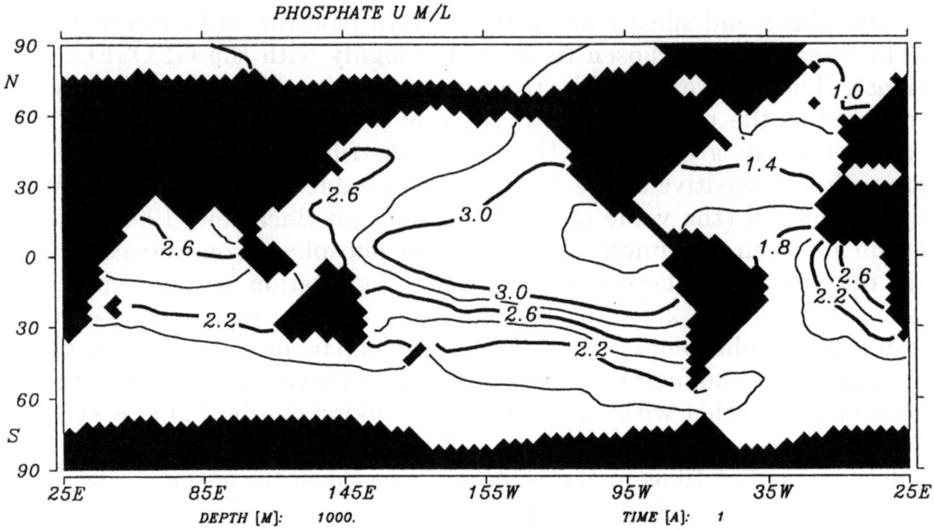

Figure 6a: Phosphate in 1 km depth

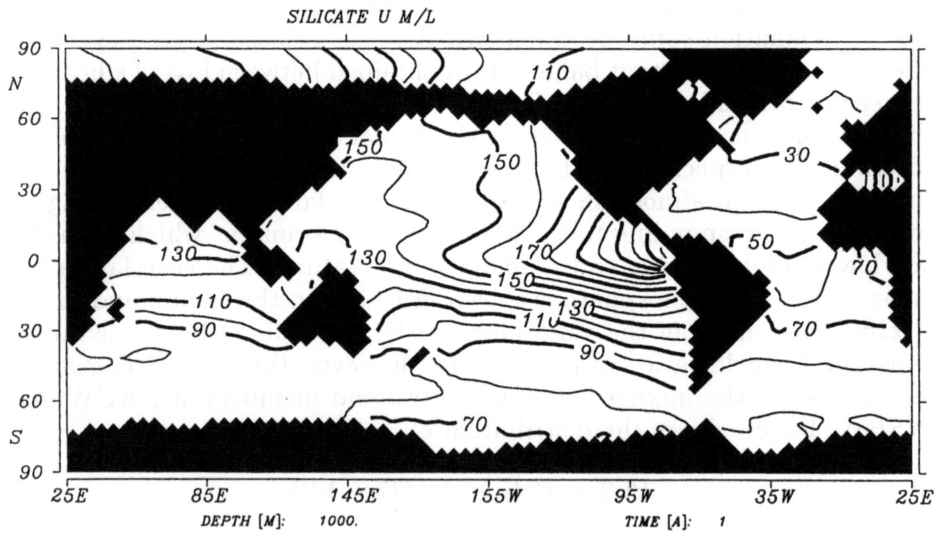

Figure 6b: Silicate in 1 km depth

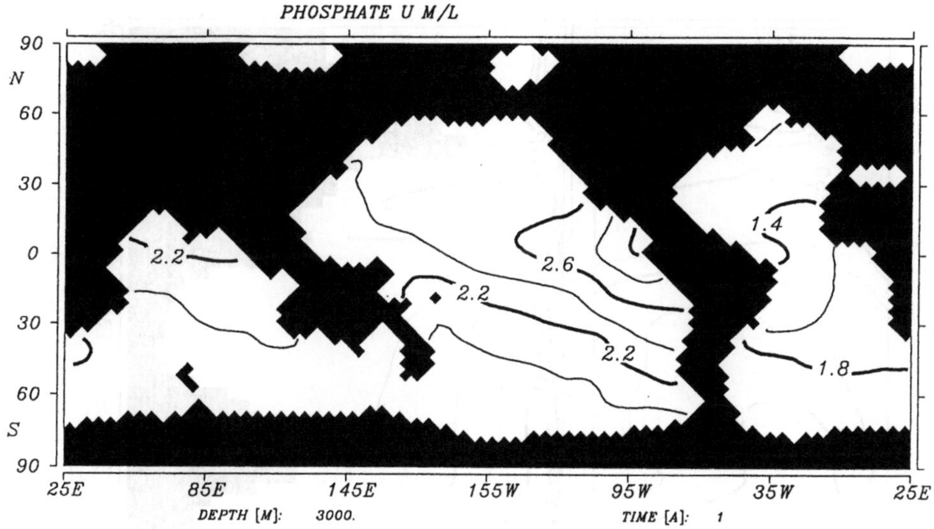

Figure 6c: Phosphate in 3 km depth

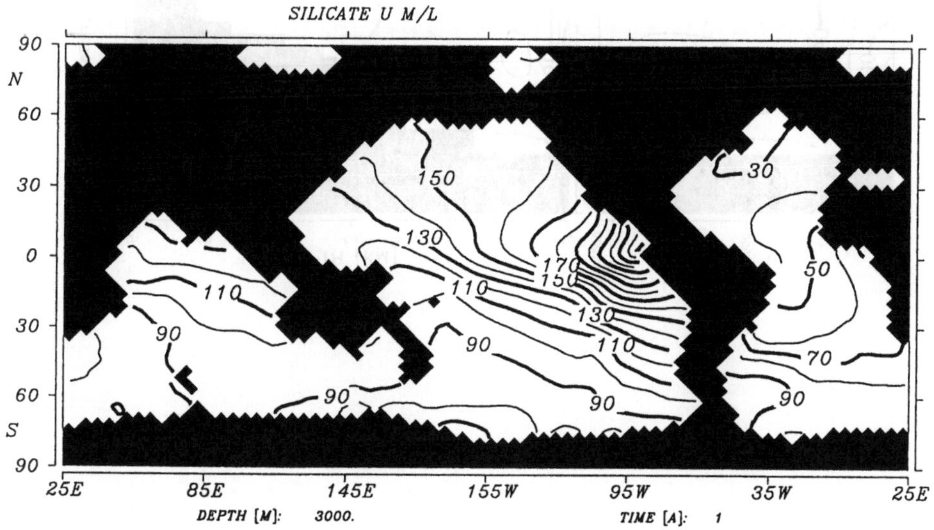

Figure 6d: Silicate in 3 km depth

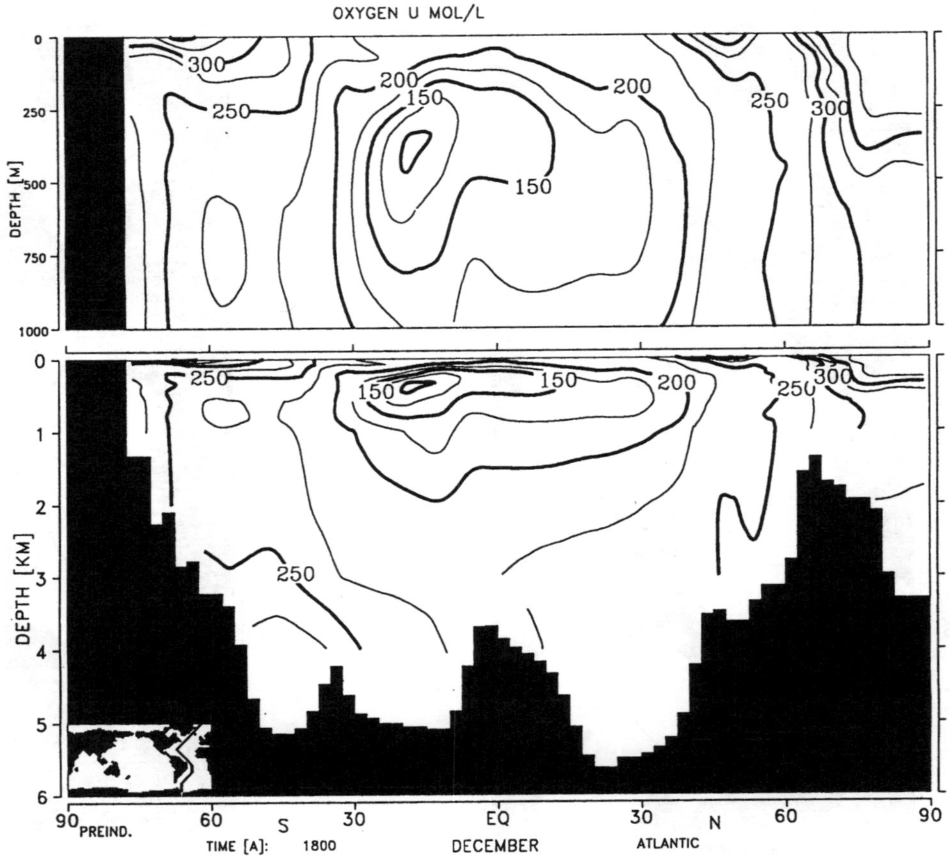

Figure 7a: Oxygen in the western Atlantic

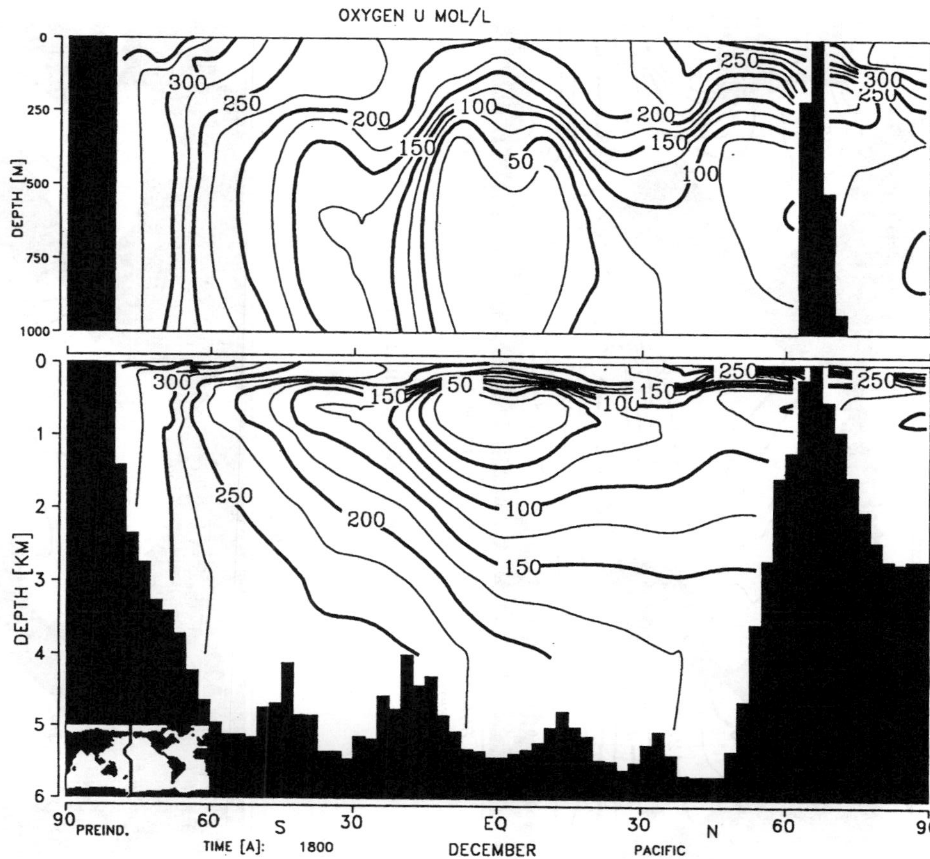

Figure 7b: Oxygen in the western Pacific

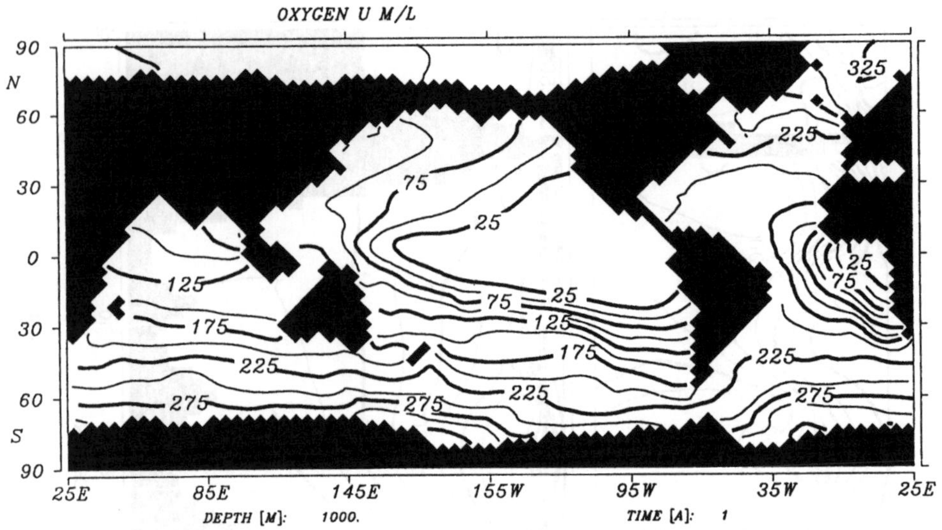

Figure 7c: Oxygen in 1 km

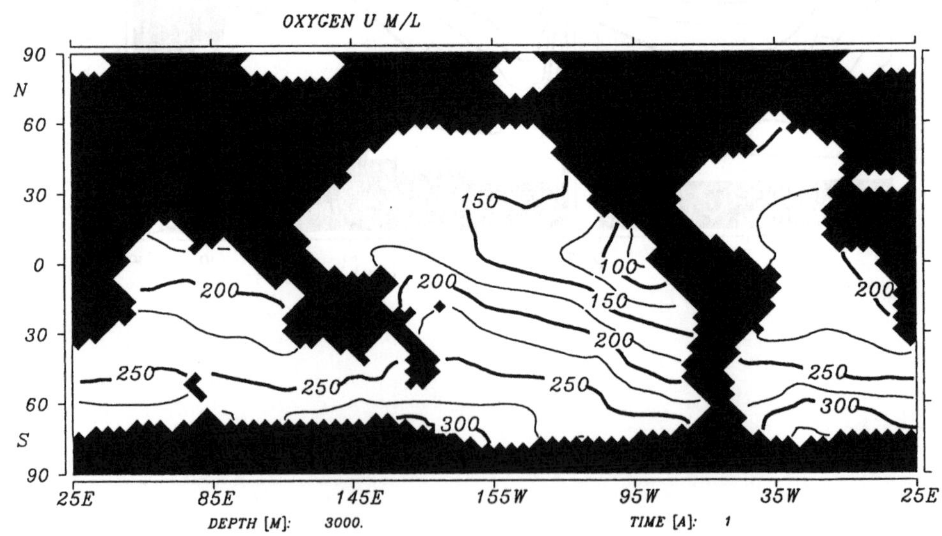

Figure 7d: Oxygen in 3 km

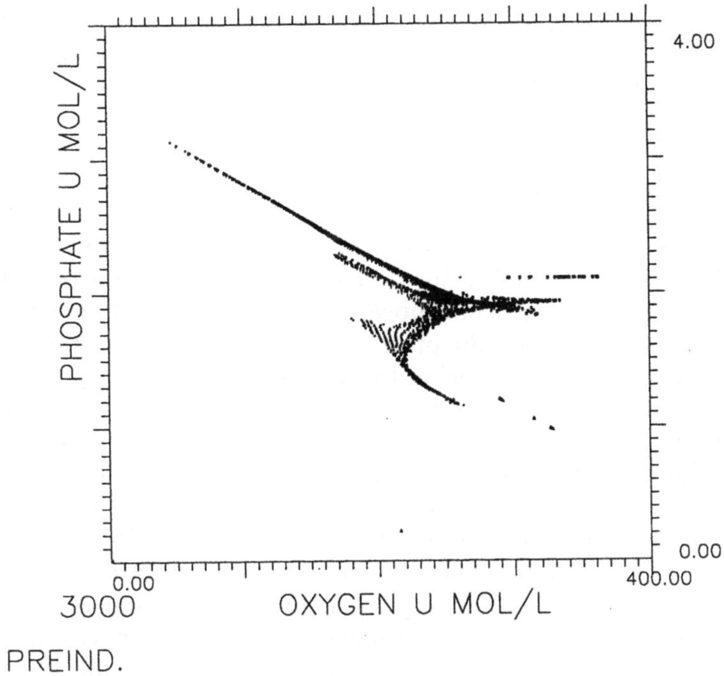

Figure 8a: scatter oxygen vs phosphate in 3 km

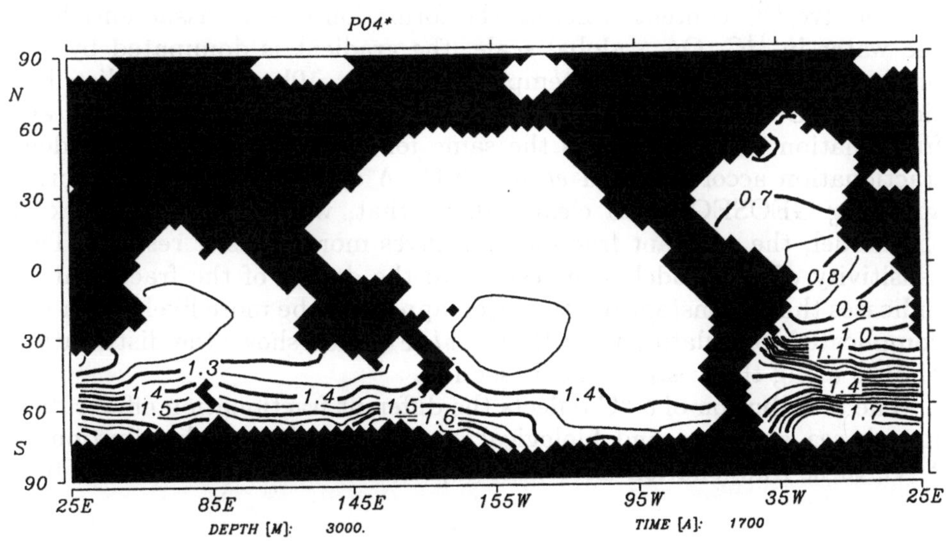

Figure 8b: $PO_4^* = PO_4 + O_2/177 - 1.95$

Water) at 0.7 and WSBW (Weddell Sea Bottom Water) at 1.8. In the Northern and central Pacific the distribution is almost uniform at values between 1.35 and 1.4, which indicates a mixing ratio of 30:70 between NADW and WSBW.

3.2 Dissolved CO_2 and Alkalinity

The inventories of carbon atoms and alkalinity were chosen to match approximately the observed values. Due to slowness of the sediment pool, no systematic tuning to all observational constraints was possible. The preindustrial pCO_2 of 280 ppm in the atmosphere was considered to be the most important criterion for the settings. Fig.9, a-d show the sections for these variables in the Atlantic and Pacific. Since the net effect of temperature on the concentration of ΣCO_2 accounts for only 50μmol/l (Maier-Reimer and Hasselmann, 1987), the gradients in the distributions are maintained primarily by the biological processes. In the global average, the model exhibits a rain ratio of calcium carbonate to organic material of 0.22. The relation in the productivity makes the gradients of ΣCO_2 about four times bigger than the gradients of alkalinity. Both distributions again reflect the conveyor belt structure with a relative depletion of Atlantic waters compared to those in the Pacific.

3.3 Carbon Isotopes

The surface map of δ^{13} C (fig. 10a) is dominated by the opposing effects of temperature and productivity on the the fractionation of isotopes: in the equatorial belt of high productivity, the temperature tends to deplete the relative ^{13}C content, whereas the formation of soft tissue enrichens the water in ^{13}C. On a global scale, the gradient is dominated by the productivity, but halved by temperature. Fig. 10b shows the distribution of δ^{13}C along the section in the western Atlantic for the constant fractionation. Fig 10c shows the same for the temperature dependent fractionation according Rau *et al* (1991). A comparison with the corresponding GEOSECS data clearly shows that, within the framework of the model, the constant fractionation gives more realistic results. This sensitivity of the model with respect to the details of the fractionation indicates that a constant fractionation appears to be more likely than the complicated formulation from Rau *et al.* Fig. 10d shows the distribution of δ^{13}C along the western Pacific section.

The distribution of δ^{13}C reflects the different fractionation effects during air-sea exchange and biological production. In the distribution of δ^{14}C, these fractionation effects are expected to be twice as strong. It is

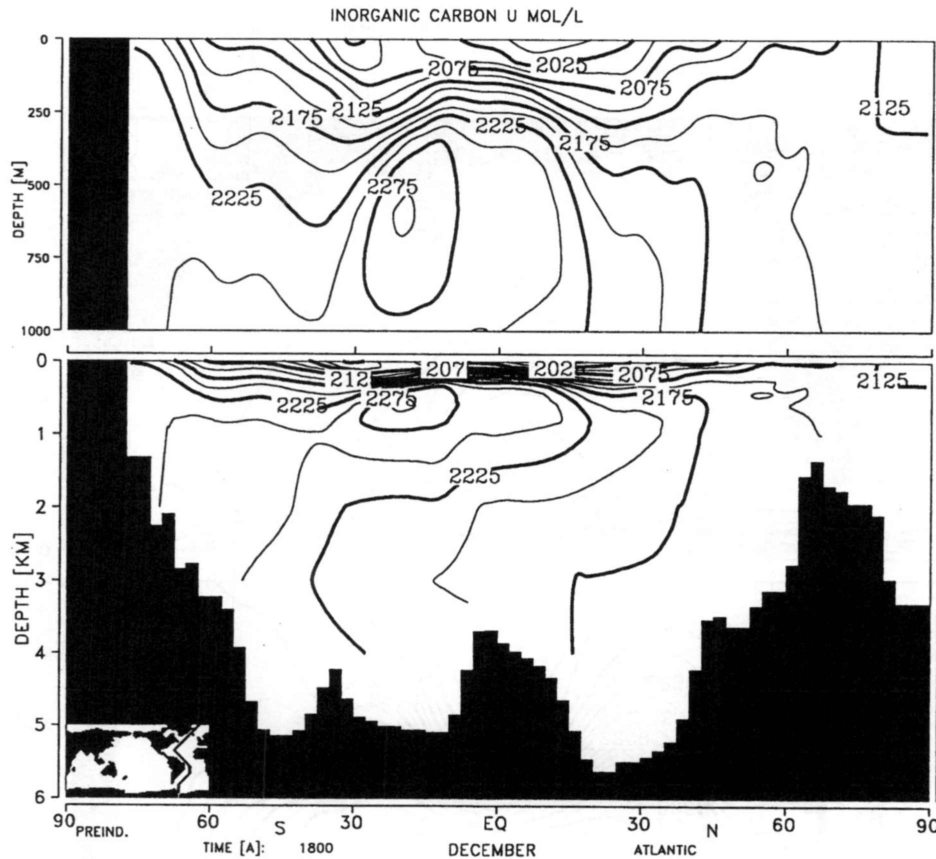

Figure 9a: Dissolved inorganic carbon in the western Atlantic

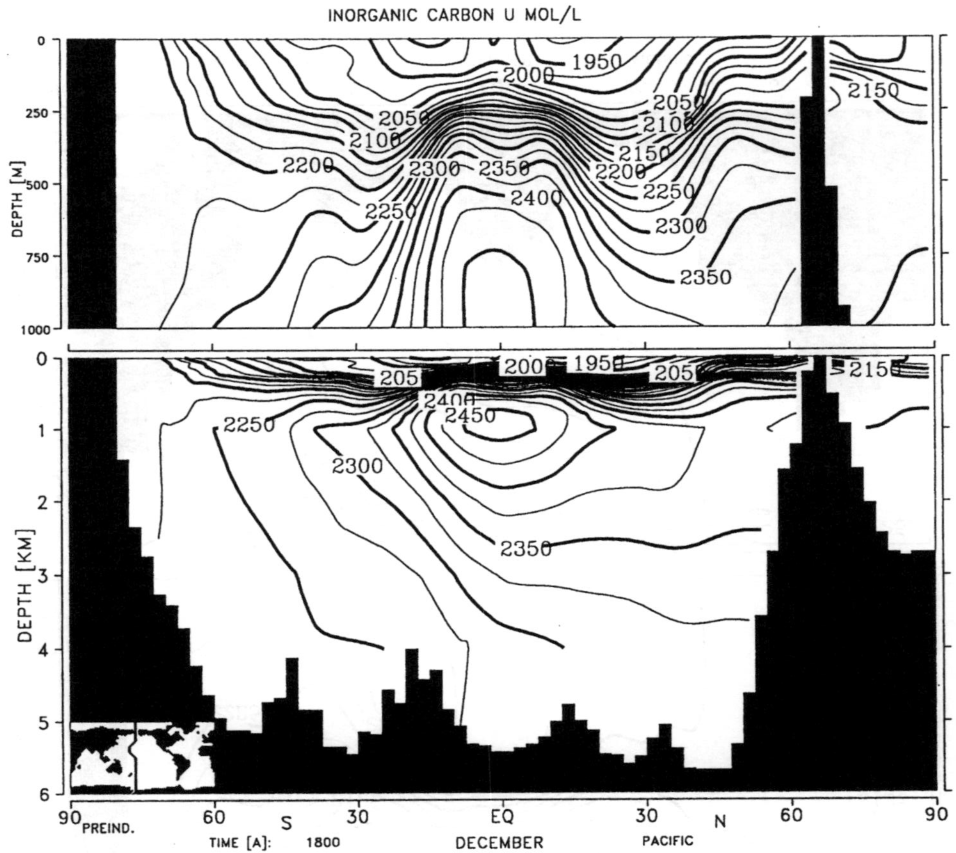

Figure 9b: Dissolved inorganic carbon in the western Pacific

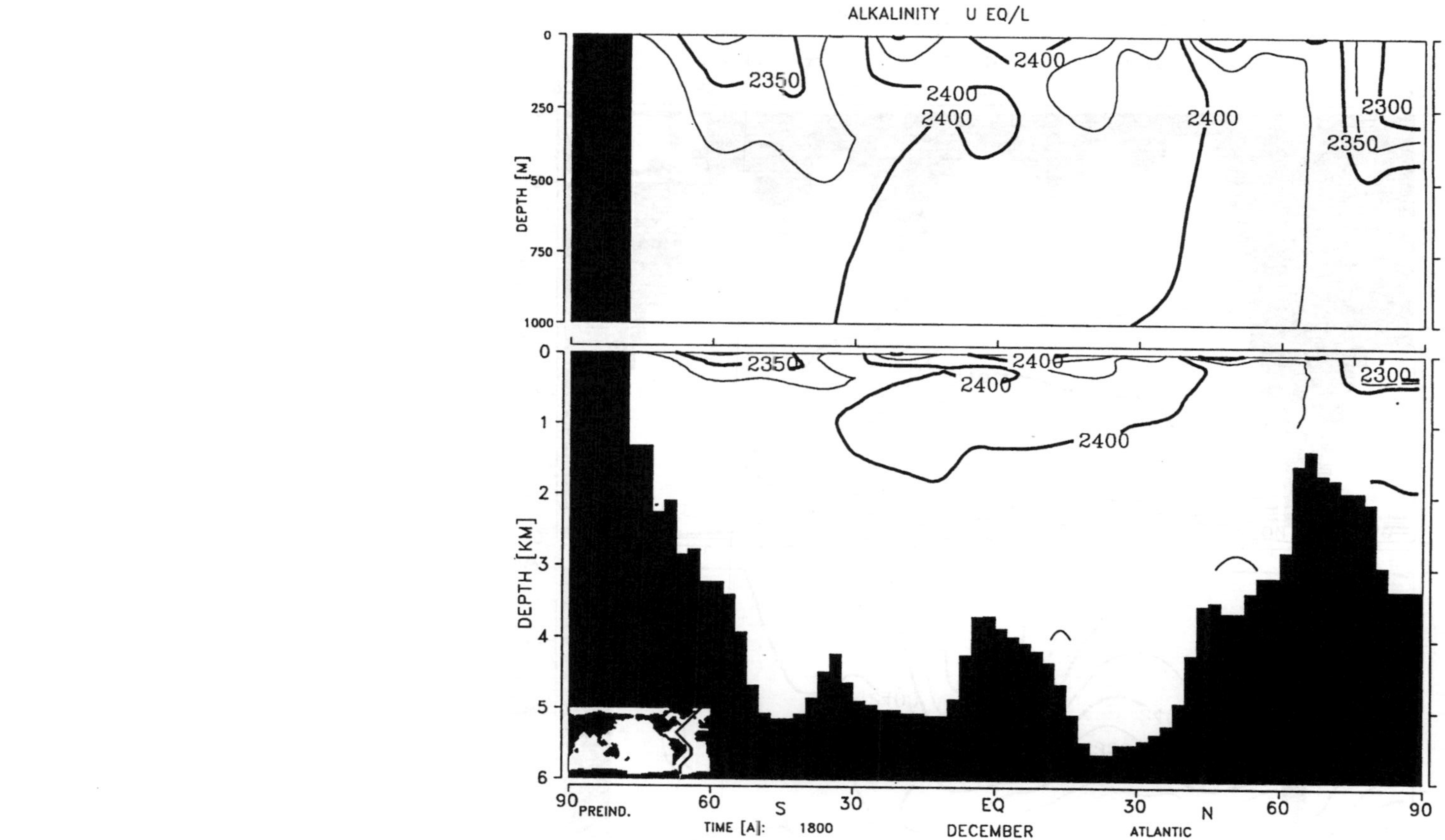

Figure 9c: Alkalinity in the western Atlantic

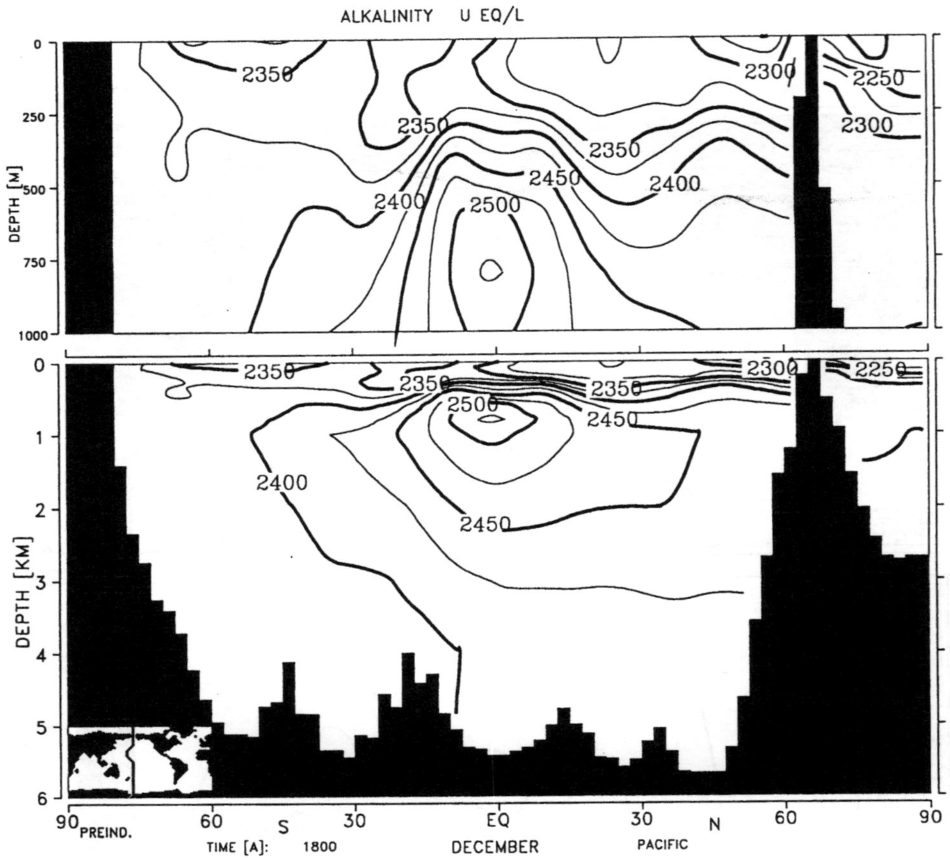

Figure 9d: Alkalinity in the western Pacific

therefore convenient to correct the distribution of $\delta^{14}C$ by twice the local $\delta^{13}C$ and to denote the result $\Delta^{14}C$. This quantity is considered to be (free from thermodynamic or biological fractionation effects) a purely physical advective tracer. The surface distribution of $\Delta^{14}C$ (fig.11a) is dominated by the effective connection with deeper waters. In the subtropical gyres, the $\Delta^{14}C$ content is almost in equilibrium with the atmosphere; in upwelling regions, the equilibration with the atmosphere is inhibited by the sluggish effective gas exchange for ^{14}C. Fig.11 b and c show the usual sections.

3.4 Carbonates

Fig.12 a,b show sections of the degree of supersaturation with respect to calcite. Here again, the conveyor belt structure of the circulation yields significant differences between the oceans. The lack of phosphate in the Atlantic yields a lower productivity (less alkalinity consumption during remineralisation) which is due to the strong lack of silicate shifted to a higher rain ratio of $CaCO_3$ (relatively more alkalinity release by calcite dissolution). Both mechanisms tend to shift the lysocline to deeper horizons. Fig. 12c shows the thickness of the sediment of calcium carbonate resulting from the rain exerted by the export productivity and the intersection of the lysocline with the topography. The basic structure is given by the 3 km isobath in the Pacific. In the Atlantic, where the position of the lysocline is deeper, the sedimentation covers a larger percentage of the sea floor.

3.5 Surface Partial Pressure

The model pCO_2 difference between sea surface and atmosphere (Fig.13) shows a strong high in the equatorial Pacific, a high in the equatorial Atlantic off the coast of Africa, and lows at high latitudes. Lower values are seen in the western Pacific and western Atlantic than further east, as the gyre surface flow carries warm water northward where it rapidly cools. These structures are all seen in data of pCO_2 in surface ocean water (Keeling 1968, Takahashi *et al.* 1983), which, additionally, reflect the changes of the air-sea difference of pCO_2 by the increased CO_2 content of the atmosphere in the last century (cf. MRH). The distribution of pCO_2 shows a pronounced seasonality resulting from a delicate balance of physical and biological effects. In spring, the biological pump begins to act with the onset of the plankton bloom at reduced turbulence and increased temperature. In some regions of the subpolar oceans, the seasonality is exactly inverse to the temperature effect. Up to now, the

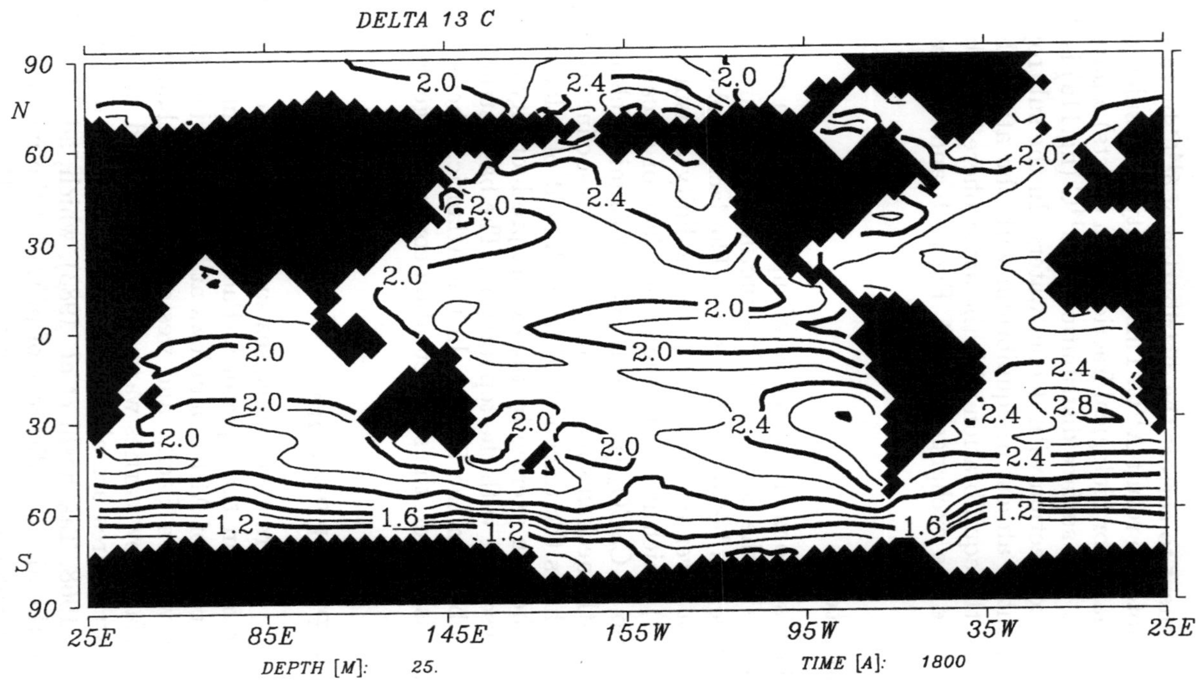

Figure 10a: Surface distribution of $\delta^{13}C$

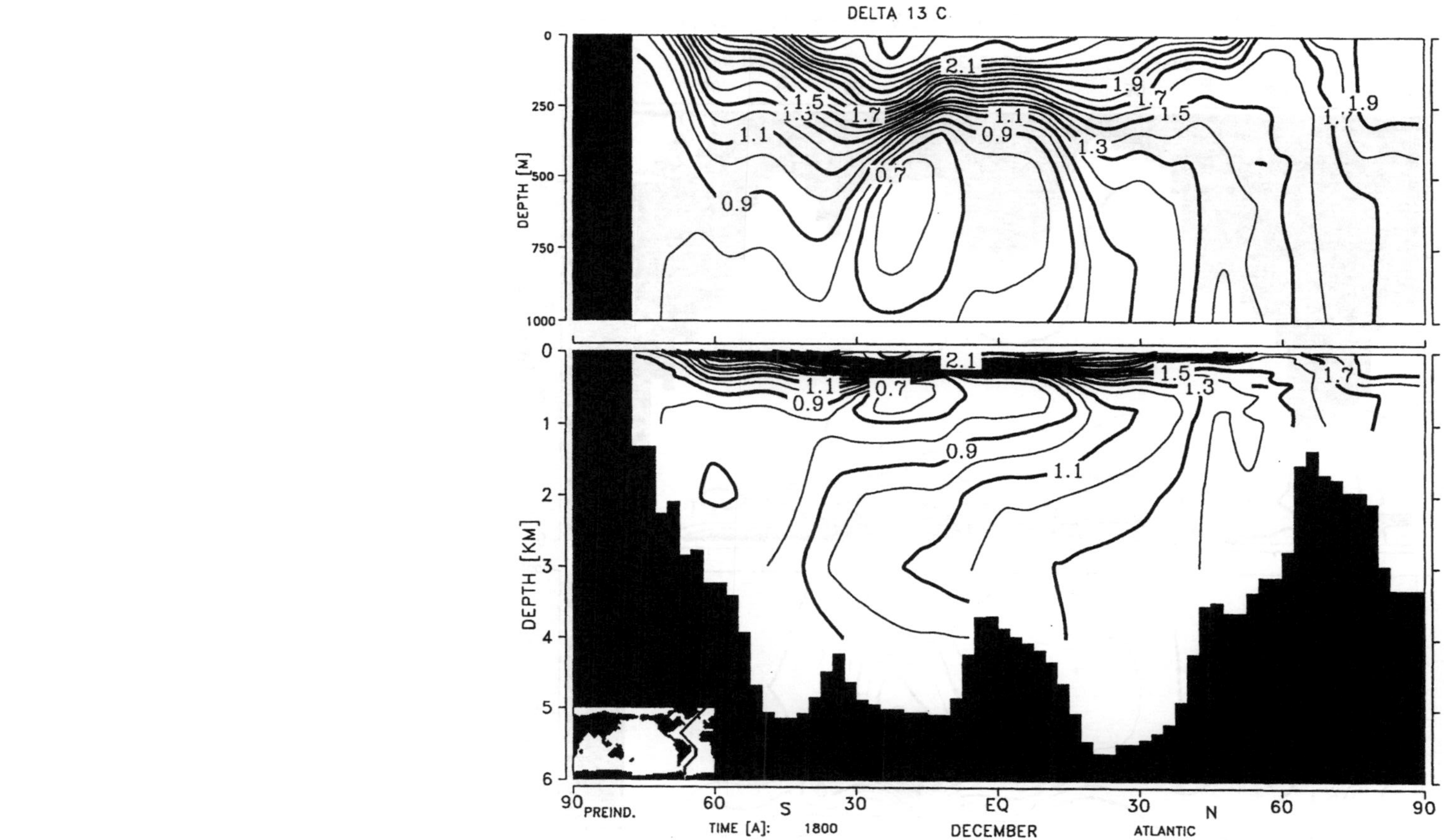

Figure 10b: $\delta^{13}C$ in the western Atlantic for constant fractionation

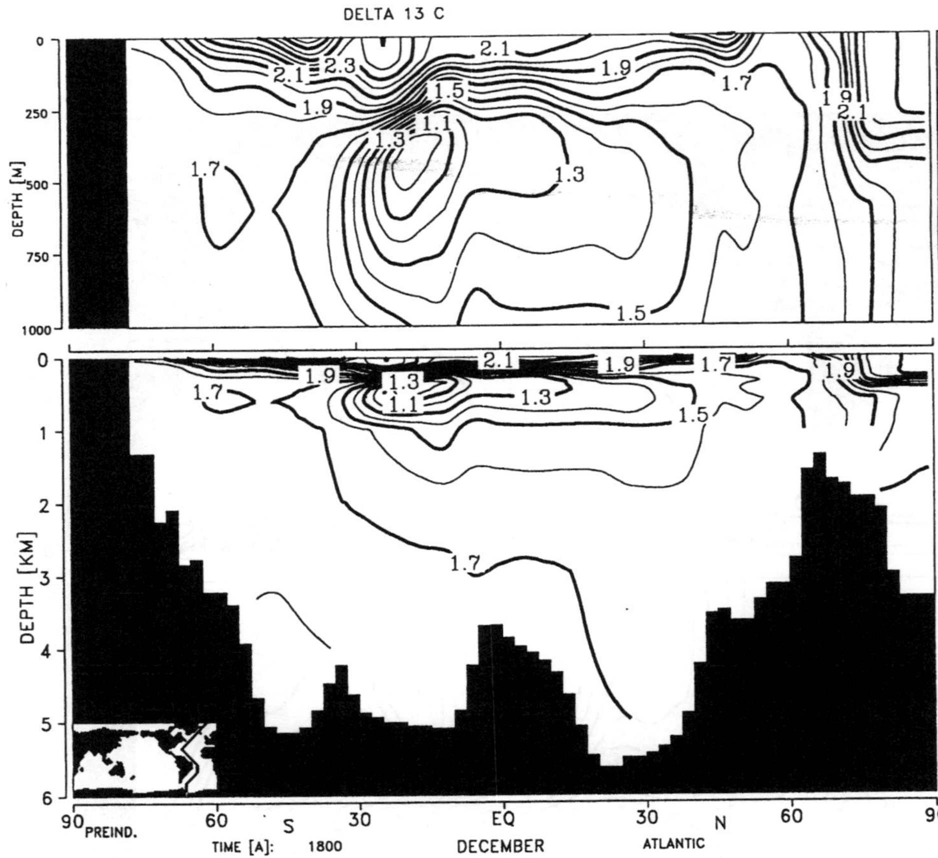

Figure 10c: $\delta^{13}C$ in the western Atlantic for Fractionation according Rau *et al* (1991)

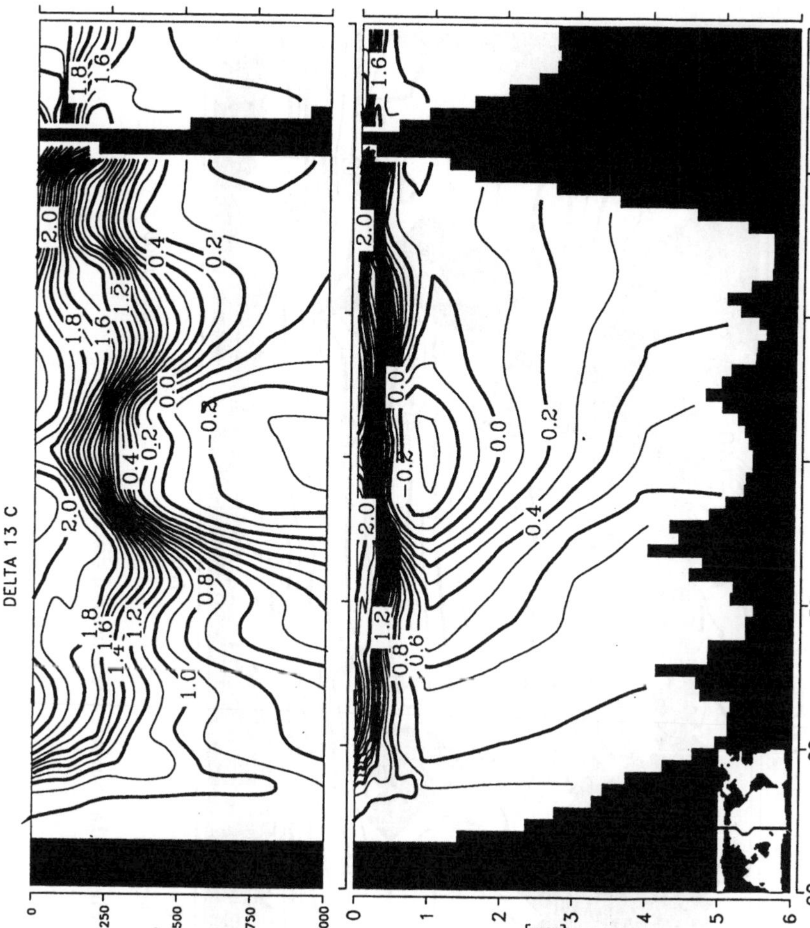

Figure 10d: $\delta^{13}C$ in the western Pacific

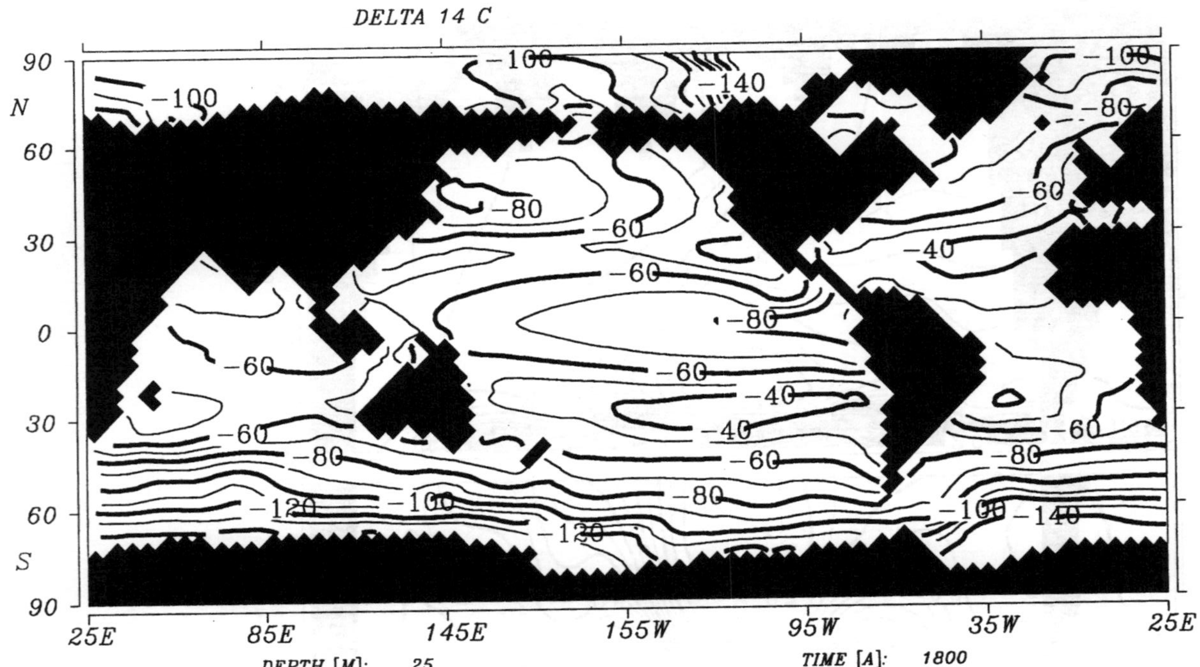

Figure 11a: Surface preindustrial distribution of $\Delta^{14}C$

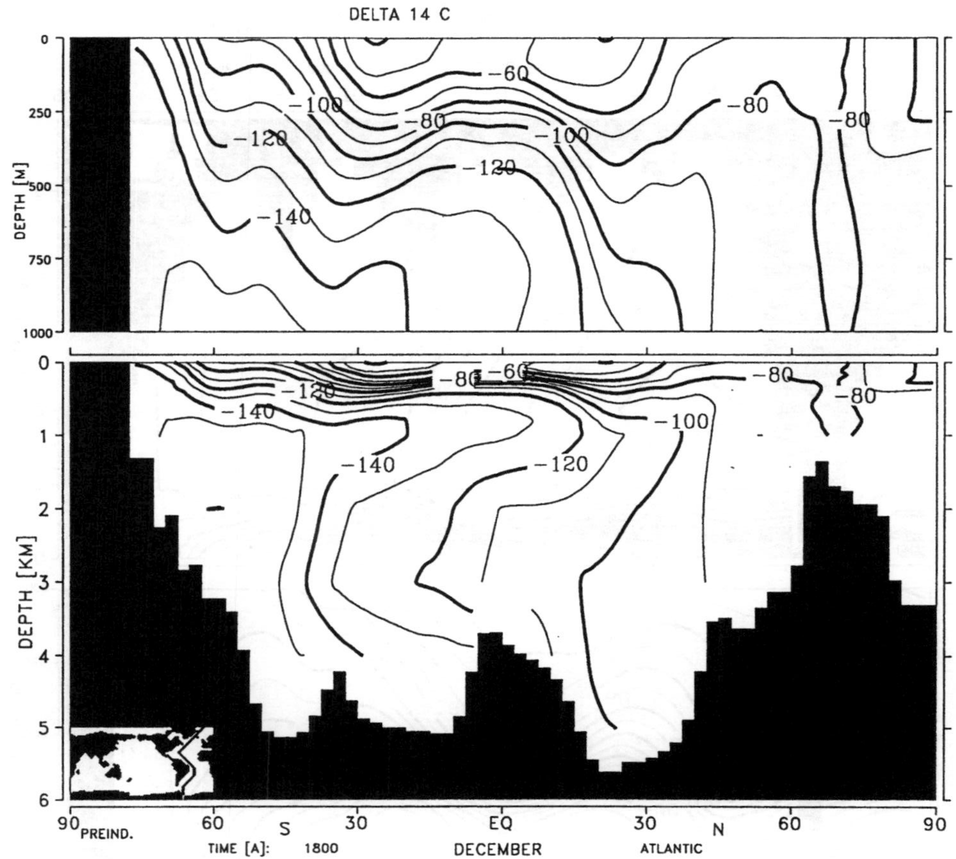

Figure 11b: Preindustrial distribution of $\Delta^{14}C$ in the western Atlantic

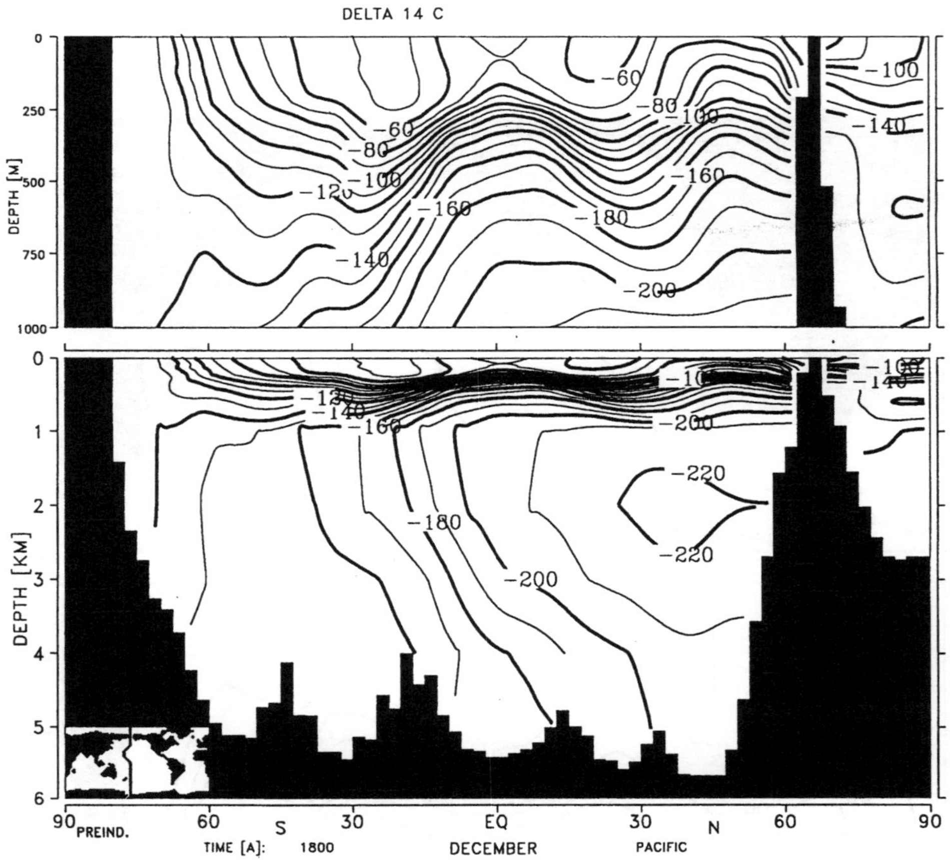

Figure 11c: Preindustrial distribution of $\Delta^{14}C$ in the western Pacific

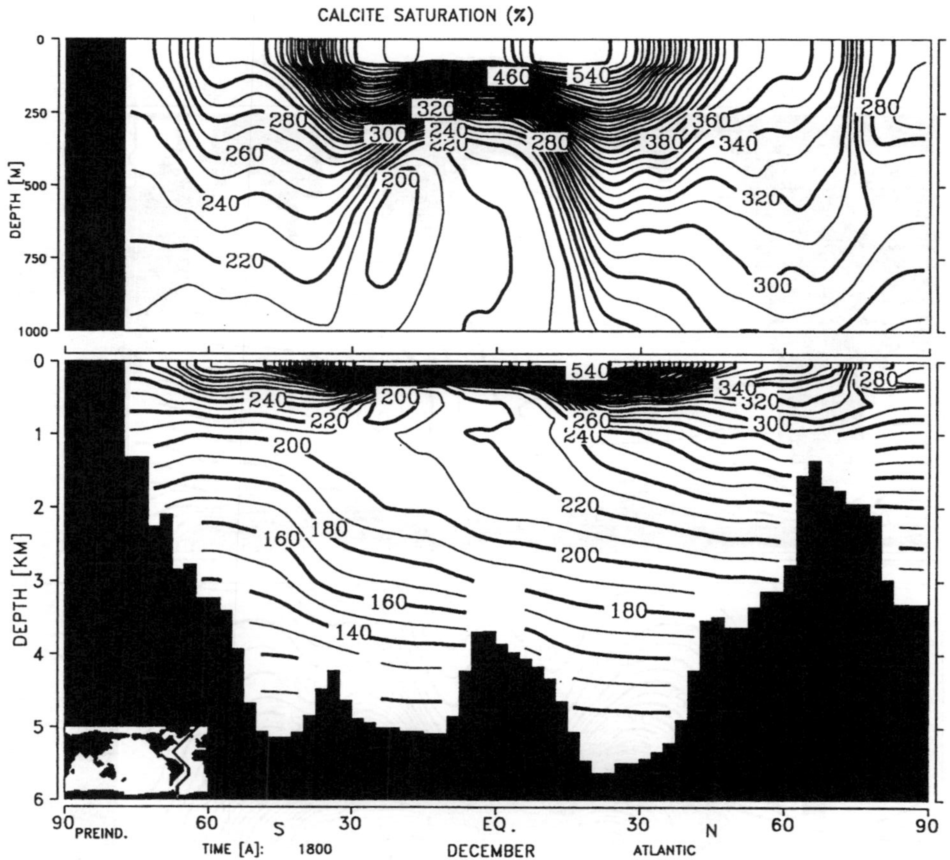

Figure 12a: Degree of saturation with respect to calcite in the western Atlantic

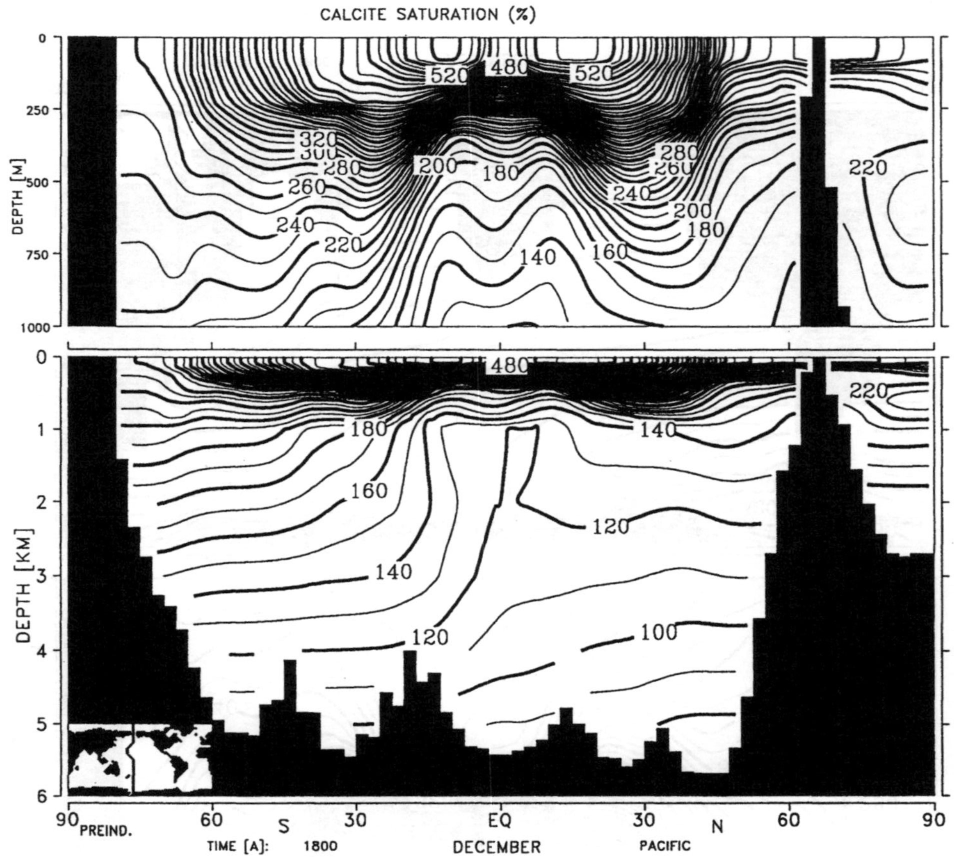

Figure 12b: Degree of saturation with respect to calcite in the western Pacific

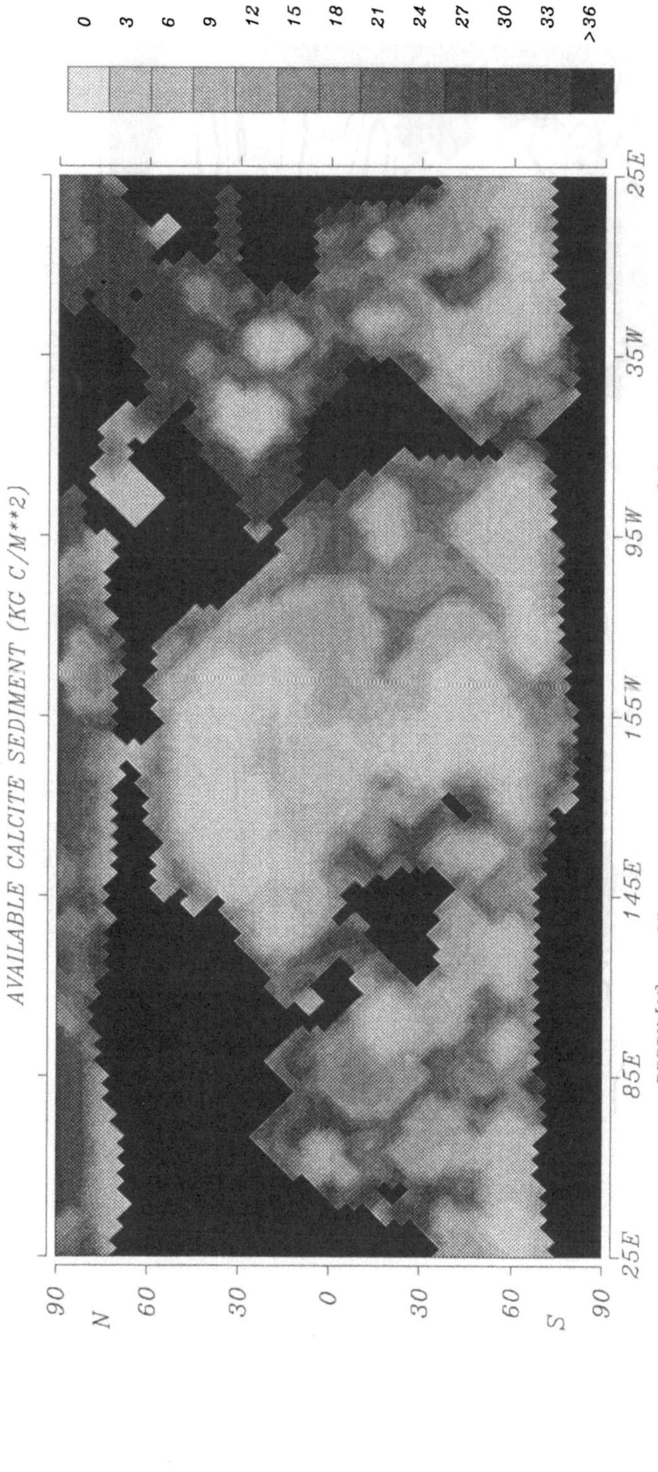
AVAILABLE CALCITE SEDIMENT (KG C/M**2)
DEPTH [M]: 25.
TIME [A]: 1975
25E
85E
145E
155W
95W
35W
25E
90
N
60
30
0
30
60
S
90
0
3
6
9
12
15
18
21
24
27
30
33
>36

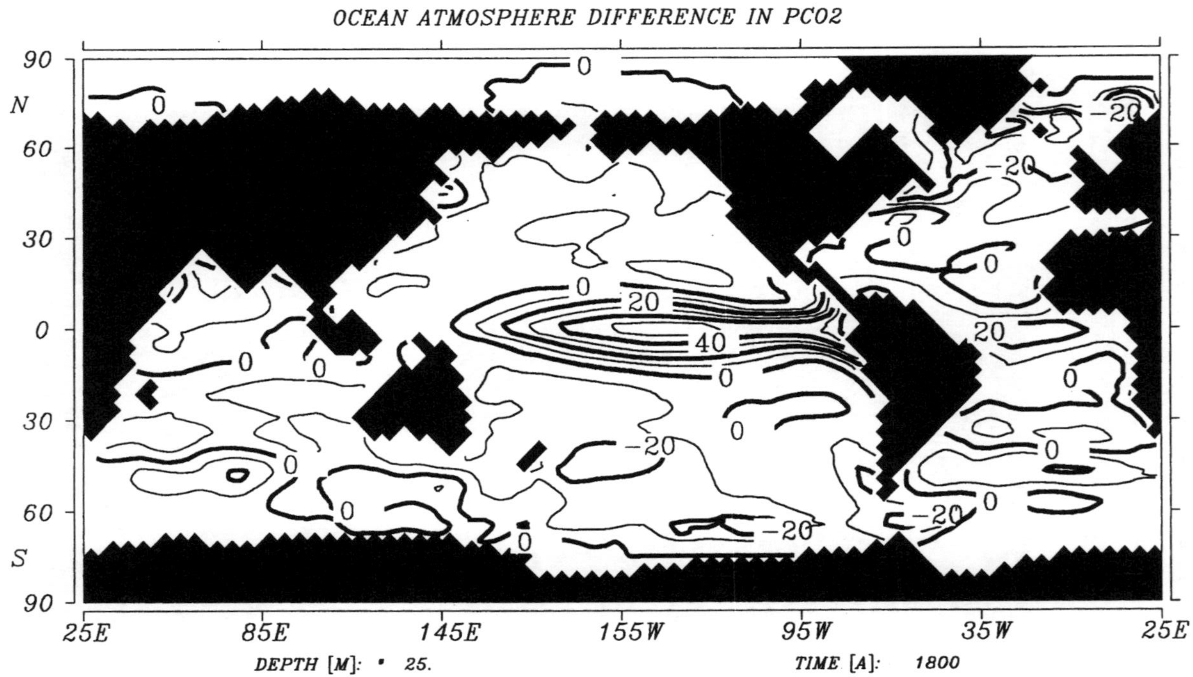

Figure 13: Annual mean sea-air difference of pCO_2

published data on this seasonal cycle of pCO_2 are not sufficient to derive globally critical tests for the details of the formulation of productivity in the model. Great efforts are underway to derive a global figure from observations.

4 Summary

The results of a radiocarbon transport model have been used to tune the thermohaline boundary conditions of a physical Ocean general circulation model. After this tuning, the modelled distributions of other tracers like phosphate, oxygen, silicate, alkalinity, and dissolved carbon dioxide exhibit realistic large scale patterns. The tracer model is not perfect and certainly not in a final state, but it is thought to be realistic enough that experiments can be carried out with the model.

Appendix: Chemical Equilibria and alkalinity

Many chemical reactions can be written in the form:

$$A^+ + B^- \rightleftharpoons AB \tag{1}$$

In a strongly diluted system, the reaction $\rightharpoonup$ is proportional to the product of the concentrations $[A^+][B^-]$, while the reaction $\leftharpoondown$ is proportional to the concentration $[AB]$. An equilibrium is established, when both reactions have the same strength , i.e.

$$[A^+][B^-] = K_{AB}[AB], \tag{2}$$

with K_{AB} the dissociation constant for the considered reaction ("law of mass action"). Relevant reactions for the marine carbon system are:

$$[H^+][OH^-] = K_W[H_2O] \tag{3}$$
$$[H^+][HCO_3^-] = K_1[CO_2] \tag{4}$$
$$[H^+][CO_3^{--}] = K_2[HCO_3^-] \tag{5}$$
$$[H^+][B(OH)_4^-] = K_B[B(OH)_3][H_2O] \tag{6}$$

Boron is not involved in biogeochemical processes, but it interacts with the carbon system by communicating with the same $[H^+]$ pool. Since $[H_2O]$ is almost constant, in the usual definition it is incorporated in the numerical values of K_B and K_W. Formally, an additional equilibrium

$$[H_2O][CO_2] = K_0[H_2CO_3]$$

should be considered. Since this K_0 is not precisely known from measurements, the notation in eq.(4) refers to the sum of $[CO_2]$ and $[H_2CO_3]$.

The exchange with the atmosphere is governed by Henry's law, that the partial pressure is proportional to the concentration:

$$pCO_2 = \alpha[CO_2] = \alpha \frac{K_2}{K_1} \frac{[HCO_3^-]^2}{[CO_3^{--}]} \tag{7}$$

At the other end of the reaction chain, Calcium has to be considered:

$$[Ca^{++}][CO_3^{--}] \rightleftharpoons [CaCO_3] \tag{8}$$

when the left side of (7) is below the pressure dependent solubility product K_{SP}, crystalline $CaCO_3$ will be dissolved, while the formation of $CaCO_3$ in supersaturated water is not observed in the open ocean. Up to now, there are only speculations that the presence of magnesium or organic compunds may suppress this reaction flow. The solubility product for calcite increases with pressure, roughly proportional to exp(z/4km), whereas the dissociation into $CaCO_3$ increases with temperature, and thus decreases with depth. The horizon where $[CaCO_3]$ equals the solubility product is named lysocline. With increasing pCO_2 from the emission of fossil fuel CO_2, the concentration CO_3 will reduce, and sedimented $CaCO_3$ can be dissolved. In computations of any changes of the system, the complications which arise from the simultaneous appearance of $[H^+]$ in different equations can be reduced by considering the alkalinity.

For physicists, the alkalinity is a rather unfamiliar and confusing quantity; a confusion which is enhanced with the realization that more than twenty different definitions appear in the literature (Rakestraw, 1949). It is, however, a routinely measured quantity. From the modeler's point of view it could have been invented as a dirty trick for simplification of the model. In pure water, we have equal numbers of $[H^+]$ and $[OH^-]$ ions. The familiar pH-value

$$pH = -log_{10}[H^+] \tag{9}$$

indicates the character of the solution to be acid ($pH < 7$) or alkaline ($pH > 7$). It is equivalent to state that the difference

$$A = [OH^-] - [H^+] \tag{10}$$

is positive for alkaline solutions and negative for acid solutions ($K_W \approx 10^{-14}(mol/l)^2$). This definition for A ("Alkalinity") can easily be generalized to include different reactions:

A = negative sum of charges

where the sum extends over those ions that may react with the H_2O system. Obviously, A does not change in any dissociation process or by the injection of any neutral compound like CO_2. When the system undergoes variations of the physical parameters like temperature or pressure, the dissociation constants and thus the concentration of the individual ions will change, but not the alkalinity.

For practical applications (as in "Tracers in the Sea"), it is convenient to consider only the carbonate contribution which define the "carbonate alkalinity"

$$CA = [HCO_3^-] + 2[CO_3^{--}] = [CO_2](K_1/[H^+] + 2K_1K_2/[H^+]^2) \quad (11)$$

Together with the inventory of carbon atoms:

$$\Sigma CO_2 = [CO_2] + [HCO_3^-] + [CO_3^{--}] = [CO_2](1 + K_1/[H^+] + K_1K_2/[H^+]^2) \quad (12)$$

Approximate solutions for the concentrations at given invenories of alkalinity and carbon can easily be found (neglecting $[CO_2]$, which is less than 1% of ΣCO_2). Insertion of the equilibrium conditions (5,6) then yields an estimate for the pCO_2, for instance from (7):

$$pCO_2 \sim \frac{(2\Sigma CO_2 - CA)^2}{CA - \Sigma CO_2} \quad (13)$$

A rigorous treatment of all reacting ions is only slightly more complicated: Consider any reactions

$$[H^+][X^-] = K_X[HX], \quad (14)$$

or

$$[Y^+][OH^-] = K_Y[YOH] \quad (15)$$

with an arbitrary number of X and Y compounds. The alkalinity is then

$$A = [OH^-] - [H^+] + \Sigma[X^-] - \Sigma[Y^+] \quad (16)$$

With given inventories of ΣX and ΣY atoms, the contributions of these elements to the alkalinity are:

$$A_X = \frac{\Sigma X}{(1 + [H^+]/K_X)} \quad (17)$$

and

$$A_Y = -\frac{\Sigma Y}{(1 + [OH^-]/K_Y)} = -\frac{\Sigma Y[H^+]}{([H^+] + K_W/K_Y)} \quad (18)$$

(The only further contribution with a higher concentration than the measuring error in chemical standard methods ("titration alkalinity") comes from the borate system). Elimination of H_2CO_3 from (11) gives an expression for CA which again has only $[H^+]$ as unknown:

$$CA = \Sigma C \frac{(K_1/[H^+] + 2K_1K_2/[H^+]^2)}{(1 + K_1/[H^+] + K_1K_2/[H^+]^2)} \tag{19}$$

The computation of the equilibrium state now reduces to the problem of determining $[H^+]$ to yield a given value of alkalinity. This can easily be achieved by a Newton iteration, for which only the inventories and the dissociation constants have to be known. For any change in the $[H^+]$, we have

$$\frac{\partial A_X}{\partial [H^+]} = -\frac{\Sigma X}{K_X(1 + [H^+]/K_X)^2} \tag{20}$$

and

$$\frac{\partial A_Y}{\partial [H^+]} = -\frac{\Sigma Y K_W/K_Y}{([H^+] + K_W/K_Y)^2} \tag{21}$$

It is easily seen that $\partial CA/\partial [H^+]$ is also strictly negative which proves the solution to be unique.

From these derivatives it becomes obvious which constituents should be included in the definition of alkalinity: a substantial value of ΣX or ΣY is essential as well as a dissociation constant in the order of magnitude of $[H^+]$.

Despite their high concentration in seawater, strongly dissociating compounds like $NaOH$ or HCl will not contribute to changes in the composition of the H^+OH^- system. A special confusion in the definition for alkalinity arises in the treatment of phosphate ions:

$$[H^+][HPO_4^{--}] = K_P[H_2PO_4^-], or \tag{22}$$
$$[H^+]^2[HPO_4^{--}] = K_P^*[H_3PO_4] \tag{23}$$

(This contribution is of rather theoretical interest, since the concentration is smaller than the modern accuracy of measurements) Different authors include either $+[HPO_4^{--}]$, or $-[H_2PO_4^-]$, or the difference $[HPO_4^{--}] - [H_3PO_4]$ (Dickson 1981 and Goyet, this volume) in the definition of alkalinity. For the more important aspect of changes of the contributions to the alkalinity, all three definitions are obviously identical.

In the model, the alkalinity changes are taken from the definition

$$A = [HCO_3^-] + 2[CO_3^{--}] + [B(OH)_4^-] + [OH^-] - [H^+] \tag{24}$$

Biological production affects the system in basically two ways: In the formation of organic material, for every phosphorus atom, 122 atoms C are taken out from the water, mainly in the form $[H_2CO_3]$. In addition to this, 17 compounds $[NO_3^-]$ are used; This belongs to a strong acid, the corresponding removal of $[H^+]$-ions increases the alkalinity. Both changes tend to reduce the pCO_2 In the formation of $CaCO_3$ shells, the reduction of the alkalinity is twice the reduction of the ΣCO_2; this process tends to increase the pCO_2. In the global average, $CaCO_3$ and organic material ("soft tissue") are formed in the relation 1:4; the global effect on the pCO_2 is a reduction by appr. 200 ppm.

References

ARCHER, D.,1991 Modeling the Calcite Lysocline. *J. Geoph. Res.,* **96,** C9, 17037 - 17050.

ATLAS,E. and R.M.PYTKOWICZ, 1977 Solubility behavior of apatite in seawater. *Limnology Oceanography* **22,** 290-300.

BACASTOW, R. and E.MAIER-REIMER Ocean - circulation model of the carbon cycle, *Clim. Dyn.,* **4,** 95 - 125, 1990.

BACASTOW,R. and E.MAIER-REIMER Dissolved organic carbon in modeling oceanic new production, *Global Biogeochem. Cycles,* **5,** 71 - 85, 1991.

BAINBRIDGE,A. GEOSECS ATLANTIC EXPEDITION, Sections and profiles. *Washington appr.1980.*

BERGER, W.H.,K.FISCHER, C.LAI, and G.WU Ocean Productivity and organic carbon flux, Part I: Overview and maps of primary production and export production. *SIO Ref. 87-30. UCSD, La Jolla, 1987.*

BOLIN,B. (ED.) Carbon Cycle Modelling, *390 pp, J.Wiley & sons,1982.*

BROECKER, W.S. and T.TAKAHASHI Neutralization of fossil fuel CO_2 by marine calcium carbonate. in: N.R.Anderson and A.Malahoff (eds): The fate of fossil fuel CO_2 in the Oceans. *Plenum Press, 197 7.*

BROECKER, W.S. and T. TAKAHASHI The relationship between lysocline depth and in situ carbonate ion concentration. *Deep Sea Res.* **25,** 65 - 95, 1978.

BROECKER, W.S. and T.H.PENG Tracers in the Sea, *691 pp. Eldigio press, Palisades,1982*

BROECKER,W.S., T.H.PENG, G.ÖSTLUND, and M.SUIVER The distribution of bomb radiocarbon in the ocean. *J. Geoph. Res.,* **90,** 6953 - 6970, 1985.

BROECKER,W.S., S.BLANTON,W.M.SMETHIE, JR, and G. OSTLUND Radiocarbon decay and oxygen utilization in the deep Atlantic Ocean, *Global Biogeoch. Cycles* **5,** 87 - 117, 1991.

BRYAN, K. A numerical method for the study of the circulation of the World *Ocean. J. Comp. Phys.* **4,** 347 - 376, 1969.

CRAIG,H., W.S.BROECKER and D.SPENCER GEOSECS PACIFIC EXPEDITION, Sections and profiles, *Washington, 1981.*

CULBERSON, C.H. and R.M.PYTKOWICZ, 1968 Effect of pressure on carbonic acid, boric acid and th pH in sea water. *Limnol. and Oceanogr.* **13,** 403 - 417

DICKSON,A.G. An exact definition of total alkalinity and a procedure for the estimation of alkalinity and total inorganic carbon from titration data. *Deep Sea Res.,* **28A,** 609-623, 1981

DUGDALE,R.C. Nutrient limitation in the sea: dynamics,identification and significance. *Limn. and Oceanogr.* **12,** 685-695, 1967.

EDMOND, J.M. and J.M.T.M.GIESKES On the calculation of the degree of saturation in sea water with respect to calcium carbonate under in situ conditions. *Geochimica et Cosmochimica Acta* **34,** 1261 -1291, 1970.

FASHAM, M.J.R., H.W.DUCKLOW, and S.M. MCKELVIE A nitrogen-based model of plankton dynamics in the oceanic mixed layer, *J. Mar. Res.,* **48,** 591 - 639, 1990.

FIADEIRO, M.E. Threedimensional modeling of tracers in the deep Pacific Ocean II: Radiocarbon and the circulation. *J. Mar. Res.* **40,** 537 - 550, 1982.

HEINZE, C., E. MAIER-REIMER, and K. WINN Glacial pCO_2 reduction by the World Ocean: Experiments with the Hamburg carbon cycle model. Paleoceanogr. 6, 395 - 430, 1991. Keeling,C.D.: Carbon Dioxide in surface waters. 4. Global distribution. *J.Geoph. Res.* **73,** 4543-4553, 1968.

KOBLENTS-MISHKE,O.L. Primary production. In: Vinogradov,M.E. (ed): Oceanology. *Biology of the Ocean, vol. 1 (in Russian). Nauka, Moscow, 1977.*

KROOPNICK, P.M. The distribution of $\delta^{13}C$ of ΣCO_2 in the world oceans. *Deep Sea Res.,* **32,** 57 - 84, 1985.

LEVITUS,S. Climatological Atlas of the World Ocean. *NOAA prof.paper 13, Rockville,Md., 1982*

LISS,P.S. and L.MERLIVAT Air-Sea gas exchange rates: Introduction and Synthesis. in: P.Buat-Menard (ed) The Role of Air-Sea Exchange in Geochemical Cycling *.D.Reidel, pp 113-127, 1986*

MAIER-REIMER, E. and R. BACASTOW Modelling of geochemical tracers in the ocean. In: *M.Schlesinger (ed.) Climate - Ocean Interaction pp.233 - 267, Kluwer Acad. Press, Dordrecht, 1990*

MAIER-REIMER,E., U.MIKOLAJEWICZ, and K.HASSELMANN Mean Circulation of the Hamburg LSG OGCM and its sensitivity to the thermohaline surface forcing. *J. Phys. Oceanogr. 1992 (in press)*

MAIER-REIMER,E. and K.HASSELMANN Transport and storage of CO_2 in the ocean - an inorganic ocean-circulation carbon cycle model. *Climate Dynamics* **2,** 63-90,1987.

MARTIN, J.H., R.M.GORDON, and S.E.FITZWATER Iron in Antarctic waters. *Nature* **345,** 156-158, 1990

MOOK,W.G., J.C.BOMMERSON, and W.H.STAVERMAN Carbon isotope fractionation between dissolved bicarbonate and gaseous carbon dioxide. *Earth and Planetary Science letters* **22,** 169 - 176, 1974.

MOORE III, B. and A.BJORKSTROM Calibrating Ocean Models by the constrained Inverse Method. *In: J.R.Trabalka and D.E.Reichle (eds): The changing Carbon Cycle A Global Analysis. Berlin, Heidelberg 1986, pp.295-328*

NAJJAR, R.G., J.L.SARMIENTO, and J.R.TOGGWEILER Downward transport and fate of organic matter in the Ocean: Simulations with a General Circulation Model. *Global Biogeochem. Cycles* **6,** 45 - 76, 1992.

NYDAL,R. and K.LOVSETH Tracing Bomb ^{14}C in the Atmosphere 1962-1980. *J. Geoph. Res.* **88 C6,** 3621 - 3642, 1983.

OESCHGER,H., U.SIEGENTHALER, U.SCHOTTERER, and A.GUGELMANN A box diffusion model to study the carbon dioxide exchange in nature. *Tellus* **27,** 168-192,1975.

ÖSTLUND, H.G. and M.STUIVER, GEOSECS Pacific radiocarbon. *Radiocarbon* **22,** 25-53, 1980

RADACH,G. and E.MAIER-REIMER The vertical structure of phytoplankton growth dynamics - a mathematical model. *Mem. Soc. Roy. Sc. Liege, 6e serie,* TOME **VII,** 113-146, 1975

RAKESTRAW, N.W., 1949 The conception of alkalinity or excess base of Sea water. *J.Mar.Res.* **7,** 14-20

RAU,G.H., P.N.FROELICH, T.TAKAHASHI, and D.J.DES MARAIS Does sedimentary organic d C record variations in Quaternary Ocean [SCO_2(aq)]?. *Paleoceanogr.* **6,** 335 - 347 ,1991.

REVELLE,R. and H.E.SUESS Carbon dioxide exchange between atmosphere and ocean and the question of an increase of atmospheric CO_2 during the past decades. *Tellus* **9,** 18-27,1957.

ROTTY, R.M. Distribution of and changes in industrial carbon dioxide production. *J. Geoph. Res* **88 C2,** 1301-1308,1983.

SARMIENTO, J.L., J.C.ORR, and U.SIEGENTHALER A Perturbations Simulation of CO_2 Uptake in an Ocean General Circulation Model. *J Geoph.Res.* **97 C3,** 3621 - 3646, 1992.

SIEGENTHALER, U. Uptake of Excess CO_2 by an Outcrop - Diffusion Model of the Ocean. *J. Geoph. Res.* **88 C6,** 3599 - 3608, 1983.

SUESS,E. Particulate organic carbon flux in the oceans - surface productivity and oxygen utilization. *Nature,* **288,** 260-263, 1980.

STUIVER, M. and H.G.ÖSTLUND, GEOSECS Atlantic radiocarbon. *Radiocarbon* **22,** 1-24, 1980.

SUGIMURA,Y. and Y.SUZUKI A high-Temperature Catalytic Oxidation Method for the Determination of non-Volatile Dissolved Organic Carbon in Seawater by Direct Injection of a Liquid Sample. *Marine Chem.* **24,** 105-131, 1988.

TAKAHASHI,T., W.S.BROECKER, and S.LANGER Redfield ratio estimates based on chemical data from isopycnal surfaces. *J. Geoph. Res.,* **90 C4,** 5907-5924, 1985.

TAKAHASHI,T., D.CHIPMAN, and T.VOLK Geographical, seasonal, and secular variations of the partial pressure of CO_2 in surface waters of the North Atlantic Ocean: *thr results of the North Atlantic TTO program in: Proceedings of Carbon Dioxide Research Conference: Carbon Dioxide, Science and Consensus CONF 820970 U.S.Dept. Energy, Washington , 1983*

TOGGWEILER, J.R., K.DIXON, and K.BRYAN. Simulations of Radiocarbon in a coarse - resolution World Ocean Model I. Steady State Prebomb Distributions. *J. Geoph. Res.* **94, C6,** 8217 - 8242, 1989.

WEISS,R.F. Carbon Dioxide in water and sea water: the solubility of a nonideal gas. *Marine Chem.* **2,** 203-215, 1974.

WEISS,R.F. The solubility of nitrogen, oxygen, and argon in water and sea water. *Deep Sea Res.* **17,** 721-735,1970.

WILLIAMS,P.M. and E.R.M.DRUFFEL Radiocarbon in dissolved organic matter in the central North Pacific Ocean. *Nature,* **330,** 246-248, 1987.

WOODRUFF,S.D., R.J.SLUTZ, R.L.JENNE, and P.M.STEURER A comprehensive ocean atmosphere data set. *Bull.Am.Soc.* **68,** 1239 - 1250, 1987.

WROBLEWSKI,J.S. A model of phytoplankton plume formation during variable Oregon upwelling. *J.Mar. Res.* **35,** 357-394, 1977.

INDEX

Springer-Verlag and the Environment

We at Springer-Verlag firmly believe that an international science publisher has a special obligation to the environment, and our corporate policies consistently reflect this conviction.

We also expect our business partners – paper mills, printers, packaging manufacturers, etc. – to commit themselves to using environmentally friendly materials and production processes.

The paper in this book is made from low- or no-chlorine pulp and is acid free, in conformance with international standards for paper permanency.

MIX
Papier aus verantwortungsvollen Quellen
Paper from responsible sources
FSC® C105338

If you have any concerns about our products,
you can contact us on
ProductSafety@springernature.com

In case Publisher is established outside the EU,
the EU authorized representative is:
Springer Nature Customer Service Center GmbH
Europaplatz 3, 69115 Heidelberg, Germany

Printed by Libri Plureos GmbH
in Hamburg, Germany